Urban Informality

Ahmed M. Soliman

Urban Informality

Experiences and Urban Sustainability Transitions in Middle East Cities

Foreword by Hernando De Soto and Nezar Al Sayyad

Springer

Ahmed M. Soliman
Architectural Department
Faculty of Engineering, Alexandria
University
Alexandria Governorate, Egypt

ISBN 978-3-030-68987-2 ISBN 978-3-030-68988-9 (eBook)
https://doi.org/10.1007/978-3-030-68988-9

This Springer imprint is published by the registered company Springer Nature Switzerland AG
The registered company address is: Gewerbestrasse 11, 6330 Cham, Switzerland

Foreword by Hernando De Soto

Today, the world population is approximately 7.7[1] billion, and it is expected that within the next 30 years, between 2.5 and 3 billion people will be added to the urban population worldwide; with the highest growth projected to be in less developed regions. The absolute number of slum dwellers in the developing world had reached more than 830 million in 2018. In the second half of the twentieth century, and at the beginning of the New Millennium, cities in the Global South, and in particular, the cities in the Middle East witnessed an arbitrary urbanization process on their peripheries of vast areas urbanized "from below."

Thus, urban informality became the dominant feature of city landscapes in the Middle East. This phenomenon has international organizations and various academics institutions all over the world trying to understand *why the urban poor are acting in their environment informally rather than formally*.

My book *The Mystery of Capital* was released in the year 2000 at the height of a global economic and stock market boom. The publicity for the book was reminiscent of the English-language launch of my earlier book, *The Other Path*, in 1986. *The Mystery of Capital* rejected all previous research and scholarship, observing what the poor five-sixths of humanity actually did and learned that they have vast assets. Turning to the obvious question:

Why then are southern countries, especially Middle Eastern countries, so underdeveloped? Why can't they turn real estate assets into liquid capital—the kind of capital that generates new wealth? For me, this is "the mystery of capital."

My quest to answer this question lead me to a novel answer. The present global crisis is the same kind of crisis that advanced nations suffered during the Industrial Revolution, when they themselves were Third World countries teeming with black markets, pervasive mafias, widespread poverty and flagrant disregard of the law. The Western nations, I argued, created the key conversion process 150 years ago, and their economies began to soar into wealth without their ever realizing what they had

[1]This is the figure for July 2019 which I obtained from two different sources.

done. I explained how this unwitting process, hidden deep in thousands of pieces of property law throughout the West, came to be, how it works, and how today it can be deliberately set up in developing and former Soviet nations.

Therefore, *The Mystery of Capital* was a message of hope, written in a period of global optimism, providing a way for all the poor developing and former Soviet nations of the world "to soar into wealth." All that was needed was for their governments and elites to focus on one key policy—providing legal title to real estate currently held under a wide range of "extra-legal" forms, including squatting, illegal subdivision, indigenous and communal laws, and many forms of commons and cooperatives. If this "dead" extra-legal property could be made legal, the real property involved could be transformed into "liquid capital," and this new capital could power a broad-ranging economic development process that might enable the world's poor to achieve prosperity.

Professor Soliman's book brings new insights on the transitions and policies that have introduced new dynamics in urban informality, to such an extent that today we need to reformulate the framework and the analytical tools applied to them. He argues that urban informality has changed in nature; its form, its topologies, and its practices have been transformed by radical transitions in the socioeconomic and political contexts as well as by the effect of development policies, of regularization or "integration" imposed or encouraged by states or by international organizations. He recalls reconsidering the matter of urban land and, more broadly, the new urban configurations: capital dynamic, land liberalization, social movements and social capital, the appearance of new actors (like NGOs and international donors), citizen participation, and mode of governance. The tensions between four modes of production, reproduction, consumptions, and distribution of goods and services in the city and its practices are examined. In this book, *Urban Informality: Experiences and Urban Sustainability Transitions in Middle East Cities*, the author makes a valuable contribution to the theme by introducing governance of urban sustainability transitions on urban informality theory to the development of sociotechnical and spatial systems. Consequently, this unique treatise focuses on those components that direct integrated urban informality within the urban components of spatial systems in the Middle East. The book addresses the resilient gap in existing knowledge and related approaches through the application of planning instruments as a necessary measure in redressing the urban informality question typical of the Middle East's landscapes.

The contents of the book are well structured in terms of the reigning perceptions of urban informality in the Middle East and the building blocks consisting of definitions, terminology, methodology, analysis, and conclusions. The book includes a critical assessment of the urban informality philosophy and its interface with related theories from both urban and regional perspectives. Professor Soliman highlights the current transitions of urban informality and how the recent waves of urbanization and globalization have changed the nature of urban informality in four dimensions—by (i) determining dominant areas of land development; (ii) discourses about social organization through lookout opportunities for social inclusion; (iii) understanding the diversity and nature of the states; and (iv) transforming the drivers of sustainable transitions at various spatial levels.

The book draws upon the experience and intensive work in three countries; Egypt, Lebanon, and Jordan. The transregional contexts and research traditions specific to these areas provide particularly useful entry points for investigation between informality and socioeconomic and political transitions within the Middle East. Addressing the issue of state power and market forces in the housing process is also discussed because similar issues could be observable in other countries of the Global South as well as of the Global North.

The two main questions are what is the significance of "secure land rights," and what is the importance of land value in the development process, especially after the Arab Spring. Consequently, this volume examines the relationship and the ways that the residents within informal settlements, government, and professionals are communicating with each other in the formulation of urban informality. Gaining experience from this special relationship, it becomes apparent that all actors have a certain role to play in the formalization process of land for the benefit of the economic development on the one hand, and for securing the housing needs for the vast majority of the urban poor on the other.

Therefore, Professor Soliman's book focuses on the clarification of three areas: a synopsis of urban informality in the Middle East, ongoing research innovation on governance of sustainability transitions, and how to deal with urban informality in the context of sustainability transitions. These areas overlap and crosscut with two other dimensions. The first dimension is comprised of six axes which are: urban sustainability transitions, urbanization process, the paradigm of informality, state power and informality, land rights, and social exclusion/marginality. The second dimension is tensions between modes of production, reproduction, consumptions, and distribution of goods and services in the city and its practices. The book attempts to clarify the importance of urban informality processes on the built environment, and how to alter urban informality in the context of ongoing sustainability transitions in the Middle East. As urban sustainability strategies currently play a key role for the development of cities, this book sheds light on urban informality within sustainability strategies in three cities.

The SDG 11 "Tracking Progress Towards Inclusive, Safe, Resilient and Sustainable Cities and Human Settlements," Target 1 "By 2030, ensure access for all to adequate, safe and affordable housing and basic services and upgrade slums." However, the leaders of Middle Eastern countries committed to confirm with the SDGs till the year 2030. Accordingly, Middle Eastern cities have been facing a triple transformation: densification of popular urban tissues, a proliferation of spontaneous housing, and considerable urban informality. Simple and legible urban structures are replaced by fragmented configurations, with suburban, subaltern, and peri-urban spaces offering the image of a still incomplete patchwork in cities of the Middle East.

You can find more on this in my book *The facts are in: The Arab Spring is a massive economic revolution*, where I addressed a nagging question among business leaders in the region: Why are Arab entrepreneurs not meeting their potential for producing prosperity—for themselves and their countries?

Both academics and practitioners share the need to formally address the challenge of integrated urban informality in the Middle East with spatial context through the adoption of applicable planning theory based on neoliberal influences and developments. Professor Soliman points out that neoliberal planning serves as the delivery instrument within socioeconomic and political transitions. In focusing on the failure of neo-liberal planning theory and practice deriving mainly from inherent vitiation of green growth and spatial justice in the Middle East, the critical thinking and analytical abilities of the writer are brought to bear. In particular, the writer addresses compliances to neoliberal participatory planning within certain national spatial systems of Egypt, Lebanon and Jordan and its implications for the transformation in the transition systems. It is essential to rethink the impact of the transition of the current and future socioeconomic, political, and ethnic diversity, if sustainable development in economic, social spaces, and spatial development is the focus.

The book proposed a new direction in the application of socioeconomic and political transitions on urban informality, and the way its related theory and planning instruments are developed. It rethinks the traditional planning and development orthodoxies and approaches in terms of an alternative theory with supporting territorial planning framework instruments to promote spatial integration, and to understand the dynamic process of socioeconomic and political transitions to govern the urban informality in a sustainable way. Professor Soliman establishes the development of alternative spatial planning theory within the Middle East, a region that historically and otherwise has been dominated by the influences and impacts of colonial legacies as far as spatial planning and development are concerned. This book contributes to the development of international theory in line with the existing socioeconomic and political transitions on urban informality in the Middle East, as well as system innovations and transitions of spatial development. Professor Soliman focuses on system innovations between distinctive socio-technical configurations, encompassing not only new technologies but also corresponding changes in markets, user practices, policy, and cultural discourses, and governing institutions. I believe the work is a ground-breaking publication as it concerns understanding the governance of urban sustainability transitions on urban informality and planning in the Middle East.

President of Institute for Liberty and Democracy, Lima, Peru
August 2019

Hernando De Soto

Foreword by Nezar Al Sayyad

In the early 1970s, Keith Hart—researching in the context of Africa—introduced the concept of the "urban informal sector" to distinguish it as set of activities that exist outside of what was then considered normal regulated economic structures of the city. It would take another three decades before Ananya Roy and I would coin the term "urban informality"[2] to describe a larger system of informal norms and processes that govern the use of space and make possible new forms of social behavior and political power for many of the world's urban poor. The commodification of informal housing within this structure is a well-studied phenomenon which allowed us to explore the relationship between the new emerging actors and spaces behind the production of this built environment. These included grass-root activists, community organizations, and religious groups that consolidated the active roles of these agents in the provision of urban services.

So what is this urban informality, and who are these urban informals? Today, whether they are called "urban marginals," "urban disenfranchised," or "urban poor" (and these terms are often used interchangeably although they may mean different things), the current era of global restructuring has greatly increased the number of such people. This is particularly true in the context of the Middle East, where it has also led to an explosion in the range of their activities. But urban informality may neither be a totally new analytical concept nor a new urban process. So, what is it then that makes the idea of urban informality relevant or important to study at this time? The answer to this question is very complex.

This book Urban Informality: Experiences and Urban Sustainability Transitions in Middle East Cities sets the context for the discussions on the urban informality in the Middle East at the end of the second decade of the twenty-first century. It highlights the current themes and practices of urban informality and shows how the recent waves of urbanization and globalization have changed over time. The book seeks to contribute to the current debates by interrogating two interrelated

[2]Roy, A. and Al Sayyad, N. (2004)

questions with regard to the Middle East: what constraints and opportunities emerge for urban informality within its larger context; how and to what extent can informality be governed at the urban level. By exploring these two questions from different angles, this book seeks to respond to a noticeable gap in the urban informality literature particularly because of its focus on three Arab countries in the Middle East. It draws upon field work and experiences in Egypt, Lebanon, and Jordan. It is also one of the first volumes to explore the connection between urban informality and urban sustainability transitions. The book is transregional in method and scope with its focus on three countries with aspiring cities that represent the diversity of cities in the Global South beyond the Middle East. It goes on to flesh out three aspect of urban informality in the Middle East: explaining its content, connecting it to sustainability discourses, and interrogating the ongoing innovations in the governance of sustainability transitions.

It is well known that much of the urban growth of the new century will be in the developing world especially in the so-called informal settlements. So today, what may be needed most in thinking about urban informality is a shift in our analytical frameworks. Thus, the current era of liberalization and globalization can be seen as giving rise to new forms of informality. Assessing the situation today, we have learned that urban informality does not simply consist of the activities of the urban poor, or is a particular status of labor, or marginal economies, or illegal land subdivisions. Rather, it is an organizing logic in and of itself as this book illustrates.

We have also learned that none of the phenomena associated with urban informality can be understood outside the context of globalization, liberalization, and structural adjustment. Despite the Internet and the spread of information technology, patterns of urban behavior and exchange at the beginning of the twenty-first century in many ways resemble those common in other times in history. Indeed, one may argue that these patterns that we term informalities has existed since the Middle Ages in different forms, particularly in the Middle East. Such a mode of urbanism, made of segregated enclaves often dominated by poverty, militarization, religious ideologies, and the politics of patronage is what informality has become is all about in this global turn. This book with its unique focus on these three countries provides a valuable contribution to this argument and to our understanding of this economic, socio-political, and cultural system we call "urban informality".

Emeritus Professor of Architecture and Planning, University of California, Berkeley, Berkeley, CA, USA

Nezar Al Sayyad

Preface

The relevance of this book draws from the engagement with sourcing new Southern perspectives on urban and regional planning. For a long time, there has been a resilient gap in knowledge regarding the determination of appropriate planning instruments for redressing urban informality, which has plagued the built environment of cities of the Global South.

During a two-day symposium on urban informality in Spring 2001 at Berkeley University, USA, I have met a distinguished group of scholars and practitioners from three world regions. At this symposium, Prof. Bromley criticized De Soto's *Mystery of Capital* in which I was engaged in his work in Egypt. At this symposium, I presented my paper entitled "Tilting at Sphinxes: Locating Urban Informality in Egyptian Cities" and received valuable feedback from all participants, especially Ray Bromley, as well as from Professors Al Sayyad and Roy who were organizing the symposium. Between 2001 and 2003, I was finishing writing up my book entitled *A Possible Way Out: Formalizing Housing Informality in Egyptian Cities*, and I was thinking to publish another volume to cover the Middle East region.

During autumn 2003, the idea of the current book cropped up in my mind during a valuable discussion between me and Prof. Ray Bromley who was visiting, as a visiting professor, the Faculty of Architectural Engineering at Beirut Arab University. During that time, I was acting as a dean of this faculty; we had a valuable discussion on urban informality issues and I proposed an outline for a book and its contents, and the initial title of the book was "Experiencing Urban Informality on the Global South" to combine between theory and practice as new productivity for understanding urban informality. I returned to Egypt at the end of 2004; since then, I had engaged for several years as a consultant and as a director of a private firm with the GOPP and UN-Habitat in setting up General Strategic Urban Plans for a few small cities in Egypt. During my practice work on these cities, I discovered that most of the cities had a housing surplus of more than fifty percent more than their actual needs. Moreover, in mid-2014, I engaged in cooperation with UN-Habitat, Cairo office, on a land readjustment project in Banha city, north of Cairo, which allowed me to grasp the importance of the grassroots and social network in putting theory into practice. It was the first project, as a land readjustment program, to implement on the ground in Egypt. However, there was a debate between me and the representatives from the GOPP and UN-Habitat about why people are going into urban informality to satisfy their housing needs. It was a big debate with many concerns being raised. Therefore, I decided to investigate the urban informality phenomenon from a wider perspective. The Arab Uprising conquered the Middle East region and opened the debate of housing and land rights for all strata of the society; hence, it was the time to dig deeply into these issues to understand urban informality transitions at the regional level.

During that period, the dynamics in planning theory were exacerbated by the ascendance of neoliberalism as global economic orthodoxy since the 1990s. This reality unleashed changes epitomized in the neoliberal planning theory. Neoliberal planning serves as a delivery instrument in the neoliberal socio-economic and political economy. After more than five decades of experimentation, neoliberal planning theory is called to serious question due to indications of its poor performance in the delivery of sustainable growth and spatial justice in the Global South. With the beginning of the new millennium, there was intensive innovation of theory and practice of governance of urban sustainability transitions to explore different dynamics, urban challenges, and breakthroughs in accelerating sustainability transitions in urban areas across the globe.

In mid-2016, I submitted an initial book proposal to Springer with the title of "Experiencing Informality: Politics, Economics, and Culture in Middle East Cities". For various reasons, I did not succeed to go on the right track for publication with Springer Nature. At the end of July 2018, I contacted. Mrs Margaret Deignan, Senior Publishing Editor, Earth Sciences, Geography & Environment, at Springer Nature, who helped to put me on the right track of the publication process. Mrs Deignan and I exchanged ideas and proposals for the book and she advised me to update and visit planning innovation on urban sustainability transitions and to link this with urban informality. The proposals were sent to external reviewers and the series editors of the book series Theory and Practice of Urban Sustainability Transitions; we discussed ideas, and feedback on the reviewers' comments were raised. It was a great idea to do so, and I visited the new insights into how cities address the sustainability challenges they face not by returning to old paradigms, or to put new wine into an old bottle, but by searching for new and innovative methods and instruments that are based on shared principles of a transitions approach. Therefore, we agreed on the final title of the book and the modification that was made to the manuscript. Moreover, we agreed that the book will be under the series Theory and Practice of Urban Sustainability Transitions. More than 8 months of exchanging of ideas and debates took place between me and the series editors as we, Loorbach, Margaret, and I, had agreed to publish the manuscript under a separate title. During my stay at the North Coast, west of Alexandria city, from August 2018 till the submission date in July 2020, the initial manuscript was updated and modified, and intensive work, according to the series editors' comments, was done to contribute urban sustainably transitions on urban informality in the Middle East.

Therefore, this study focuses on the clarification of urban sustainable transitions on urban informality. The study started from the thoughts of the Cairo School of Urban Studies, and modified to fit with the nature of the study and formulated what is called the Alexandria School of Urban Sustainable Transitions (ASUST). The study is formulated through three folds which are crosscutting with the other two dimensions. The three folds are: a synopsis of urban informality in the Middle East, ongoing research innovation on the governance of sustainability transitions, and how to deal with urban informality in the context of sustainability transitions. The three folds are crosscutting and overlapped with the other two dimensions: the six axes and the four modes of tensions. The six axes explore: urban sustainability transitions, the urbanization process, the paradigm of informality, state and power, land rights, and social exclusion. The scope of tensions is crosscutting between the three folds and the six axes. The scope of tensions has four modes in which capital accumulation plays a prime element: modes of production, reproduction, consumptions, and distribution of goods and services in the city and its practices. The output of the integration between the three folds and the two dimensions is the determination of dominant areas of developments: areas of land development; discourses about the social organization through lookout opportunities for social inclusion; understanding of the diversity and nature of the states; and transformation of the drivers of sustainable transitions at various spatial levels. The book, therefore, approaches the question of urban informality as the "hotspots" or a "black hole" or "a site of critical analysis" or "a site of transitions" within a city, popular expressions mainly based on the tension and pressure created between landscape and regime and the transformative pressure of niches on regimes. The conclusion is that urban informality has changed in nature; its form, its topologies, and its practices have been transformed by radical transitions in the socio-economic and political context, and the tensions between the four modes in the city and its practices.

The content of the book is a timely response that postulates a theoretically and partially compelling Southern perspective of planning in the Middle East context. Given the precarious disposition of Middle East's political transitions and planning experience, the book proposes a new direction, which rethinks neoliberalism as a meta-theory of planning in the Middle East, especially after the Arab Uprising. It pioneers an original school of thought that presents a general theory and practice of planning for the Egyptian environment, as well as for the Middle East, and other cities of the Global South in the twenty-first century.

Alexandria, Egypt Ahmed Soliman
20/12/2020

Acknowledgment

I acknowledge the divine providence which propelled the delivery of this work and the team spirit of my wife, Manar, and children, Aya and Yahya. I acknowledge Prof. Hernando Do Soto, Prof. Nezar Al Sayyad, and Mr. Geoffrey Payne for writing the forewords and a back cover for this book; I appreciate their valuable support and productive help. I am very grateful to Mrs. Margaret Deignan at Springer Nature Switzerland AG for her time, energy, and continuous support, for her constructive comments throughout the process of preparation, and for the acceptance of the book's manuscript. The anonymous reviewers' comments and the feedback by the series editors of the book series, Theory and Practice of Urban Sustainability Transitions, helped improve the manuscript; I appreciate their valuable comments. Distinct thanks go to Prof. Derk Loorbach and Mr. Satoru Mizuguchi for their valuable comments during the various drafts of the manuscript. Special thanks go to Dr. Christian Witschel, Editorial Director, who took the matter of publication very seriously. Also, the helpful follow up from Mrs Malini Arumugam, Project Coordinator at Springer Nature, and Ramabrabha Selvaraj, Project Manager at Spi, who was responsible for initial and final preparation of the project coordination of the book project, is accredited. The work of the language editor, Kenneth J. Quinn, is much appreciated, who has improved the language and brought it to a good standard. The support of my colleagues, especially Prof. Mohsen Zahran and Prof. Mohsen Brada in the Architectural Department, Faculty of Engineering, Alexandria University, and in the Faculty of Urban and Regional Planning, Cairo University, respectively, are much appreciated. The valuable comments on the introductory and concluding chapters from my colleagues are very much appreciated. The valuable efforts by the production department of Springer Nature are much appreciated; without their skills and intensive care, this work would not be a reality. The valuable comments and opinion of Prof. Ray Bromley are much appreciated. Last but not least, credit for the reproduction of all drawings in this book goes to my son, architect Yahya, whose work I appreciate. I accept responsibility for any omissions and errors.

Contents

Abbreviations

AGM	Amman Greater Municipality
ASUST	Alexandria School of Urban Sustainable Transitions
ANC	Al Alamein New City
ASEZA	Aqaba Special Economic Zone
CAPMAS	Central Agency for Public Mobilization and Statistics
CBOs	Community Based Organizations
CDR	Council for Development and Reconstruction
CO_2	Carbon Dioxide
DFID	Department for International Development, UK
DIY	Do-It-Yourself
DLS	Department of Land and Survey
Drift	Dutch Research Institute for Transitions
ESCWA	A region containing 13 countries and the 6 Gulf Cooperation Council member states
FAO	Food and Agricultural Organization
GAM	Greater Amman Municipality
GCC	Gulf Cooperation Council, contains six Gulf countries
GCR	Greater Cairo Region
GDCD	General Directorate of Civil Defense
GDP	General Domestic Production
GIS	Geographical Information Systems
GIZ	German Technical Cooperation
GLTN	Global Land Tool Network
GNP	General National Production
GOPP	General Organization for Physical Planning
GTZ	German Technical and Financial Cooperation
HCHSV76	Habitat Conference on Human Settlements in Vancouver in 1976
IIHS	Indian Institute for Human Settlements
ILO	International Labour Office

IRFED	Institut de Recherche et de Formation en Vue du Développement Harmonisé (a mission of French experts (priests) who were commissioned by President Chehab to evaluate strategies for Lebanon's development)
ISDF	Informal Settlements Development Fund
ITS	Informal Tended Settlements
JMRC	Jordan Mortgage Refinance Company
JRP	The Jordan Response Plan 2015
KSI	Dutch Knowledge Network on Systems Innovation and Transition
LE	Egyptian Pound
LR	Land Readjustment
MB	Muslims Brothers
MDGs	Millennium Development Goals
MENA	A branch of the World Bank group that includes the Middle Eastern and North African countries
MLP	Multi-levels Perspective
MNCs	Multinational corporations
MP	Multi Phase
MUIP	Monitoring of Urban Inequalities Programme
N-AERUS	Network-Association of European Researchers on Urbanization in the South
NCAC	New Capital Administrative City
NGOs	Non-Governmental Organizations
NSHP	National Social Housing Programme
NUA	The New Urban Agenda
OECD	The Organization for Economic Co-operation and Development, Paris
PILaR	Participatory and Inclusive Land Readjustment
PSUP	Participatory Slum Upgrading Programme
3RP	The Regional Refugee & Resilience Plan
SDGs	Sustainable Development Goals
SDG	Sustainable Development Goal
SNM	Strategic Niche Management
Solidere	A Lebanese joint-stock company in charge of planning and redeveloping Beirut Central District
STRN	Sustainability Transitions Research Network
S&T	Science and Technology
SUD	Sustainable Urban Development
TLM	Transitions Land Management
TT	Technical Team
UN	United Nations
UNCED	United Nations Conference on Environment and Development
UNDP	United Nations Development Programme
UNHCR	United Nations High Commissioner for Refugees
UNICEF	United Nations Children's Fund

UNRWA	United Nations Relief and Works Agency
USAID	United States Agency for International Development
USD	United States Dollars
WHO	World Health Organization
VNR	Voluntary National Review
YNHP	Youth National Housing Program

Part I
Objectives, Substantive Issues, and Structure of This Book

Chapter 1
Scientific, Knowledgeable Background, and Book Architecture

Abstract Urban informality has become the main phenomenon of contemporary urbanization in the Global South. This is due to rapid arbitrary urbanization processes, combined with the rapid population growth that characterize the sociotechnical landscape confronting most of the cities of the Global South, as well as cities of the Middle East. Overall, the urban informality transition is a multidisciplinary and a multidimensional process occurring simultaneously at different levels as a response to informal urbanization and to the transformation of the socioeconomic, socio-spatial, and political milieu of the Global South and, especially, of Middle Eastern cities. This chapter outlines this volume's structure and elaborates the current urban informality transitions and how the recent waves of urbanization and socioeconomic, socio-spatial, and political transitions have changed the complex system of transitions in the urban informality sustainability of the Middle East's cities. Urban informality lacks good services, structures, and institutions, causing persistent environmental and socioeconomic degradations. Various local and international policies and development aid have tried to remedy these but failed to do so. Thus, there is an urgent need to develop planning innovations, vision/discourse, new models/structures/organization, and new practices to realize a transition from informal unsustainable to formal/or semiformal sustainable settlement. The study elaborates a new model for ameliorating urban informality transitions towards sustainability to contribute to the sustainable development process of a given environment.

Keywords Urban informality · Sustainable transitions · Governance of urban sustainability transitions · Globalization · State power · Access to land · Social exclusion · Social inclusion

1.1 Introduction

Flying over greater Cairo, Egypt, the city with more than a thousand minarets, it is easy to recognize the degrees of deterioration and breakdown of urbanity. The belt of urban informality surrounding the perimeter of the Greater Cairo Region (GCR), is

A. M. Soliman, *Urban Informality*, https://doi.org/10.1007/978-3-030-68988-9_1

becoming a common phenomenon in most Egyptian cities. This cancer of urban informality has been infesting not only Egyptian cities but also most of the cities of the Middle East (Soliman 2012a), as well as the Global South. This phenomenon has multiplied on a large scale and has been characterized by growing complexity and diversity over the last seven decades.

In the second half of the twentieth century and at the beginning of the new millennium, cities in the Global South, or what was previously called the Third World[1], and in particular, the cities of the Arab-Muslim world in the Middle East[2] witnessed arbitrary urban socio-spatial growth as their vast transitional and peripheral areas urbanized "from below" (Maconachie 2007; Gerges 2015; de Satgé and Vanessa 2018; Roy and AlSayyad 2004; Rocco and Ballegooijen 2018). Urban informality, illegal housing development, unplanned occupancy, self-built

---

[1]The term "Third World" had its principal origins in the search for an alternative to the polarized politics of the immediate post-war years. A path between the capitalist and communist protagonists of the Cold War is the Third. Thus, the First World was the capitalist West, the Second World was the Soviet Bloc, and the Third World comprised the rest. After the collapse of the Soviet Union and the emergence of globalization and networked societies, the terminology has been questioned. The North and South terminology emerged, while the United Nations is still using "developed" and "developing" countries. Throughout the present study, the "North" or "developed" and the "South" or "developing" countries is used, but there is no exact line to separate the two, unless one were to propose a particular level of income, level of GDP and GNP per person, and the absence of informality. Always the higher level exists in the North the lower in the South. For further details, see, for example, D. Drakakis-Smith (2000) and M. Castells (2010).

[2]Among the many definitions of "the" "Middle East", it is perhaps most easily defined in terms of the peoples who speak Arabic, Persian or Turkish, the main Semitic historic languages of the *Qur'an*. The Middle East's geographic and cultural region is located in southwestern Asia and in north and north-eastern Africa. The geopolitical term the Middle East, first coined in 1902 by United States naval officer Alfred Thayer Mahan, originally referred to the Asian region south of the Black Sea between the Mediterranean Sea to the west and India to the east. The Middle East was seen as the area in between (Dumper and Stanley 2007; Cammett et al. 2015). In modern scholarship, the term refers collectively to the Asian countries of Bahrain, Cyprus, Iran, Iraq, Israel (and the Palestinian-occupied Gaza Strip and West Bank), Jordan, Kuwait, Lebanon, Oman, Qatar, Saudi Arabia, Syria, Turkey, the United Arab Emirates, Yemen, and the African country of Egypt. A broader, more cultural definition might include the Muslim countries of Morocco, Algeria, Tunisia, Libya, Sudan, Afghanistan, and Pakistan. MENA is a branch of the World Bank group that includes the Middle East and North Africa Region (MENA) and covers a wide array of countries from Morocco in the west to Iran in the east. Each one of the countries has a long and rich history and strong individual characteristics. Others (Amineh 2007; Elsheshtawy 2004) called "The Greater Middle East", in addition to the previous countries, the countries of Afghanistan and Pakistan; and the Central Eurasian countries (i.e., the five Central Asian republics and the three new states in the south Caucasus). A region-wide strategy is bound, by its very nature, to generalize that do not always do justice to the specific situation of a particular country. Nevertheless, several similarities exist that constitute a good basis for general principles and overall guidance, which then needs to be adapted in what must remain country-specific strategies. It is important to keep this caveat in mind while reading the present study. On the other hand, the ESCWA region contains 13 countries, the six of the Gulf Cooperation Council (GCC) which are Bahrain, Kuwait, Oman, Qatar, Saudi Arabia, and the United Arab Emirates, and others with more diversified economies: Egypt, Iraq, Jordan, Lebanon, Palestine, Syria, and Yemen. For further discussion of other definitions and arguments, see (Khalidi 1998).

structures, illegal conversions of agricultural land, peri-urban areas, land taking, temporary appropriation of public spaces (Antonio et al. 2020) are among the current terminologies applied to the urban fabric of the Middle East's cities that have created various urban problems on the periphery of those cities. Long denied, ignored, or suppressed, these quarters appear as an accursed aspect of a resolutely modernist urban project conducted since the independence of these countries. Roy (2011) emphasized in her work on post-colonial urbanism how the study of cities can be enriched through a renewed engagement with postcolonial studies, but all the cities of the Middle East are not categorized within this classification. Most of the Middle East's cities have their own traditional settings and are very rarely classified as postcolonial urbanist as they are no longer offer "subaltern resistance". Cairo, for example, has had a unique urbanism through its history, and so, too, Beirut, a city of seven civilizations. Amman as a Tribal city. Tripoli, Damascus, and Baghdad were nearly destroyed after the Arab Spring. New cities along the Suez Canal in Egypt, namely Suez, Port Said and Ismailia, were built during the construction of Suez Canal which officially opened on 17 November 1869. These cities were built based on the concept of garden cities, and social cohesion was a dominant consideration. It seems that postcolonial urbanism is no longer in the form of "subaltern resistance" in the Middle East, but it may be so in Latin America and India. Except for a few, northwest Africa's cities are rather 'a new way of urbanity' (Soliman 2017).

Due to the socioeconomic instability, weak state apparatus, less transparency in legal proceedings and enforcement of legal frameworks and relatively high levels of economic and social inequality, most of governments of the Middle East do not overcome the problems of informal urbanization in which urban informality flourished and which caused various urban problems. However, many housing policies have emerged (Soliman 2012b) that do not meet the basic needs of the low-income groups, nor do they encourage the necessary transition to "formalize informality" as urban informality continued to spread on the periphery of the major urban centers of the Global South.

However, the need and drive for urban sustainability transitions (Kabisch et al. 2017) and the environmental consequences of urbanization are apparent (Vasenev et al. 2018). Urban sustainability transitions are nonlinear and complex processes of change, driven by deeper transformations through persistent societal challenges in which fundamental and structural changes in urban systems are addressed (Frantzeskaki et al. 2017a, b). Sociotechnical transitions examine how material, social, and cultural changes of socioeconomic-spatial dynamics interact in transitions towards sustainable development. Sociotechnical is an approach of urban sustainable transitions, and its components of the multilevel perspective of landscape, regime, and niche correlate with urban informality transition; this occurs through changing the goals, opportunities, potentialities, and areas of urban development, formally or informally. Little attention is given to the role of spaces and places in transitions which makes one ill-prepared to understand and explain its geographically uneven development (Coenen et al. 2012). Thus, the urban informality, due to informal urbanization, cannot be addressed by incremental improvements, but by its resilience and adaptability which require shifts to new kinds of systems, shifts which are called *'urban informality sustainability transitions'* (Soliman 2020).

This volume seeks to contribute to this examination by answering two interrelated questions: What sorts of relaxations, tensions, and pressures among the sociotechnical and socio-spatial systems for urban informality transitions emerge within an urban context?; and to what extent can urban-informality transitions be sustainable at the urban level and how? By exploring these two questions from a sociotechnical perspective, this volume seeks to respond to a notable gap in the urban informality transitions literature.

This book develops an analytical framework for the serious investigation of the historical specificities of the Global South's urban informality within two different Mediterranean countries and within an inland territory. The retrospect of urban informality draws useful lessons to investigate how to drive this phenomenon to integrate or to create a sustainable built environment. Investigation of urban informality in the Middle East begins with transregional comparisons between the southern and the eastern coasts of the Mediterranean, or between Egypt, Lebanon, and Jordan. Its starting point is the research on urban informality in three different environments; Egypt has endured tremendous socioeconomic transformation, and the other two countries are confronting a huge influx of refugees and social diversity, and they have encountered the political conflicts that occurred in the Middle East with the emergence of a free-market system. It is very rare, however, to find comparative and analytical research within three different environments in the light of socioeconomic structural composition, socio-spatial structure, political systems, sizes of the population, and size and type of territory and terrain. Also crucial is the need to deeply rethink the assorted explanations that would lead to the integration of urban informality into urban sustainability transitions.

This chapter comprises five sections. This first section serves as an introductory slice, which is followed by the emergence of urban informality. The third highlights a reflection of urban sustainability transitions on urban informality. The fourth explores the contribution of urban informality in the transition process, while the fifth describes presents the book's structure, which contains four parts.

1.2 Emergence of Urban informality

Slums or informal settlements, as a sociotechnical landscape, are caused by a range of interrelated factors including informal urbanization, weak governance—particularly in the area of policy, planning and land management, migration and the movement of people related to urban densification, disaster, conflict and economic vulnerability, climate change, long term poverty and the lack of affordable housing. These have led to the appearance of an emerging urbanism, known as informal, urban informality or indiscriminate forms of urbanity. Today, most cities of the Global South are surrounded by belts of urban informality, or a black hole in cities (Harris 2018), or "a site of critical analysis" (Banks et al. 2020), which constitute between 30% and 40% of housing production in these cities. This urban informality causes serious urban, social, and environmental problems which has led into the deterioration of the built environment. Issues of stark socioeconomic inequalities, social exclusion, extreme poverty, high unemployment, marginality, pollution, environmental degradation, inadequate infrastructure, poor environment conditions, climate change (Satterthwaite et al. 2020), as well as unsustainable environmental footprints beyond the city boundaries that are linked to the spread of urban informality. Indeed, more than one million urban dwellers now live in slums or informal settlements, and the absolute numbers increase every year (UN-Habitat 2018). Other studies indicated that more than a fifth of the world's urban population is living in informal settlements (Satterthwaite et al. 2020).

In the literature, there are various thoughts and policies that act as sociotechnical niches towards the transformation of slums and the appearance of urban informality in cities. In 1954, W. Arthur Lewis (1954) identified informality as presenting a dualistic model of interaction between the modern and traditional sectors in under-developed countries. But there were, of course, earlier rounds of discussion about informal sector, notably the Kenya report debates of the early 1970s. In 1971, in Hart's study of urban migrants to Accra, Ghana, Hart then proposed dividing the economy of developing counties into two sectors, formal and informal. However, in recreating and renaming Lewis's categories, Hart spoke of income opportunities rather than sectors, and he defined informality simply as "self-employment." His distinctions were subsequently defined more precisely by the International Labor Office (ILO). It could be said that the slum phenomenon was explored a long time ago, indeed since the Industrial Revolution (Geddes 1915).

The early search on informal housing and land markets was initiated in Latin America in the 1960s and involved a rich diversity of political scientists, architects, anthropologists, sociologists, and many other experts. Charles Abrams (1964) provided a wealth of first-hand experience that improved our understanding of the phenomenon of informal housing in three continents: Latin America, Asia, and Africa. He studied the relationship between informal rapid urbanization, urban land, and shelter, and concluded that a great deal could be done to improve conditions in urban informal housing through better planning and management of urban resources and through greater political courage in dealing with predatory interests.

The extensive experience of John Turner's six years of research in Lima, Peru, reported in the article "Dwelling resources in Latin America", appeared as a special issue of Architecture Design in August 1963. Later, Turner's novel idea (Turner 1976), was presented in "Housing by People" at the first UN-Habitat Conference on Human Settlements in Vancouver in 1976. He advocated a radical change of relationships between people, as a niche innovation, and government, as a sociotechnical regime, in which government ceases doing what it does badly or uneconomically—building and managing houses—and concentrates on what it has the authority to do: ensure equitable access to resources which local communities and people cannot provide for themselves. Architecture for the Poor (Fathy 1973), a classic work by Hassan Fathy, pioneered appropriate technology for building in

Egypt, especially by working to reestablish the use of traditional mud construction as opposed to Western building designs, material configurations, layouts, and relied on a self-help housing technique. This was the first to emphasize a radical change or transition between the government and local community for housing production to provide equitable access to resources and services on one hand and to sustain natural resources on the other. Hidden potential transitional pathways were to construct housing for low-income groups in a sustainable way, but the relationships or potential pathways between the modes of informal housing systems of production/reproduction, distribution, consumption, and other forms of government institutions, capital, and community concerned were not stipulated.

Since then, many scholars and international organizations have studied (Ward 1982b) the main causes of the spread of urban informality in the cities of the Global South, wherein many concepts have emerged. In spring 2001, a two-day symposium was held at University of California entitled "Housing and Urbanization in Developing Countries". From this emerged an important volume edited by Roy, A. and Al Sayyad, N. (2004) entitled "Urban Informality: Transnational Perspectives from the Middle East, Latin America, and South Asia". The present author participated in this symposium and contributed chapter. Prior to this symposium, an indispensable book entitled "The Mystery of Capital: Why Capitalism Triumphs in the West and Fails Everywhere Else" by Hernando De Soto (2000) argued that the importance of property ownership, as a part of sociotechnical regime, is the key to ending poverty—but that it will work only if the poor can use their property to generate further wealth. In other words, it is necessary to change dead capital into a live capital so as to contribute to the development process in a given environment.

A key assumption underlying the work of all those who advocated a reconsideration of the meaning of informality has been that poor countries must diversify their economic base away from dependence on primary production through manufacturing. In this regard, scholars have claimed that small enterprises generate more jobs and require less capital investment per job than larger enterprises. The debates were about the nature of the informal sector (Chen and Carré 2020) and of dual economies, or "invisible hand of the market" (Pittfor 2019) and did not directly tackle the socio-spatial aspects of urban informality and the emerging forms of urbanism. All of the above studies have tried to link urban informality with the progress of informal economic development, level of urbanization, socioeconomic transitions, availability of affordable land, level of intervention of international organizations on the informal housing systems, and the political willingness of government to meet the basic needs of the bottom strata of a society. None of the above scholars did not try to link urban informality and the ongoing transitions of socioeconomic, socio spatial, and political circumstances, nor they did not try to investigate how to manage urban informality transitions through the interconnected aspects that might led to the formalization of urban informality in a sustainable way.

For most of the metropolises in the Global South, as well as cities of the Middle East, the beginning of the twenty-first century saw the spread of urban informality and the possibility of its integration and adaptation with the urban fabric (Hosseinioon 2018, 2019; Soliman 2019). This beginning was accompanied with a standstill and frozen social, economic, political, and environmental transitions all over the Global South as twentieth century's planning model was extended into the twenty-first century.

In the context of economic globalization, the absorption and containment of unplanned zones of urban development have become imperative as a minimal response to the extensive liberalization of the use value and exchange value of land and to the redistribution of resources and power "in the city." Hence, urbanization has become a driving force for socio-spatial development, as well as a source of development with the power to change and improve lives (UN-Habitat 2018a, b). This fundamental long-term transformation will involve the realignment of urban society, its technologies and infrastructures, urban cultures, and lifestyles, as well as governance and institutional frameworks. The diversity, complexity, duality, and contingent interactions between informality and formality are the main urban challenges that would lead to a new and more nuanced appreciation of the production of urban spaces across cities of the Global South. The New Urban Agenda[3] (NUA) (UN-Habitat 2016b) emerges as a tool to harness 'the transformative power of urbanization". Among the main challenges of the rapid urbanization process in the Global South are the arbitrary socio-spatial development combined with the spread of the urban informality on the periphery of cities and also how to handle the transitions of the arbitrary development to meet the continuous process of transitions of socioeconomic and political milieu that occurred in cities of the Global South.

Two opposing perspectives on the political meaning of urban informality have been established (Rocco and Ballegooijen 2018; Harris 2018). The first, the 'emancipatory perspective', frames urban informality as a practice that fosters autonomy, entrepreneurship, and social mobility. The other perspective, a more critical one, sees informality predominantly as a result of political exclusion, inequality, and poverty. But it would argue that urban informality is a response to rapid informal urbanization and socioeconomic, socio-spatial, and political transformations in rapidly developing settings. In other words, urban informality is *"a reproduction of housing production to a status quo"* and *"a site of transitions"* of the continuous transitions of socioeconomic and political milieu. It therefore does not necessarily relate to poverty (Roy 2009), rather it relates to capital accumulation, where real estate assets are considered the most secure future investments in rapidly developing settings (Soliman 2012a; Harris 2018; De Soto 2000). Subsequently, the scene of informal urbanization took place under the spread of urban informality, which led to the change of the thinking of the scholars towards housing policies.

---

[3]The New Urban Agenda (NUG) represents a paradigm shift based on the science of cities; it lays out standards and principles for the planning, construction, development, management, and improvement of urban areas along its five main pillars of implementation: national urban policies, urban legislation and regulations, urban planning and design, local economy and municipal finance, and local implementation. It is a resource for every level of government, from the national to local; for civil society organizations; the private sector; constituent groups; and for all who call the urban spaces of the world "home" to realize this vision (UN-Habitat 2016b).

It is noted that the urban informality has changed in its nature, in its form, in its practice and in its topologies (Sims 2010; Soliman 2019; Roy and AlSayyad 2004; d'Alençona et al. 2018; Devlin 2018; Marx and Kelling 2019; Obremski and Carter 2019; Boanada-Fuchs and Vanessa 2018). The socioeconomic-spatial transitions of urban informality have produced three fundamental forms: slums, informal settlements, and inadequate housing (United Nations 2018). Others classified housing typology in three forms: semi-informal settlements, squatting housing and hybrid or ex-formal housing. The three typologies further identified twelve subtypes within the three main types, and a total of 13 minor variants were also identified within some of the more complex subtypes (Soliman 2004; Grashoff and Yang 2020; Boanada-Fuchs and Fuchs 2018). This typology depends on socioeconomic-spatial transitions over time, as a result of a new planning innovation, in which it changed types according to transition pathways.

In practice, urban informality is in transition as a result of changing sociotechncal landscape pressures (migration, informal urbanization, climate change, demographic changes, digitalization, etc.) and emerging new visions, models, tools, mechanisms, and practices (as sociotechnicl niches). The way informal settlements have developed and are managed is structurally changing: from self-organized and driven by economic forces to empowered community collective coordination of sustainable and more just socioeconomic systems. Also, urban informality, as a sociotechnical landscape, has been transformed by radical transitions and tensions and pressure of "a sociotechnical regime" in the socioeconomic and political context on one hand, and the effect of regularization or "integration", as "sociotechnical niches", imposed or encouraged by the states or by the international organizations (Harris 2018; Hosseinioon 2019; Fawaz 2017), on the other hand.

However, urban informality from transitions perspective, as *"a site of transitions"*, is defined as the outcome of the chaotic urbanization associated with capital—whatever the amount, sources, and sorts, formally or informally. It is generated by urban regimes (public and private, whatever their legal form and economic status), and its path-dependency, utilizing the main collectives resources within the market (land, labor, material, etc., as well as, the legislative process that controls the operation of these resources) in facilitating and controlling land provision for housing and economics activities, interacting and integrating within the desires of the grassroots, who create new planning innovations, as sociotechnical niches, for socio-spatial struggling for *producing, reproducing, distributing, and consuming* of goods and services at a reasonable cost, in a reasonable time and in sustainable way. Yet the grassroots are creating micro-level innovations, as niches, to build integrated systems from below for sustainability. In short, urban informality is operating as *"a site of transitions"* to cope and integrate with the tremendous ongoing transitions of the socioeconomic and political milieus.

These transitions and policies have introduced new dynamics in the urban informality (such as socioeconomic-spatial transitions), to such an extent that it is today necessary to reformulate new models and analytical tools (as niches) applied to them. New fundamentals of "an urban regime" have to be reconsidered: the matter of urban land and, more broadly, the new urban configuration, and the capital dynamic

(Piketty 2014; Amin 1976) or circular economy (Pomponi and Moncaster 2017) or a green new deal to overcome the triple crunch (Pittfor 2019).

However, land liberalization (McAuslan 2013) and the mode of governance (Gupta et al. 2015) constitute the urban regime, while social movements, social capital (Dennis 2005), the appearance of new actors (like NGOs, international donors), citizen participation (Dominelli and Moosa-Mitha 2014; Wittmayer et al. 2019) constitute "a sociotechnical niche". The continuous transition of urban informality, as a sociotechnical landscape, has amplified the tensions and pressures between modes/transitions and pathways of production, reproduction, consumption and distribution of goods and services. A shocking transition attacked the Middle East in the form of "The Arab Spring" in which democracy and major socioeconomic changes (or regimes) have introduced new ways to handle, enhance, or formalize urban informality as a primary concern for human rights and social inclusion, as well as rights to the city as a main principle for human rights. Property rights or legal land adjustments and reform and reconciliation are today's realities of "a regime" to ensure better practices and responsible performance by various states in the Middle East (Soliman 2012a), e.g., see the Egyptian case's chapter. In terms of access to land, as a part of a regime, property rights are evolving to become the rights of being secure in a place. This means redefining the structure of land legalization procedures, as transition pathways, within the state and its responsibilities. The question of the illegality of land and real estate today concerns most of the inhabitants of regional metropolises (Payne 2002), as well as those of secondary cities. It can therefore no longer be contained in an unacknowledged fringe or "*a container for storing the human rights*."

Furthermore, from the perspective of the overall control desired by the local powers, the passage from a policy of containment to a policy of transitions has become vital and represents an inescapable reversal. Urban informality is in transition and being considered as an integral part of the urban matrix (Devlin 2018), constituting a major part of the built environment on the ground. Transitions management of urban informality requires the introduction of a multidisciplinary and multidimensional approach. Also, socioeconomic-spatial transitions and the issues of land and real estate, as well as with the sociological examination of ways to develop and manage cities, are essential. Moreover, one has to be able to grasp these new modes of hybridization of urban informalities (Harris 2018) and its territories and their transitions' management, as much based on their similarities as based on the expression of their differences and their political, cultural, and socioeconomic manifestations.

However, urban informality, as a transitive phenomenon in the form of *"a site of transitions"*, in a broad sense is attached to unusual and various kinds of arrangements, networks, activities, production, reproduction, the distribution and consumption of goods and services that produce 'a new way of life' (Al Sayyad 2004) or 'ways of life' living in a 'culture of poverty' (Lewis 1966), or living in a "culture of diversity". Many scholars view informal housing, or urban informality, exclusively as a new type of built environment, or 'a new urban pattern', 'a new urbanity' (Soliman 2015a), or a new forms of self-governance management and local experimentation intended to reconfigure the production of urban informal housing[4]. In general, the grassroots have formulated a method or tools of change, as a sociotechnical niche, for meeting their own needs, while others, such as a regime, do not understand or try to understand this process of transitions. The goal is how to formalize the informal sector, to find a compromise between the grassroots and decision makers, or at least to integrate the urban informality into a formal context in a sustainable way. However, urban informality involves a broad range of diverse illegal activities, or illegal pathways, involving various aspects of life, rather than a strategy (d'Alençona et al. 2018), and is processed and interacts legally and illegally with the prevailing transitional conditions of a given environment. This means that an urgent need to develop a new vision/discourse, new models/structures/organizations, and new practices is crucial to realize a transition from informal unsustainable to "formalized sustainable settlements".

1.3 Reflection of Urban Sustainability Transitions on Urban Informality

Modern societies face fundamental sustainability problems in several domains, such as energy, water, transportation, informal urbanization, and the triple crunch of financial, climate change, and the coming global energy (Pittfor 2019). In recent years, planning innovations, in the form of system innovations and transitions, have taken place (Elzen et al. 2004; Geels and Schot 2010; Markard et al. 2012; Koehler et al. 2017). It has therefore been suggested that societies need to fundamentally restructure systems of production, reproduction, consumption, and distribution of goods and services by initiating so-called sustainability transitions, aiming at sustainability for the future of the development of urban agglomerations on the planet. The emerging field of transitions is characterized by a wide variety of topics, approaches, and methodologies (Elzen et al. 2004; Geels and Schot 2010; Markard et al. 2012; Koehler et al. 2017).

Transition concepts are an analytical framework to analyze dynamics and patterns from one condition to another in specific societal systems (Loorbach 2020). Transitions towards sustainability is consistent with the general understanding of sociotechnical transitions, which are conceptualized as major changes in technological, organizational, and institutional terms for both production and consumption of goods and services. Sociotechnical transitions involve a broad range of actors (Grin et al. 2010) and typically unfold over considerable timespans (e.g., 25 years and more), have the flexibility to be locked in and locked out, and can be coproduced by both technological and social dynamics (Wittmayer et al. 2019). It assumed that the

[4]Housing informality reflects the informal process, as a new trend, of housing production by the bottom strata of society, while urban informality reflects a context of various informal socioeconomic and spatial activities and a dual system of the socio-political arena.

Global North undergoes long periods of relative stability and optimization that enable societal systems to be transferred in a sustainable way, but in the Global South this seldom happened, and it is time to try new planning innovations to improve the fragmented situation and to cope with the multi-sectoral dynamics of transitions.

One of the central approaches that describes and analyzes such complex transformational processes is the multi-level perspective (MLP) (Geels 2005; Geels and Schot 2010; Grin et al. 2010). The model envisions sociotechnical transitions to unfold through developments on three analytical 'levels': innovative practices (niche experiments), structure (the regime), and long-term, exogenous trends (the landscape) (Schot 1998; Rip and Kemp 1998; Geels 2005).The sociotechnical landscape, on the macro level, forms a broad exogenous environment that as such is beyond the direct influence of regime and niche actors, e.g., urbanization process, demographics changes, way of life, etc. A regime, on the meso level, is a dominant set of structures, culture, and practices which are shared by groups of actors. The existing regimes at the meso level often slow down the processes of change, but the power of regimes may also be utilized to bring about or not a transition. Niches, on the micro level, may develop by the grassroots and/or by the bottom strata of the society , such as new initiatives, new techniques and new forms of culture and management. Niches may also emerge within a regime and not only on the micro level (Grin et al. 2010). Regime innovation may be to replace or absorb niche innovation and the opposite seldom happened. Within niches, there are learning processes regarding innovations, new practices, and behaviors. There is a process of variation and selection on this micro-level resulting in path dependencies (locked-in or locked-out, or windows of opportunity or windows of necessity or a window of status quo), which may lead to the exclusion of other paths.

Urbanization, as a sociotechnical landscape, has multiple social, political, cultural, environmental and economic dimensions that profoundly influence social development and innovation (Frantzeskaki et al. 2017a, b). The questions therefore are how to alter urban informality transitions, how to make these areas sustainable and resilient, how it will be governed in a sustainable way, and how it will be converted to prosperity. The sustainability and transformations of urban informality, in particular in cities of the Middle East, and the practice of everyday life (Lefebvre 1971) and social security, as a regime (Gough et al. 2008) that constitutes "a new pattern of urbanity" (Soliman 2012a, b), or "a site of transition", as a niche, are predicated upon and conditioned by governance and political economy. Also, a regime in the form of corporate governance, corporate social responsibility, corporate international donors and sustainability are not interchangeable but are inevitably related and all must appear (Crowther et al. 2018) at the right times within the right place. While the grassroots, as the main component of a niche, are competing, adjusting, and resilient in all aspects of changes (Hosseinioon 2019) systematically and sustainably to comply with transformations. This has led to establishing or modifying or altering the existing urban informality, according to the availability of resources, on an incremental basis for example the process of establishing an informal settlement at scattered, collective, and consolidated transitions bases. Such

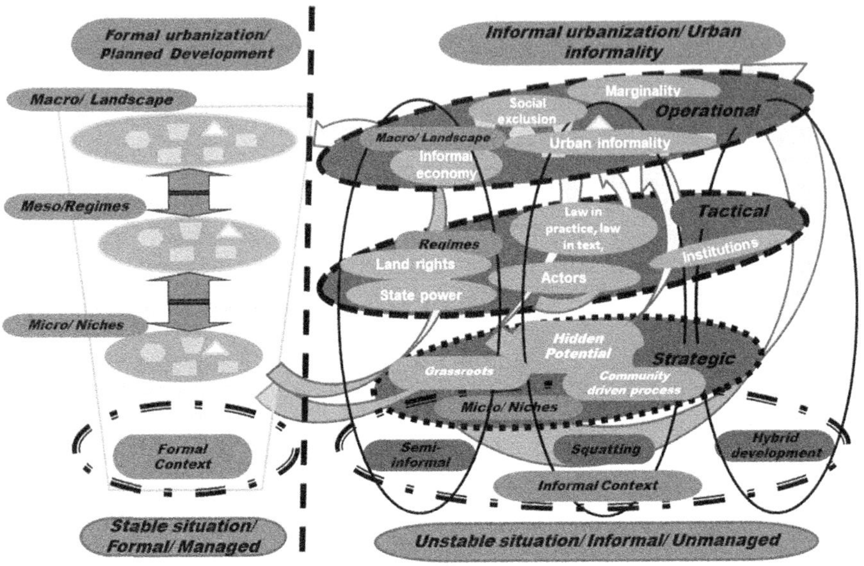

Fig. 1.1 Diagram of the relationship between urban informality, as a sociotechnical landscape, and the multilevel perspective and how it operates. (Source: The author and adapted from Grin et al. 2010)

structural systemics of the local grassroots and realignments within urban contexts can be referred as a part of urban sustainability transitions (Frantzeskaki et al. 2017b; Elzen et al. 2004).

Figure 1.1 illustrates the multilevel Perspective (MLP) as a reflection of sociotechnical system. As indicated on the left-hand side, it constitutes three structural levels: niche, regime, and landscape (Grin et al. 2010; Geels 2010; Geels and Schot 2007). On the right-hand side, the MLP illustrates urban informality with its three typologies of semi-informal settlement, squatting settlement, and hybrid development. At the niche level, the three informal typologies are innovated by the grassroots to satisfy their socioeconomic and housing needs. Informal settlements are formulated through hidden potential pathways (Soliman 2019) of the grassroots through a "community driven process" (Soliman 2017) as an innovation for housing policy. Moreover, the sociotechnical regime has a major impact on activities, political willingness, socio-spatial changes, peoples' behavior, capital accumulation, and social amenities for the bottom strata of the societies in the Middle East. Land rights, state power, and various official institutions constitute the regime in which tensions and pressure in the socioeconomic and political context are imposed. On the socioeconomic landscape, urban informality emphasizes alignments and interacting processes and the importance of tensions and pressures, which create windows of opportunity for transition. The process of hybrid settlement is a way of formalization through lock-out methods of legalizations, whatever these methods may be, followed by transitions from informality to formality. In the case of semi-informal and squatting settlements, security of tenure and regularity frameworks are legal channels for development that would result in a systematic urban informality sustainability (see, for example, (Soliman 2017)).

1.4 Contributing to Urban Informality Transitions

Achieving better understanding of the governance of urban informality sustainability transitions presents a main agenda for the SDGs and the NUA. Several studies have argued the need for policy instruments to combine various policy instruments in so-called transformative learning (Kim and Merriam 2011), including mixed methods (King 2009; Madsen and Cook 2010), and experimental studies using arts-based methods (Butterwick and Lawrence 2009) to reach a practical and a theoretical determination of an issue. However, the study employs a transformative framework together with mixed policy and art-based methods to cope with the diverse case studies to better understand transformative learning, and they reflected the epistemological perspectives that researchers bring to these studies. These transformative, mixed methods, and arts-based methodologies are a framework of practical systems that directly engages the culturally diverse countries with a focus on increased social justice (Mertens 2012) for accessing urban informality sustainability transitions. This approach identified three basic acceptance systems that constitute a paradigmatic viewpoint: ontology (the nature of reality), epistemology (the nature of knowledge and the relationship between the knower and that which would be known), and methodology (the appropriate approach to systematic inquiry) (Mertens 2012).

This book sets the context for the discussions of the societal system of urban informality transitions in the Middle East. A system is here defined as a network or correlation between three pillars, institutions (states), material artifacts (capitals) and knowledge and inspiration (societies) that articulates a transition development process. The societal system in the Global North went through long periods of relative stability and optimization that were followed by relatively short periods of radical change. But in the Middle East, stability is very rare and new planning innovations are needed to overcome a fragmented situation. In the Middle East, one of the most important contexts of the urban transition is the peri-urban interface or urban informality where there is a connection of formal and informal institutions, duality of economy, duality of poverty (poverty in cash and poverty in kind), inequality, and environmental degradation (Allen 2014; Sims 2010). Formal/informal land-market forces are often most closely associated with such transitions (Soliman 2017). The central assumption is that urban informality transitions produce new spaces and places, and this alters the existing ones, not only physically but also in all aspects of socioeconomic and political transitions, and would reflect the urban sustainability transitions on the quality and the level of production, reproduction, consumption and distribution of goods and services which would "constitute prosperity" . This fundamental long-term transformation of urban informality involves the realignment of urban society in all aspects of life. In effect, such readjustments are shockwaves, nonlinear and complex processes of sustainable transition of societal system, driven by deeper transformations compete/cope either with such transformations or what is actually happening on the ground.

The question is how to alter the dynamic of urban informality in a way to contribute to urban sustainability transitions, as well as to integrate it within the urban context. The study assumes that the underlying patterns and dynamics of historical urban informality transitions can be used as bases for hypothesizing possible future urban patterns and possible interventions. It is not the intention to examine differences of planning theories between the Global North and South, nor to make a comparison between traditional planning in the Middle East and colonization planning, nor to make a comparison between formality and informality. The question posed is, instead, why it is possible to apply approaches from the Global North in the Global South/Middle East? It should be noted that most scholars who have researched urban informality in the Global South have or are being educated in the western societies, e.g., John Turner, Charles Abrahams, Geoffrey Payne, Forbes Davidson, Mike Davis, et al. It is very rare to find scholars who are from the Global South speaking to the issues involved.

The focus of this book is the Arab–Muslim world in the Middle East. It has the potential to explore new perspectives in urban sustainability discussions in the Middle Eastern countries, which is closely linked to urban sustainability in Europe (Devlin 2018; Harris 2018). The book draws upon experience and intensive work in three countries: Egypt, Lebanon, and Jordan. The transregional contexts and research traditions specific to these areas provide particularly useful entry points to investigate informality and socioeconomic and political transformation in the Middle East. Addressing the issue of state power and market forces in the housing process is also discussed because similar issues could also be observable in other countries of the Global South, as well as those of the Global North. Therefore, this volume examines the relationships and the ways that the residents within informal settlements, the government and professionals are communicating with each other in the formulation of the informal land markets that constitute a part of the regime. Gaining experience from this special relationship, it becomes apparent that all actors have a certain role to play in the formalization process of land tenure (UN-Habitat 2016c; Soliman 2017; Grin et al. 2010). The main question is: What the significance of "secure land rights"?

The unique feature of the book is its transregional focus on three distinct states: Egypt, Lebanon, and Jordan. First, the book takes account of the specific effect of the three countries' population that marks the level of urbanization. Second, it investigates of the effect of globalization on the process of urban informality within three different contexts. Third, the three countries have hosted an enormous number of refugees. Fourth, the book sheds light on the many ways that the three states have attempted to intervene in informal areas to progress towards formalization. This is evident in the thematic organization of case-studies where the exploration of a substantive question is pursued.

Egypt, Lebanon, and Jordan have given rise to burgeoning cities in the so-called Global South challenge of the Middle Eastern countries. The three countries have different socioeconomic structure, vary geopolitically, and all are involved to some extent with the confrontation with Israel. Each of the three countries was affected in one way or another by the so-called Arab Spring. The three countries have continuously sought organizational and management innovations in traditional bureaucracies. In the social sciences, five major challenges have been dominant in defining the parameters and perspectives for investigating urban informality transitions in the three countries. These challenges are grouped into three sets: the regime set as the political economy of globalization and the influx of refugees; the landscape set as indiscriminate rapid urbanization, demographic context, and urban informality; and the niche set that comprises the various innovations that have been applied in those cities. Examples of innovations include, for example, the Cairo school of Urban Studies and its modification to the Alexandria School of Urban Sustainable Transitions. Lebanon and Jordan bear a new Marxist pedigree and are thus overdetermined in their privileging of capitalism as the only mechanism and class struggle as the only resolution to urban problems. Egypt, in a transition era, has witnessed the transformation of political economy structures, socioeconomic transformation, and great upheavals of urban proletarians. The book, therefore, approaches the question of urban informality as the 'hotspots' or "black holes" or "a site of critical analysis" or "a site of transitions" within a city. These are based on a popular expression concerning the tension and pressure created between landscape and regime, and the transformative pressure of niches on regimes.

However, during the last few decades, the Middle Eastern countries have witnessed incredible changes, external and internal, in the urban, political, and socioeconomic milieu. Subsequently, new transitions of socioeconomic, political, and environmental orders have emerged[5]. Globalization has spread and put extra strain on the region. The world financial crisis[6] of 2008 (Harvey 2016) and the world health crisis of Corona virus (Covid-19) with its sharp drop in share prices have put further pressure on these countries. Thus, the international economic market after

[5]These changes, for instance, are: the invasion of Iraq; the changing presidency of Syria; the assassination of Rafiq El Hariri on 14 February 2005 in Lebanon and Anwar Sadat on 6 October, 2001, in Egypt; the death of King Hassan of Jordan; the death of Yasser Arafat; the death of Saddam Hussein of Iraq; and the continuing conflict between Israel and Palestine (Soliman 2012a). The invasion of Iraq in 2003 initiated by George W. Bush was the beginning of the remapping the Middle East region. During December 2010 and January 2011, the 'Arab Spring,' or Arab awakening/revolts, or Jasmine Revolution, were the most influential events that caused significant changes in the Middle Eastern countries at the time (Talani 2014; Fergany 2016). In early 2019, peoples' movements changed the political situation in Algeria and Sudan. On the global level, the aftermath of the attack on the World Trade Center on the 11 September 2001 posed a continuing challenge, and it still is for the Middle Eastern countries.

[6]The price of oil reached its peak in June 2004, rising to $42 per a barrel, while in September 2008 it approached $140 and dropped in October 2008 below $5 after the international economic crisis. Due to Covid-19 crisis, the barrel price dropped below $25 in April 2020, matching the price in October 1973.

Covid-19 will be changed, and a new economic era will have begun. Therefore, the roles of the states in the Middle Eastern countries, and their relationship with civil societies, are starting to be remodeled in their policies towards the rest of the world and towards their own civil societies and political economies (Fergany 2016; Jamshidi 2014). Thus, shedding light on the recent changes of socioeconomic and political transitions at national and international level is essential. The nexus of these changes influences the built environment of the cities of the Middle East, particularly the spread of the urban informality.

As a result of the speedy urban expansion of the cities of the Middle East, like Cairo, Beirut, and Amman, innovative planning thoughts have emerged. There is a shift of planning from the 'Chicago School' of urban sociology (CSUS) to the 'Los Angeles School' of Postmodern Geography (LASPG) to absorb this growth (Roy 2014), and a recent shift to the Cairo School of Urban Studies (CSUS) concerning the importance of social movement. The latter school strives to capture the innovations of a generation of Cairo-centered scholars and critical thinkers. It is now time to introduce Southern mode of thinking, modifying the CSUS initiated in 2005, when reflecting upon the experience gained from the local and international scholars and as a reaction to grasp the current rapid informal urban agglomeration and the speedy social-movement transitions. The CSUS explored six crucial issues: cities and citizenship, elite domination, urban social movements, state forms, public policy, and subaltern politics in the globalized Middle East (Singerman and Amar 2006). The CSUS worked against the grain of the two dominant dualistic paradigms: the neo-liberal notion of dictatorial regimes locked in opposition to liberal middle class civil societies, and the neo-imperialist 'Clash of Civilizations' vision that imagines the world split between elite Westernized secularist modernizers and backward Islamist fundamentalists. After the Arab Spring, especially in Egypt—after the 30 June 2013 revolution—the regional structure has changed wherein Egypt is taking a lead to modernize and develop the country after the clearance of the "Muslim Brotherhood" (in Arabic, *al-Ikhwan al-Muslimin*, or simply *al-Ikhwan*) and other Islamist groups who tried to apply of the neo-imperialist "clash of civilization". These changes might result in a Fourth-Deep Transition as the outcome of the recent social movements, as has happened in Sudan, Algeria, and Paris (Yellow Jacket protesters), and the protests that began in Minneapolis on May 26, 2020, following the death of George Floyd, all of have been followed by Global Anger, not only in the Global North but also in the Global South. These social movements occur in complex urban systems during a time of rapid urban agglomeration changes and socioeconomic, political, technological, and global transformations. They are asking for a better urban future and enhancement of the standard of living. The appearance of urban informality, as the main components of societal technical systems, seeks to satisfy housing needs and other socioeconomic amenities for the bottom strata of a society. The CSUS emphasized a cosmopolitanism that has often been embedded in transnationalism, normative, universalist, and imperialist discourses (Singerman and Amar 2006). But the CSUS was introduced before the Arab Spring, Covid-19, and the collapse of *Al-Ikhwan Al-Muslimin,* and many deviations have occurred at the local and international levels, so a small shift to this school should transfer from urban studies to urban sustainable transitions studies. The latter is more comprehensive, in a continuous sustainable transitions process, resilience to cope with fluctuations of transformative circumstances. It has a wide scope to acclimate with international forces, and includes urban studies as well as all urban challenges in cities to cover all aspects of transitions that are going on. Let us call this the "Alexandria School of Urban Sustainable Transitions (ASUST)", as the author is working at Alexandria University and has experience in rapidly urban transitions in the Middle East.

Therefore, urban sustainability transitions, as a new planning innovation (Geels et al. 2017) from the ASUST Perspective, is introduced as a research arena to achieve better governance of urban sustainability transitions involving urban informality (as explained in detail in Chaps. 3, 4, 5, 6, and 7). Frantzeskaki (Frantzeskaki et al. 2017b) argued that, to achieve sustainability transitions, strategies for the co-evolution of material and institutional transformations in sociotechnical and socio-ecological systems are essential. However, the ASUST builds on these propositions and relies on directing natural resources towards the needs of current and future generations, involving the grassroots to development processes, inspiring people's innovations regarding the circular economy, encouraging social networks, familiarizing people to ongoing transitions of informality, and to protecting/guiding cities of the Middle East from the current arbitrary/chaotic urbanization process to a more sustainable transitions profile. The idea is to link the theory and practice of urban sustainability transition, as well as to manage and oversee the sustainability transitions from urban informality. Currently, the states of Middle East have committed to justice and urban socio-spatial control in the new globalized era.

The ASUST relies on five assumptions regarding the people/social movement: Individuals move from rural to urban areas in response to economic incentives; people move to urban areas for social and political reasons; individuals move in response to information and communication networks incentives; people displacement and ethnic conflicts may also lead to changing of socio-spatial settings; and finally, an internal migration/mobility within the city is critical for a better environment (for further details see Chap. 3).

As illustrated in Fig. 1.2, the ASUST perspective of multiple and continuous processes conceives of transition as an interference of Policy Strategy Credibility, Consistency of Dimensions, Policy Processes (Coherence of processes), Systemic change (four perspectives), Systematic directions of Transitions, and Instrument mix Comprehensiveness. First, policy strategy is the general theme of the book, pertaining; (i) to link the theory and the practice of urban informality transitions and apply them to the discussion of the governance of sustainability transitions concerning the Middle East; (ii) to integrate urban informality into formality transitions to open windows of opportunities for constructive roles in sustainable development; (iii) to modify the compact pattern that dominated the Global South/Arab cities; and (iv) to enhance the imported western urban planning that do not suite the needs of those cities for which new planning innovations are needed to overcome a fragmented situation. This volume explores the emerging transformative pressures and tensions, as well as alternative futures that might be possible beyond 'more of

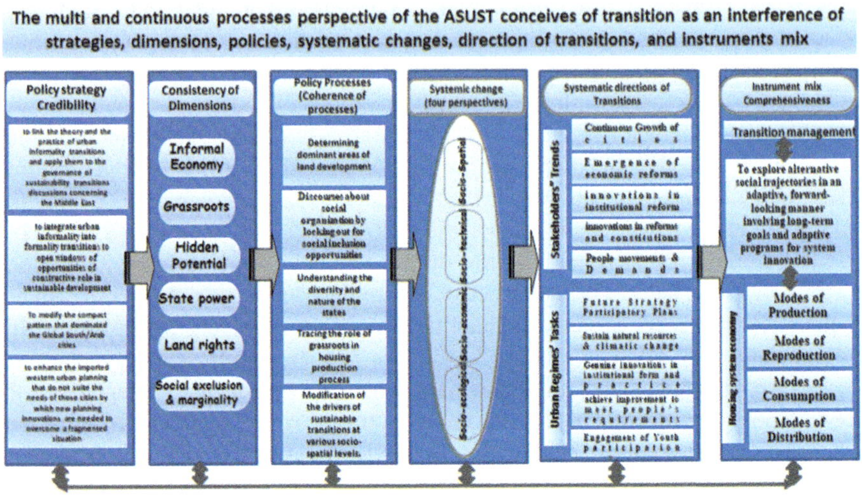

Fig. 1.2 Diagram illustrating the ASUST perspective as multiple and continuous processes for transition. (Source: The author)

the same' or 'optimizing' along the dominant regime pathway. It tries to link urban informality with urban sustainability transitions as a quest for new value systems and for a better way of life.

Second, Fig. 1.3 illustrates the consistency of dimensions in which this study focuses on the clarification of three elements cross-cutting with other two dimensions. The three elements are a synopsis of urban informality in the Middle East, ongoing research innovation on governance of sustainable transitions, and how to deal with urban informality in the context of sustainable transitions as the target. The three elements crosscut and overlap with the other two dimensions: the six axes and the mode of tensions. The first dimension is the six axes which are: (1) exploring informal economy; (2) the grassroots process; (3) the paradigm of hidden potentialities; (4) state and power; (5) land rights; and (6) social exclusion and marginality. The second dimension comprises the modes of informal housing systems of production/reproduction, consumptions and distribution of goods and services in the city and its practice. The output of the integration between the three elements and the two dimensions (six access and the mode of tensions) is the determination of production/reproduction, distribution, and consumption of goods and services and areas of the level of social networks, as a sociotechnical landscape, in which the grassroots, as a sociotechnical niche, play a great role in transitions process. This would lock out opportunities and hidden potentialities for sustainability by understanding and driving the diversity and nature of a sociotechnical regime and by transforming the drivers of urban informality transitions on various socio-spatial levels.

The conclusion is that the urban informality has changed in its nature, its form, its topologies, and its practices on the ground. This would lead into the transformation

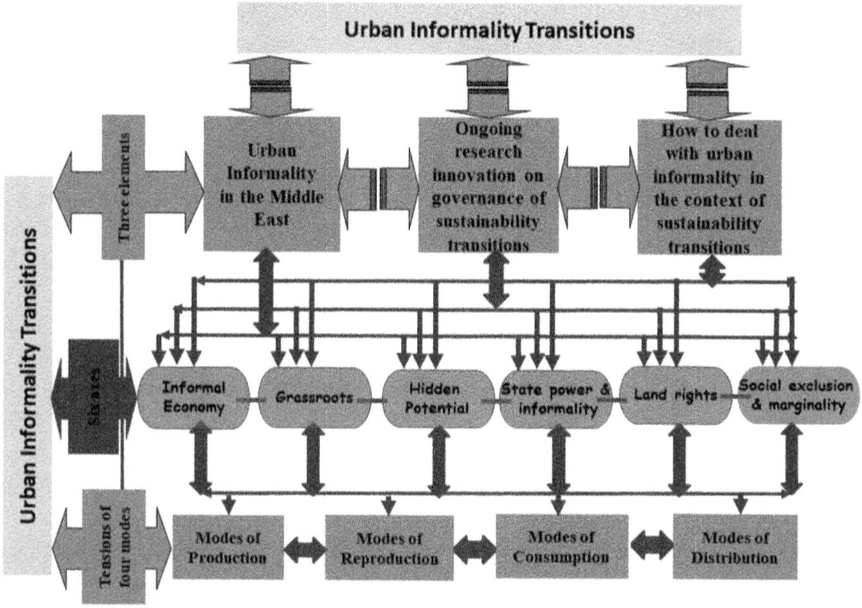

Fig. 1.3 Flowchart depicting the framework behind the present study. (Source: The author)

of the socioeconomic, political context, as a regime, and the stabilization of the tensions, and socio-spatial as a landscape, between the modes of informal housing systems of production, reproduction, and distribution and consumption of goods and services in the city. Urban marginalized informality (e.g., the landscape described here) has developed historically but under regime pressures became increasingly and persistently problematic, for example, large-scale societal tensions, pollution, political disability, and economic crises leading to increasing destabilization. The grassroots within a regime start to worry and to address the crisis with more traditional tools/approaches. But also, the grassroots start to come up with new ideas, visions, and approaches.

Third, Fig. 1.2 illustrates the policy processes in which elements of the sociotechnical system are interrelated and interdependent (Markard et al. 2012) to achieve a transition. The innovation in urban informality system may take the form of a process of socio-spatial economic transformation that may requires one generation or more. It reviews the history and current urban informality transitions and how the recent waves of urbanization and globalization have changed the nature of urban informality in five dimensions: (i) by determining dominant areas of land development; (ii) through discourses about social organization by locking out for social inclusion opportunities; (iii) by understanding the diversity and nature of the states; (iv) by tracing the role of grassroots in housing production process; (v) and by modifying of the drivers of sustainable transitions on various socio-spatial levels. The idea involved is one of trying to link theories and practice, by integrating and correlating between urban informality transitions within urban sustainability transitions, and how can be governed on the ground.

Therefore, this book examines urban informality processes, as a landscape, in which various variables and uncertainties such as social exclusion, marginality, and informal economy influence the transformation process. Other variables or uncertainties, such as the land delivery system, state power, grassroots, actors, regulations, and the law, constitute the regime. A niche reflects the hidden potentials of the grassroots and the mechanisms of informal land markets. The integration and correlation, positive or negative, between niches and regimes produce the landscape either formal and/or informal development. The output of the urban informality on the ground is goods and services in which various pressures and tensions form among the actors. The question is how to illuminate these pressures and tensions to sustain such a process? On the one hand, the actual attempts at the transformation of the built environment, as a landscape (informal economy, urban informality, arbitrary growth, socio-spatial transfer, social exclusion, marginality) and, on the other hand, the internal and external or specific power, as a regime structure, in the urban context (regularization policies, projects for development or improvement of the environment, new forms of urban management).. The dynamics of the actors and the renewal of action systems, as a niche, sustainability transitions of urban informality rendered possible or not by the necessary is the dominant task of the book to "formalize and integrate of the urban informality within the urban context".

Fourth, systemic change is proposed to include socio-spatial perspective as a fourth dimension for transitions to sustainability, while the other three perspectives are: socioeconomic, socioecological, and sociotechnical perspective (Loorbach et al. 2017) . This will be discussed in detail in Chap. 2. As illustrated in Fig. 1.4, urban informality, as a landscape, is a continuous socioeconomic-spatial process, and it correlates and links to the power of the three pillars: the state, society, and capital. The output of this integration is the modes of informal housing systems of production/reproduction, consumption and distribution of goods and services. This socioeconomic-spatial process is passing through three stages of transformation: the formulation, consolidation, and transitions of urban informality in which three actions of pressure, relaxation, and tensions emerge amid the landscape, regime, and niche, respectively. On the other hand, socio-spatial dimensions of sustainability transitions have been largely ignored in the transitions literature (Smith et al. 2010; Coenen et al. 2012). It is important to include the socio-spatial transformation as a mirror to reflect the final output of the modes of informal housing systems that help to achieve a process of sustainable urban informality transitions.

It would help to have a closer look into the transitions of urban informality from a wider perspective to gain practical experience on the various mechanisms that accelerated and associated the solidarity of the built environment for urban informality. This will reflect the future of planning in attempting to combine or to integrate informality and formality and to find a way to formalize the informality in a sustainable way in the light of a gray and uncertain future. Four significant considerations are explored: ***What people do, what people are able to do, what people are prepared to do, and what people are willing to do.*** These questions are

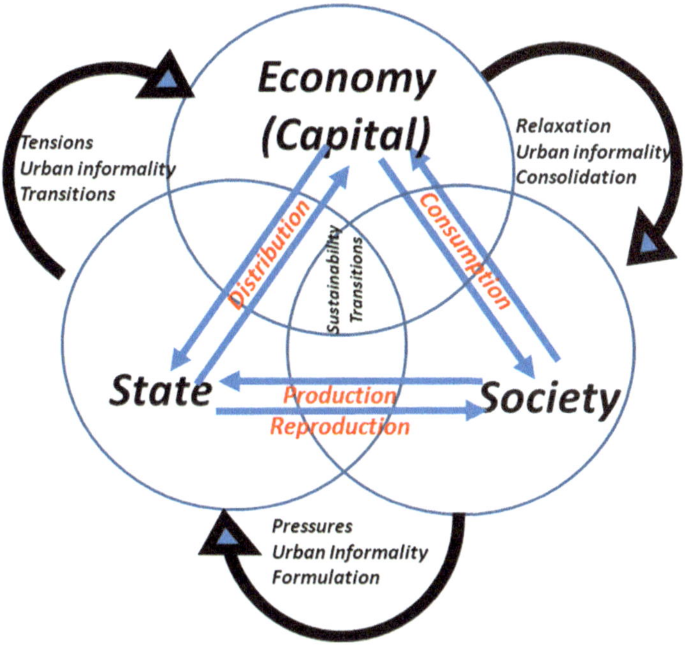

Fig. 1.4 The relationship between the process of urban informality, as a landscape, socio-spatial growth, the modes of informal housing systems, and the three pillars, as well as how it operates. (Source: The author)

based on the analysis that current regimes do not support people to work together for common sustainable development (Loorbach 2020). Thus, the book proposes including the socio-spatial perspective into the ASUST to identify relevant elements, path-dependency, persistent problems, and desired transitions. The framework is the over-all vision and direction for transitions in the short term; it is about experimentation and collaboration with research, community, and policy (e.g., in transition arenas and experiments). The ASUST relies on the correlation and integration of the three pillars, three elements, two dimensions, and socio-spatial perspective for urban informality sustainability transitions which is very rarely observed in the literature and in the research arena (Wieczorek 2018; Mguni 2015) (see Figs. 1.2, 1.3, and 1.4 and 2.1)

Fifth, systematic directions of transitions are composed of two instructions: trends of transitions and tasks of transitions. The former explores five trends: continuous growth of cities; emergence of economic reforms; innovations in institutional reforms; innovations in reforms and constitutions; and people movements and their demands for goods and services. The latter examines various tasks of transitions from five angles: future logical strategy participatory plans; sustaining natural resources and climatic change; genuine innovations in institutional form and practice, achieving improvement to meet people's requirements; and engaging the participation of youths in the developmental process (as explained in Chaps. 3, 4, 5, 6, and 7).

Finally, instrument-mix comprehensiveness is composed of two instruments: transition management and the modes of informal housing systems of goods and services. Transition management could be added to the ASUST as a promising model for sustainable development, allowing societies to explore alternative social trajectories in an adaptive, forward-looking manner involving long-term goals and adaptive programs for system innovation (Kemp et al. 2005). Transition management builds on the argument that the interconnected, complex and global character of current challenges, such as climate change or sprawling urbanization and growing social inequalities, requires a radical change in the basic systems providing societal needs for energy, water or shelter (Schot and Kanger 2016) because housing is a central instrument of social policy and social transition. Urban informality, as a landscape in the form of "a site of transitions", is examined in transformation processes through capitalistic production, reproduction, distribution and consumption of various goods and services. By enhancing a networked society, as an active urban regime, stimulated by increased digitalization, informatization, through the great acceleration in resource use, through rescaling of technology, but also by changing the power of state and grassroots; and by transforming the drivers of change at various socioeconomic-spatial levels. In short, the concept of the ASUST conceives of transition as an interference in processes at three levels: innovative practices (niche experiments), structure (the regime), and long-term, exogenous trends/forces (the landscape) (Schot and Geels 2008; Schot 1998; Rip and Kemp 1998; Geels 2005). The three levels represent functional relationships between actors, grassroots, structures and working practices that are closely linked.

1.5 Book Presentation and Chapter Contributions

The book focuses on urban context transformation to empirically and theoretically understand sustainability transitions on urban informality through a sociotechnical perspective. It is based on the ASUST and the issue of the three pillars, three elements, the two dimensions, and the socio-spatial perspective for urban informality sustainability transitions as previously discussed. The book is composed of four parts as follows:

1.5.1 Part One: Objectives, Substantive Issues, and Structure of This Book

Part one, the current introductory chapter, sheds light on the main outline of the book, draws intention to the main themes of the study, gives the reader an overview of the main arguments, and focuses on the structure of the book.

1.5.2 Part Two: Urban Informality Transitions in Middle Eastern Cities

Part two includes six chapters that revolve around the academic basis of urban sustainability transitions discussions, urbanization in the era of globalization, the paradigm of urban informality, state, society, economy, and urban informality, land rights, governance and urban informality, along social exclusive versus inclusive, marginality, and urban informality. Urbanization, informal economy, and marginality, and social exclusion constitutes the landscape, while state power, land rights, constitute the regime level. The innovation of the grassroots in establishing the built environment of urban informality constitutes the niche. This structure gives the reader the chance to grasp and explore the various forces that accelerate the urban informality transitions. This part is not in sequence, but it proves useful to structure them as follows:

1.5.2.1 The Sustainability Transitions, Governance, and Urban informality

The sustainability transitions research[7] is the recognition that many environmental problems need new planning innovations to tackle these urban dynamics challenges. These problems cannot be addressed by incremental improvements, but require shifts to new kinds of systems, shifts which are called 'sustainability transitions' (Markard et al. 2012). The concept of transition has been studied for decades in several disciplines, e.g., in biology and population dynamics, in economics, in sociology, in political science, in science and technology, and in systems science

[7]In recent years, planning innovations have taken place aimed at sustainability for the future of urban agglomerations development on the planet. Intensive research programs on system innovations and transitions were established in 2005, led by the Dutch Knowledge Network on Systems Innovation and Transition (KNSI) at the Dutch Research Institute for Transitions (Drift), The Netherlands (Grin et al. 2010). Also, the Sustainability Transitions Research Network (STRN) was inaugurated in 2009 at the 1st European Conference on Sustainability Transitions in Amsterdam with the purpose of creating a new interdisciplinary academic community (Koehler et al. 2017). STRN became a core network to attract contributors to the development of sustainability transition studies.

(Grin et al. 2010), but very rarely are socio-spatial transitions touched upon. Transitions in sustainability consist of shifts or 'system innovations' between distinctive sociotechnical configurations, encompassing not only new technologies but also corresponding changes in markets, user practices, policy, and cultural discourses and governing institutions (Geels et al. 2008).

Because the sustainable-transitions (ST) approach emphasized three interrelated angles (complex systems analysis, a sociotechnical perspective, and a governance perspective (Grin et al. 2010), the study focuses on those interlinked angles as transdisciplinary (Friedmann 2005; Bums and Friedmann 1985) and multidisciplinary field of studies (Loorbach and Oxenaar 2018). The term transition is broadly used in many scientific disciplines and refers to a nonlinear shift from one dynamic equilibrium to another. The term sustainability transitions is increasingly used to refer to the large-scale societal changes that are deemed necessary to solve "grand societal challenges" (Loorbach et al. 2017).

As sustainability transitions are quests for new value systems to assist the needs of the current and future generations (Turnheim et al. 2015), urban informality could play a great part in sustainability development. Urban marginalized informality (as a landscape) has developed historically but existing regime pressures became increasingly and persistently problematic. Urban informality transitions reflects all the activities of co-operation, tensions, pressures, and conflicts, within and between the state and citizens of urban informality, whereby they go about organizing the use, consumption, production/reproduction, and distribution of human, natural, and other resources, and social life. However, a closer look shows that the formulation of urban informality transitions from a wider perspective is essential to gain experience in the various mechanisms that accelerate and are associated with the solidarity of the built environment. The essential assumption is that societal systems go through long periods of relative stability and optimization that are followed by relatively short periods of radical change, but in the Global South relative stability is very rare, thus, new planning innovations are needed to overcome the socio-spatial fragmented situation.

Transition management is presented as "a strategy to deal with environmental degradation by stimulating sustainable development as a specific aim of policy making" (VROM 2001). Transition management is a promising model for sustainable development, allowing societies to explore alternative social trajectories in an adaptive, forward-looking manner involving long-term goals and adaptive programs for system innovation (Kemp et al. 2005). The issue of the governance of urban sustainability transitions on urban informality in the developing countries is very rare in the literature and in the research arena (Wieczorek 2018; Mguni 2015) Sustainability transitions management builds on the ASUST perspective relies on housing as a central instrument of social policy and social experience and equity. It also relies on the interconnection, and complex of current challenges such as sprawling urbanization, growing socioeconomic inequalities, requires a radical change in the basic societal and economic systems which is providing societal needs for goods and services.

However, this chapter attempts to explain how to apply the transition concept on urban informality from the ST point of view and relies on the ASUST, as a new concept, that might help decision makers within the Middle East to govern these settlements in a sustainable way. But this chapter also expects patterns in processes and dynamics of transitions to be found across the diversity of urban informality in the Global South. Therefore, this chapter attempts to trace the various tensions and pressures imposed by decision makers and other institutions (as a regime) in pushing low-income groups into the city's peripheries and how those groups are reacting to create an environment suitable to their requirements and needs.

1.5.2.2 Urbanization in the Era of Globalization and the Arab Spring

Urbanization, as a landscape, is indeed one of the most significant trends of the past, present, and future centuries, providing the foundation and momentum for the global transition. Currently, about 54% of the world population resides in urban areas (UN-Habitat 2016a). Urbanization is not handling population growth alone, but it has a major impact on socioeconomic and cities transformations. Cities are taking the lead to address many of the global challenges of the twenty-first century, including poverty, inequality, exclusion, unemployment, formal/informal urban sprawl, environmental degradation, and climate change. The transformative force of urbanization is likely to be greater in developing countries (UN-Habitat 2016b), with possible implications for harnessing the positive nature of urbanization. Will the urbanization process, as a driving force or an enabler power, come to terms with the urban informality sustainability transitions in the Global South?

Patterns of urbanization in the Middle East are varied in the form of suburbanization, peri-urbanization, informal urbanization, displaced urbanization, urban sprawl, or plotting urbanism as an ordinary process of urbanization (Karaman et al. 2020) and have constituted a significant trend over the last seven decades. There is an urgent need to rethink the past and present patterns of urbanization and to tackle the following challenges: first, competing jurisdictions between cities, towns, and surrounding peri-urban areas to attract suburban development; second, the true costs to the economy and to society of fragmented land use and road networks spatial development; third, the factual conflict between inclusion and exclusion of the urban poor, and between social capital and the growing capitalism; fourth, commodification of housing and land, which might accentuate socioeconomic differentiations between property-owners, who often live in the same area, and their tenants; fifth, how to cope with climatic change to sustain a healthy planet; and sixth, how to come up with affordable alternatives to accommodate the additional 2.5 billion people that would seem likely to reside in cities by 2050.

Globally, a billion new houses are needed by 2025 to accommodate 50 million new urban dwellers per year; costs are estimated at $9–11 trillion by 2025 (UN 2016), so how much land is needed, and ***where, how, and who will provide the necessary funds***. The huge amount of needed housing will accelerate both rapid urbanizations, formally and informally, and amplify urban informality in the Global

South. In all countries, the influx of immigrants into the cities from rural zones markedly accentuates the cultural diversity, in ethnically diverse countries like Lebanon, and the ethnic diversity and character of the county's nationalism in countries such as Jordan. Informal rapid urbanization coinciding with the longest period of political and socioeconomic transition in the Middle East has created a huge demand for housing plots and the spread of urban informality on the peripheries of the Middle East's cities.

According to (United Nations 2018), in Arab countries "urban growth rates will remain higher than total population growth rates in the foreseeable future... urban growth has been the result of rural-to-urban migration as well as high fertility and declining rates of mortality." The UN notes that "Many cities are now going through a critical phase of development, marked by dwindling resources, increasing poverty, and serious environmental degradation." This is an apt description of the situation of the population shifts in Middle East countries that are subject to various circumstances and depend on various incentives. It varies from time to time because what happened in the past does not necessarily occur now. Also, the rhythm of movement between countries, between and within one country, and within a city, becomes faster than before, as it occurred during the short time in which urban informality flourished. The phenomenon of population mobility is also characterized by a planned move rather than an arbitrary move due to the availability of information and knowledge that became available through the network (Castells 2010), and other sources of data for a certain area. Globalization has also given rise to major population shifts between various regions within a given area and shifts between countries. On the other hand, there have been extensive migratory movements of refugees from hunger and war: in Lebanon during the civil war and in Egypt, and in Jordan during the defeat in the 1967 war, and the three countries have a significant number of refugees due to the Syrian crisis.

It seemed that a new era was opening in the political history of the Middle East, beginning in late 2010 and early 2011. Cities of the Middle East witnessed an urban insurgency raised in the "new social movements" or Arab Spring of community collectives from slum neighborhoods. Muslim Brotherhood (in Arabic, *al-Ikhwan al-Muslimin, or simply al-Ikhwan*) and more privileged students group (Fergany 2016; Singerman and Amar 2006) began to cooperate, demanding fair elections, affordable shelter, job opportunities, decent life, equity, and respect for human rights. After a controversial beginning under the aegis of marginality theory (Murshed 2002), analyses of political and social mobilization by diverse urban movements and constituencies are very much in fashion today (Caroline and Yatzimirsky 2014; Parnell and Oldfield 2014). Arbitrary urbanization and rapid population growth of urban and rural agglomerations in the Middle East have increased the demand for housing, especially for low-income groups, increased social exclusion, and decreased social capital. It is argued that a real shift from old wisdom to a new system of innovation through the co-evolution of material dimensions and institutional transformations of sociotechnical systems is crucial to create a sustainable environment.

1.5.2.3 The Paradigm of Urban Informality: Laws, Norms, and Practices

During 1960s and 1970s, many scholars questioned the spread of slums on the peripheries of cities in developing countries (Abrams 1964; Ward 1976; Stokes 1962; Gans 1962, 1971; Seely 1959, 1971; Turner and Fichter 1972; Turner 1976). The role of self-help housing applied to upgrading and sites and services programs was the main policy in tackling the housing shortage in the South. This phenomenon spanned the Global South with some experiences from developed countries (Hart 1973; Rakowski 1994; Perlman 1976; Dwyer 1975; Ward 1982b; Roy 2004). Latin American urbanization was the fertile arena for understanding the nature and mechanisms of squatters' areas (Roy and Al-Sayyad 2004) and opened the debate for how to come to grips with this problem. There are various terminologies, definitions, and typologies for such areas across the Global South, which are neither exact nor concise (Gilbert and Gugler 1982; Harris 2018; Grashoff and Yang 2020). Each country has a local generic name for them (Abrams 1964; Ward 1982a, 1982b; d'Alençona et al. 2018; Rocco and Ballegooijen 2018; Marx and Kelling 2019).

Informal rapid urbanization coinciding with the longest period of political and socioeconomic transitions in the Middle East have created a huge demand for housing plots and the spread of urban informality, as a landscape, on the periphery of cities (Drakakis-Smith 2000; Soliman 2012; Hosseinioon 2019). On the other hand, the social and cultural representations of the urban informality, often stigmatized by decision makers and the media (as a regime), when they are not simply ignored by the local authorities, are yet to be analyzed. The notion of urban informality (as a landscape) often reaches the point of defining populations excluded from the dominant representations. It is therefore also necessary to understand how the inhabitants of informal areas themselves encode and internalize the judgments and norms (as a niche) that are applied to them, as well as the way "the other" (regime) regards them. Thus, the pressures between the state (as a regime) and the grassroots (as a niche) is the cornerstone to integrate urban informality into sustainable development (as a landscape). To name urban informality also leads the researcher to represent and classify it (Roy 2005). Moreover, how do we account for the recovery of these areas, for their extent and their limits, in the methodology and in the attempts at definition, and in the modes of cartographic and statistical representation as well?

Scholars (Edensor and Jayne 2011; Soliman 2012b) and practitioners analyzing community-level collective action, as a niche, have become increasingly interested in how relationships based on trust, reciprocal exchange, and social networks, i.e. social capital, affect outcomes. The idea that development plans as such could be directly 'implemented' reflects a very traditional conception of a plan as a spatial blueprint, which would steadily be translated into the built environment on the ground (Healey 2003). The relationships between the three independent systems: state and market (as a regime), urban informality (as a landscape), and social network (as a niche), are correlated. The latter reflects strong social ties, which would formulate local finance customs, as well as working on creating, directing, and sustaining cultural identity. It also has the power to enforce various pressures on the regime to satisfy their social amenities. Finally, one must consider the hybridization of legal and land systems governing the right of land use that can well be the driving force behind the urban dynamics in the urban informality. Do the day-to-day adjustments, the individual arrangements (between occupant, holders of property titles, authorities) concerning the right of use of land and constructed property not call into question the often-simplifying idea of the absence of regulation in the urban informal and irregular quarters?

1.5.2.4 State Power, Society, Economy, and Urban Informality

In recent years, theoretical and policy-relevant linkages among the states, as a part of sociotechnical regime, towards civil society, economy, urban informality, and urban conflict, have changed. A clear understanding of the recent transformation of socioeconomic and political transitions is introduced to stimulate the varied function of states, as regimes, and its influence on socioeconomic transitions. Conflict may combine different forms and conflict within the forms themselves, as well as the extent to which each generates insurgency or patronage and may vary with time and circumstance. Understanding the nature and the concept of the states (as a prime actor of a regime) and their changing roles towards various strata of society became essential. The power of the state changed according to the circumstances of economic transformation rather than social transformation, and recently changed towards health transformation (due to Covid-19 pandemic). The weakness or powerlessness of the states in the Middle East is due to their economic status transitions, and the influences of the affluent groups and international organization on the states' decisions.

The policies regarding the rehabilitation, restructuring, and conforming of urban informality pursued by the states with the aid of the international organizations must also be elucidated. These policies, as sociotechnical niches, are in effect decided upon in political and economic contexts often marked by populism and authoritarianism and by attempts to decentralize/democratize that are not without effect on the nature of the urban projects pursued. How does one analyze the political choices or pressures made in urban informality transitions? Do they respond to real social demands? How is the latter identified, understood, and represented? Can one simply facilitate or induce democratization from below, or bottom-up through the "authorized or imposed participation of the inhabitant", or the grassroots as an innovator, in the projects at the urban informality transitions? What are the modalities of implementation, what is the impact of urban policies, as a sociotechnical regime, intended to integrate and pacify, but which can also control, manage, marginalize, or indeed exclude? What is the legitimacy of the large voluntarism urban sites, as a sociotechnical landscape, which can be decontextualized? What city model, as an alternative niche, does one intend to develop through these interventions in the informal quarters? Understanding the nature and transformation of urban informality and the visions towards securing their future has become a prime concern for the states in the Middle East.

The multi-diversities of the state's modes and the state's roles on housing production represent a crucial instrument for urban sustainability transitions. Moreover, political informality in housing processes has created an environment that encouraged the urban informality transitions that generated huge differences in terms of the approach, scope, scale, and liability of housing process, even though increasingly international peer-to-peer exchange and networking play an important role in the future formation of urban-informality transitions. The changing attitude and action of the states (covert or overt or both), as a regime, towards urban informality are the outputs of the transitions of the socioeconomic and political arenas in the Middle East. This transitions process is the outcome of the changing international policy, as an exogenous regime, towards urban informality and looking out for the lowest strata of the society, creating comprehensive, sustainable and organized urbanization for development and poverty reduction, with the power to change and improve lives.

1.5.2.5 Land Rights, Governance, and Urban Informality

Worldwide, access to land, as a part of a regime, is critical to the welfare of rich and poor people alike and the ability to make effective use of it. Since land is a key asset for the urban poor, land policies are of fundamental importance for economic activity, poverty reduction, sustainable management, and the well-being of households. During the most recent decades, many countries in the Middle East have undertaken land reform in one guise or another, aimed at enhancing the security of property rights. Also, the provision of property rights to residents of low-income settlements became a central theme for a social network as the best way is to accelerate their incorporation into the overall housing system or land-market system. This paradigm has become so significant within multilateral agencies that housing and land issues are mentioned in the Millennium Development Goals (MDGs) in relation to Part 1 of Goal 7 and Goal 11 of the Sustainable Development Goals (SDGs) (United Nations 2018). This has emerged as a method to harness 'the transformative power of urban informality'. An ongoing debate in studies on land, natural resources, and housing is that which concerns the role of institutional form in relation to economic performance and sustainability. Mainstream (economic) theories of development presuppose a straightforward relationship between the two with certain institutional forms—such as formal, private, and titled property rights—regarded as imperative for economic growth. Who drives the formation and adaption of the "increasing versus decreasing" land delivery system for urban informality?

In particular, it is also necessary to discover which norms are at the source of the legalization procedures or of the conforming of the informal areas, to observe how, between occupants—contracting parties, landowners, public powers (as stakeholders of a regime), etc. Various systems of costs, values, and representations are confronted, as a landscape, with the application of the law, as a regime, within the machinery of regularization. It would be relevant to go more deeply into the relationship between the law that confers ownership and the right of occupant use, or again, the effects of the procedures of regularization or adjustment of the right of land use when these procedures ignore the prevailing legal system based on an admixture of customary and positive law. In the Middle East countries, religion acts as a cornerstone of the right to land tenure and ownership of land. In Islamic law, there is a formal hierarchy of sources of law. The two foundational and primary sources of Islamic Law (*Shari'a*) are the *Qur'an* and the *Sunna* (Sait and Lim 2006). In addition, the classical law relating to property and land rights underwent several periods of influence—Ottoman, colonial, and post-colonial/modern periods. The environment of the Middle East contains various tribes, who have the power to pressure the state and which they have their own customs concerning land tenure. These patterns have spurred tensions between local tribes and the state (Al Naber and Nolle 2016) or between the regime and niche. Also, diverse paths of land and property rights exist in many cities of the Middle East.

Most recent research does not examine the diversity of either land tenure or legalization processes being introduced in informal housing areas. However, this section aims to show that land tenure and security are diverse and complex due to several different mechanisms of state intervention, market forces, multi-actors, and land acquisitions. There are several factors associated with informal residential developments on both agricultural and desert land, the government either directly or indirectly making a major contribution. The process of securing land titles or rights to land may be subject to corruption, with the land title being given to those who either have a strong connection with politicians or have the capital to cover the cost of the legislative process. This has led to the study of the nature of the relationship between the three pillars, as a regime, multi-driven processes, as niches, and urban space, as a landscape, by which the management of transitions land is formulated in a sustainable way only through a more encompassing and democratic political process. Also, strengthening the administrative capacity of land registers and political willingness are preconditions to establishing efficient transitions land management and systems of property rights, while tenure security is achieved by accretion of various rights over time.

Grassroots groups, as niches, such as these across the Middle East, Africa, Asia, and Latin America, are ready to engage at a higher level in implementing pro-poor land policies (Soliman 2017). Land interventions are based on an exclusive, top-down approach that fails to involve the grassroots communities they are meant to serve. This is one of the major reasons that land management remain so poorly implemented and why implementation tools and regulations, as a weak regime, are often ineffective. It is argued (Soliman 2017) that the interdependent nature of stakeholders and/or users (where stakeholders are the individuals or collectives within a network) is a key distinguishing characteristic of a social network, while stakeholders' structures, connectivity, markets forces, and the distribution of power, as a regime, are key components of transitions land management (Layton 2009; Jackson 2010). The Global Land Tool Network (GLTN) (UN-Habitat 2012) recognizes that grassroots engagement is a necessary element in any land tool design and implementation.

1.5.2.6 Social Exclusive Versus Inclusive, Marginality, and Urban Informality

Both social exclusion and governance, as a regime, are issues high on the agenda of international agencies, as well as of many governments in the South. The SDGs, particularly Goal 11 and other urban indicators, alongside the 2016 New Urban Agenda, offer a renewed opportunity for the global community to confront several emerging global urban challenges. Among these are the urgent need to halt uncontrolled urban sprawl, reverse the growth of urban slum populations, institute smart, safe, and efficient urban transport systems, and improve urban environments by creating safe public spaces. Clearly, achieving Goal 11 targets by 2030 will require a focus on a range of cross-cutting, routine, and persistent challenges. Goals 1, 5, and 11 stress the need to invest in cities to address socioeconomic inequalities, social exclusion, extreme poverty, high unemployment, poor environmental conditions, and drivers of climate change (UN-Habitat 2018). This fundamental long-term transformation will involve the realignment of urban society, its technologies and infrastructures, urban cultures, and lifestyles, as well as governance and institutional frameworks.

In the Middle East's cities, most of the suburbs or peri-urban or hybrid urbanization areas, as a landscape, (Bums and Friedmann 1985) are occupied by the lowest strata of the society and characterized by overcrowding, the worst environment, lack of services, and substantial distance from job opportunities. These suburbs/peri-urban contain not only informal residential areas but also an informal economy sector that constitutes a high proportion of the national domestic production of a city (Chen and Carré 2020). Major cities in the Middle East have become loci of contested space, as a sociotechnical landscape, with growing numbers of "marginalized people", or "structurally irrelevant people" (Castells 1997) or invisible urban enclaves (Sibley 1995), or "a deviant situation" (Singerman, D.(ed.) 2011) or "urban disenfranchised" (Bayat 2004) or "socio-spatial struggle for survival" (Soliman 2017) who are now claiming their rights to the city as well. Therefore, the division of cities into many cities within a city became noticeable, where affluent groups occupied the most privilege quarter, while the lowest strata of the society occupied the worst locations or locations far from various social services and job opportunities, for example, western and eastern Beirut, western and eastern Amman, and old and modern Cairo.

The conjunction of conditions linked with urban poverty, violence, and ethnic and migrants' concentration is spatially expressed in "invisible" urban enclaves. Very often, these places—despite their scale—in cities of the Middle East are not marked on city maps and are categorized by the majority as "illegal." Many areas of such characteristics exist in Cairo, Beirut, and Amman. These places become the signifiers of the socially constructed and demonized image of the "other" (Sibley

1995), and cities divided where urban areas have been segmented into two sectors, formally and informally, or legal and illegal, or maybe more than two divisions (Sims 2010; AlSayyad 2006). Also, ethnicity is playing a great part in the division of the cities in the Middle East (Soliman 2008). Yiftachel (2009) also addressed the scant attention given to ethnicity as an element of urban materiality. He argues that 'ethnocracy' produces 'gray' spaces that only partially incorporate the ethnically marginalized and which lie between the legal and the illegal of formal planning systems.

Marginality, inequality, and social exclusion, as a product of sociotechnical landscape, became the dominating feature of the urban fabric of cities of the Middle East. Marginality combined with social exclusion reflects the ways in which some people, usually seen as individuals or parts of distinct social groups or categories of people, have been excluded or pushed to the periphery of physical and economic growth or political development. In other words, the regime pressures these group by pushing them out of the organized areas. The rich and above middle-income classes settled on a traditional pattern, formal areas, while the urban poor, who mainly immigrated from rural areas, have organized themselves and formulated their own pattern, living as deviants (Singerman, D.(ed.) 2011). So, marginalization became a "socio-spatial type" and a way of struggle to the rights of the city.

In the solidarity of passive networks, as a niche, the people-movement (Singerman 1995) became politicized. However, people operating in the networks may interact with local elites and government bureaucrats to obtain publicly subsidized goods and services. Other forms of exchange in these informal modes of participation—such as seeking marriage partners or employment opportunities, migrating to seek better jobs, or exchanging gifts—do not entail engagement with the authorities but rather play out among neighbors and community members (Singerman 1995), and sometime the state may take the role of informality on behave of the citizens.

1.5.3 Part Three: Cross-Regional Scene

Part 3 includes three chapters that center on experimentation treats the cross-regional scene of urban informality transitions in three Middle East cities: Cairo, Beirut, and Amman. These cities reflect the phenomenon of urban informality transitions in Egypt, Lebanon, and Jordan respectively.

1.5.3.1 A Puzzle of Urban Informality Transitions in Egypt

The urbanization process in Egyptian cities had taken place in the form of belts of urban informality built on agricultural land, or a belt of poverty or a belt of misery surrounding the Egyptian cities. This is a stone on the chest of Egypt. The current arbitrary urban pattern of Egyptian cities has accumulated over time due to arbitrary urban and housing development policies, the instability of socioeconomic and political transitions, and the spread of urban informality (Bayat 2004; Soliman 2004a, b; 2015a, b; Dorman 2007; Kipper and Fischer 2009; Singerman and Amar 2006; Sims 2010). The phenomenon of urban informality in Egypt is as old as the establishment of the *Heliopolis* district near Cairo in 1907, and may even be older than that.

Approximately 50% of the urban and rural agglomeration of Egyptian now live in this so-called *ashwaiyyat* or housing informality which has become the dominant urban fabric of Egyptian cities. The result is a growing divergence in Egyptian cities: between the *old* and the *new* parts of the urban agglomeration, a "Great Divide" between "establishing", "struggling", and "emerging" urban spaces. Urban informality transitions in Egypt do not represent a need for housing so much as a method for liquid capital and social capital accumulation, and it is a matter of collateral investment to create wealth and to secure residents' future from the fluctuation of the economy within and outside the country. The government policy's failure to provide affordable, viable housing for the urban poor has led many people to invent a policy, as a niche, to build homes—semi-legally or illegally—on privately-owned agricultural lands.

In 2016, it is estimated that there are around 10 million housing units are empty and unoccupied. These statistics reinforce the argument that the puzzle of urban housing in Egypt is a problem not of scarcity but of distortions of both urban development and the housing market caused by an accumulation of ill-conceived and inadequate policies, as a regime, that has led over time to a mismatch between supply and demand and to severely curtailed private sector investment in housing construction. A recent study estimated that informal housing varies between 40% and 60% of the urban housing stock of 27.1 million physical housing units (UN-Habitat and Arab Republic of Egypt 2015). The total area of urban informality was at least 416.58 thousand feddans constituting over 39% of the total built-up area in Egypt in about 2016 (CAPMAS 2016; ISDF 2016).

With the absence of clear housing and planning policies, urban informality spread in a way that has eaten 1.2 million feddans of the best agricultural land since 1984 until now. Recently, the state issued a temporary law of reconciliations number 17 of 2019 amended by-law number 1 of 2020 put in force from July to September 2020. This law aims to set up a reconciliation agreement between two parties: the owners of informal housing units/informal buildings and the state. This agreement must be accompanied by certain fees to be paid to the state that will give a formal building license to the offender as de facto recognition. The building license does not provide security of land tenure, but rather it prevents any threat of demolishing the informal unit/building. It is assumed that there are around 2.0 million violations in which the state will collect around 40.0 billion Egyptian pounds for these reconciliations. There are no clear tools or mechanisms on how to legalize urban informality by which the state created a new taxonomy of informality as "a hybrid formal status". This law will sustain the status quo or laissez-faire of the informality without any proper development or upgrading for the urban fabric of informality. The urban informality morphology will still remain because they surround the major urban centers. In short, urban informality operates as "a site of transitions" to cope and integrate with the tremendous ongoing transitions of socioeconomic and political milieu. Urban informality transitions are a comprehensive component for the relationship between the challenges of the political economy, socio-spatial transitions, and social capital and social exclusion. Urban informality transitions have evolved and changed through five eras. These were Nasserism (1952–1970), a period of an extensive intervention; Sadatism (1970–1981), a period of active but ad-hoc intervention; Mubarakism (1981–2011) a period of fluctuating intervention, the post Arab Uprising (2011–2014) when the state was absent, a period of arbitrary intervention that retreated from interventionist policies and when housing markets fluctuated constantly, and the current intervention (2014–2020): a period of exhaustive and reconciliation intervention. These transitions rely on certainty and uncertainty aspects, or pathways, by which three trends have emerged.

These transitions occurred according to three trends: *relaxation, tensions, and pressure* in sociotechnical systems. These trends interact and interrelate with regime systems (e.g., institutions' management, political transitions, planning regulation, land delivery system, and capital forces), landscape (e.g., housing production, population growth, population mobility, social groups, 'radical' change in terms of scope of change traditions, norms etc.), and niche (e.g., a community-driven process and a time period that witness transitions).

On the sociotechnical regime level, state intervention in the housing field has aligned with the interests of the politically powerful, the affluent group and the poor, depending on the pressures and tensions that were imposed by the lowest strata of the country. The state became involved in housing due to humanitarian reasons, functionalism, social control, political control, and human-rights considerations. Thus, the state assumed a new triple role—of pressures, relaxations, and reconciliation—widening its control over society and providing funds to improve informal settlements. This is not to overplay the role of the state and the certainty of success of such tools. Therefore, many actors are involved, interests inevitably conflict, and struggles ensue with unpredictable outcomes by which systemic innovations occur by emergent and nonlinear dynamics (Geels et al. 2016; O'Brien 2015).

On the sociotechnical niche level, to absorb the people's anger, the state responded to this huge demand for housing in four relaxations policies: turning a blind eye to illegal construction on existing residential buildings; continuing the *laissez-faire* attitudes towards illegal residential development on agricultural land; adapting a relaxed attitude by sustaining the *status quo* of the urban informality; and finally by issuing a temporary law for reconciliation between the state and violators. However, the role of the state has changed from an enabler, supporter, or provider into a developer and a contractor in the housing market.

On the sociotechnical landscape level, the Egyptian case highlights the relationship between state power and housing production, and it shows the effects of the new attitude of the state to housing policies and practice. It also draws out contributions of understanding socioeconomic and political transitions on urban informality, on one hand, and it offers the possibilities and limitations in governing sustainability transitions at the level of the city, on the other hand. In absolute terms, the progress made on housing the urban poor has certainly not been sufficient to offset the growth of urban informality in Egypt, but the continuous transitions process of urban informality have secured the immediate needs for affordable housing and job opportunities.

1.5.3.2 The Pocket of Urban Informality Transitions in Lebanon

Lebanon has a 200-km coastline and a total area of 10,452 km^2, running north-east – south-west along the eastern shores of the Mediterranean Sea. Lebanon is a highly urbanized country with more than 87% of its population living in urban areas and 64% living in large urban agglomerations (Beirut and its suburbs, Tripoli, Saida, Zahle, and Tyre). The primary characteristic of these five million inhabitants involves their belonging to 18 different Christian and Muslim sects (Harris 2009). Urbanization processes in Lebanon have been occurring without any guiding strategies or plans, merging the cities into single large agglomerations, threatening arable lands and biodiversity, creating transportation and traffic problems, and increasing the challenge of infrastructure and services provision. The urbanization of the country was of such explosive and uncontrollable proportions (Martin 2015) that migrants began pouring into the periphery of the capital populated by former Armenian refugee camps and neighborhoods, as well as Palestinian refugee camps. Lately, after the Syrian crises in April 2011, Syrian refugee camps became the borderline where the camps and the cities met.

The chapter on Lebanon serves to link three social sciences: political transitions, sectarianism changes, and urban development transitions that have had a major influence on urban informality transitions. This chapter connects these three concepts within the research paradigm of 'sustainability transition'. It is shown that the influx of Syrian refugees has influenced the ongoing change of the dominant socioeconomic and political-intensive development path and addresses the causes of local and regional changes on the urban fabric of Lebanese cities and the formulation of urban informality. The chapter focuses on the recent socioeconomic, political, and spatial transitions, with emphasis on Beirut city, in order to explain the extent to which the religions/ethnics, along with political with the demise of pre-war safety net mechanisms, has affected the urban development that is remapping the country(UN-Habitat 2018d).

The formulation and spread of urban informality in Lebanese territory became the main challenge for the Lebanese government and international agents, so it is not only a matter of arbitrary spatial development. Rather it became a global challenge to host the growing mass of Syrian refugees. The chapter fosters longer-term proactive strategies and policies and specific measures to realize two policy goals: 'sustainable development' alongside and contributing to a 'sustainable peace', as the possible result of a large-scale transition of the systems of production, consumption, and governance. Since the civil war, urban expansion has occurred in the form of informal areas developed around major cities and towns. These informal areas are inhabited by the Lebanese who were displaced from their areas of origin or by refugees, mainly Palestinian refugees and smaller refugee communities (such as the Kurdish).

Urban informality transitions have occurred according to three trends: *relaxation, tensions, and pressure* in sociotechnical systems. On the sociotechnical regime level, the recent pressures of transitions, either locally or internationally, has resulted in the spread of urban informality, changed the land uses within Beirut city, and created what is called the "Misery Belt" around Beirut. Before the outbreak of war in Syria, Lebanon had an estimated four million inhabitants. State power, Syrian refugees, and social segregation and ethnicity have created tensions between the local citizens and refugees in Lebanon. Policymakers, politicians, international donors, and practitioners followed relaxation attitudes to overcome sustainable transitions on urban informality in Lebanon and have to acknowledge a potential strategy that accommodates low-income groups of host communities and foreign refugees.

On the sociotechnical niche level, the arbitrary physical urbanization process has directed pressure on the structure and practice of Lebanese territory, either directly or indirectly, through a diversified system of citizen participation, community development, and ethnicity enrolled/engagement. As a result, there thus arises a triple process of status quo of urban segregation—on the one hand, that of the various ethnic minorities with respect to the dominant ethnic group and that of various refugees with respect to the host community and, on the other hand, that of the different ethnic minorities amongst themselves.

On the sociotechnical landscape level, regions of urban informality currently host approximately 1.5 million refugees. This has put further tensions on the country with the highest number of refugees per capita in the world. Lebanon, as such, represents an empirically central case to understand the dynamics of the Syrian refugee crisis (Nassar and Stel 2019; Anholt and Sinatti 2020). The perception of Syrian refugees as predominantly a security threat and as pressure on the regime (rather than as a humanitarian problem) has, for instance, been one of the foundations under the 'no camp-policy': large concentrations of refugees were to be avoided for fear of them becoming 'terrorist hotbeds' (Turner 2015). Thus, the Lebanese case is mapping the 'political economy' of (in)formality, in other words: explicitly exploring who benefits from ambiguous governance modes in what way, should then become part of the policies, programs, and projects that seek to help refugees and host communities. Furthermore, the recent massive explosion in Beirut harbor in August 2020 puts yet further burden on the state.

1.5.3.3 Hills of Urban Informality Transitions in Jordan

The Jordanian chapter sheds light on the transformation of socioeconomic, political and spatial contexts on the spread of urban informality during the recent few decades. The Hashemite Kingdom of Jordan has a geographical area of 89,318 km^2. The total population of the kingdom is 10.05 million (including Syrians and Syrian refugees) (Hashemite Kingdom of Jordan 2017), with an average density of 73.5 people per km^2, and an annual population growth rate of 2.3%. The percentage of the population in urban areas is around 82.6%. The kingdom consists of twelve governorates, with major cities of Amman, Zarga, Irbid, Russeifa, Wadi al-Seer, Ajloun, Agaba, Madaba, Salt, and Ard Ramtha (UNDP and GDCD 2008).

Jordan has witnessed phases of tremendous socioeconomic, political, and physical transitions since its existence until now. These transitions' phases are the post-independence phase (1918–1948), the post-foundation of Israel phase (1948–1973), the rapid-boom phase (1973–1999), and the rapid-regional transitions phase (1999 to the present). During these phases of transitions, an enormous number of Palestinian and Syrian refugees and other nationalities overwhelmed the country, asking for goods and services for their survival. In 2017, it was estimated that the total population of Jordan was 10,053,000 of which 3,077,716 are non-Jordanians, constituting 30.61% of the total population of the country (Hashemite Kingdom of Jordan 2017). According to the government's estimate, Jordan hosts 1.4 million Syrians, of whom 646,700 are refugees and 750,000 lived in the country before the Syrian crisis beginning in March 2011. Currently, Syrians represent around 14% of the national population.

The transitions processes in Jordan witnessed the three trends of *relaxation, tensions, and pressure* in sociotechnical systems. On the sociotechnical regime, Jordan has first to overcome the crisis of foreign refugees to meet the immediate needs of Syrian and Palestinian refugees both in and out of camps, as well as vulnerable Jordanians affected by the crisis. Second, it needs to restore and reinforce municipal services and infrastructure degraded as a result of the sharp demand increases in critically affected sectors, in particular, solid-waste management, housing, environment, energy, and transport. Third, it is critical (i) to cope with rapidly expanding employment and livelihood opportunities (ii) to strengthen the coping capacities of vulnerable Jordanians, who have been impacted by the crisis, (iii) to address social imbalances and strengthen social cohesion in Jordanian communities hosting large numbers of refugees, and finally, (iv) to regulate the arbitrary urban growth and to eliminate spreading of urban informality to cope with tremendous socioeconomic transition in a sustainable way (Hashemite Kingdom of Jordan 2014).

In the sociotechnical niche, the three types of informal housing areas have emerged as innovations by the grassroots. These are: illegal land subdivision (semi-informal areas or *ahya' al fuqara*—housing of the poor), illegally occupied land (squatting areas or Bedouin areas), and hybrid areas occupied by Jordanian and non- Jordanian foreign refugees' camps (*sakan 'ashwa'i* or arbitrary housing). The latter has put further pressure on the Kingdom, causing many urban and socioeconomic challenges to tackle the rapid urbanization processes, scarce natural resources and a lack of financial resources, on one side, and the needs of the influx of foreign refugees on the other.

On the sociotechnical landscape, due to the various transitions and transformation that Jordan has passed through, rapid informal expansion dominated the urban growth in Jordanian territory. The influx of Syrian, Lebanese, Iraqis and Palestinian refugees have resulted in tremendous physical and social problems. This is seen in crowded, sub-standard housing in poorly planned and serviced neighborhoods that pose major risks of abuse and conflict, but also high likelihoods to deteriorate the urban environment. The influx of refugees has also contributed to the collapse of the housing market in the country and has encouraged, or at least been associated with, the spread of urban informality on the periphery of the major Jordanian cities. From a social point of view, the Syrian crisis has increased *tensions* between Syrian refugees and affected Jordanian host communities in the country. Furthermore, informal rapid urbanization and informal housing growth extended planning and zoning ordinances to newly developed areas which were beyond the financial ability of the poor who were pushed towards living in crowded or sub-standard housings (Al-Hasanat 2003; Nuseir et al. 2003), or in informal settlements, many of whom are settled in the eastern sector of Amman, while the rich are living in western Amman. In Jordan, spatial inequalities are not only a forerunner of social and economic divisions; these, in turn, cause further inequalities and other forms of exclusion and marginalization. The result is a growing divergence and pressures in Jordanian cities, Amman and the other 'Ammans' within the city, the formal city, and the informal city found in "the Jordanian urban fabric" (Khirfan 2019).

The chapter concludes that Jordanian national identity is multi-scalar, merging Arab supranationalism with Jordanian and Palestinian identities, i.e., multiple scales of nationalism (Culcasi 2016). The case of Jordan indicates 'unpicking' and 'unpacking' networks of actors at niche and regime levels (as noted by (Genus and Coles 2008)). This is indicated by the presence of the international refugees, and the multi-scale of nationalism through a transparent planning strategy in cooperation with various stakeholders. This sort of cooperation is quite clear in following urban planning strategy based on transition management through a participatory approach by involving local and international stakeholders.

1.5.4 Part Four: Epilogue

The fourth, and last, part presents an epilogue on urban informality transitions in Middle East cities. It emphasizes implementing principles of governance of sustainability transitions (Frantzeskaki et al. 2017a, b) on urban informality at the national level. Urban informality, *as a site of transitions*, is integrated and interrelated with various changes that are occurring within a given environment. A model of integration interlinking various taxonomies of urban informality is reached. However, the challenge for highly developed cities is how to transform existing urban informality to environmental sustainability without compromising the socioeconomic benefits that they provide. The three taxonomies of urban informality—bottom-up, technocratic, and hybridization—are in a dynamic process, and may change their categories over time from one typology to another, also reflecting on the population's behavior and action in each alternation. In order to adjust or modify the existing condition for each taxonomy of urban informality, one has to understand exactly four thresholds of people's transitions; ***what people are willing to do, what people are able to do, what people do, and, what people are prepare to do*** in their own settings. The first threshold concerns *what people are willing to do,* i.e., their reactions towards their settings that affect how planning happens. These deserve more attention than they have received. Once willingness has been introduced, the crossing of thresholds depends on the numbers of people involved in informal practices, their physical location, and especially their manner of social organization. A second threshold concerns the emergence of embedded social legitimacy and *what people are able to do*, a development with practical significance for those involved and also, potentially, for planning practices and wider politics. A third threshold concerns *what people do* when regular/irregular practices are crossed, becoming covert, an unavoidable matter of debate and policy. A fourth threshold concerns *what people are prepared to do*. This is the least predictable step: it depends on the prevalence and physical visibility of changing legal development into legal development and vice-versa, but on much else besides. The four modes of urban informality can be texted with the Multi-Level Perspective through potential socio-spatial transition pathways by which management of urban sustainability transitions constitutes the cornerstone of this process according to the prevailing environment.

References

Abrams, C. 1964. *Man's Struggle for Shelter in an Urbanizing World*. Cambridge, MA: MIT Press.

Al Naber, M., and F. Nolle. 2016. The Politics of Accessing Desert Land in Jordan. *Land Use policy*: 492–503.

Al Sayyad, N. 2004. Urban Informality as a "New" Way of Life. In *Urban Informality: Transnational Perspectives from the Middle East, Latin America, and South Asia*, ed. A. Roy and N. Al Sayyad, 7–30. Lanham: Lexington Books.

———. 2006. Whose Cairo? In *Cosmopolitan: Politics, Culture, and Urban Space in the New Globalized Middle East*, ed. D. Singerman and P. Amar, 539–542. Cairo: American Unversity Press in Cairo.

Al-Hasanat, S. 2003. Housing Information System. *Sixth Annual Seminar on Housing Sector Assessment*. Amman: Housing and Urban Development Corporation, 107–113.

Allen, A. 2014. Peri-Urbanisation and the Political Ecology of Differential Sustainability. In *A Routledge Handbook on Cities of the Global South*, ed. S. Oldfield and S. Parnell, 21–36. London/New York: Routledge.

Amin, A. 1976. *Unequal Development: An Essay on the Social Formations of Peripheral Capitalism*. Sussex: The Harvester Press Limited.

Amineh, M. Parvizi. 2007. *The Greater Middle East in Global Politics : Social Science Perspectives on the Changing Geography of the World Politics*. Leiden, The Netherlands: Brill.

Anholt, R., and G. Sinatti. 2020. Under the Guise of Resilience: The EU Approach to Migration and Forced Displacement in Jordan and Lebanon. *Contemporary Security Policy* 41 (2): 311–335.

Antonio, J., C.M. Lara-Hernandez, C.M. Coulter, and A. Melis. 2020. Temporary Appropriation and Urban Informality: Exploring the Subtle Distinction. *Cities* 99. https://doi.org/10.1016/j.cities.2020.102626.

Banks, N., M. Lombard, and D. Mitlin. 2020. Urban Informality as a Site of Critical Analysis. *The Journal of Development Studies* 56 (2): 223–238.

Bayat, A. 2004. Globalization and the Politics of the Informals in the Global South. In *Urban Informality: Transnational Perspectives from the Middle East, Latin America, and South Asia*, ed. Ananya Roy and Nezar AlSayyad, 79–102. Lanham: Lexington Books.

Boanada-Fuchs, A., and F.B. Vanessa. 2018. Towards a Taxonomic Understanding of Informality. *IDPR* 40 (4): 397–420.

Bums, L.S., and J. Friedmann. 1985. *The Art of Planning: Selected Essays of Harvey S. Pedoff.* New York: Plenum Press.

Butterwick, S., and R.L. Lawrence. 2009. Creating Alternative Realities: Arts-Based Approaches to Transformative Learning. In *Transformative Learning in Practice: Insights from Community, Workplace, and Higher Education*, ed. J. Mezirow, E.W. Taylor, et al., 35–45. San Francisco: Jossey-Bass.

Cammett, M., I. Diwan, A. Richards, and J. Waterbury. 2015. *A Political Econmoy of the Middle East*. Boulder: Westview Press.

CAPMAS. 2016. *A Study for Development and Upgrade Informal Areas in Egypt (in Arabic)*. Cairo: CAPMAS.

Caroline, M., and S. Yatzimirsky. 2014. *Megacity Slums: Social Exclusion, Space and Urban Policies in Brazil and India*. London: Imperial College Press.

Castells, M. 1997. *The Power of Identity*. Malden: Blackwell Publishers Inc.

———. 2010. *The Rise of the Network Society*. 2nd ed. West Sussex: Wiley.

Chen, Marty, and Françoise Carré. 2020. *The Informal Economy revisited: Examining the Past, Envisioning the Future*. Oxon: Routledge.

Coenen, L., P. Benneworth, and B. Truffer. 2012. Toward a Spatial Perspective on Sustainability Transitions. *Research Policy* 41: 968–979.

Crowther, D., S. Seifi, and A. Moyeen. 2018. Responsibility and Governance in Achieving Sustainability. In *Approaches to Global Sustainability, Markets, and Governance*, ed. D. Crowther, S. Seifi, and A. Moyeen, 1–18. Singapore: Springer.

Culcasi, K. 2016. Warm Nationalism: Mapping and Imagining the Jordanian Nation. *Political Geography* 54: 7–20.

d'Alençona, P.A., H. Smith, and E.A. de Andres. 2018. Interrogating Informality: Conceptualisations, Practices and Policies in the Light of the New Urban Agenda. *Habitat International* 75: 59–66.

de Satgé, R., and W. Vanessa. 2018. *Urban Planning in the Global South*. Cham: Palgrave Macmillan.

De Soto, H. 2000. *The Mestry of Capital: Why Capitalism Triumphs in the Westand Fails Everywhere Else*. London: Blaack Swan.

Dennis, R. 2005. *Marginality, Power, and Social Structure: Issues in Race, Class, and Gender*. London: Elsevier Ltd.

Devlin, R.T. 2018. Asking 'Third World questions' of First World informality: Using Southern Theory to Parse Needs from Desires in an Analysis of Informal Urbanism of the Global North. *Planning Theory* 17 (4): 568–587.

Dominelli, L., and M. Moosa-Mitha. 2014. *Reconfiguring Citizenship: Social Exclusion and Diversity Within Inclusive Citizenship Practice*. Surrey: Ashgate Publishing Limited.

Drakakis-Smith, D. 2000. *Third World Cities*. 2nd ed. London: Routledge.

Dumper, M.R.T., and B.E. Stanley, eds. 2007. *Cities of the Middle East and North Africa: A Historical Encyclopedia*. Santa Barbara: ABC-CLIO, Inc.

Dwyer, D.J. 1975. *People and Housing in Third World Cities*. London: Longman.

Edensor, E., and M. Jayne, eds. 2011. *Urban Theory Beyond the West: A World of Cities*. London: Taylor & Francis.

Elsheshtawy, Y. (ed.). 2004. *Planning Middle Eastern cities : An Urban Kaleidoscope in a Globalizing world*. London: Routledge.

Elzen, B., F. Geels, and K. Green. 2004. *System Innovation and the Transition to Sustainability: Theory, Evidence and Policy*. Gloucestershire: Edward Elgar Publishing Limited.

Fathy, H. 1973. *Architecture for the Poor: An Experiment in Rural Egypt*. Chicago: The University of Chicago Press.

Fawaz, M. 2017. Planning and the Refugee Crisis: Informality as a Framework of Analysis and Reflection. *Planning Theory* 16 (1): 99–115.

Fergany, N. 2016. *Arab Revolution in the 21st Century? Lessons from Egypt and Tunisia*. Hampshire: Palgrave Macmillan.
Frantzeskaki, N., V.C. Broto, L. Coenen, and D. Loorbach. 2017a. Urban Sustainability Transitions: The Dynamics and Opportunities of Sustainablity Transitions in Cities. In *Urban Sustainability Transitions*, ed. N. Frantzeskaki, V.C. Broto, L. Coenen, and D. Loorbach, 1–19. Oxon: Routledge.
———. 2017b. *Urban Sustainability Transitions*. New York: Routledge.
Friedmann, J. 2005. Globalization and the Emerging Culture of Planning. *Progress in Planning* 64: 183–234.
Gans, H. 1962. *The Urban Villagers; Group and Class in the Life of Italian-Americans*. New York: Free Press of Glencoe.
———. 1971. The West End: An Urban Village. In *Internal Structure of the City*, ed. L.S. Bourne, 300–308. Oxford: Oxford University Press.
Geddes, P. 1915. *An Introduction of the Town Planning Movement and to the Study for Civics*. London: Williams & Norgate.
Geels, F. 2005. The Dynamics of Transitions in Socio-Technical Systems: A Mulit-Level Analysis of the Transition Pathway from Horse-Drawm Carriages To Automobiles. *Technology Analysis and Strategic Management* 17 (4): 445–476.
———. 2010. Ontologies, Socio-Technical Transitions (to Sustainability), and the Multilevel Perspective. *Research Policy* 39: 495–510.
Geels, Frank W., and Johan Schot. 2007. Typology of Sociotechnical Transition Pathways. *Research Policy* 36: 399–417.
———. 2010. The Dynamics of socio-Technical Transitions: A Sociotechnical Perspective. In *Transitions to Sustainable Development; New Directions in the Study of Long Term Transformative Change*, ed. J. Grin, J. Rotmans, and J. Schot, 11–93. New York: Routledge.
Geels, F.W., M.P. Jacobsson, and S. Hekkert. 2008. The dynamics of sustainable innovation journeys. *Technology Analysis & Strategic Management 20* (5): 521–536.
Geels, F.W., F. Berkhout, and D.P. van Vuuren. 2016. Bridging Analytical Approaches for low-Carbon Transitions. *Nature Climate Change* 6: 576–583.
Geels, et al. 2017. *A research agenda for the Sustainability Transitions Research Network*. Amsterdam: STRN.
Genus, A., and A.-M. Coles. 2008. Rethinking the Multi-Level Perspective of Technological Transitions. *Research Policy* 37 (9): 1436–1445.
Gerges, F.A. 2015. *Contentious Politics in the Middle East: Popular Resistance and Marginalized Activism Beyond the Arab Uprisings*. New York: Palgrave Macmillan.
Gilbert, A., and J. Gugler. 1982. *Cities, Poverty and Development: Urbanisation in the Third World*. Oxford: Oxford University Press.
Gough, I., G. Wood, A. Barrientos, and P. Bevan. 2008. *Insecurity and Welfare Regimes in Asia, Africa and Latin America: Social Policy in Development Contexts*. New York: Cambridge University Press.
Grashoff, and Yang. 2020. Towards Critique and Differentiation: Comparative Research on Informal Housing. In *Comparative Approaches to Informal Housing Around the Globe*, ed. U. Grashoff, 1–21. London: UCL Press, University College London.
Grin, J., J. Rotmans, and J. Schot. 2010. *Transitions to Sustainable Development: New Directions in the Study of Long Term Transformative Change*. New York: Routledge.
Gupta, J., K. Pfeffer, H. Verrest, and M. Ros-Tonen, eds. 2015. *Geographies of Urban Governance: Advanced Theories, Methods and Practices*. Cham: Springer.
Harris, R. 2009. Reflections on Lebanon. In *Lebanon: Liberation, Conflict, and Crisis*, ed. Barry Rubin, 9–24. New York: Palgrave Macmillan.
———. 2018. Modes of Informal Urban Development: A Global Phenomenon. *Journal of Planning Literature* 33 (3): 267–286.
Hart, K. 1973. Informal Income Opportunities and Urban Employment in Ghana. *Journal of Modern African Studies* 11 (1): 61–89.
Harvey, D. 2016. *The Ways of the World*. London: Profile Books Ltd.

Hashemite Kingdom of Jordan. 2014. *Jordan Response Plan 2015 for the Syria Crisis*. Amman: Ministry of Planning and International Cooperation.
———. 2017. *Jordan in Figures*. Amman: Department of Statistics.
Healey, P. 2003. Collaborative Planning in Perspective. *Planning Theory* 2 (2): 101–123.
Hosseinioon. 2018. Measuring Urban Resilience to Natural Disasters for Iranian Cities: Challenges and Key Concepts. In *Urban Disaster Resilience and Security: Addressing Risks in Societies*, ed. F. Fiedrich and A. Fekete, 71–90. Cham: Springer.
———. 2019. Urban Resilience and Informality: Effects of Formalisation in Golestan, Iran. In *Urban Resilience for Risk and Adaptation Governance: Theory and Practice*, ed. G.B. Morató, O.C.N. Tollin, and M.R.C. Jordi, 11–128. Cham: Springer.
ISDF. 2016. *Slums Areas in Egypt*. Cairo: ISDF.
Jackson, L. 2010. *Influences on Financial Decision Making Among the Rural Poor in Bangladesh*. Unpublished Ph.D. thesis. Sydney: School of Marketing, The University of Western Sydney.
Jamshidi, M. 2014. *The Future of the Arab Spring Civic Entrepreneurship in Politics, Art, and Technology Startups*. Oxford: Elsevier Inc.
Kabisch, N., H. Korn, J. Stadler, and A. Bonn, eds. 2017. *Nature-based Solutions to Climate Change Adaptation in Urban Areas: Linkages Between Science, Policy and Practice*. Cham: Springer.
Karaman, O., et al. 2020. Plot by Plot: Plotting Urbanism as an Ordinary Process of Urbanisation. *Antipode*: 1–30.
Kemp, R., D. Loorbach, and J. Rotmans. 2005. Transition Management as a Model for Managing Processes of Co-Evolution Towards Sustainable Development. *The International Journal of Sustainable Development and World Ecology*: 1–15.
Khalidi, R. 1998. The "Middle East" as a Framework of Analysis: Re-mapping a Region in the Era of Globalization. *Comparative Studies of South Asia, Africa and the Middle East* 18 (1): 74–81.
Khirfan, L. 2019. Planning and Unplanning Amman: Between Formal Planning and Non-traditional Agency, In *The Palgrave Handbook of Bottom-Up Urbanism*, ed. M. Arefi and C. Kickert, 201–218. Cham, Switzerland: Springer International Publishing.
Kim, S.J., and S.B. Merriam. 2011. Methodological Challenges in Studying Transformative Learning. In *Adult Education Research Conference*, 364–370. Toronto: New Prairie Press.
King, K. 2009. Workplace Performance–PLUS: Empowerment and Voice Through Professional Development and Democratic Processes in Health Care Training. *Performance Improvement Quarterly* 21 (4): 55–74.
Kipper, R., and M. Fischer. 2009. *Cairo's Informal Areas: Between Urban Challenges and Hidden Potentials*. Cairo: GTZ.
Koehler, J., F. Geels, F. Kern, E.. Onsongo, and A. J. Wieczorek. 2017. *A Research Agenda for the Sustainability Transitions Research Network*. Technical. Amsterdam: STRN Working Group, STRN.
Layton, R.A. 2009. On Economic Growth, Marketing Systems, and the Quality of Life. *Journal of Macro Mark* 29 (4): 349–362.
Lefebvre, H. 1971. *Everyday Life in the Modern World*. New York: Harper & Row Publishers.
Lewis, W. Arthur. 1954. Economic Development with Unlimited Supplies of Labor. *Manchester School of Economics and Social Studies* 22 (2): 139–191.
———. 1966. Culture of Poverty. *Scientific American* 215 (4): 19–25.
Loorbach, interview by A. Soliman. 2020. *Series Editor* (March 15).
Loorbach, D., and S. Oxenaar. 2018. *Counting on Nature: Transitions to a Natural Capital Positive Economy by Creating an Enabling Environment for Natural Capital Approaches*. Rotterdam: Dutch Research Institute for Transitions (DRIFT), Erasmus University.
Loorbach, D., N. Frantzeskaki, and F. Avelino. 2017. Sustainability Transitions Research: Transforming Science and Practise for Societal Change. *Annual Review of Environment and Resources* 42: 599–626.
Maconachie, R. 2007. *Urban Growth and Land Degradation in Developing Cities: Change and Challenges in Kano, Nigeria*. Hampshire: Ashgate Publishing Limited.

Madsen, S.R., and B.J. Cook. 2010. Transformative Learning: UAE, Women, and Higher Education. *Journal of Global Responsibility* 1 (1): 127–148.
Markard, J., R. Raven, and B. Truffer. 2012. Sustainability Transitions: An Emerging Field of Research and Its Prospects. *Research Policy* 41 (6): 955–967.
Martin, D. 2015. From Spaces of Exception to 'Campscapes': Palestinian Refugee Camps and Informal Settlements in Beirut. *Political Geography* 44: 9–18.
Marx, Colin, and Emily Kelling. 2019. Knowing Urban Informalities. *Urban Studies* 56 (3): 494–509.
McAuslan, P. 2013. *Land Law Reform in Eastern Africa, Traditional or Transformative? A Critical Review of 50 Years of Land Law Reform in Eastern Africa,1961–2011*. Oxon: Routledge.
Mertens, D.M. 2012. Transformative Mixed Methods: Addressing Inequities. *American Behavioral Scientist* 20 (10): 1–12.
Mguni, P. 2015. *Sustainability Transitions in the Developing World: Exploring the Potential for Integrating Sustainable Urban Drainage Systems in Sub-Saharan Cities*. Copenhagen: PhD submitted to University of Copenhagen.
Murshed, S.M. 2002. *Globalization, Marginalization and Development*. London: Routledge.
Nassara, J., and N. Stel. 2019. Lebanon's Response to the Syrian Refugee Crisis – Institutional Ambiguity as a Governance Strategy. *Political Geography* 70: 44–54.
Nuseir, A., M. Tuffaha, and S. Momani. 2003. *Housing Strategies and Its Impact on Housing Sector. Housing and Urban Development Corporation*. Amman: Sixth Annual Seminar on Housing Sector Assessment, 215–222.
O'Brien, K. 2015. Political Agency: The Key to Tackling Climate Change. *Science* 350: 1170–1171.
Obremski, Helena, and Claudia Carter. 2019. Can Self-Build Housing Improve Social Sustainability Within Low-Income Groups? *TPR* 90 (2): 167–193.
Parnell, S., and S. Oldfield. 2014. *The Routledge Handbook on Cities of the Global South*. Abingdon/Oxon: Routledge.
Payne, G., ed. 2002. *Land, Rights and Innovation: Improving Security for the Urban Poor*. London: ITDG.
Perlman. 1976. *The Myth of Marginality: Urban Poverty and Politics in Rio de Janeiro*. Stanford: California University Press.
Piketty, T. 2014. *Capital in the Twenty-First Century*. London: Th e Belknap Press of Harvard University Press.
Pittfor, A. 2019. System Change, Not Climate Change. *Internationa Politices and Society*.https://www.ips-journal.eu/regions/global/article/show/system-change-not-climate-change-3712/.
Pomponi, F., and A. Moncaster. 2017. Circular Economy for the Built Environment: A Research Framework. *Journal of Cleaner Production* 143: 710–718.
Rakowski. 1994. Introduction: Land and Housing Development in Middle Eastern and North Africa Cities. *International Journal of Urban and regional Research* 12 (1): 1–7.
Rip, A., and R. Kemp. 1998. Technological Change. In *Human Choice and Climate Change, Volume 2*, ed. S. Rayner and E.L. Malone, 327–399. Columbus: Battelle Press.
Rocco, R., and J.V. Ballegooijen, eds. 2018. *The Routledge Handbook on Informal Urbanization*. New York: Routledge.
Roy. 2004. The Gentleman's City: Urban Informality in the Calcutta of New Communism. In *Urban informality: Transnational Perspectives from the Middle East, Latin America, and South Asia*, ed. A. Roy and N. Al Sayyad, 147–170. Lanham: Lexington Books.
———. 2005. Urban Informality: Toward an Epistemology of Planning. *Journal of the American Planning Associarion* 71 (2): 147–158.
———. 2009. The 21st-Century Metropolis: New Geographies of Theory. *Regional Studies* 43 (6): 819–830.
———. 2011. Postcolonial Urbanism: Speed, Hysteria, Mass Dreams. In *Worlding Cities : Asian Experiments and the Art of Being Global*, ed. A. Roy and A. Ong, 307–335. Chichester: Blackwell Publishing Limited.

———. 2014. Worlding the South: Toward a Post-colonial Urban Theory. In *The Routledge Handbook on Cities of the Global South*, ed. S. Parnell and S. Oldfield, 9–20. London: Routledge.
Roy, A., and N. Al-Sayyad. 2004. *Urban informality: Transnational perspectives from the Middle East, Latin America, and South Asia*. Lanham: Lexington Books.
Sait, S., and H. Lim. 2006. *Land, Law and Islam: Property and Human Rights in the Muslim World*. London: Zed Books Ltd.
Satterthwaite, D., D. Archer, S. Colenbrader, and D. Dodman. 2020. Building Resilience to Climate Change in Informal Settlements. *One Earth* 21: 143–156.
Schot, J. 1998. The Usefulness of Evolutionary Models for Explaining Innovation: The Case of the Netherlands in the Nineteenth Century. *History of Technology* 14: 173–200.
Schot, J.W., and F. Geels. 2008. Strategic Niche Management and Sustainable Innovation Journeys: Theory, Findings, Research Agenda, and Policy. *Technology Analysis and Strategic Management* 20 (5): 537–554.
Schot, J., and L. Kanger. 2016. *Deep Transitions: Emergence, Acceleration, Stabilization and Directionality Science Policy Research Unit (SPRU)*. Sussex: University of Sussex.
Seely, J. 1959. The Slum: Its Nature, Use and Users. *Journal of the American Institute of Planners* 25 (19): 7–11.
———. 1971. The Slum: Its Nature, Use and Users. In *Internal Structure of the City*, ed. L.S. Bourne, 464–474. Oxford: Oxford University Pres.
Sibley, D. 1995. *Geographies of Exclusion: Society and Difference in the West*. London: Routledge.
Sims, D. 2010. *Understanding Cairo: The Logic of a City Out of Control*. Cairo: America University Press in Cairo.
Singerman, D. 1995. *Avenues of Participation: Family, Politics, and Networks in Urban Quarters of Cairo*. Princeton: Princeton University Press.
———. 2011. The Siege of Imbaba, Egypt's Internal "Other", and the Criminalization of Politics. In *Cairo Contested: Governance, Urban Space, and Global Modernity*, ed. Diane Singerman, 111–144. Cairo: The American University in Cairo Press.
Singerman, D., and P. Amar. 2006. *Cairo Cosmopolitan: Politics, Culture, and Urban Space in the New Globalized Middle East*. Cairo: The American University in Cairo Press.
Smith, A., J. Vob, and J. Grin. 2010. Innovation Studies and Sustainability Transitions: The Allure of the Multi-level Perspective and Its Challenges. *Research Policy* 39 (4): 435–448.
Soliman, A. 2004a. *A Possible Way Out: Formalizing Housing Informality in Egyptian Cities*. Lanham: American University Press.
———. 2004b. Tilting at Sphinxes: Locating Urban Informality in Egyptian Cities. In *Urban Informality in an Era of Liberalization: A Transnational Perspective*, ed. Ananya Roy and Nezear Al Sayyad, 171–208. Lanham: Lexington Books.
———. 2008. Diversity of Ethnicity and State Involvement on Urban Informality in Beirut. *Theoretical and Empirical Researches in Urban Management, Year 3, Number 9*: 15–32.
———. 2012a. Housing and the State in the Middle East. In *International Encyclopedia of Housing and Home*, ed. Susan J. Smith, 346–354. Amsterdam: Elsevier.
———. 2012b. The Egyptian episode of self-build housing. *Habitat International* 36 (2): 226–236.
———. 2015a. Collective Planning Process: A Driving Seat for Formalising Urban Informality in Egypt. *International Journal of Urban Sustainable Development*, June 1–24.
———. 2015b. The Crucible of Housing the Bottom of the Pyramid in Egypt. In *Housing the Urban Poor in Developing Countries*, ed. B. Aldrich and R. Sandhu, 165–189. Jaipur: Rawat Publications.
———. 2017. Land Readjustment as a Mechanism for New Urban Land Expansion in Egypt: Experimenting Participatory Inclusive Processes. *International Journal of Urban Sustainable Development* 9 (3): 313–332.
———. 2019. So It's Always a Chance: Community-Led Solutions to New Urban Expansion. In *New Cities and Community Extensions in Egypt and the Middle East*, ed. S. Attia et al., 159–179. Cham: Springer.

———. 2020. Understanding Sustainability Transitions for Urban Informality in the Middle East. *International Journal of Urban Sustainable Development*. https://doi.org/10.1080/19463138.2020.1728692.
Stokes, C. 1962. A Theory of Slums. *Land Economic* 38 (3): 187–197.
Talani, L. S. 2014. *The Arab Spring in the Global Political Economy*. Hampshire, UK: Palgrave Macmillan.
Turner, J. 1976. *Housing by People*. London: Marion Boyars Publishers Ltd.
———. 2015. Explaining the (Non-) Encampment of Syrian Refugees: Security, Class and the Labour Market in Lebanon and Jordan. *Mediterranean Politics* 20 (3): 386–404.
Turner, J., and R. Fichter. 1972. *Freedom to Build: Dwellers Control of the Housing process*. New York: Macmillan.
Turnheim, B., F. Berkhout, F. Geels, A. Hof, A. Mcmeekin, and B. Nykvist. 2015. Evaluating sustainability Transitions Pathways: Bridging Analytical. *Global Environmental Change* 35: 239–253.
UN. 2016. *Habitat III Issue Papers: 20 – Housing*. New York: UN.
UNDP and GDCD. 2008. *Disaster Risk Management Profile (Amman–Jordan)*. Amman: General Directorate of Civil Defense (GDCD).
UN-Habitat. 2012. *Tapping the Potential: The Role of Grassroots in Land Policy Implementation*. Technical report. Nairobi: UN-Habitat.
———. 2016a. *Urbanization and Development: Emerging Futures*. World Cities Report 2016. Nairobi: United Nations Human Settlements Programme (UN-Habitat).
———. 2016b. *The New Urban Agenda*. Nairobi: UN-Habitat.
———. 2016c. *Evaluation of Participatory Review of Egyptian Planning and Related Urban Development Legislation to Support Sustainable Urban Development, April 2016*. Nairobi: UN-Habitat.
———. 2018a. *DG 11 Synthesis Report 2018: Tracking Progress Towards Inclusive, Safe, Resilient and Sustainable Cities and Human Settlements*. SDG 11 Synthesis Report High Level Political Forum 2018. Nairobi: Un-Habitat.
———. 2018b. *Prosperity for All: Enhancing the Informal Economy Through Participatory Slum Upgrading*. Nairobi: UN-Habitat.
———. 2018d. *National Urban Policies Program in Lebanon: Diagnosis Report 2018*. Nairobi: Un-Habitat
UN-Habitat and Arab Republic of Egypt. 2015. *Egypt Housing Profile*. Cairo: Arab Republic of Egypt.
United Nations. 2018. *Department of Economic and Social Affairs, Population Division*. New York: United Nations.
Vasenev, V.I., E. Dovletyarova, Z. Cheng, and R. Valentini, eds. 2018. *Megacities 2050: Environmental Consequences of Urbanization*. Cham: Springer.
VROM. 2001. *Where There's a Will There's a World*. The Hague: Ministry of Housing: Spatial Plannning Environment (VROM).
Ward, B. 1976. *The Home of Man*. Harmondsworth: Penguin Books Ltd.
———. 1982a. Informal Housing: Conventional Wisdoms Reap-Praised. *Built Enviroment* 8 (2): 84–94.
———. 1982b. *Self-Help Housing: A Critique*. London: Alexandrine Press, Mansell Publishing Co.
Wieczorek, Anna J. 2018. Sustainability Transitions in Developing Countries: Major Insights and Their Implications for Research and Policy. *Environmental Science and Policy* 84: 204–216.
Wittmayer, J., J. Backhaus, F. Avelino, B. Pel, T. Strasser, I. Kunze, and L. Zuijderwik. 2019. Narratives of Change: How Social Innovation Initiatives Construct Societal Transformation. *Futures* 112. https://doi.org/10.1016/j.futures.2019.06.005.
Yiftachel, O. 2009. Theoretical Notes on 'Gray Cities': The Coming of Urban Apartheid? *Planning Theory* 8 (1): 88–100.

Part II
Urban Informality Transitions in Middle Eastern Cities

Chapter 2
Governance and Sustainability Transitions in Urban Informality

Abstract At the beginning of the new millennium, innovative urban planning focused on urban sustainability transitions to tackle the major urban challenges, one of which is urban informality. In this unprecedented era of increasing urbanization, and in the context of Habitat III and the connection between the New Urban Agenda (NUA) and the 2030 Agenda for Sustainable Development (SDGs), urban informality has become the main topic to make the planet sustainable. This chapter emphasizes the recent transformation of a new planning innovation as a shift from thinking as a 'pearl of planning wisdom' to more comprehensive sustainable transitions. It draws on the roots of sustainable development to enrich the debate on the importance of trends in governance of sustainability transitions to managing the current urban challenges. It first briefly considers the recent debate on urban sustainability transitions as a new approach capable of facing the urban challenges of urban informality. Then, the debate on the management of the ways in which urban sustainable informality emerges. It concludes that urban informality in the cities of the Global South has to be engaged in a more multi-disciplinary and multi-dimensional approach. Urban informality, from the perspective of urban sustainability transitions, interrelates and interlinks with socioeconomic, socio-spatial, and political transitions in each environment.

Keywords Sustainable development · Sustainability transitions · Urban governance · Transition management · Sociotechnical systems · Multilevel perspective · Co-design and learning · Grassroots level

2.1 Introduction

In recent years, planning innovation has become an important issue in urban planning debates. Over time, planning innovation has witnessed several concepts according to a socioeconomic, spatial, and political transformation of the Global North, as well as the South, and to the rapid development of technology. This chapter traces the evolution of planning innovation before and after Habitat III up to now to examine the effect of theories on urban planning practice in the Middle East. The aim

A. M. Soliman, *Urban Informality*, https://doi.org/10.1007/978-3-030-68988-9_2

is to open an academic discussion on the theory and practice of governance of sustainability transitions and its correlation with urban informality in the Middle East. Two themes are raised in this chapter: one is to trace the recent academic debate on sustainability transitions (ST); and the second is to question whether the Middle Eastern cities are suitably equipped to tackle the challenges of urban informality in a sustainable way. The chapter aims to link theory and practice to a confrontation of urban informality in the built environment of the Middle East. It explores the different dynamics, challenges, and breakthroughs in accelerating ST of urban informality across Middle Eastern countries. It addresses the question: To what extent does urban planning require a deep understanding of the context in which it proposes to intervene on urban informality and how should this understanding shape what planners do?

This chapter briefly investigates the roots of sustainable development as a new approach, which was already underway by the 1970s, to engage with the major challenges facing the planet. Ward and Dubos (1972) examined a close connection between wealth distribution and the conservation of planetary resources, while Perloff (Bums and Friedmann 1985) questioned the quality of the urban environment to sharpen the meaning of the concept and substantially covered matters aligned with the treatment here.

With the beginning of the Third Millennium, a new research field questioned the sustainability development to guide natural resources towards the needs of current and future generation, and to protect the planet from climatic changes, biodiversity degradation, and spatial and natural hazards. Afterwards, many academic institutions in Europe developed an approach linking the Theory and Practice of Urban Sustainability Transition. This approach was then expanded further to address managing and governance of the ST.

The question of urban challenges some long-held assumptions in sustainable transitions where both theory and practice have tended to smooth over this kind of sensitivity in favor of concepts and practices that are blind to place and held to be valid in the Middle East when such is not always the case. Two assumptions are touched upon: First, is it possible to borrow new planning innovations from the Global North and apply them in the Global South/Middle East?; and, second, is it conceivable to link between the theory and practice of governance of ST and its correlation to urban informality in the Middle East? This chapter is structured as follows. Sections 2.2 and 2.3 cover these issues and illustrate the recent debate on ST. Section 2.4 examines the governance of ST and Sect. 2.5 outlines a framework as a possible paradigm on governance ST on urban informality. A short conclusion closes the chapter.

2.2 The Roots of Sustainable Development

Planning is unique, and its uniqueness stems partially from the inability to define in a single, narrow definition that which fits in every instance (Faludi 1973); (Paris 1982; Friedmann 2011). Planning has become an inherently transdisciplinary field of study (Bums and Friedmann 1985; Friedmann 2005) composed of multidisciplinary perspectives (Loorbach and Oxenaar 2018). Friedmann (2014) exploring transactive planning, radical planning, the concept of the "good city", civil society, rethinking poverty, quality of environment, and the diversity of planning cultures, through which emerges a coherent and compelling story about how the evolution of thinking about planning over several decades has helped to shape its practice. Faludi (Faludi 1973) argues that 'Theory of Planning', rather than 'Theory in Planning', should constitute the core subject of planning theory. It reflects the instruments and techniques used in ('planning'). By definition, 'Theory in Planning' helps planners to understand their area of concern. It is also identified as a substantive theory. 'Theory of Planning' refers to the very nature and rationality of planning as a form of state action and its role in societal development. It helps planners to understand themselves and their operating methods. Theory of planning is synonymous with procedural theory because planning is not a single entity that could fit into one category, nor can planners fully agree on what planning really is.

Contrary to popular belief, not all planning theories were constructed prior to their application on actual grounds. In fact, many of the theories being studied presently are a result of close observation of cities as they were transformed under political, social, economic, and spatial circumstances (Friedmann 2014). Theories of planning, however, mean different things to different people. Practitioner planners tend generally to view theories as useless in their practical endeavors. Conversely, planning academicians tend to view, and heavily rely upon, theories as an integral part of the planning profession. Thus, the interrelation, interaction, and the nexus between theory and practice of urban planning became the main concern of academic thinkers of today.

Since the industrial revolution that took place in Britain from the mid-eighteenth century up until now, cities continue to morph in shape for numerous reasons (Friedmann 2011) and various circumstances (Hall 1998). When in 1915 Patrick Geddes published 'Cities in Evolution', he was trying to fight against the social and environmental chaos and evil of the spontaneous sprawl of the city commencing with the Industrial Revolution. He was promoting a certain aesthetic quality of the city space and, at the same time, he was linking social evolution to spatial design and quality of the environment as in "The Three Ecologies". Geddes's book was also the first publication to understand transition as a shift the emphasis from a developmental paradigm to an evolutionary one, in which small changes can lead to big effects. In other words, in a more or less intuitive manner, he was trying to investigate the possibility of a method/theory that would link cities' morphologies to the process of their functioning in a sustainable way.

During the last seven decades, a number of 'theories of the city' have been postulated in the context of urban planning, ranging from rational planning models (Knight 1921) and political economy models (Harvey 2009) through equity planning (Hall 1998), the advocacy planning model (Lynch 1960), and sustainable/smart/green planning models (Loorbach and Oxenaar 2018) which in turn have informed the way in which we understand cities. For David Harvey (2009), a city is 'a complex dynamic system in which spatial form and social process are in continuous interaction with each other', and the historically, culturally constructed institutional framework and capital within urban areas are what matters in formulating planning models (Harvey 2016). Others (Castells 1983; Blokland and Savage 2008) saw the city as a fragmented social-spatial reality, 'a Dual City', formal and informal, brought about by technological change which created a conflict between a 'space of flows' and a 'space of places', and formulated a social capital model. On the other hand, cities in the Middle East are transferable and fragmented due to the speedy process of socioeconomic and political transitions and might be more than "a dual city", which are creating "a new pattern of urbanity" (Soliman 2017) as "a new resilience planning" to cope with immediate transitions. Recently, systems of cities, in the North and South, became more complicated because of the complex systems of interaction and interrelated disciplines originating from the debate over the complexity of globalization (see Blokland and Savage 2008; Parnell and Oldfield 2014).

Amos Rapoport in his influential book *Human Aspects of Urban Form* (Rapoport 1977) arrived at a concept for analyzing social and urban changes allowing for cultural differences at all levels of definition, choice of devices and rules (hidden/non-hidden) (tangible/non-tangible) for maintaining good qualities in the built environment. His conclusion underlines that the nexus between man, various activities, natural resources, and the environment is the final product of the urban fabric and the dynamics of the transitions of human settlements. Barbara Ward, or Lady Jackson, in her classical book (Ward 1989) *Progress for a Small Planet* explored three topics that dominate discussions of the global environment: pollution; the consequences of the affluent running ever faster through finite resources; and the growing tensions between rich and poor. She suggested a global strategy for meeting the basic needs of the disadvantaged and showed how the vast inequalities between countries can be reduced. She used the phrases "inner limits" and "outer limits" to refer to the inner limits of the human right to an adequate standard of living and the outer limits of what the Earth can sustain. Ward and Dubos are the co-authors of *Only One Earth* (Ward and Dubos 1972), written for the 1972 UN Stockholm Conference on the Human Environment (Earth Summit 1) and the subsequent Conference on Trade and Development (1974) in which a major concern was the sustainability development. Since the first Habitat Conference on Human Settlements in Vancouver in 1976 (HCHSV76), urban sustainability has become a necessity and moral duty for all nations as an exciting journey into "future sustainable built environments". In the HCHSV76, Ward has warned the nations of the mushrooming population in the Global South and the increasing scarcity of natural resources on the planet, as well as the increasing gap between rich and poor nations and individuals alike.

Since the early 1980, most international organizations, governments, scholars, and thinkers became more aware of sustainable development for the Earth's cities. With the emergence of sustainability as a concept in the 1987 Brundtland's Report (Brundtland 1987) to, covering long-term environmental strategies and international solidarity and cooperation, more efficient actions are being taken by human actors with respect to the environment and the expectations and aims of the global society. There has been a paradigm shift towards 'sustainable development' in human development policy (Fiedler 2014), plus attention to how to govern sustainable development (Monkelbaan 2019), and how to reduce the negative impact on environment of economic growth (Brundtland 1987).

Drakakis-Smith (2000) examined sustainability from the urbanization point of view. He makes a clear distinction between 'sustained growth' and 'sustainable urbanization'. In both, a broader objective of sustainability aims to meet present needs 'without compromising the ability of future operations to meet their own needs.' Sustainability also emphasizes the interlinked nature of the individual components of rapid urbanization. This means that, as far as possible, urban management policies need to consider the implications of change in one part of the system on the remainder. It is also true that economic growth must bring returns to the individual citizen as much as it brings returns to the firm or the state.

Sustainable development is seen by many (Mitlin and Satterthwaite 2004; Hall 1998); (Lefebvre 1991); (Beall and Fox 2009; Loorbach et al. 2016; Bicknell et al. 2009; Meadowcroft 2000; Maheshwari et al. 2016) as aimed at "promoting the human well-being, meeting the basic needs of the poor and protecting the welfare of future generations (intra- and inter-generational justice), preserving environmental resources and global life-support systems (respecting limits), integrating/distributing economics and environment in decision-making, and encouraging popular participation in development processes". With the current situation of Covid-19, health sustainability is another dimension to be taken into account for protecting well-being in urban informal settlements of current and future generations (Abusaada and Elshater 2020).

As cities witness tremendous changes in population, the rapid transformation of the urban fabric and fluctuating economic growth became a necessary and justifiable development objective for the urban transitions process. Drakakis-Smith (2000) stated that urban sustainability must satisfy the following requirements: equity, social justice and human rights, basic human needs, such as shelter and health care, social and ethnic self-determination, environmental awareness and integrity, and awareness of linkages across both space and time, i.e., not seeking gain at the expense of someone elsewhere in the world or of the generations to come.

The management of sustainable cities required the interrelationship between economic, social, cultural, and political processes to balance between "losers" and "winners" in which narrative of change is existed (AL-Sayyad 2004; Wittmayer et al. 2019). There are four main issues raised: sustain natural resources; meet the current needs for an existing population without harming that which is to come;

sustain economic growth, sustain health care, and determine how to manage; and govern these sorts of transitions in a sustainable way. In general, sustainability is already understood as a goal that must accompany all transformations that occur in each environment. During the 1990s, "sustainability" emphasized social, economic, and environmental well-being, or what is often referred to as the 'triple bottom line' approach to sustainable development in the academic literature (Elkington 1997). More formally, 'sustainable urban development' (SUD) has been defined as 'development that improves the long-term social and ecological health of cities and towns' (Wheeler 1998).

Since the HCHSV76 in Istanbul in 1996, the adoption of the Millennium Development Goals (MDGs) in 2000 and the United Nations Conference on Housing and Sustainable Urban Development (Habitat III), in October 2016, in Quito, Ecuador, eradicating slum dwellings has been the prime concern globally. Also, the international organizations agreed that urban poverty needs to be recognized as multidimensional deprivation. On the other hand, the Sustainable Development Goals (SDGs) (UN-Habitat 2018) aimed at tracking progress towards inclusive, safe, resilient and sustainable cities and human settlements. Goal 11, Target 1, indicates that by 2030 ensuring access for all to adequate, safe, and affordable housing and basic services and the upgrading of slums should be prioritized. By 2050 the world urban population is expected to nearly double, making urbanization one of the twenty-first century's most transformative trends (UN-Habitat 2018). Therefore, there are increasing considerations of how to cope with these urban transitions in a new systematically sustainable way.

As reported by Kimberly Etingoff (2017), urban areas consume more than 66% of the world's energy and generate more than 70% of global greenhouse gas emissions. With the world's population expected to reach 10.875 billion by 2100 (UN 2019), nearly 90% of whom will live in urban areas, a critical question for planetary sustainability is how the size of cities affects energy use and carbon dioxide (CO_2) emissions. How the rights and entitlements of the urban poor are expanded and protected constitutes a key issue within the debate on sustainable cities. If the population of urban poor continues to increase at such rapid rates, and if the capacity of the growth of the informal sector to absorb this increase is nearing the saturation point, then more sympathetic and constructive policies must appear soon. With the increasing demand for housing for the urban poor and the spread of urban informality in the cities of the Global South, informal urbanization will continue to formulate the urban fabric of those cities (Soliman 2019).

There is a need to envisage a systemic transition in the existing built environments, not just to zero carbon emissions, but across the entire ecological footprint of our cities and the regions within which they are embedded, simultaneously promoting economic security, social health, and resilience (Rotmans 2006). However, the New Urban Agenda (NUA) (UN-Habitat 2018) responded to urban challenges' problems and drew a roadmap for building cities that can serve as engines of prosperity and centers of cultural and social well-being, while protecting the environment. Thus, the leaders of the planet agreed on eight task forces of the NUA. It is a vision of pluralistic, sustainable, and disaster-resilient societies that fosters green economic growth. It aims to pave the way to making cities and human settlements more inclusive, ensuring that everyone can benefit from urbanization while paying attention to those in vulnerable situations (UN-Habitat 2009). One of the main tasks of the NUA is to ensure that all citizens have access to equal opportunities for goods and services, while facing no discrimination, including access to a affordable shelter in urban areas, and to tackle the problem of urban informality.

2.3 Understanding Sustainability Transitions

The planet is currently confronted by a quadruple challenge: environmental degradation, climatic change, social inequality (Kanger and Schot 2019), and the crisis of Covid-19. These challenges are fundamentally linked to the First Deep Transition from both the "Western" mode of thinking (e.g., LaFreniere 2007; Lamba 2010), and the Latin American mode of thinking (Perlman 1976; Moore 2015). Both recognize the functioning of societal subsystems, such as poverty, social exclusion, marginality, and duality of economy through the emergence of a macro-level selection environment called 'industrial modernity' (e.g., Piketty 2014; Moore 2015). The Second-Deep Transition (Kanger and Schot 2019) refers to the creation and expansion of a wide range of sociotechnical systems of sustainable transitions for the provision of transport, energy, food, housing, healthcare, communications, etc., or tackling social inequality. The Third Deep Transition involves climate change which leads to environmental degradation. The Fourth Deep Transition is the Covid-19 crisis which cripples the whole planet and has infected more than 130 million.

The four deep transitions have led many academic institutions to innovate a new planning concept for tackling these urban challenges. One of these attempts is transition and transition management introduced in 2001 in the Fourth Dutch National Environmental Policy Plan (NMP4), which stressed four sustainable transitions of energy, biodiversity, agriculture, and mobility transitions. Also, transition management was presented as "a strategy to deal with environmental degradation by stimulating sustainable development as a specific aim of policy making" (VROM 2001) and to achieve the requirements of the MDGs. Later, the Sustainability Transitions Research Network (STRN) was inaugurated in 2009 at the 1st European Conference on Sustainability Transitions in Amsterdam to create a new interdisciplinary academic community (Koehler et al. 2017). The STRN became a core network to attract contributors to the development of sustainability transition studies. Since then, enormous numbers of research studies have introduced the question of governance of ST from different perspectives.

Sustainability Transitions (ST) are defined by many scholars (Coenen et al. 2012; Frantzeskaki et al. 2017a, b; Grin et al. 2010) as systems innovations that influence geographical settings through changes of sociotechnical, economic technical, and political configurations in a specific location (international, national, and local). The term sustainability transitions is increasingly used to refer to large-scale societal changes, deemed necessary to solve "grand societal challenges" (Loorbach et al.

2017). ST are long-term, multi-dimensional, and fundamental transformation processes. ST are quests for new value systems to change the behavior of human and the organization of natural resources in a sustainable way to meet the needs of the current populations and fulfill the requirements of future generations. One particularity of ST is that guidance and governance often play a specific role. ST inform the direction of the transition, as well as the direction of the socio-spatial development. Thus, the main output of ST is a shift from a certain mode of sociotechnical to another in a sustainable way, or a shift from a certain socio-spatial setting to another (Grin et al. 2010).

The ST are the recognition that many environmental problems and urban challenges are undergoing continuous change, such as climate change, loss of biodiversity, resource depletion, and many others, but they very rarely recognized the transformations of the spread of the urban informality in cities of the Global South (Soliman 2020). These problems cannot be addressed by incremental improvements but require shifts to new kinds of systems, shifts that are called 'ST' (Markard et al. 2012). Thus, ST are understood as 'involving a broad range of actors', a broad range of forces and as involving 'shifts in power' in which "a fundamental change in structure (e.g., organizations, institutions), culture (e.g., norms, behavior, way of life) and practices (e.g., routines, skills)" flourish (Loorbach and Rotmans 2010).

The concept of transition has been studied for decades in several disciplines, e.g., biology and population dynamics, economics, sociology, political science, science and technology studies, and in systems sciences (Grin et al. 2010), but very rarely are socio-spatial transitions touched upon (Truffer et al. 2015). It is argued that transition concepts are an analytical framework to analyze dynamics and patterns in specific societal systems (Loorbach, Series Editor 2020) to achieve sustainable development, but not what sort of dynamics and patterns because most studies viewed transitions from the economic point of view and ignored the socio-spatial transitions in the built environment (Koehler et al. 2017). The term transition is broadly used in many scientific disciplines and refers to a nonlinear shift from one dynamic equilibrium to another (Loorbach et al. 2017). Transition in sustainability is shifts or 'system innovations' between distinctive sociotechnical configurations, encompassing not only new technologies but also corresponding changes in markets, user practices, policy, cultural discourses, and governing institutions (Geels et al. 2008). A more detailed definition is that a transition is a gradual, continuous process of '...change where the basic character of society (or a complex sub-system of society) transforms' (Rotmans et al. 2001) in which the final outcome is a new ingredient or elements or innovations of production and consumption (Markard et al. 2012) that help to achieve sustainable development. An interesting broader definition of transitions was explored by Leftwich (Leftwich 2010) who proposed "all the activities of co-operation and conflict, within and between societies, whereby the human species goes about organizing the use, production and distribution of human, natural and other resources in the production and reproduction of its biological and social life". Therefore, the harvest of a transition is the equity of the distribution, production, co-production, and consumption of goods and services among different strata of a society for better and sustainable development or, in short, sustainability transitions.

These transitions occur according to several variables: co-evolution and multiple changes in sociotechnical systems, multi-actor interactions between social groups, 'radical' change in terms of scope of change, and time periods that witness transitions (Geels and Schot 2007). The trend of transitions and transformation builds on a more mature body of research and knowledge, grounded in diverse scientific disciplines. The failure of ST implies both that it is inevitable in the longer term that destabilization/disruption and systemic change will happen, and that systemic change is needed. Therefore, the process of transitions is a transformative or a shift (positively or negatively) from a certain condition or a composition or ingredients to another. It is argued (Loorbach et al. 2017) that the main driver behind the emergence of transitions research has been the search for new insights and ideas to understand how to steer clear from unsustainability lock-in and how to mobilize and empower disruptive innovations and transformative capacity from the system toward desirable ST.

The following section examines the trend of systemic innovation of sustainability challenges, sociotechnical systems and barriers to change, and the multilevel perspective on ST.

2.3.1 Systemic Innovation of Sustainability Transitions Challenges

The trend of transitions and transformation builds on a more mature body of research and knowledge, grounded in diverse scientific disciplines. Various models developed in this field aim to explain how transitions unfold, fold and how to govern them (Koehler et al. 2017), and these challenges require system innovation, i.e., deep-structural changes of the sociotechnical configurations underlying the respective sectors (Markard et al. 2012; Truffer and Coenen 2012). The failure of ST implies both that it is inevitable in the longer term that destabilization/disruption and systemic change will happen and that systemic change is needed. Therefore, the process of transitions is a transformative or a shift (positively or negatively) from a certain situation or a composition or ingredients to another. It is argued (Loorbach et al. 2017) that the main driver behind the emergence of transitions research has been the search for new insights and ideas to understand how to steer clear of unsustainability lock-in and how to mobilize and empower disruptive innovations and transformative capacity from the system toward desirable ST.

Transitions are fundamental changes and transformations that take place in the societal system and on earth. There is a great volume of research looking at the role of institutions in shaping transition policies, and these institutions shape the transitions processes (Fuenfschilling and Truffer 2014). There is an immediate need to move on from historical lessons or analyses of transitions in the making to become more explicit on how to develop policy-relevant scenarios and toolboxes based on interdisciplinary and multi-disciplinary knowledge generated by transition scholars.

The transitions approach does not offer a blueprint or a guidebook on how to govern such processes. The transitions approach offers conceptual frames for better understanding the sectoral dynamics of transitions and the co-evolution of the different societal systems and subsystems over the course of a transition (Kemp et al. 2005; Voß et al. 2009).

Also, it is argued that transitions are conceptualized as societal processes of fundamental change in the structure, culture, and practices of a societal system (Frantzeskaki and de Haan 2009). Transitions as processes of 'degradation' and 'breakdown' versus processes of 'build up' and 'innovation' (Gunderson and Holling 2002) have been witnessed in history (Geels 2004), and the question emerges how to handle it at the present time and in the future.

The EEA (2019) and Loorbach et al. (2017) identified three systemic innovations to categorize ST that present different ways to understand ST. These are sociotechnical, socio-ecological, and socioeconomic transitions. Although these approaches share a similar interest in the nonlinear process of transitions, they come from different disciplinary and epistemological backgrounds and bring different types of insights and methods to ST research. First, the sociotechnical approach employs insights from innovation studies, social movement theory, geography, and institutional theory. It provides a rich body of historical case studies to explain the dynamics of transitions in key production–consumption and market systems (Geels 2020). Within the sociotechnical research, "transitions" initially refer to large-scale transformations within a society or important subsystems, during which the structure of the societal system fundamentally changes (Rotmans et al. 2001) or sectoral societal change. Sociotechnical systems are the combination of human and non-human aspects that create functional configuration that works to change the nature of capitalism as the nature of society interacts or/and interlinks. Second, socio-ecological transformation addresses shift in complex environmental systems and combines this knowledge with increasing contributions from the social sciences that aim to explain deliberate transformations to sustainability through social change. Finally, socioeconomic transitions include diverse disciplines, such as political economy, sociology, philosophy, and positive psychology, contributing to destructive patterns of social shapeshifting, consumerism, and materialism (Pel and Kemp 2020; Wittmayer et al. 2019; Moore 2015) or to create a new sustainable mode of production, co-production, distribution, and consumption of goods and services.

On the other hand, the sociotechnical system covers a broad range of socioeconomic, political, and spatial transitions, and it criticized on four levels. One concerns the 'bias' towards (sustainable) technologies and their development from "its socio-institutional embedding" (Swilling et al. 2015). This observation may be driven by the initial writings on the intervention towards creating markets where green technologies are adopted and reshape the incumbent regimes (e.g. Rip and Kemp 1998). Another criticism refers to missing links with ecological aspects, material, and the energy-intensity of economic activities and human-nature relationships. By focusing purely on society–technology interactions, the sociotechnical approach is found of little use for understanding transitions of predominately agricultural systems on which poor people in the least developed countries are so reliant. Third, the sociotechnical system relies on changes in technologies and capitalism and ignored the influence of the transformation on the socio-spatial development in cities. Finally, transitions of sociotechnical systems may not achieve the planned stabilization or may even fail entirely (Chang et al. 2017). One of the very salient weaknesses in ST is related to the way of treating spaces and places in sociotechnical systems studies (Coenen et al. 2012). The argument is that the challenges posed by sustainability requirements for the development of new products, methods of distribution, means of consumption, technologies and lifestyles, necessitate a more elaborate conceptualization of the formation and transformation of sociotechnical systems and its spatial dimensions. This is because socio-spatial dimensions or "geography of ST" (Coenen et al. 2012) ignore ST in most studies in which the different socio-spatial contexts are largely treated as "empty containers" (Truffer and Coenen 2012), or "a container full off a culture of arbitrary socio-spatial growth and a new pattern of way of life".

However, the three systemic innovations of transitions rely on stable and well-established societies in the developed world, and do not recognize the transformation of the rapid, informal spatial growth of cities, nor the political instability, or weaker official institutions in the Global South. Most of countries in the Global South contain weaker state apparatus, less efficient bureaucracies, higher levels of political and economic instability, less transparency in legal proceedings and enforcement of legal frameworks and relatively high levels of economic and social inequality (Pel and Kemp 2020). Furthermore, these countries are typically characterized by less advanced industrial processes, a rapid informal urbanization, reliance on extended family ties and clientelism, and employment in the informal sector (Viotti 2001; Bell 2007). Therefore, 'in practice', urban informality, as a main output of the informal urbanization process, has been transformed by radical transitions and tensions in the socioeconomic and political context. Also, the effect of improper development policies, of regularization or "integration" imposed or encouraged by the states or by the international organizations, such as found in Egypt, Lebanon, and Jordan, are the main obstacles for transitions. Thus, urban informality transitions in sociotechnical system are nonlinear processes that are the product of a dialogue or sectoral dynamic of actions, or interactions between role makers (regime) and game players (niche), without which there can be no existential urban fabric (landscape). However, it is not necessary that players involved in transitions are attached to only one level, e.g., regime actors or niche actors, since in reality actors could engage on all levels (Jørgensen 2012; Frantzeskaki et al. 2018a, b).

The emergence of the urban informality phenomenon in cities of the Global South at its core, for example, required a complementary development of land delivery system, regulations, services (e.g., social amenities, infrastructures), user practices, etc. Because land is the main component of housing production and other land uses, many scholars (De Soto 2000; Mitlin and Satterthwaite 2004; Payne 2002) have paid great attention to formalizing and regulating the ownership and security of land tenure. Thus, the spatial transformation of land in cities of the Global South is the cornerstone of shaping the built environment in which the various activities are integrated and interlinked to provide an environment for generating the production, co-production, distribution, and consumption of various goods and services.

Even though the socio-spatial transitions do not change just the very structures of existing sociotechnical system, they also affect related societal domains, such as living, housing, and working activities. In the same way, ST are long-term, nonlinear, and multi-dimensional processes (in various aspects of life) through which the establishment of the socio-spatial systems shift to more sustainable modes of production, co-production, distribution, and consumption of various commodities. Thus, the urban informality transitions are possible outcomes of changes in "the institutional rectangle" of the realms of market, government, science and technology, and the way of living of civil society and their mutual alignment. The dominant means to spread urban informality is meeting the residents' requirements of various goods and services that change the way of life and create a new pattern of urbanity, as "*a new way of transitions*", and this can take roughly one or two generations (25–50 years) to be achieved (Kemp et al. 2007).

Going back to Turner's concept (1976) of housing as *"housing is not what it is"*, but *"what it does in people's lives"*, it could be argued "*housing is not what it is rather what it has done to the urban fabric of a given context*" (Soliman 2017). Nearly 50 years have passed since John F. C. Turner's work (Turner 1976) in the barriadas of Peru, which inspired him to develop his housing theory, influencing housing policies in Lima, and other cities in developing countries, but his concept has depleted the socio-spatial fabric of the built environment of most cities that adopted his concept. However, until recently, socio-spatial dimensions of ST have been largely ignored in the transitions literature (Smith et al. 2010; Coenen et al. 2012). Therefore, it is important to include the socio-spatial transformation as a mirror to reflect the final production/production and consumption of urban informality transitions process. So far, these different strands of transitions research have developed relatively in parallel and have not been integrated, leaving open several questions on how civil society and government exercise power in the socio-spatial transitions and how this relates to, and differs from, the way in which low-income groups exercise power in the formulation of their built environment.

However, a definitive statement of what space or urban fabric means for transitions is probably not the most pressing and interesting aspect of the geography of ST in which the search for new growth models are questioned (Bridge, et al. 2013; Cooke 2008; Van den Bergh 2013; Truffer and Coenen 2012). A greater awareness and conceptual clarification of the role of urban informality in shaping the urban fabric of cities in developing countries are attracting many scholars (Soliman 2020). Urban informality transitions are constituted socio-spatially and unpacking this configuration that helps to understand better the underlying processes that give rise to these patterns. This requires both contextual analysis of the particular settings (spaces, space with space, and places) and its integration with concerned people in which transitions are embedded and evolve, while at the same time paying attention to the geographical connections and interactions (i.e., the socio-spatial relations) between that place, space, other spaces, and people.

Building on earlier work by Truffer and Coenen (2012), three main dimensions of ageography of transitions are reached: socio-spatial embedding with sociotechnical and issues of power. However, the sociotechnical and socio-spatial transitions

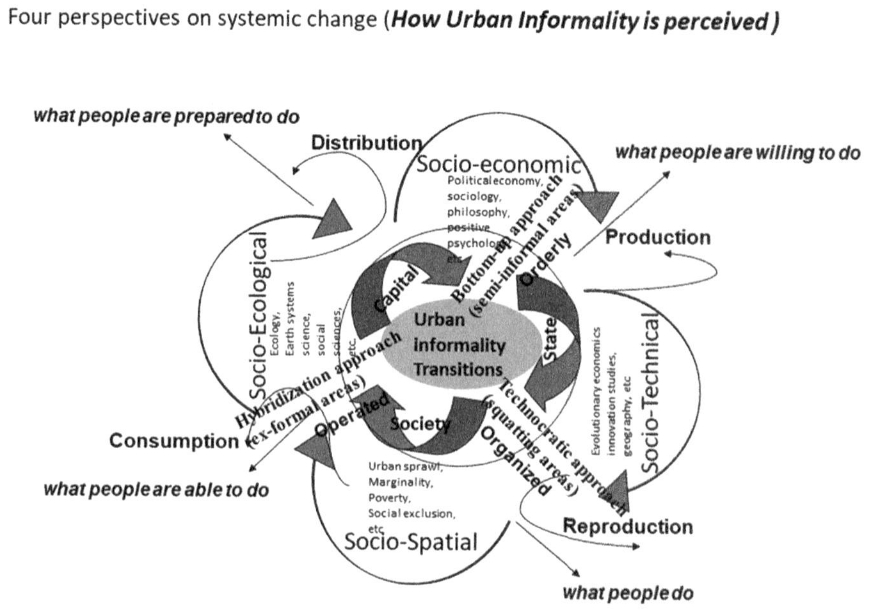

Fig. 2.1 Diagram of the interaction of four perspectives on systemic change (*How Urban Informality is perceived*). (Source: adopted by the author from (Loorbach et al. 2017))

(Socio-Spatial-Technical Transitions) (SSTT) are configured by the dominating power, which have two folds. First, SSTT focuses on precisely the kinds of trans-boundary production, co-production, distribution, and consumption of goods and services, such as energy, mobility, and housing. It offers a clear framework for understanding the dynamics of change arising from the interactions of societal systems with the drivers of change of the socio-spatial context at the macro, meso and micro levels. In doing so, it provides a foundation for mapping the implications for policy intervention in transitions, as well as links, and connects with various aspects of transitions (political, capital, biodiversity, natural resources, etc.). Given this, the discussion draws on the different framings of systemic change (sociotechnical and socio-spatial transitions), for example, in highlighting themes such as interactions between systems, social practices, market-based instruments, production, reproduction, distribution, consumption of goods and services, and various social amenities.

As illustrated in Fig. 2.1, transitions to sustainability emphasize three interrelated systemic framings of change: socio-economic, socioecological, and sociotechnical perspective (Loorbach et al. 2017), and the study focuses on those interlinked perspectives as transdisciplinary (Friedmann 2005; Kemp et al. 2007) and multidisciplinary field of studies (Schot and Kanger 2016). Figure. 2.1 depicts an attempt to include socio-spatial transitions as a fourth approach to a change in systematic innovations in order to include the geographical transitions as a main component influencing the shape and the pattern of a given urban fabric. All changes of chaotic land-use patterns and excessive urban sprawl are vital for modern economies and societies to achieve end-use functions, such as residential development, mobility, industrial processes, and residential development. All these activities are configured in organizational chains that represent different subsystems of the urban fabric. These subsystems inject the production, co-production, distribution and use of various commodities or activities. The arbitrary socio-spatial transitions of urban informality in the Global south are linked and interrelated to capital, state power, empowerment of the grassroots, and market mechanisms. The four modes of production, co-production, consumption and distribution of goods and services, determine the dominant areas of transitions of socioeconomic, socio-spatial, and political transitions.

However, the SSTT transitions aim to respond to the deficiencies of, and thus complement, the empowerment of the grassroots by digging into its historical transitions to explore how these shifts take place, thereby providing insights into contemporary social sustainability issues (Shirazi and Keivani 2019). It relies on responding to the four dimensions; *what people are willing to do, what people are prepared to do, what people do, and what people are able to do*. A more detailed definition is that a SSTT is "a gradual, continuous process of socio-spatial change where the basic pattern of the urban fabric (or a complex of sub-systems of a city) transforms", and is a rapid process in changing the natural resources. In a simple term, the SSTT are looking backward for a retrospect of a certain socioeconomic and spatial situation and looking forward to a prospect for it in a sustainable way in which the everyday life of citizens are shaped and created "a new way of life" (AL-Sayyad 2004) or "a new urban pattern" (Soliman 2017) or "a site of transitions". This transformation has a great impact in shaping socially, economically, and politically the built environment in which it is absorbing the needs generated by the rapid population growth and meeting the speedy urbanization process. There are various diverse lock-ins, lock-out, and systemic interactions to achieve the rapid and transformative change in the SSTT systems. One is the appropriate regularity collective framework of land to suite the requirements of the low-income groups for meeting the increasing demand of housing plots in cities of the Global South.

2.3.2 The Multi-level Perspective on Sustainability Transitions

The systemic changes of contemporary environmental problems are often called 'sociotechnical transitions', because they involve alterations in the overall configuration of various activities that entail technology, policy, markets, consumer practices, infrastructure, cultural meaning, and scientific knowledge (Elzen, et al. 2012; Geels 2014). The multilevel perspective (MLP) has emerged as a fruitful middle-range framework for analyzing sociotechnical transitions to sustainability (Grin et al. 2010; Geels 2010). Based on historical evidence of past transitions, the multilevel perspective suggests that transitions involve interacting processes between three analytical levels: innovative practices (niche experiments), a sociotechnical structure

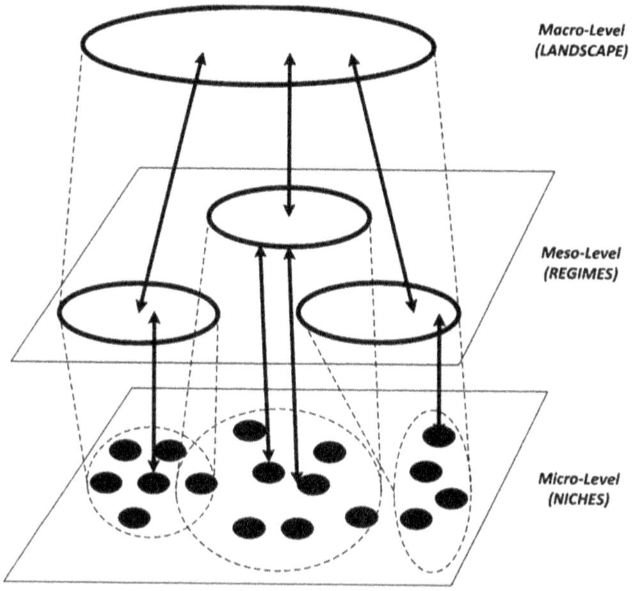

Fig. 2.2 Map illustrating the MLP process and its levels. (Source: The multi-level perspective of sustainability transitions (Geels and Schot 2007))

(the regime), and long-term, exogenous trends (the landscape) (Schot and Geels 2008; Schot 1998; Rip and Kemp 1998; Geels 2002, 2005; Köhler et al. 2019).

As shown in Fig. 2.2, the MLP of transitions comes about through interactions between processes on three levels: (a) niche-innovations build up internal momentum, through learning processes, price/performance improvements, and support from powerful groups; (b) destabilization of the regime creates windows of opportunity for niche innovations; and (c) changes on the landscape level create pressure on the regime (Geels and Schot 2007). The central level comprises sociotechnical regimes; sets of rules and routines that define the dominant "way of doing things." Regimes account for path-dependence, stability and are often locked-in, which hinders radical change. Regimes are stabilized by the sociotechnical landscape, a 'broad exogenous environment' that, as such, is beyond the direct influence of actors (Grin et al. 2010), and what time consumes. The landscape encompasses such processes as urbanization, urban informality processes, demographic changes, and wars or health crises (such as Covid-19) that can put pressure on regimes making them vulnerable to more radical changes. Regimes transform on condition of the availability of alternatives that can fulfil the same societal function. The emergences of innovations within niches are protected spaces outside, or on the fringe of, the established regime, where new entrants or relative outsiders (such as start-up family companies, housing development, and small social networks) can experiment without direct exposure to mainstream market pressure and institutional forces. In the case of the cities of the Global South, a community-driven process is created as niche innovations to fulfill the requirements of housing low-income groups, regardless of regulatory framework of official institutions. Alternatives are developed in niches, protected spaces that facilitate experimentation with novelties in the Middle Eastern cities. Sometimes, such niche spaces are created by public policy interventions-markets created for environmental technologies and services by regulatory or public procurement demand. In this case, niche and regime are overlapped and might combine or integrate to form a sociotechnical agglomeration or niche might disappear and leave the sociotechnical arena for the regime's act.

In the context of the MLP, system transformation is driven by change agents and occurs in the outcome of mutually reinforcing contextual, landscape pressures, internal regime destabilization processes, and scaling up of innovations developed in niches. Smith et al. (2010) argue that the evolution of the MLP in the context of innovation studies and sustainable development raises some challenging analytical and practical issues. These are niche dynamics, unlocking regimes, socio-spatial aspects of transitions, methods to map transitions, politics of transitions, and interlinkage with the dynamics of governance. It is argued that if a ST perspective provides a problem focus, then the MLP provides a helpful framework for organizing the broad interdisciplinary analysis that is needed. There is considerable scope for innovation studies to contribute further, just as was done in the genesis of the MLP framework.

On the other hand, the multi-phase (MP) concept describes a transition over time as a sequence of four alternating phases: (i) the pre-development phase from dynamic state of equilibrium in which the status quo of the system changes in the background, but these changes are not visible; (ii) the take-off phase, the actual point of ignition after which the process of structural change picks up momentum; (iii) the acceleration phase in which structural changes become visible; and (iv) the stabilization phase where a new dynamic state of equilibrium is achieved. The final shared concept is that of co-design and learning (Grin et al. 2010), and sometimes called learning by doing, and doing by learning. This means that knowledge is developed in a complex, interactive design process with a range of stakeholders involved through a social process of learning (Grin et al. 2010; Grin and Loeber 2007), and participant observation (Alexander et al. 1975).

Figure 2.3 illustrates the process of urban informality transitions which combines ML, MP, and SSTT. It distinguishes two directions: first multi-phases perspective is perpendicular to multi-level perspective; second, multi socio-spatial phases intersects multi-level perspectives, each of which has three phases within complex transition processes, characterized by different processes, actors, and barriers. The two directions have three phases which provide the logical framework for exploring the management of transitions process, as indicated later in this chapter. The horizontal direction has three variables/barriers: emergence, diffusion, and reconfiguration. The multi socio-spatial phase is a perpendicular axis to the multi-phases direction which has three stages: initial, intermediate, and consolidation. The first phase, scattered expansion, involves the emergence of radical innovations at the niches level. Scattered expansion takes place when the government erected some buildings, as a political informality, for its own employees on the periphery of the

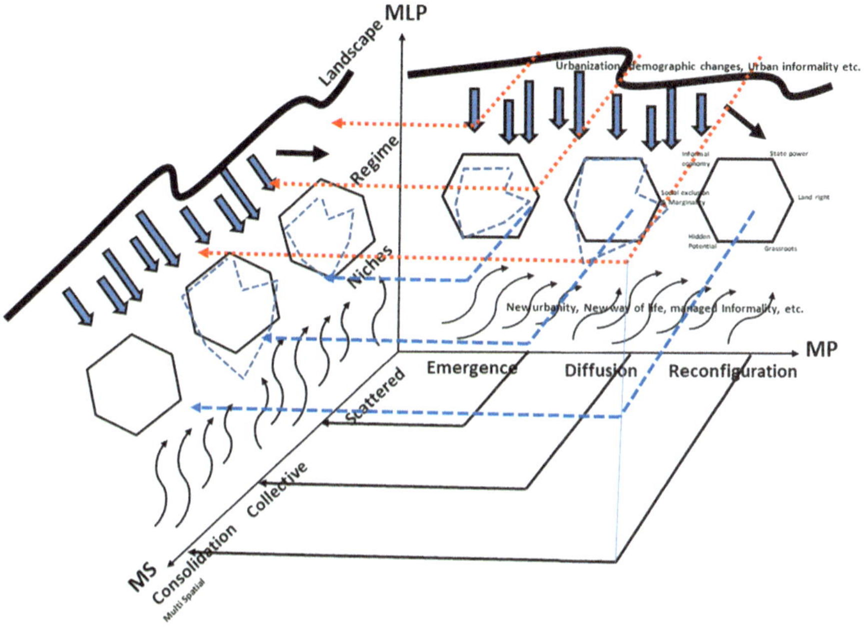

Fig 2.3 Diagram distinguishing the multi-level perspective intersecting with multi-phases and multi socio-spatial phases. (Source: The author)

urban centers, and this was recognized by outsiders as the initial development of the residential areas. These residential buildings were constructed on a main road to gain access to infrastructure and services, such as electricity, water, and sewerage. These attracted other people to develop housing on adjacent plots, and they followed the government steps.

The second multi socio-spatial phase is a collective expansion that began when the main services were developed within or close to informal residential areas, thus improving sanitary conditions or when a main road was constructed on the urban periphery. Changing housing systems and a stronger demand for services increased the desirability of suburban living. As a result, lower-income groups moved increasingly away from crowded central locations towards less desirable parts of the city and used a different residential building style. The fact that the public authorities installed various services within a certain peripheral area gave the impression that these areas would soon be developed as residential areas. As a result, while some government sponsored expansion of the peripheral areas took place, more private residential buildings were built on nearby vacant plots. As demand for housing grew, the peripheral areas became the main means by which low-income groups could acquire land for housing. The scale of such expansion in various settlements has varied greatly in the three cities; Cairo, Beirut, and Amman—much influenced by the level of public transport or informal private transport linking the settlement to zones of employment.

The third multi socio-spatial phase consolidates expansion and generally has two stages. The first stage is when the settlement becomes recognized by speculators, and more illegal land sub-division takes place and more buildings become established. Demand for apartments grows and small developers/contractors start to build more houses to cater to the newcomers. The informal areas become affordable to low-income groups, which encourages people to move out of crowded, inner-city rented flats into a newly established residential area. The advantages of the informal areas are relatively low land prices (compared to prices within the urban center) and the availability of vacant land accessible to various facilities and job opportunities. Both these factors encourage more people and investors to invest in housing production. This has helped low-income groups change their status from tenant to owner occupier. The second stage is characterized by vertical expansion. With the increasing price of land, new stories are added to the existing buildings in which densification is taking place.

This process is linked to the continuous changes of social and economic statuses of the residents, as well as continuous transitions at the national level and interrelated to the level of pressure of regime, or *laissez-faire* policy of the regime at a certain time for certain purposes which created a window of opportunity for housing production. Thus, urban informality transitions represents not only informal housing production but also a method for liquid capital and social capital accumulation. Urban informality transitions in the Middle East does not represent a need for housing so much as it is a matter of collateral investment to create wealth and to secure residents' future from the fluctuation of the economy within and outside the country. The current urban informal pattern of cities in the Middle East has accumulated over time due to arbitrary urban and housing development policies, and the spread of urban informality (Bayat 2004; Soliman 2004a, b; 2015a, b; Dorman 2007; Kipper and Fischer 2009; Singerman and Amar 2006; Sims 2010). Thus, unban informality offers a relevant entry point to avoid the urban divide and intra-city inequality and to create a new pattern of urbanity. Policymakers, as part of the regime, have much on their plate right now, and economic recovery plans cannot move faster than efforts to address the current urban informality transitions on the ground. But, as states shift their focus to recovery, the choices that countries make will define what tomorrow looks like and whether we are better able to manage future global crises.

2.4 Sustainability Transitions Management

Just as ST attracted many scholars, a call for the governance of ST that is built around 'provisional, flexible, revisable, dynamic and open approaches that include experimentation, learning, reflexivity, and reversibility' (Grin et al. 2010) is taking place. Governance transitions imply a less state-centric view of politics and pay attention to issues of negotiation, deliberation, and self-governance (Kemp et al. 2007). Governance transitions are about the structured ways and means by which the divergent preferences of interdependent actors are translated into policy choices depend on transparency, accountability, and trust to assign values so that the plurality of interests of societal sector are transformed into coordinated action and the compliance of actors is achieved.

Various approaches have been developed to direct transitions that aim to produce analyses of transitions, but also prescriptive advice on how to steer transitions, including work on Transition Management (Rotmans et al. 2001; Loorbach 2010), Strategic Niche Management (Kemp et al. 1998; Hoogma et al. 2002), and Reflexive Governance (Voss et al. 2006; Voss and Bornemann 2011). The challenges is how to steer transitions in desirable directions, but also of how to do so within timescales that help avoid dangerous environmental change (e.g., Sovacool 2016) and/or devastation of the urban fabric of a given environment. Classic work on governance (Kooiman 2003) defines governing as "the totality of interactions, in which public, as well as private actors, participate, aimed at solving societal problems or creating societal opportunities, attending to the institutions as contexts for the governing interactions, and establishing a normative foundation for all those activities." Experiments are one tool that can be used to implement transitions in practice (Hoogma et al. 2002), and there is a great need to move beyond experimentation as a key transition governance tool (Koehler et al. 2017). There is a lack of research on how intermediaries can interact in the niche–regime interface, destabilize incumbent regimes, and operate in later phases of transitions (Ingram 2015). There are many issues to be investigated such as traditional policies, forward-looking analysis, widening the use of quantitative methods, extending transition studies to include multiple levels of governance, further development of the analysis of transition experiments in several directions, and evaluation of the output of transitions processes.

The Global Campaign on Urban Governance (GCUG) seeks to contribute to the eradication of poverty through improved urban governance (UN-Habitat 2018). Urban governance is defined as a specific coalition of city-level forces assembled within a multilevel governance context in pursuit of a particular urban vision (Swilling and Hajer 2017). Urban informality is formulated and configured through self-organization and social networks to achieve well-being and to access affordable shelter and services. The traditional view is that urban governance refers to "the broad constellation of social, political and economic forces that mold the process of urban development" (Wittmayer et al. 2018). Lefebvre (1991) argued that urban governance unfolds across geographical scales because urbanization processes include cities, regions, cross-border agglomerations, as well as supranational hierarchies.

Kemp et al. (2007) distinguish governance transition and transitions management. A governance approach is based on insights from governance and complex systems theory as much as upon practical experiment and experience (Loorbach 2007). The former, the governance perspective to ST suggests, in line with sustainability theories and approaches, that governance and policy require a radically different set of guiding principles in the context of ST. A search is required for steering mechanisms and tools that coordinate societal and political processes in a participatory and deliberative fashion, while engendering commitment to sustainability values (Frantzeskaki and Loorbach 2012).

Transition management is a promising model for sustainable development, allowing societies to explore alternative social trajectories in an adaptive, forward-looking manner involving long-term goals and adaptive programs for system innovation (Kemp et al. 2007). Meadowcroft (2007) describes transition management as follows: "The theory has a modular structure, with several elements being combined to produce the whole". Particular components include: the image of the transition dynamic with the distinct stages of the transition process; a three-level analytical hierarchy of 'niche', 'regime' and 'landscape' that provides a framework for understanding transition processes; a basket of future-oriented visioning devices (goals, visions, pathways. and intermediate objectives); a practical focus for activities (arenas and experiments); and a broad 'philosophy of governance' that emphasizes decision-making in conditions of uncertainty and "the gradual adjustment of existing development pathways in light of long-term goals."

In transition management, the governance process is a cyclical process of development phases at various scale levels (Loorbach 2007). The transition management cycle theorizes that four different types of governance activities can be distinguished when observing actor behavior in the context of societal transitions: strategic, tactical, operational, and reflexive (Frantzeskaki and Loorbach 2012; Wittmayer et al. 2018). These activities exhibit specific characteristics (in terms of the type of actors involved, the type of process they are associated with, the type of product they deliver, and the level of consumption) which make it possible to (experimentally and exploratively) develop specific instruments that have the potential to govern transition processes (Frantzeskaki and Loorbach 2012). A fifth activity as a *Commitment of package of actions,* by which action, policy sets, cost, and effectiveness accelerate the empowering of niches, could be added.

Figure 2.4 illustrates the four activities of strategic, tactical, operational, and reflexive, in addition to a fifth activity of commitment of package of actions. The model provides a practical tool for transition management in the urban informality context. Strategic level governance activities focus on the long term and the overall system, taking into consideration the mechanisms and the incremental development process of urban informality. Tactical level governance activities focus on the mid- and long-terms and on parts of the system. Actors designing actions with a tactical scope are concerned with the overall direction of the system, rather than with specific goals unfolding over the mid-term (5–15 years), and tend to be in charge of program development, agenda-setting, and regulation. Operational level governance activities focus on the short and long terms and concrete activities—they are part of the constant flow of everyday decisions and actions by professionals and citizens alike. The reflexive level is inherently about learning and consequently adapting activities on the different levels and between the different levels of the mechanisms of urban informality. The commitment-of-package-of-actions level is about reducing the gap of the tensions between the regime and people concerned. It offers transparency and a bridge of trust between all stakeholders involved in the transition management.

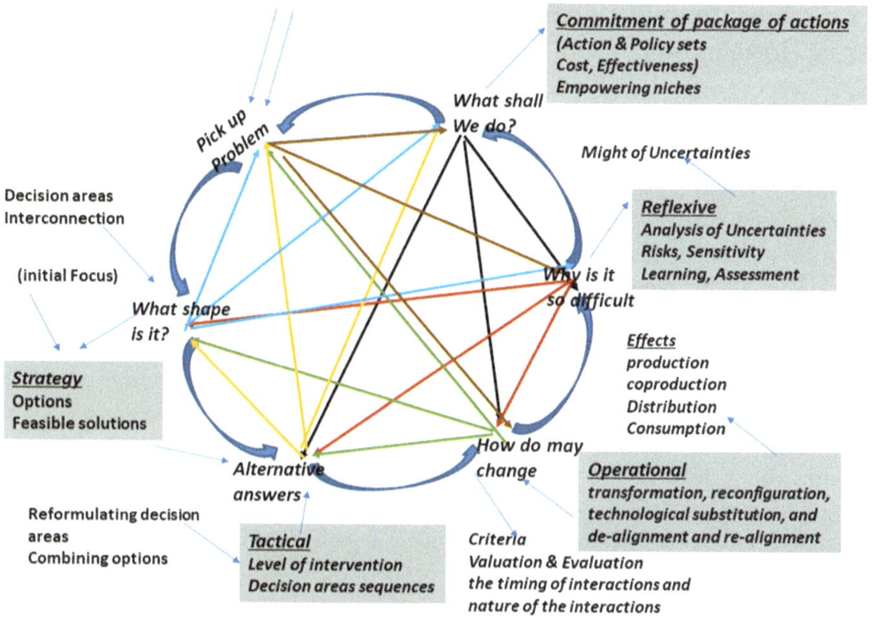

Fig 2.4 Scheme of the transition management process in urban informality. (Source: the author adopted from (Frantzeskaki et al. 2018a)

The model also relies on using combinations of the two criteria of the timing of interactions and nature of the interactions in which four different transition pathways are developed (Geels and Schot 2007). These are transformation, reconfiguration, technological substitution, and de-alignment and re-alignment. An additional fifth proposition addresses a possible sequence of transition paths, i.e., how transitions may start with one path, but shift to others. The linkage between these typologies with the Multi-Level Perspective (MLP) is outcomes of alignments between developments on multiple levels. These transition typologies have been developed to conceptualize the different pathways that transitions may take depending on the configuration and timing of the interactions between landscape factors, niche-level innovations, and sociotechnical regimes (See Geels et al. 2016; Geels and Schot 2007; Wieczorek 2018; Mguni 2015).

The issue of the transition management of urban ST on urban informality in developing countries is very rarely emphasized in the literature and in the research arena (Wieczorek 2018; Mguni 2015). The ST literature documents a rapidly growing and influential field of research (Markard et al. 2012; Köhler et al. 2019). This model builds on the argument that the interconnected, complex, and global character of current challenges, such as climate change, sprawling urbanization, or urban informality and growing social inequalities, requires a radical change in the basic systems providing societal needs for energy, water, or shelter (Schot and Kanger 2016).

2.5 Sustainability Transitions Management of Urban Informality

Faludi (1970) elaborated upon a tentative list of the elements of the planning environment as follows: the level and place of development; norms and value; political system and administrative structure; institutional structure; cleavages in society; and specific features of a society. Thus, the starting point for sustainable transitions management on urban informality is the recognition that many environmental problems, such as pollution, lack of hygienic sanitary systems, resource depletion (clean water, clean methods of transportation, security, and safety), and physical problems, such as high densities, compact urban fabric, and institutional and legalization problems, are grand challenges. Transition management relates to sustainable consumption and production patterns in sociotechnical systems such as electricity, heat, buildings, mobility, and agro-food. These problems cannot be addressed by incremental improvements or upgrading but require shifts to new kinds of systems, shifts which are called 'ST management' (Grin et al. 2010; Markard et al. 2012), but the question remains how do ST stimulate urban informality transitions.

Soliman (2012a) stated that housing informality, as a commodity or a product or a consumption, does not operate in isolation. Rather it depends on the domination of capital, labor, a willingness of the state, collective resources, and cooperation among all stakeholders. He recognizes the local people and their capability for handling financial investment, managing, and maintaining the physical environment, controlling the local resources, and their competence in participating in the national economy. The formulation processes of urban informality are created through social networks and cultural norms depending on the organizational bases that dictate those rules and the means among the residents. Also, the state institutions play an important role to support, directly/indirectly or not support, what people do, the way people manage their environment, and regulate the prevailing law to be for the benefit of the collective resources generated by the urban informality.

However, urban informality is the result of continuous urban transitions that resulted from a cooperation or an arrangement or relationship among a group of actors (whatever the level and the type of actors involved), the result and the functions of capital (whatever the amount and sort), state involvement (whatever the level of intervention), and the nature of the built environment being produced. Urban informality promotes sustainability, where value-for-cost is maximized, thereby allowing residents the opportunity to control the built environment, and where people are encouraged to invest in shared amenities. In short, urban informality is the echo of the duality of socioeconomic structure and the absence of the state involvement in urban informality processes. The question is how to alter urban informality transitions management, how to make these areas sustainable, and how they will be governed. The sustainability and transformations of urban informality, in particular in cities of the Middle East, in the practice of everyday life (Lefebvre 1971) and social security (Gough et al. 2008) that constitute a new urbanity (Soliman

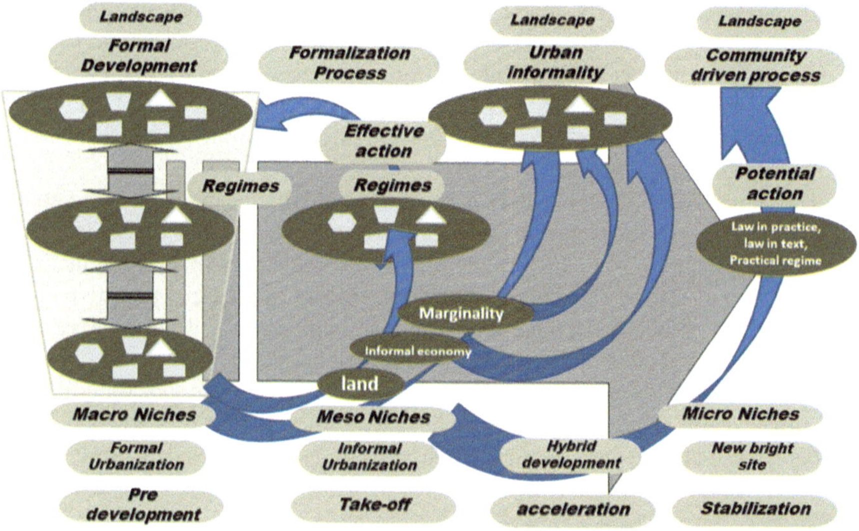

Fig. 2.5 Scheme of the linkage between sociotechnical systems, the MLP, the MP, and urban informality transitions. (Source: adapted from (Grin et al. 2010; Soliman 2019))

2012) are predicated upon and conditioned by governance, production, reproduction, distribution and consumption of housing, economic activities, and social amenities for the bottom strata of the societies in the Global South.

Linking the MLP with the MP, the MLP emerge through interactions between processes at three structural levels (Geels 2010; Geels and Schot 2007; Grin et al. 2010) of niches, regime, and landscape perspective. The MP constitutes four distinct phases (Grin et al. 2010): the pre-development, the take-off, the acceleration, and the stabilization phases. As illustrated in Fig. 2.5, the model is based on the linkage between sociotechnical systems, the MLP, and the MP through the interaction and correlation among the three levels of niche-innovations, regime, and landscape, while the MP comprises four phases of transition: predevelopment, takeoff, acceleration, and stabilization. It continues to process in two directions, vertically, and horizontally, in which interactions occur between the MLP and the MP through various pathways. It highlights socio-cognitive aspects of the regime's rules and their changes that potentially shape sociotechnical transitions.

This model provides X-curve transition phases and pathways (Loorbach and Oxenaar 2018) through the interactions between various levels of the MLP and the MP to create an enabling, sustainable urban management (new residential areas; upgrading of existing informal areas, enhancement social amenities; and guiding new informal development of virgin areas, etc.) for the acceleration of desired transitions. The final landscape of this process is four outputs: formal development, formalization of informal areas, upgrading of existing urban informality, and, finally, encouraging community-driven processes in newly established areas. In each phase of the transition, vertically and/or horizontally different strategies and instruments can be used to create a sustainable urban fabric for the uptake of natural dynamic transitions: population growth, production, reproduction, consumption, capital dynamics, etc. The learning concept is the output of gaining pieces of knowledge generated from the several phases of the progress and from the mechanisms of different pathways.

This model has the advantages that, first, it comes with the dynamic of the duality of economic and social exclusion in the Global South. Second, it is a flexible process that meets changes beyond the formal/informal and regulatory framework, including also behavioral, cultural, and practical changes caused by the rapid transitive urbanization. Third, it is an adaptable model to cope with outer and inner forces (economic, social, and political) that might affect the emergence of societies. Fourth, it copes with natural disasters, local conflict, poverty degradation, spreading of informal urbanization, fluctuation of the market, etc. Fifth, it is elastic, so it works with a diversity of actors to be able to challenge incumbent ideas and interests and to creatively adapt methods and tools to different contexts requiring specific skills and capacities.

Therefore, this model will investigate five main dimensions, as a reflection to the Alexandria School of Urban Sustainability Transitions (ASUST), to the governance of ST on urban informality in the Global South, especially the Middle East. These are: urbanization dynamics perspective (Drakakis-Smith 2000); relationships between state, market, and society; the paradigm of urban informality; laws, norms, and practices and their expression in life-styles (Lefebvre 1971); land rights and property transitions perspectives (Payne 2002), and social exclusion and right to the city. Work on transitions and system changes has expanded under different terms, e.g., regime transformation (Avelino et al. 2016), technological revolutions, technological transitions (Geels 2002), system innovation (Elzen et al. 2012) (Geels 2005), transition management (Rotmans et al. 2001), capital transitions (Piketty 2014) community based organization transition (Soliman 2012b) and, finally, people-movement trends.

Even though many other pathways may exist in the proposed model, Geels and Schot (2007) develop six pathways; reproduction process; landscape pressure on the regime, de-alignment and re-alignment path; technological substitution; the reconfiguration pathway and sequence of transition pathways, many other pathways may exist in the proposed model. These innovations trigger further adjustments in the architecture of the niches, the regime, and the landscape. Also, the MP development and its interaction and correlations with the MLP might create many pathways according to various circumstances for governance sustainable transitions on urban informality. This would yield a range of dynamic patterns that combine in different ways to produce multi pathways.

In general, urban informality in the Global South is the output of the domination of capital and its reproduction (De Soto 2000; Bromley 2004), the size of production (Turner and Fichter 1972; Fathy 1973), the level of consumption (Davis 2006; Sims 2010; Perlman 1976; Ward 2004; Soliman 2017), and societal transitions (Mitlin and

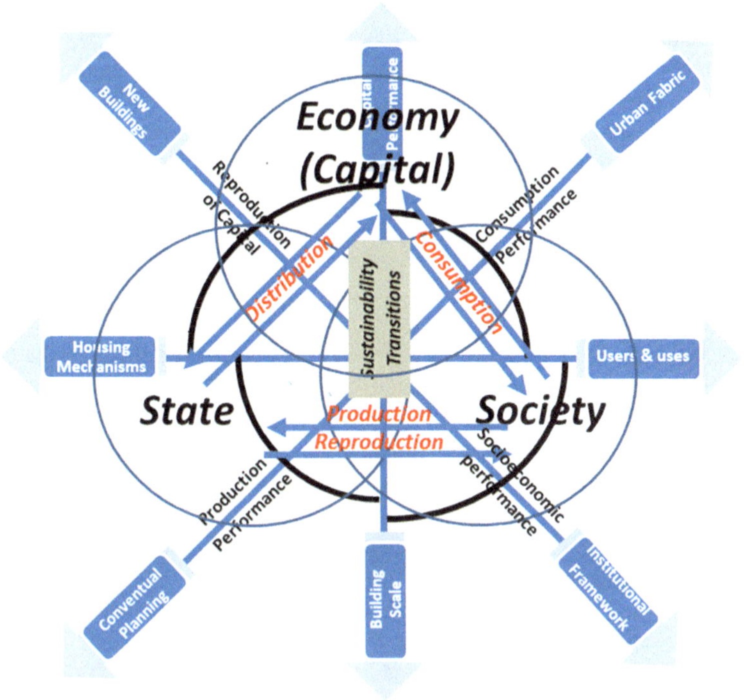

Fig. 2.6 Scheme of the linkage between four perspectives on the urban informality transitions in cities. (Source: The author)

Satterthwaite 2004; Hatina 2007; Roy and AL-Sayyad 2004; Harvey 1985, 2016; Soliman 2019).

As illustrated in Fig. 2.6, there are four perspectives on the urban informality transitions in cities. These range from: production performance, including housing mechanisms, conventional planning and building scale; reproduction of capital perspectives, including interrelated and linkage between capital performance, new buildings, and housing mechanisms; consumption perspectives, including building scale, institutional framework and users and uses lifetime trends; and socioeconomic perspectives including users and uses, urban fabric, and capital performance as the ‘urban sustainable community’ discourses. Knowledge mapping of the built environment in cities must also recognize the complexity. In this context, it is important to realize that sustainable development in urban informality can apply at various levels of building, neighborhood, city, regional, national, and international scales. Frequently the model has treated the built environment not only as spatially connected and complex, but also a socioeconomic complex that generates goods and services for the urban fabric. This sociotechnical connectivity relates to the complexity of infrastructure, spaces and places and communities together with how urban form and function relate. In this sense, a focus purely on buildings leads to a lack of strategic focus. Moreover, as Bai et al. (2010) suggest, there is frequently an inherent temporal ('not in my term') mismatch of spatial ('not in my patch') and institutional ('not my business') scales between urban decision making and global environmental concerns, where urban decision-makers are often constrained within short timescales, the immediate spatial scale of their jurisdictions, and within 'nested' governmental hierarchies.

2.6 Conclusion

What is suggested here, then, is the logic of directing the flashback of the topic of urban informality as the main challenge facing the Global South, as well as an in-depth look at the ST management. The development of sustainable management along transitions lines is the key to achieving the major objectives to sustain the built environment associated with urban informality. It is clear that ST management of urban informality to do so or not to do so will have a great deal to do with whether the polarization of societies in the Middle East will be increased or diminished within the next decade or two.

The linkage between ST and urban informality is the cornerstone of this chapter. This chapter attempts to clarify four themes; a brief synopsis of the roots of sustainable development; ongoing research innovation on ST; ST management; and how to deal with urban informality in a sustainable way. The basic concepts underlying the ST debate, which has been going on for a few years now, notably center on the management of ST. There are many different conceptions of ST in the literature, and there are even more understandings of different styles of governance/ management. Our "hub and spokes" provide a structured illustration of the different dimensions attributed to "ST" in the literature.

Three important points are highlighted in this context. First is the unifying element to ST that concentrates on climate changes, energy consumption, deterioration of natural resources, etc., and very rarely are the ongoing rapid changes in the urban fabric of cities in the Global South touched upon. Second, different strands of the literature have focused on different facets of the relationship between state intervention, on the one hand, and societal autonomy, on the other, ignoring the influence of market forces and political transferees on the ongoing changes. Accordingly, the main objective is organized according to a conception of ST and how urban informality could be sustainable. Third, it is appropriate to return to the roots of sustainability and shows that our approach brings in new insights and/or new wine in old bottles.

The chapter demonstrates that many of the existing debates in the literature are inherently inclusive to the Global North as they mix different explicit and implicit dimensions differently than the ones in the Global South. Also, it is very rare to question urban informality and its negative impact on the urban context. In order to avoid such analytical problems, the chapter suggests a two-step analysis that is derived from the ASUST to test the configuration of urban informality within ST.

In the first step, the relationship between urban informality and ST is examined in relation to their policy and politics with proper ties in correlation and interaction of various forces. Only on this basis, a second step could then draw meaningful cross-linkages between governance in the form of institutional structures, actor constellations, ST, urban informality, and resulting linkage instruments. As part of the first step, a sequence of correlations is formulated between four dimensions (the concept of transitions, sociotechnical level (macro, meso and micro), levels of intervention (market forces, government's policies, NGOs, and society), and sustainable transitions dimensions. The integration of urban informality with urban context in more detail is presented a planning process of three modes of sustainability (complex systems, sociotechnical perspective, and governance perspective) in the policy dimensions by which physical, economic, political and social networks adaptation are reached. These four dimensions are distinguished along with two variables: the type of instruments applied (legally binding legislation or soft law) and the approach to implementation (flexible or rigid). Further pathways for the other dimensions could follow suit.

The second step focuses on a land issue as the main component of informal residential development and its relationship with the informal economy and marginality. Relying on the MLP and the MP, four possibilities emerge: formal, informal, hybrid, and new bright sites, each of which has certain treatments and adaptations. This sort of adaption has to be within the context of ST dynamics, including locked-in/out regimes that are challenged by changing contexts, ecological stress and societal pressure for change, as well as experiments and innovations in niches driven by entrepreneurial networks and creative communities and proactive administrators.

It is essential to understand the art of ST and it is being managed in Western societies. Whether urban planning in the Middle Eastern cities is following a certain concept of theory of planning or has not, explicit planning thinking must be considered. This chapter has emphasized the transformation of socioeconomic and political atmospheres in which sustainability transitions are emerging. Now the Middle East, especially after the Arab Spring, has lost its way in following a certain theory of planning to tackle the perpetual challenges of urban informality.

ST management dynamics includes locked-in/out regimes that are challenged by changing contexts, ecological stress, and societal pressure for change, as well as experiments and innovations in niches driven by entrepreneurial networks, creative communities, and proactive administrators. But also included are resistance by vested interests, uncertainties about the future amongst urban populations, political instabilities, and the erosion of social services and systems of provision. On the other hand, creating a sustainable environment from the roots would alleviate future disasters and avoid future deterioration of the built environment. The early ideal cities of the Industrial Revolution proved at that time its resilient sustainability and capability for modifying a built environment that provides a better way of life in a sustainable context.

References

Abusaada, H., and A. Elshater. 2020. COVID-19's Challenges to Urbanism: Social Distancing and the Phenomenon of Boredom in Urban Spaces. *Journal of Urbanism: International Research on Placemaking and Urban Sustainability*. https://doi.org/10.1080/17549175.2, https://doi.org/10.1007/s11524-020-00438-6.

Al Sayyad, N. 2004. Urban Informality as a "New" Way of Life. In *Urban Informality: Transnational Perspectives from the Middle East, Latin America, and South Asia*, ed. A. Roy and N. Al Sayyad, 7–30. Lanham: Lexington Books.

Alexander, C., et al. 1975. *The Oregon Experiment*. Mountain View: The Center for Environmental Structure.

Avelino, F., J. Grin, B. Pel, and S. Jhagroe. 2016. The Politics of Sustainability Transitions. *Journal of Environmental Policy & Planning* 18 (5): 557–567.

Bai, X., R. McAllister, R. Beaty, and B. Taylor. 2010. Urban Policy and Governance in a Global Environment: Complex Systems, Scale Mismatches and Public Participation. *Current Opinion in Environmental Sustainability* 2: 1–7.

Bayat, A. 2004. Globalization and the Politics of the Informals in the Global South. In *Urban Informality: Transnational Perspectives from the Middle East, Latin America, and South Asia*, ed. Ananya Roy and Nezar AlSayyad, 79–102. Lanham: Lexington Books.

Beall, J., and S. Fox. 2009. *Cities and Development*. London: Routledge.

Bell, S.T. 2007. Deep-Level Composition Variables as Predictors of Team Performance: A Meta-Analysis. *Journal of Applied Psychology* 92 (3): 595–615.

Bicknell, J., D. Dodman, and D. Satterthwaite. 2009. *Adapting Cities to Climate Change: Understanding and Addressing the Development Challenges*. London: Earthscan.

Blokland, T., and M. Savage. 2008. Social Capital and Networked Urbanism. In *Networked Urbanism: Social Capital in the City*, ed. T. Blokland and M. Savage, 1–20. Hampshire: Ashgate Publishing Limited.

Bridge, G., S. Bouzarovski, M. Bradshaw, and N. Eyre. 2013. Geographies of Energy Transition: Space, Place and the Low-Carbon Economy. *Energy Policy* 53: 331–340.

Bromley, R. 2004. Power, Property and Poverty: Why De Soto's Mystery of Capital Cannot Be Solved. In *Urban Informality: Transnational Perspectives from the Middle East, Latin America, and South Asia*, ed. A. Roy and N. Al Sayyad, 271–288. Lanham: Lexington Books.

Brundtland, G.H. 1987. *The World commision on Environment and Development: Our Common Future*. Technical Report. New York: United Nations.

Bums, L.S., and J. Friedmann. 1985. *The Art of Planning: Selected Essays of Harvey S. Pedoff*. New York: Plenum Press.

Castells, M. 1983. *The City and Grassroots: A Cross-Cultural Theory of Urban Social Movements*. London: Edward Arnold (Publishers) Ltd.

Chang, Rui-dong, J. Zuo, Z. Zhao, V. Soebarto, G. Zillante, and X. Gan. 2017. Approaches for Transitions Towards Sustainable Development: Status Quo and Challenges. *Sustainable Development*. https://doi.org/10.1002/sd.1661.

Coenen, L., P. Benneworth, and B. Truffer. 2012. Toward a Spatial Perspective on Sustainability Transitions. *Research Policy* 41: 968–979.

Cooke, P. 2008. Clean Tech and an Analysis of the Platform Nature of Life Sciences: Further Reflections Upon Platform Policies. *European Planning Studies* 16: 375–393.

Davis, M. 2006. *Planet of Slums*. London: Verso.

De Soto, H. 2000. *The Mestry of Capital: Why Capitalism Triumphs in the Westand Fails Everywhere Else*. London: Blaack Swan.

Dorman, W. J. 2007. *The politics of Neglect: The Egyptian State in Cairo, 1974-198*. PhD. thesis, School of Oriental and African Studies, University of London.

Drakakis-Smith, D. 2000. *Third World Cities*. 2nd ed. London: Routledge.

EEA. 2019. *Sustainability Transitions: Policy and Practice*. Technical. Luxembourg: European Environment Agency.

Elkington, J. 1997. *Cannibals with Forks: The Triple Bottom Line of 21st Century Business*. Oxford: Capstone.

Elzen, B., M. Barbier, M. Cerf, and J. Grin. 2012. Stimulating Transitions Towards Sustainable Farming Systems. In *Farming Systems Research into the 21st Century: The New Dynamic*, ed. I. Darnhofer, D. Gibbon, and B. Dedieu, 431–455. New York: Springer.

Etingoff, K. 2017. *Sustainable Cities: Urban Planning Challenges and Policy*. Waretown: Apple Academic Press, Inc.

Faludi, A. 1970. The Planmng Environment and the Meaning of Planning. *Regional Studies* 4: 1–9.

———. 1973. *A Reader in Planning Theory*. Oxford: Pergamon Press Ltd., Headington Hill Hall.

Fathy, H. 1973. *Architecture for the Poor: An Experiment in Rural Egypt*. Chicago: The University of Chicago Press.

Fiedler, J. 2014. *Urbanisation, Unlimited: A Thematic Journey*. Cham: Springer.

Frantzeskaki, N., and H. de Haan. 2009. Transitions: Two Steps from Theory to Policy. *Futures* 41 (9): 593–606.

Frantzeskaki, N., and D. Loorbach. 2012. Governing Societal Transitions to Sustainability. *International Journal of Sustainable Development* 15 (1/2).

Frantzeskaki, N., V.C. Broto, L. Coenen, and D. Loorbach. 2017a. Urban Sustainability Transitions: The Dynamics and Opportunities of Sustainability Transitions in Cities. In *Urban Sustainability Transitions*, ed. N. Frantzeskaki, V.C. Broto, L. Coenen, and D. Loorbach, 1–19. Oxon: Routledge.

———. 2017b. *Urban Sustainability Transitions*. New York: Routledge.

Frantzeskaki, N., M. Bach, K. Holscher, and F. Avelino. 2018a. *Co-Creating Sustainable Urban Futures: A Primer on Applying Transition Management in Cities*. Cham: Springer.

———. 2018b. Introducing Sustainability Transitions Thinking in Urban Contexts. In *Co-creating Sustainable Urban Future: A Primer on Applying Transition Management in Cities*, Matthew Bach, ed. Katharina Holscher, and Flor Avelino Niki Frantzeskaki, 63–79. Cham: Springer.

Friedmann, J. 2005. Globalization and the Emerging Culture of Planning. *Progress in Planning* 64: 183–234.

———. 2011. *Insurgencies: Essays in Planning Theory*. Oxon: Routledge.

———. 2014. Austrians-in-the-World. Conversations and Debates About Planning and Development. *European Spatial Research and Policy* 21 (1): 11–22.

Fuenfschilling, L., and B. Truffer. 2014. The Structuration of Socio-Technical Regimes – Conceptual Foundations from Institutional Theory. *Research Policy* 43 (4): 772–791.

Geels, F. 2002. Technological Transitions as Evolutionary Reconfiguration Processes: A Multi-Level Perspective Case-Study. *Research Policy* 31 (8/9): 1257–1274.

———. 2004. From Sectoral Systems of Innovation to Socio-Technical Systems: Insights About Dynamics and Change from Sociology and Institutional Theory. *Research Policy* 33: 897–920.

———. 2005. The Dynamics of Transitions in Socio-Technical Systems: A Mulit-Level Analysis of the Transition Pathway from Horse-Drawm Carriages To Automobiles. *Technology Analysis and Strategic Management* 17 (4): 445–476.

———. 2010. Ontologies, Socio-Technical Transitions (to Sustainability), and the Multilevel Perspective. *Research Policy* 39: 495–510.

———. 2014. Regime Resistance Against Low-Carbon Transitions: Introducing Politics and Power into the Multi-Level Perspective Theory. *Culture Society* 31 (5): 21–40.

———. 2020. Micro-Foundations of the Multi-Level Perspective on Socio-Technical Transitions: Developing a Multi-Dimensional Model of Agency Through Crossovers Between Social Constructivism, Evolutionary Economics and Neo Institutional Theory. *Technological Forecasting and Social Change* 152. https://doi.org/10.1016/j.techfore.2019.119894.

Geels, F. W., and Johan Schot. 2007. Typology of Sociotechnical Transition Pathways. *Research Policy* 36: 399–417.

Geels, F. W., M.P. Jacobsson, and S. Hekkert. 2008. The Dynamics of Sustainable Innovation Journeys. *Technology Analysis & Strategic Management* 20 (5): 521–536.

Geels, F.W., F. Kern, G. Fuchs, N. Hinderer, K.G. Nele, J. Mylan, M. Neukirch, and S. Wassermann. 2016. The Enactment of Socio-Technical Transition Pathways: A Reformulated Typology and a Comparative Multi-Level Analysis of the German and UK Low-Carbon Electricity Transitions (1990–2014). *Research Policy* 45: 896–913.

Gough, I., G. Wood, A. Barrientos, and P. Bevan. 2008. *Insecurity and Welfare Regimes in Asia, Africa and Latin America: Social Policy in Development Contexts*. New York: Cambridge University Press.

Grin, J., and A. Loeber. 2007. Theories of Learning. Agency, Structure and Change. In *Handbook of Public Policy Analysis. Theory, Politics, and Methods*, ed. Gerald J. Miller, Mara S. Sidney, and Frank Fischer, 201–222. New York: CRC Press.

Grin, J., J. Rotmans, and J. Schot. 2010. *Transitions to Sustainable Development: New Directions in the Study of Long Term Transformative Change*. New York: Routledge.

Gunderson, L.H., and C.S. Holling. 2002. *Understanding Transformations in Human and Natural Systems*. Washington, DC: Island Press.

Hall, P. 1998. *Cities in Civilization: Culture,Technology and Urban Order*. London: Weidenfeld and Nicolson.

Harvey, D. 1985. *The Urbanization of Capital: Studies in the History and Theory of Capitalist Urbanization 2*. Oxford: Basil Blackwell Ltd.

———. 2009. *Cosmopolitanism and the Geographies of Freedom*. New York: Columbia University Press.

———. 2016. *The Ways of the World*. London: Profile Books Ltd.

Hatina, Meir. 2007. *Identity Politics in the MIddle East Liberal Thought and Islamic Challenge in Egypt*. London: Tauris Academic Studies.

Hoogma, R., R. Kemp, J. Schot, and B. Truffer. 2002. *Experimenting for Sustainable Transport: The Approach of Strategic Niche Management*. London: Spon Press.

Ingram, J. 2015. Framing Niche-Regime Linkage as Adaptation: An Analysis of Learning and Innovation Networks for Sustainable Agriculture Across Europe. *Journal of Rural Studies* 40: 59–75.

Jørgensen, U. 2012. Mapping and Navigating Transitions – The Multi-Level Perspective Compared with Arenas of Development. *Research Policy* 41: 996–1010.

Kanger, L., and J. Schot. 2019. Deep Transitions: Theorizing the Long-Term Patterns of Sociotechnical Change. *Environmental Innovation and Societal Transitions* 32: 7–21.

Kemp, R., J. Schot, and R. Hoogma. 1998. Regime Shifts to Sustainability Through Processes of Niche Formation: The Approach of Strategic Niche Management. *Technology Analysis & Strategic Management* 10: 175–196.

Kemp, R., D. Loorbach, and J. Rotmans. 2005. Transition Management as a Model for Managing Processes of Co-Evolution Towards Sustainable Development. *The International Journal of Sustainable Development and World Ecology*: 1–15.

Kemp, René, Derk Loorbach, and Jan Rotmans. 2007. Transition Management as a Model for Managing Processes of Co-evolution Towards Sustainable Development. *International Journal of Sustainable Development & World Ecology* 14: 1–15.

Kipper, R., and M. Fischer. 2009. *Cairo's Informal Areas: Between Urban Challenges and Hidden Potentials*. Cairo: GTZ.

Knight, F.H. 1921. *Risk, Uncertainty, and Profit*. Boston: Houghton.

Koehler, J., F. Geels, F. Kern, E.. Onsongo, and A. J. Wieczorek. 2017. *A Research Agenda for the Sustainability Transitions Research Network*. Technical. Amsterdam: STRN Working Group, STRN.

Köhler, J., F.W. Geels, and F. Kern. 2019. An Agenda for Sustainability Transitions Research: State of the Art and Future Directions. *Environmental Innovation and Societal Transitions*. https://doi.org/10.1016/j.eist.2019.01.004.

Kooiman, J. 2003. *Governing as Governance*. London: SAGE.

LaFreniere, G.F. 2007. *The Decline of Nature: Environmental History and the Western Worldview*. Bethesda/Dublin/London: Academic.

Lamba, H.S. 2010. Understanding the Ideological Roots of Our Global Crises: A Pre-Requisite for Radical Change. *Futures* 42 (10): 1079–1087.
Lefebvre, H. 1971. *Everyday Life in the Modern World*. New York: Harper & Row Publishers.
———. 1991. *The Production of Space*. Maiden: Blackwell Publishing.
Leftwich, A. 2010. *Redefining Politics. People, Resources and Power*. New York: Routledge.
Loorbach, D.. 2007. *Transition Management New Mode of Governance for Sustainable Development*. Ph.D. thesis. Grifthoek/Utrecht: International Books.
———. 2010. Transition Management for Sustainable Development: A Prescriptive Complexity-Based Governance Framework. *Governance: An International Journal of Policy Administration and Institutions* 23 (1): 161–183.
Loorbach, interview by A. Soliman. 2020. *Series Editor* (March 15).
Loorbach, D., and S. Oxenaar. 2018. *Counting on Nature: Transitions to a Natural Capital Positive Economy by Creating an Enabling Environment for Natural Capital Approaches*. Rotterdam: Dutch Research Institute for Transitions (DRIFT), Erasmus University.
Loorbach, D., and J. Rotmans. 2010. The Practice of Transition Management: Examples and Lessons from Four Distinct Cases. *Futures* 42 (3): 237–246.
Loorbach, D., J. Wittmayer, H. Shiroyama, J. Fujino, and S. Mizuguchi. 2016. *Governance of Urban Sustainability Transitions: European and Asian Experiences*. Tokyo: Springer.
Loorbach, D., N. Frantzeskaki, and F. Avelino. 2017. Sustainability Transitions Research: Transforming Science and Practise for Societal Change. *Annual Review of Environment and Resources* 42: 599–626.
Lynch, K. 1960. *The Image of the City*. Cambridge, MA: MIT Press.
Maheshwari, B., P. Vijay Singh, and B. Thoradeniya. 2016. Balanced Urban Development: Is It a Myth or Reality? In *Balanced Urban Development: Options and Strategies for Liveable Cities*, ed. P. Basant Maheshwari, Singh Thoradeniya, and Bhadranie Vijay, 3–16. Cham: Springer.
Markard, J., R. Raven, and B. Truffer. 2012. Sustainability Transitions: An Emerging Field of Research and Its Prospects. *Research Policy* 41 (6): 955–967.
Meadowcroft, J. 2000. Sustainable Development: A New(ish) Idea for a New Century? *Political Studies* 48: 270–387.
———. 2007. Who is in Charge here? Governance for Sustainable Development in a Complex World. *Journal of Environmental Policy and Planning* 9 (3): 299–314.
Mguni, P. 2015. *Sustainability Transitions in the Developing World: Exploring the Potential for Integrating Sustainable Urban Drainage Systems in Sub-Saharan Cities*. Copenhagen: PhD submitted to University of Copenhagen.
Mitlin, D., and D. Satterthwaite. 2004. *Empowering Squatter Citizen: Local Government, Civil Society, and Urban Poverty Reduction*. London: Earthscan.
Monkelbaan, J. 2019. *Governance for the Sustainable Development Goals Exploring an Integrative Framework of Theories, Tools, and Competencies*. Singapore: Singapore Pte Ltd./Springer.
Moore, J.W. 2015. *Capitalism in the Web of Life: Ecology and the Accumulation of Capital*. London: Verso.
Paris, C. 1982. *Critical Reading in Planning Theory*. Oxford: Pergamon Press Ltd., Headington Hill Hall.
Parnell, S., and S. Oldfield. 2014. *The Routledge Handbook on Cities of the Global South*. Abingdon/Oxon: Routledge.
Payne, G., ed. 2002. *Land, Rights and Innovation: Improving Security for the Urban Poor*. London: ITDG.
Pel, B., and R. Kemp. 2020. Between Innovation and Restoration; Towards a Critical-Historicizing Understanding of Social Innovation Niches. *Technology Analysis & Strategic Management*. https://doi.org/10.1080/09537325.2020.1750588.
Perlman. 1976. *The Myth of Marginality: Urban Poverty and Politics in Rio de Janeiro*. Stanford: California University Press.

Piketty, T. 2014. *Capital in the Twenty-First Century*. London: Th e Belknap Press of Harvard University Press.
Rapoport, A. 1977. *Human Aspects of Urban Form: Towards a Man – Environment Approach to Urban Form and Design*. Oxford: Pergamon Press Ltd., Headington Hill Hall.
Rip, A., and R. Kemp. 1998. Technological Change. In *Human Choice and Climate Change, Volume 2*, ed. S. Rayner and E.L. Malone, 327–399. Columbus: Battelle Press.
Rotmans, J. 2006. A Complex Systems Approach for Sustainable Cities. In *Smart Growth and Climate Change*, ed. M. Ruth, 155–180. Cheltenham/Northampton: Edward Elgar.
Rotmans, J., R. Kemp, and M. van Asselt. 2001. More Evolution Than Revolution: Transition Management in Public Policy. *Foresight* 3 (1): 15–31.
Roy, A., and N. Al-Sayyad (eds.). 2004. *Urban informality: Transnational perspectives from the Middle East, Latin America, and South Asia*. Lanham: Lexington Books.
Schot, J. 1998. The Usefulness of Evolutionary Models for Explaining Innovation: The Case of the Netherlands in the Nineteenth Century. *History of Technology* 14: 173–200.
Schot, J.W., and F. Geels. 2008. Strategic Niche Management and Sustainable Innovation Journeys: Theory, Findings, Research Agenda, and Policy. *Technology Analysis and Strategic Management* 20 (5): 537–554.
Schot, J., and L. Kanger. 2016. *Deep Transitions: Emergence, Acceleration, Stabilization and Directionality Science Policy Research Unit (SPRU)*. Sussex: University of Sussex.
Shirazi, M., and R. Keivani, eds. 2019. *Urban Social Sustainability: Theory, Policy and Practice*. New York/Abingdon: Routledge.
Sims, D. 2010. *Understanding Cairo: The Logic of a City Out of Control*. Cairo: America University Press in Cairo.
Singerman, D., and P. Amar (eds.). 2006. *Cairo Cosmopolitan: Politics, Culture, and Urban Space in the New Globalized Middle East*. Cairo: The American University in Cairo Press.
Smith, A., J. Vob, and J. Grin. 2010. Innovation Studies and Sustainability Transitions: The Allure of the Multi-level Perspective and Its Challenges. *Research Policy* 39 (4): 435–448.
Soliman, A. 2004a. *A Possible Way Out: Formalizing Housing Informality in Egyptian Cities*. Lanham: American University Press.
———. 2004b. Tilting at Sphinxes: Locating Urban Informality in Egyptian Cities. In *Urban Informality in an Era of Liberalization: A Transnational Perspective*, ed. Ananya Roy and Nezear Al Sayyad, 171–208. Lanham: Lexington Books.
———. 2012a. Housing and the State in the Middle East. In *International Encyclopedia of Housing and Home*, Susan J. Smith (ed.), 346–354. Amsterdam: Elsevier.
———. 2012b. The Egyptian episode of self-build housing. *Habitat International* 36 (2):226–236.
———. 2015a. Collective Planning Process: A Driving Seat for Formalizing Urban Informality in Egypt. *International Journal of Urban Sustainable Development*, June 1–24.
———. 2015b. The Crucible of Housing the Bottom of the Pyramid in Egypt. In *Housing the Urban Poor in Developing Countries*, ed. B. Aldrich and R. Sandhu, 165–189. Jaipur: Rawat Publications.
———. 2017. Land Readjustment as a Mechanism for New Urban Land Expansion in Egypt: Experimenting Participatory Inclusive Processes. *International Journal of Urban Sustainable Development* 9 (3): 313–332.
———. 2019. So It's Always a Chance: Community-Led Solutions to New Urban Expansion. In *New Cities and Community Extensions in Egypt and the Middle East*, S. Attia et al. (eds.), 159–179. Cham: Springer.
———. 2020. Understanding Sustainability Transitions for Urban Informality in the Middle East. *International Journal of Urban Sustainable Development*. https://doi.org/10.1080/19463138.2020.1728692.
Sovacool, B. 2016. How Long Will It Take? Conceptualizing the Temporal Dynamics of Energy Transitions. *Energy Research & Social Science* 13: 202–221.
Swilling, M. and Hajer, M. 2017. Governance of urban transitions: towards sustainable resource-efficient urban infrastructures. *Environmental Research Letters* 12.

Swilling, M.J., K. Musephine, and J. Wakeford. 2015. Developmental States and Sustainability Transitions: Prospects of a Just Transition in South Africa. *Journal of Environmental Policy and Planning* 5 (5): 1–23.

Truffer, B., and L. Coenen. 2012. Environmental Innovation and Sustainability Transitions in Regional Studies. *Regional Studies* 46 (1): 1–21.

Truffer, B., J.T. Murphy, and R. Raven. 2015. The Geography of Sustainability Transitions: Contours of an Emerging Theme. *Environmental Innovation and Societal Transitions* 17: 63–72.

Turner, J. 1976. *Housing by People*. London: Marion Boyars Publishers Ltd.

Turner, J., and R. Fichter. 1972. *Freedom to Build: Dwellers Control of the Housing process*. New York: Macmillan.

UN. 2019. *United Nations, Department of Economic and Social Affairs, Population Division, World Population Prospects 2019, Volume II: Demographic Profiles*. New York: United Nations.

UN-Habitat. 2009. *Urban World Bridging the Urban Divide, Why Cities Must Built Equally*. Technical. Nairobi: Un-Habitat.

———. 2018. *DG 11 Synthesis Report 2018: Tracking Progress Towards Inclusive, Safe, Resilient and Sustainable Cities and Human Settlements*. SDG 11 Synthesis Report High Level Political Forum 2018. Nairobi: Un-Habitat.

Van den Bergh, J.C.J.M. 2013. Economic-Financial Crisis and Sustainability Transition: Introduction to the Special Issue. *Environmental Innovation for Societal Transitions* 6: 1–8.

Viotti, E.B. 2001. *National Learning Systems: A New Approach on Technical Change in Late Industrializing Economies and Evidences from the Cases of Brazil and South Korea. Science*. Cambridge, MA: Center for International Development, Harvard University, Technology and Innovation Discussion Paper No. 12.

Voss, J.-P., and B. Bornemann. 2011. The Politics of Reflexive Governance: Challenges for Designing Adaptive Management and Transition Management. *Ecology & Society* 16 (2): 9.

Voss, J.P., D. Bauknecht, and R. Kemp, eds. 2006. *Reflexive Governance for Sustainable*. London: Edward Elgar Publishing.

Voß, Jan-Peter, A. Smith, and J. Grin. 2009. Designing Long-Term Policy: Rethinking Transition. *Policy Science* 42: 275–302.

VROM. 2001. *Where There's a Will There's a World*. The Hague: Ministry of Housing: Spatial Plannning Environment (VROM).

Ward, B. 1989. *Progress for a Small Planet*. London: Penguin.

Ward, P. 2004. Informality of Housing Production at the Urban-Rural Interface: The "Not-so Strange Case" of the Texas Colonias. In *Urban Informality: Transnational Perspectives from the Middle East, Latin America, and South Asia*, ed. A. Roy and N. Al Sayyad, 243–270. Lanham: Lexington Books.

Ward, B., and R. Dubos. 1972. *Only One Earth: The Care and Nauntance of a Small Planet*. Stockholm: United Nations conference on the human environment.

Wheeler, S. 1998. Planning Sustainable and Livable Cities. In *The City Reader*, ed. R.T. LeGates and F. Stout. London/New York: Routledge.

Wieczorek, Anna J. 2018. Sustainability Transitions in Developing Countries: Major Insights and Their Implications for Research and Policy. *Environmental Science and Policy* 84: 204–216.

Wittmayer, J., F. Steenbergen, and N. Frantzeskaki. 2018. Transition Management: Guiding Principles and applications. In *Co-creating Sustainable Urban Futures: A Primer on Applying Transition Management in Cities*, ed. N. Frantzeskaki, M. Bach, and P. Mguni, 81–102. Tokyo/Heidelberg/New York/Dordrecht/London: Springer.

Wittmayer, J., J. Backhaus, F. Avelino, B. Pel, T. Strasser, I. Kunze, and L. Zuijderwik. 2019. Narratives of Change: How Social Innovation Initiatives Construct Societal Transformation. *Futures* 112. https://doi.org/10.1016/j.futures.2019.06.005.

Chapter 3
Urbanization and Urban Informality in the Era of Globalization

Abstract In 2018, United Nations estimated the total population of the planet was 7.7 billion of which 55.28% are live in urban areas. One billion people live in slums and informal settlements today, representing about 30% of the world's urban population. The 2030 Agenda for Sustainable Development and New Urban Agenda focused on sustainable urbanization and human settlements. Urbanization is an unstoppable phenomenon, but it is a transformative process capable of galvanizing momentum of many aspects of the global environment. Today and in the recent past, cities of the Middle East are witnessing a tremendous transition of socioeconomic, spatial, and political milieu in which urbanization processes, formally or informally, have played a major role in shaping those cities. This chapter looks at the recent transformation of the urbanization process in the Middle Eastern cities in the light of three dimensions: spatial urban growth, demographic changes and shifts, and political mobilization and dynamics. The nexus between the urban informality, as "a new type of urbanism" or "a site of transitions" and the urbanization process is highlighted to investigate the main challenges in this regard and its impact on the cities' urban fabric. To ameliorate the deteriorating socioeconomic, spatial, and political dilemmas posed by Middle East's urbanization process, a new innovation in driving the transformation process is needed to make cities safe, inclusive, resilient, and sustainable.

Keywords Urbanization processes · Globalization · Demographic changes · Urban growth · Rural-urban migration · Political mobilization · Clientelism · Urban conflict

A. M. Soliman, *Urban Informality*, https://doi.org/10.1007/978-3-030-68988-9_3

3.1 Introduction

Since world population reached 3 billion in 1959, the rate of growth has increased, peaked, and begun to slow. Each succeeding milestone was reached more quickly than in the last few decades. It took 15 years to reach 4 billion, 13 years to hit 5 billion, only 11 years to get to 6 billion at the end of 1998. The interval to attain 7.63 billion was slightly longer, at 20 years, as the global rate of population growth has slowed (UN 2018). By 2050, the world urban population is expected to nearly double, to an average estimate of 9.77 billion, making urbanization one of the century's most transformative trends (UN-Habitat 2018b). Projections of the world population in 2050 range from 8.1 billion (if fertility rates fall to a global average of 1.7 children per woman) to 10.9 billion (if they remain unchanged). The gap of nearly 3 billion between those possibilities is greater than the combined populations of China and India today.

The urbanization process, as a sociotechnical landscape, is the output of the four principles of development. One is the increase in the proportion of the population who are living in cities. Another is the increase in the level of the urbanism leading to the appearance of many urban settlements. A third is the transformation of society because so many people and such a large proportion of the world's population live in towns and cities and follow lifestyles that are urban in origin and character (Clark 2003). The fourth is the dynamic change of global economy system, especially after Covid-19 or what is called the triple crunch of financial, climate, and the coming global energy (Pittfor 2019). The emergence of urban informality, as a common form of urbanism or as "a site of transitions", in most cities of the Global South, especially in the Middle East, has been mainly the main result of the tremendous rapid chaotic urbanization process, spontaneous urban growth, either formally or informally, an increase in rural–urban migratory movements, booming of industrialization processes, expansion of population displacement, and evolution the informal economy in the metropolitan areas. The absolute number of slum dwellers in the developing world grew from 766.7 million in the year 2000 to an estimated 827.6 million in 2010. This means that 55 million additional slum dwellers have been added to the global urban population since the year 2000 (UN-Habitat 2009), nearly 5.5 million annually, and have reached around 1 billion in 2018 (UN-Habitat 2018a).

This chapter attempts to go beyond the debate of whether urbanization is enforced by a tremendous transition of socioeconomic situations or a result of political strategy and interests. Such an analytical dichotomy reduces the complexity of reality since, beyond the fact that urbanization, formally or informally, is a natural phenomenon, it frames social norms and socio-spatial transitions, and constructs the image of the "other." Socio-spatial transition is considered a useful political tool for categorizing places in the city. These places are socio-spatial boundaries that protect and separate the majority from the minority or separate the formality from the informality and reproduce the existing power relationships. The objective of this chapter is to develop a clear understanding of urbanization process in the South, especially Middle East countries, and to illustrate (rather than to test rigorously) how propositions derived from that theory compare with a variety of available evidence. It intends, however, to focus upon the level of urbanization progress and its impact on the urban growth by which urban informality has flourished. The chapter focuses on the conceptual, empirical, and policy-relevant linkages among urbanization, rural-urban migration, socio-spatial transformation, political mobilization, and urban conflict.

It is pursued to understand the urbanization process in the Global South within the context of globalization, what it is, what generates it, and how to deal with its consequences. It is identified and met through the investigation of three assumptions relating to the current transformation prospects. One is that the higher the level of the ratio of urban population growth and economic transformation the greater the level of informal/formal urbanization, the higher the level of immigrants, and the greater the chance of creating a dual economy or accelerating the invisible hand of the market (Pittfor 2019). Approaches to data collection capable of recording varied forms of urban conflict that may arise are measured only indirectly. The second observes that the greater the level of informal urbanization the higher is the level of socio-spatial urban sprawl. Conflict may combine different forms and the forms themselves, as well as the extent to which each generates insurgency or patronage and may vary with time and circumstance. In effect, this is an effort to think differently about these questions and the variety of evidence that may hold some answers. The third is that cities are the engines of national economies and cultures, as well as the magnets for migration and the poorly managed urban growth that has dangerous consequences, where a governance dilemma develops. The management of cities is an increasingly complex process as the number of local stakeholders' swells (Tulchin et al. 2002) and is often transformed due to international donors' involvement.

The chapter is organized into six sections, beginning with the current introductory section, followed by a review of the recent trends and future scenarios for urbanization and urban transitions, with special emphasis on Middle East urbanization. Then, demographic changes and shifts, demographic dynamics, and rural-urban migration are examined. The fourth section discusses political mobilization in both a descriptive and an analytical framework; the urban conflicts in the Middle East and its relationship to urbanization and migration are then considered. An analysis of policy options designed to ameliorate the deteriorating economic, social, and environmental dilemmas posed by the Middle East's rapid urban growth concludes this part of the study.

3.2 Urbanization and Urban Transitions

Urbanization itself is an urban transition process phenomenon. Urbanization, as a sociotechnical landscape, is seen as the engine for development while others see it as an unconstrained process (Fiedler 2014) by which further pressure comes to bear on the built environment, unless it is planned in a sustainable way to cope with the present and future needs of a society and to sustain its natural resources. The old wisdom views urbanization as simply a demographic phenomenon, but others see it as a transformative process (UN-Habitat 2018a) capable of galvanizing momentum for many aspects of global development. Urban growth, as a consequence of urbanization, can be manifested in the densification of cities (growing upwards), and/or in the spatial expansion of cities (growing outwards) (Tulchin et al. 2002). Most of the urban growth in the Middle East has witnessed spatial expansion, growing upwards and outwards, due to the arbitrary urbanization process. The adoption by the international community of SDGs includes the standalone urban goal- SDG 11- to make cities safe, inclusive, resilient, and sustainable. The Sendai Framework and the NUA firmly place urbanization at the forefront of international development policy (UN-Habitat 2018b).

The concept of development perspective implies that development cannot be interpreted solely in terms of economic growth or political or social transformations, but it is the outcome of continuous various aspects of transitions in a given area. The term "development" is reflected the transitions of the major economic, political, and social objectives and values that societies strive for. Goulet (2006) distinguished the basic components or core values in the wider meaning of development, which are life sustenance, self-esteem, and freedom. Currently, health safety could be added as a fourth component as a response to Covid-19 pandemic. Life sustenance is concerned with the provision of basic needs; self-esteem is concerned with the feeling of self-respect and independence. Freedom refers to freedom from the three evils of want, ignorance, and squalor so that people are more able to determine their own destiny. Life after Covid-19 has changed, transforming the way people live, work, travel, and socialize, and as social distancing and new protection measures have become a way of life to protect people on the planet (UN-Habitat 2020). All four of these core components are interrelated, interconnected, and interchangeable regarding the ways development has occurred and implemented.

The political economy in Egypt, Lebanon, Jordan, and the Gulf States is shaped on the national and international levels and absorbs the impact of macro political and new economic forces (Carnoy and Castells 1993; Piketty 2014; Pittfor 2019) on micro institutions, such as the family and kinship, rural-urban migration, and public and private entrepreneurs. On the other hand, Piketty (2014) strongly argued that economics need to be viewed from the much broader perspective of the overall social system of a country which includes values, beliefs, attitudes to effort and risk-taking, and the class system, if development mistakes are to be avoided that stem from implementing policy based on economic theory alone. However, it is apparent that development is the vehicle for allowing an increasing realization of human potential, a driver for the urbanization process, a vehicle for capital accumulation and class struggle (Harvey 1985), and a vehicle for production, reproduction, distribution, and consumption of capital accumulation. This is achieved by widening the extent and scope of material consumption at the same time as expanding the opportunities to enable human aspirations to be fulfilled. Development as a continuous transition process, should include the achievement of poverty-alleviating spatial enhancement and improve social, health, and economic justice. The 2019 UNESCO Report (UNESCO 2019) stresses ensuring the availability and sustainable management of water and sanitation for all to eradicate poverty, build peaceful and prosperous societies, and reduce inequality. This requires making increased and effective use of a country's resource base to raise the level of real incomes and health safety and to decrease inequality through sustainable development growth.

Urbanization is characterized and even defined by fundamental transitions in the physical concentration of population, the nature and scale of economic production, spatial growth, and social structures and patterns of interaction. Urbanization has multiple social, political, cultural, environmental, and economic dimensions that will profoundly influence socioeconomic development and innovation. This fundamental long-term transformation will involve the realignment of urban society, its technologies and infrastructures, urban cultures, lifestyles, and circular economy, as well as governance and institutional frameworks. Such structural systemic realignments within urban contexts can be referred to as urban sustainability transitions, i.e., fundamental and structural transitions in urban systems through which persistent societal challenges are addressed (Frantzeskaki et al. 2017). A major fundamental and structural transitions is land transformation and its manifestation in countries with different political systems questions the city and the city government as the central actor. Examples range from shifts towards urban planning to the role of space and place in transitions which leaves us ill-prepared to understand and explain its geographically uneven development (Coenen et al. 2012). Globally, one in eight persons lives in slums. In total, around a billion people live in slum conditions today, of whom 881,080,000 slum dwellers are estimated to be living in developing countries (IIHS 2016). The number is expected to increase to 2 billion by 2030 unless major transitions are made to the present policies and practices of urban management (UN-Habitat 2003).

Frantzeskaki et al. (2017) indicates that sustainability transitions require two fundamental issues: possible strategies for governing urban transitions and the co-evolution of material and institutional transformations in sociotechnical and socioecological systems. Possible strategies vary across the cities of the world according to their socioeconomic situations, their governances, and their capabilities to handle this sort of transitions. The second issue concerns the co-evolution of material and institutional transformations. The latter requires intensive institutional reforms, especially in cities of the Global South, and requires the simplification of land regulations. While the co-evolution of material, either goods or services or both, would require investments and certain type of technology to achieve the required transitions. In the Global South, levels of socioeconomic and political instability, less transparency in legal proceedings and enforcement of legal frameworks, and relatively high levels of economic and social inequality are considered the main urban challenges to applying such transitions. On the other hand, in setting up transitions' strategy in the Global South, the maximization of certainties and the minimization of uncertainness should be considered. However, urban sustainability transitions will involve the realignment of urban society's technologies and infrastructures, culture, lifestyles, and circular economies, as well as governance and institutional frameworks in which persistent societal challenges are addressed regarding how shifts from one situation to another can occur in a sustainable way.

In almost all countries, the influx into the cities of emigrants from rural zones markedly accentuates the cultural diversity and ethnic diversity in already ethnically diverse nation such as Lebanon. Globalization and anti-globalization, as in 'enclosing the commons' (Harvey 2016), have also given rise to major population shifts between different parts of a given region, and migration between countries (such as Lebanon and Jordan) in which urban-fabric developments have changed. This is so despite the fact that international migratory movements are complex in their patterns and do not follow the stereotyped views held by public opinion but have influenced the urban fabric developments of the hosted countries (e.g., Lebanon and Jordan). On the other hand, there are extensive migratory regional movements of refugees from hunger and war, in Lebanon during its Civil War, in Egypt during the Suez Canal crisis of 1956 and defeat in the 1967 War, in Jordan and Syria during the defeat in the 1967 War, and in Iraq during US-led invasion under G.W. Bush. After the Arab Spring, a mass of regional movements of refugees moved from Syria, Libya, Iraq, and Yemen to Europe and Arab neighborhood countries (mainly Egypt, Lebanon, and Jordan).

Globalization has even begun to dominate the vocabulary of urban policy-making and politics, with scholarly articles devoted to the study of how cities market themselves in global terms (Duffy 1995) or how politicians symbolically use globalization in city electoral campaigns (Stokes et al. 2013) along with people using the internet during the Arab Spring (Castells 2015). It seems that all are globalized now. The title of David Clark's book (Clark 2003) *Urban World/Global City* merely takes this logic to its most extreme, collapsing global cities and an urban world into each other as linguistic shorthand for the modern condition, and proposes that globalization is grounded in contemporary social and economic formations that are the most recent products of a capitalist mode of capital accumulation in the form of the super-rich (Forrest et al. 2017). Globally, a billion new houses will be needed by 2025 to accommodate 50 million new urban dwellers per year. Costs are estimated at $US 9–11 trillion by 2025 (UN 2016a). Egypt needs around a trillion Egyptian pounds[1] to meet housing demand by 2030, so: *How much land is needed, where, how, and who will provide the necessary funds?* This immense factor needed housing will accelerate both rapid urbanizations, formally and informally, continuous spatial transitions, and the amplification of urban informality in the Global South.

In the last five decades, five very important trends have attracted the attention of international and local organizations concerned urban policies within cities in the South. One is an extension of what has been witnessed since the early years of the post-WWII period that saw the continuous growth of cities in terms of both their population and their physical pressure on the natural resources. The other, occurring since the mid-1970s and early 1990s in many countries in the South, is the

[1]One US$ is equal to 15.8 Egyptian pounds.

emergence of novel proposals for economic reforms at the local and international levels. The third trend came after 9/11 and involves the introduction of innovations in institutional reform and practice, new reforms, and the readjustment of regulations and laws of governments in the South, especially the Middle East countries. The fourth trend is the Middle East uprising which has abated , where new reforms and constitutions are in place and huge international urban hubs are currently being created. The fifth involves people movements and demands for social justice in some Arab countries and other countries in Europe (for example, the Yellow Jackets in France). All these trends have accelerated the process of the transition of urban regimes, niches, and landscape developments. However, as indicated in Chap. 1, the ASUST examines these five trends, as well as looks at the role of the three pillars in transitions processes and how they will assist the emergence of sustainability transitions.

These new trends are linked to the Habitat Conference of Vancouver (1976), the Earth Summit, UNCED in Rio de Janeiro (1992), the Habitat Summit in Istanbul (1996), and they align with the United Nations Millennium Development Goals (MDGs). The recent debate of Sustainable Development Goals (SDGs), Habitat III, and New Urban Agenda (NUA) emphasize the rights of the cities and the bridging of urban divides. It includes ways of bridging urban income divides and reducing inequality and poverty in cities, as well as fostering equal access to shelter, water and sanitation, and services in cities without slums. Subsequent to Habitat III and the NUA Declaration in October 2016, the international agencies also began reporting on Millennium Development Goal 7, Target 11 (now Target 7d), by designing global monitoring and implementation strategies that include slum indicators and a Monitoring of Urban Inequalities Program (MUIP). The objective of the MUIP is to enhance the evidence-based formulation of urban development policies by systematically disaggregating data between urban slums and non-slum areas through the implementation of Urban Inequalities Surveys (UN-Habitat 2010). The aims are, first, the responsibility of urban regimes to prepare future spatial strategy plans and adopt human settlements policies to cope with the rapid transitions of socioeconomic and urban fabric development. The urban regimes have to ensure that the urban poor have the right to adequate shelter, health, water, and energy. Second, they need to introduce genuine innovations in institutional form and practice and attempt to respond to at least some of the pressures of growth and expansion emerging within the built environment. Third, within the global information era, they must achieve significant improvement and meet the requirements of the lives of people in the South to have the best and the cheapest goods and services originating from anywhere in the world. Fourthly, a youth supplement is looking at the urban challenges faced by youth, one of the largest single demographics in the Global South. Fifth, there is an emerging need to convince international leaders to find possibilities for, and constraints, to adapting to climate change in urban areas in low and middle-income nations (Bicknell et al. 2009). Finally, after 9/11, the financial crisis of 2008, the Arab Uprising, and the Covid-19 crisis, they must impose or introduce new reforms and readjustments in sociotechnical landscape developments and rearrangements of urban regimes of the North and the South alike. The Middle

East countries are adopting new measures to thwart terrorists, develop educational systems, enhancement health systems, and regulate arbitrary developments to cope with the rapid transformation.

However, the urbanization processes comprise complex interplays of socioeconomic, political, technological, geographical, and cultural factors. Studies of urbanization are intended to describe these changes, interpreting their significance and understanding the socioeconomic and political processes that brought them about. Chadwick (1971) indicated that urbanization and civilization are very definite economic and technological processes, and these suggest that development (i.e., economic and thus technological) implies urbanization or change, by and large, from a rural to an urban way of life. On the other hand, Habitat III indicates that urbanization has become a driving force, as well as a source of development with the power to change and improve lives. It achievable through the promotion of a model of urban development that integrates all facets of sustainable development to promote equity, welfare, and shared prosperity (UN 2016b). SDGs Goal 11 views urbanization not simply as a demographic phenomenon but also as a transformative process capable of galvanizing momentum for many aspects of global development. Urbanization is an unstoppable phenomenon (UN-Habitat 2018b). On the other hand, Satterthwaite (2014) asks whether the SDGs address the deficiencies in the MDGs. He concludes the SDGs need to be much more explicit about how all of the national governments and international organizations are committed to this achievement. This includes both what these organizations demand and what they can contribute to meeting the goals and targets. It also includes how urban regimes can guide or control the continuous rapid growth of cities that are the places where broader societal transitions impact (e.g., mobility transitions, energy transitions, housing transitions, etc.).

As illustrated in Fig. 3.1, in the last three decades Cairo, Beirut, and Amman have expanded and urbanized at least threefold, their population has nearly doubled, and urban informality has become the main theme for their urban fabric developments, surrounding the periphery of those cities in what is called "a belt of poverty" or "a belt of misery" or "a stone on the chest". Also, as the informal economy has flourished and gated communities have been established within the cities, they comprise a formal old core city, a new expansion city, an informal city, and super-rich gated communities. For example, this means there are a Cairo and many other Cairos within metropolitan Cairo itself (Al Sayyad 2006), and so Beirut and many other Beiruts, and Amman and many other Ammans (Potter et al. 2009; Khirfan 2017).

As illustrated in Fig. 3.2, urbanization as a fundamental transformative process is the engine and the vehicle for a development process integrated and interacting with transformations in sociotechnical and socioecological systems. This has been influenced by four major factors: state performance, access to urban determinants, realignment of urban society, and the effectiveness of capital. In a broad sense, each has various components that determine or compose the performance of the various variables. On the other hand, all the factors are integrated and correlated, working as a complete package to achieve a practical implementation, formally or informally or both, of real transformations in sociotechnical and socioecological systems of a

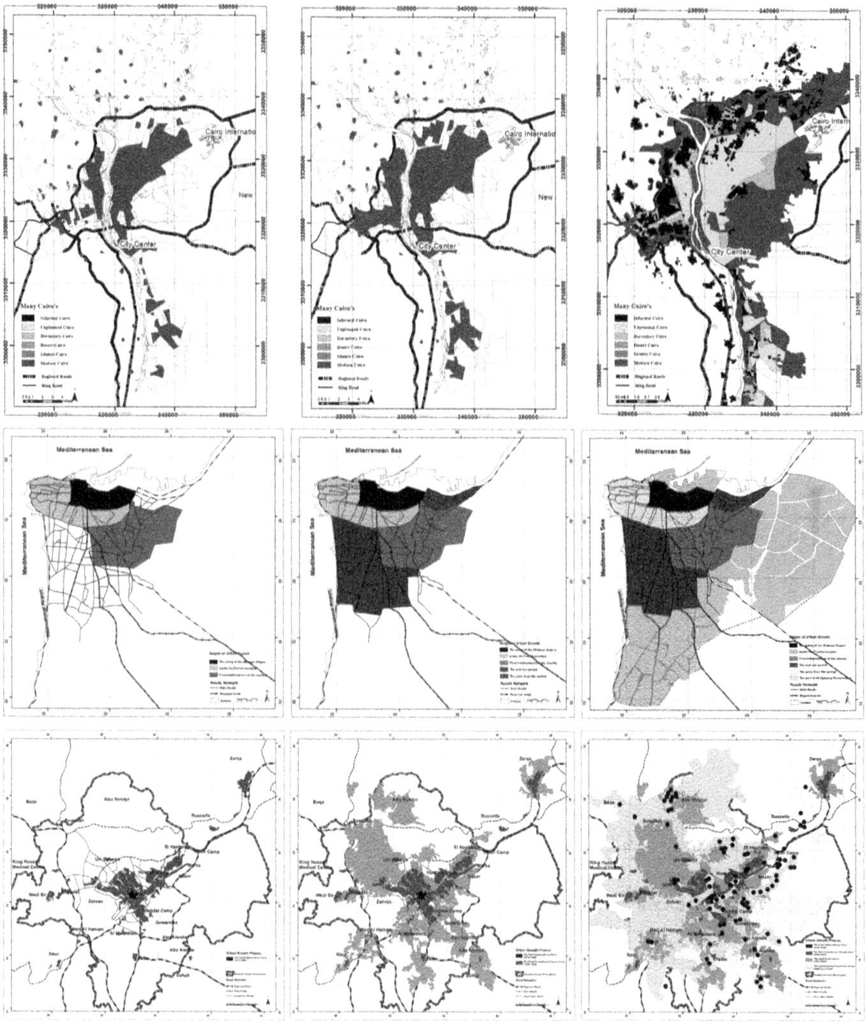

Fig. 3.1 Maps indicating the growth of Cairo (top), Beirut (center), and Amman (bottom) in the last six decades. (Source: The author)

given environment. In all stages of urbanization processes, grassroots play essential role in speeding or delaying the transformative process of a given environment. This depends on the level of conflict between the state and a society, including ways to institutionalize agonistic participatory practices. It also would require that new urban governing spaces be coproduced through participative transformation requiring experimentation and innovation in re-designing urban knowledge architectures (Richardson et al. 2018), taking into consideration political mobilization in a given setting. Also, the transformative process could not be pursued without mobilizing

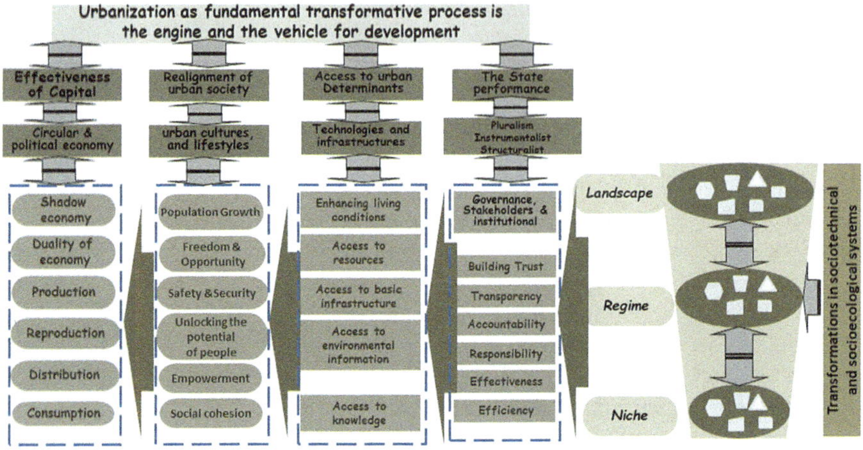

Fig. 3.2 Diagram showing the correlation between urbanization as fundamental transformative processes and sociotechnical and socioecological systems. (Source: The author)

various types of financing instruments, such as social and natural capital, alongside governance as a necessary prerequisite for achieving sustainable development. Realignment of urban society in preparing urban cultures and lifestyles is essential for cooperation in transitions process, such as controlling the demographic dynamics, guiding rural-urban migration, unlocking the potentialities of people, empowering the role of women and youth, etc. This would integrate diverse sources of knowledge and perspectives, 'glocally' forging connections between local urban challenges and broader global, national and regional macro-developments, and enabling pragmatic decisions in a context of complexity and uncertainty (Frantzeskaki et al. 2018a, c).

To sum up, at the heart of theory of cities is the elaboration of new spaces of transaction and transitions. These spaces open up conventional designations of scale and configure new "places" for "spaces" in an economy oriented around densities of knowledge creation and the generalization of control. It also unleashed developments in the internet revolution that extend from "the space of places to the space of flows" (Castells 2015). Notions of world cities point to new economic capacities, social transitions, and infrastructures that construct, assemble, and channel flows of information, goods, and influences (Castells 2010, 2nd edition); (Taylor et al. 1995). On the other hand, there is the knowledge economy as a dominating dimension for population movements that became an important element in orienting the development process and capital accumulation in a given area. However, the wave of urban informality is directed towards the informal knowledge economy, which is currently flourishing across cities of the Middle East.

3.3 Demographic Changes and Shifts

There is no doubt that probably no problem may be more threatening to the Earth's environment than the proliferation of the human species. As a reflection of urbanization processes, a variety of studies have emphasized population growth in the urban areas in the light of demographic development processes and ignored, to a certain extent, the socioeconomic, spatial, and political transformation within a given area. On the other hand, the planet, human beings, and natural resources are three components that formulate the Earth's environment, and, for all of history, man's struggle to survive and enhance the living environment. Therefore, studying the concomitant demographic changes is essential to understand the process of urbanization and socioeconomic transitions that accompanied this process. The following subsections discuss the demographic changes and causes of rural–urban migration.

3.3.1 *Demographic Dynamics*

The world population reached 6.1 billion in 2000, up from 2.5 billion in 1950, and 4.4 billion in 1980. The latest world population estimate, for July 2020, are 7.794 billion (UN 2020). World population is projected to grow to about 8.5 billion in 2030, to 9.77 billion in 2050, and eventually to stabilize between 10.5 and 11 billion. World population growth reached a peak of 2% per year in 1965 and has since declined to 1.19% in 2010 (Table 3.1). The growth rate is projected to decline to 0.98% and 0.87% in 2020 and 2050, respectively.[2] While one in every five persons on Earth survives on less than $1 a day, during the 1990s the share of people suffering from extreme income poverty fell from 30 to 23% (UNDP 2003).

Most of the population growth is occurring in poorer developing countries. African and Middle East nations are expanding at the fastest rate. While the period of 1950–1975 saw population growth evenly divided between the urban and rural areas of the world, the period since has seen the balance tipped dramatically in favor of urban growth. In 2018, for the first time in history, over half of the world's population lived in urban areas, and by 2050 this will have risen to 66%, while the bulk of this growth will be taking place in the Middle East nations (UN 2018). Whereas 30% of the world population lived in urban areas in 1950, the proportion of urban dwellers rose to 47% by 2000 and is projected to attain 60% by 2030. Almost all the population increase expected during 2000–2030 will be absorbed by the urban areas of the less developed regions, whose population will likely rise from approximately 2 billion in 2000 to just under 4 billion in 2030. The urban population of the more developed regions is expected to increase more slowly, rising from 0.9 billion in 2000 to 1.0 billion in 2030.

[2]All the data in this section are based on the World Economic and Social Survey, various issues (United Nations publications)

Table 3.1 Global regional population sizes, rates of change, and population densities

	Estimates and projections (000)					Rate of change (percent)				Population density (people/Km2)	
Region, subregion, country or area	2010	2015	2020	2030	2050	2010–2015	2015–2020	2020–2025	2025–2030	2015	2030
WORLD	6 958 169	7 383 009	7 795 482	8 551 199	9 771 823	1.19	1.09	0.98	0.87	56.8	65.7
Sub-Saharan Africa	845 136	969 234	1 106 573	1 418 333	2 167 652	2.74	2.65	2.54	2.43	44.3	64.8
AFRICA	1 049 446	1 194 370	1 352 622	1 703 538	2 527 557	2.59	2.49	2.36	2.25	40.3	57.5
Eastern Africa	346 987	399 458	457 440	587 330	888 129	2.82	2.71	2.57	2.43	59.9	88.1
Middle Africa	131 351	153 743	178 959	237 771	384 005	3.15	3.04	2.91	2.78	23.7	36.6
Northern Africa	204 310	225 136	246 049	285 204	359 905	1.94	1.78	1.56	1.39	29.0	36.7
Southern Africa	59 016	63 420	67 595	74 786	85 800	1.44	1.28	1.09	0.93	23.9	28.2
Western Africa	307 781	352 614	402 579	518 446	809 719	2.72	2.65	2.57	2.49	58.1	85.5
ASIA	4 194 425	4 419 898	4 623 454	4 946 586	5 256 927	1.05	0.90	0.75	0.60	142.4	159.4
Eastern Asia	1 595 829	1 635 150	1 663 619	1 678 653	1 586 491	0.49	0.35	0.17	0.01	141.4	145.2
South-Central Asia	1 768 531	1 892 013	2 009 437	2 216 422	2 476 228	1.35	1.20	1.06	0.90	183.2	214.6
Central Asia	63 156	68 705	73 821	81 973	94 431	1.68	1.44	1.16	0.94	17.5	20.9
Southern Asia	1 705 374	1 823 308	1 935 616	2 134 449	2 381 797	1.34	1.20	1.06	0.90	284.9	333.5
South-Eastern Asia	597 328	634 610	669 016	727 991	797 649	1.21	1.06	0.92	0.77	146.2	167.7
Western Asia	232 738	258 124	281 382	323 521	396 560	2.07	1.73	1.49	1.30	53.7	67.3
EUROPE	737 164	740 814	743 390	739 456	715 721	0.10	0.07	−0.02	−0.08	33.5	33.4
Eastern Europe	294 541	293 244	290 776	281 413	258 519	−0.09	−0.17	−0.28	−0.38	16.2	15.6
Northern Europe	100 314	103 097	105 863	110 635	117 583	0.55	0.53	0.48	0.41	60.6	65.0
Southern Europe	153 943	152 441	151 553	148 825	140 123	−0.20	−0.12	−0.17	−0.19	117.7	114.9
Western Europe	188 366	192 032	195 197	198 584	199 496	0.39	0.33	0.19	0.15	177.0	183.1
LATIN AMERICA AND THE CARIBBEAN	597 562	632 381	664 474	718 483	779 841	1.13	0.99	0.85	0.71	31.4	35.7

Caribbean	41 725	43 310	44 679	46 887	48 258	0.75	0.62	0.53	0.43	191.6	207.5
Central America	160 557	172 635	184 127	204 496	231 563	1.45	1.29	1.13	0.97	70.4	83.4
South America	395 280	416 436	435 667	467 100	500 020	1.04	0.90	0.76	0.63	23.8	26.8
NORTHERN AMERICA	342 937	356 004	369 159	395 453	434 655	0.75	0.73	0.71	0.66	19.1	21.2
OCEANIA	36 636	39 543	42 384	47 683	57 121	1.53	1.39	1.23	1.12	4.7	5.6
Australia/New Zealand	26 490	28 414	30 233	33 448	38 898	1.40	1.24	1.07	0.95	3.6	4.2
Melanesia	8 983	9 933	10 909	12 905	16 754	2.01	1.87	1.74	1.62	18.8	24.4
Micronesia	503	519	541	589	656	0.62	0.84	0.84	0.84	163.7	185.7
Polynesia	659	677	701	742	813	0.53	0.71	0.52	0.60	83.7	91.7

Source: (UN 2020) United Nations, Department of Economic and Social Affairs, Population Division (2018). World Urbanization Prospects: The 2019 Revision

Table 3.2 Population (in thousands) of urban and rural areas at mid-year and the percentage that is urban (2018)

	Population of urban and rural areas at mid-year (thousands) and percentage urban, 2018			
Region, subregion, country or area	Urban	Rural	Total	Percentage urban
WORLD	4219817.32	3413002.01	7632819.33	55.29
More developed regions	993837.09	269362.59	1263199.68	78.68
Less developed regions	3225980.23	3143639.42	6369619.65	50.65
AFRICA	547602.18	740318.34	1287920.52	42.52
South Sudan	2534.06	10384.99	12919.05	19.61
Northern Africa	123644.17	114140.51	237784.68	52.00
Algeria	30510.02	11498.04	42008.05	72.63
Egypt	42437.53	56938.21	99375.74	42.70
Libya	5183.36	1287.59	6470.96	80.10
Morocco	22602.74	13589.06	36191.81	62.45
Sudan	14380.22	27131.31	41511.53	34.64
Tunisia	8038.43	3620.75	11659.17	68.95
Total	123152.29	114064.96	237217.26	60.25
Western Asia	195049.69	77248.71	272298.40	71.63
Bahrain	1399.13	167.87	1566.99	89.29
Iraq	27723.81	11615.95	39339.75	70.47
Israel	7811.92	640.92	8452.84	92.42
Jordan	9010.37	893.43	9903.80	90.98
Kuwait	4197.13	0.00	4197.13	100.00
Lebanon	5398.41	695.10	6093.51	88.59
Oman	4083.21	746.74	4829.95	84.54
Qatar	2671.55	23.30	2694.85	99.14
Saudi Arabia	28133.14	5421.21	33554.34	83.84
State of Palestine	3848.42	1204.36	5052.78	76.16
Syrian Arab Republic	9903.24	8381.17	18284.41	54.16
Turkey	61554.69	20362.18	81916.87	75.14
United Arab Emirates	8255.64	1285.97	9541.62	86.52
Yemen	10595.26	18320.02	28915.28	36.64
Total	184585.90	69758.22	254344.12	80.56
Total Middle East	310272.26	194208.17	504480.43	

Source: (UN 2018) United Nations, Department of Economic and Social Affairs, Population Division (2018). World Urbanization Prospects: The 2018 Revision

As illustrated in Table 3.2, the level of urbanization is considerably lower in the less developed regions, where 40% of the population lived in urban areas in 2000. This proportion is noticeably higher than it was in 1950 (18%) and is expected to rise to 56% by 2030. Thus, by 2030, the less developed regions will reach a level of urbanization like that exhibited by the more developed regions in 1950. There are marked differences in the level and pace of urbanization among the major areas constituting the less developed regions of the world. Latin America and the Caribbean as a whole are highly urbanized, with 75% of its population living in urban settlements in 2000, a proportion higher than that of Europe. Moreover, this proportion is twice as high as the one estimated for Africa or Asia. With 37% of their populations living in urban areas in 2000, Africa and Asia are considerably less urbanized and, consequently, are expected to experience rapid rates of urbanization during 2000–2030. It is expected that by 2030, 53 and 54%, respectively, of their inhabitants will live in urban areas. Furthermore, by 2030, Asia and Africa will each have higher numbers of urban dwellers than any other major area of the world, and Asia will account for 54% of the urban population of the world, up from 48% in 2000 (UNDP 2003).

Rural-urban migration and the transformation of rural settlements into cities are important determinants to the high population growth expected in urban areas of the less developed regions over the next 30 years. In combination with the universal reduction of fertility levels that are expected to occur in the future, these changes will lead to the eventual reduction of the rural population of the less developed regions whose growth rates will become negative in 2025–2030 for the first time. Therefore, it is expected that the rural population of the less developed regions will reach a peak around 2025 and then begin to decline, just as the rural population of the more developed regions has done since 1950.

By far the better known of the four trends is the demographic expansion of cities in developing countries. It is argued (Stren 2002) that urbanization is one of the powerful emerging realities of the late twentieth and early twenty-first centuries. Consider the following factors. One is that over the next three decades virtually all the population growth in the world in both developed and developing countries will take place in urban areas. Another is that urban areas in developing regions are growing five times faster than urban areas in the developed world. Third, between 2000 and 2030, urban areas in developing regions will average 2.3% growth per year, while the growth of rural populations will be just 0.1% per year. Finally, by 2030, 79% of the world's urban population will be living in developing countries. As illustrated in Table 3.3, the total population of the Middle East will increase from 491.98 million in 2010 to 820.73 in 2050, or by 66.82%.

As detailed in Table 3.3, more than 64% of the population of the Middle East countries lives in cities in 2018. In the year 2025, the rate of urbanization of the region as a whole could reach 72%. In Egypt, Jordan, and Lebanon, the rate of urbanization will reach 55.6, 95.3, and 93.4%, respectively. The population of the urban agglomerations of the three countries will grow from 95.52 million inhabitants in the year 2010 to 173.02 million in 2050, of which more than 55% will live in urban areas. As for the expected growth of the cities themselves, taken individually, several transformations will occur that might results in higher that the estimated figure. It is estimated that Egypt alone will reach 180.0 million by the year 2050. This will influence socioeconomic, political, administrative, and religious transitions in which urban regimes, urban-fabric developments, and urban dynamics will occur within the national territory. However, within the present context of globalization and liberalization of the economy, urban growth risks, and subsequently the spread

Table 3.3 Projected population of the Middle Eastern countries 2010–2050

Country	Capital City	Area [sq. km]	Pop growth rate 1990–1995	2010 Projection	2020 Projection	2030 Projection	2040 Projection	2050 Projection
Algeria	Algiers	2,381,740	2.3	36.11	43.33	48.82	53.24	57.43
Bahrain	Manama	665	2.1	1.24	1.69	2.o1	2.2	2.32
Egypt	Cairo	1,001,450	1.9	84.1	102.94	119.7	137.06	153.43
Iran	Tehran	1,648,000	1.7	74.56	86.72	88.86	91.89	93.55
Iraq	Baghdad	437,072	2.8	30.76	41.5	53.29	66.75	81.49
Israel	Jerusalem	20,770	2.2	7.176	8.921	9.98	11.28	12.57
Jordan	Amman	92,300	3.1	7.18	10.2	11.12	12.68	14.18
Kuwait	Kuwait	17,820	3.1	2.99	4.3	4.87	5.32	5.64
Lebanon	Beirut	10,400	1.8	4.242	5.07	5.3	5.39	5.41
Libya	Tripoli	1,759,540	2.4	6.19	6.66	7.34	7.82	8.12
Morocco	Rabat	446,550	1.8	34.41	37.07	40.87	43.71	45.66
Oman	Muscat	212,460	3.4	3.546	4.953	5.89	6.34	6.75
Qatar	Doha	11,437	1.8	1.78	2.79	3.23	3.53	3.77
Saudi Arabia	Riyadh	1,960,582	3.4	27.42	42.922	39.48	42.77	45.05
Syria	Damascus	185,180	2.6	21.066	18.92	26.6	30.79	34.02
Tunisia	Tunis	163,610	1.4	10.64	11.9	12.82	13.43	13.88
Turkey	Ankara	780,580	1.7	72.32	83.83	88.41	92.98	95.62
United Arab Emirates	Abu Dhabi	82,880	2	8.27	9.81	11.05	12.2	13.16
Yemen	Sanaa	527,970	2.4	23.6	30.24	36.81	42.98	48.3
Sudan	Khartoum	11,741,006	2.7	34.38	43.54	54.84	67.35	80.38
Total				491.98	597.306	669.28	749.71	820.73

Source: (UN 2018) United Nations, Department of Economic and Social Affairs, Population Division (2018). World Urbanization Prospects: The 2018 Revision

Table 3.4 Changes in the urbanization rates of the Middle East countries (1950–2030)

	1950	1960	1970	1980	1990	2000	2010	2020	2030
Algeria	22.3	30.4	39.5	43.5	51.4	57.1	62.2	67.5	71.7
Morocco	26.2	29.2	34.6	41.3	48.4	55.5	61.7	66.7	71.0
Tunisia	31.2	36.0	44.5	51.5	57.9	65.5	71.3	75.2	78.4
Libya	18.6	22.7	45.3	69.3	81.8	87.6	89.7	90.9	92.0
Egypt	31.9	37.9	42.2	43.8	43.6	42.7	44.0	48.2	54.4
Jordan	35.9	50.9	56.0	60.2	72.2	78.7	80.1	82.2	84.4
Israel	64.6	77.0	84.2	88.6	90.3	91.6	93.3	93.9	94.6
Palestine	37.3	44.0	54.3	61.1	64.0	66.8	70.0	73.5	76.9
Lebanon	22.7	39.6	59.4	73.7	84.2	89.7	92.1	93.1	93.9
Syria	30.6	36.8	43.3	46.7	48.9	51.4	55.4	60.6	65.6
Turkey	21.3	29.7	38.4	43.8	61.2	65.8	69.9	73.7	77.0

Source: United Nations 2019

of urban informality, will become highly selective, concentrating on certain agglomerations, or even in certain regions, to the detriment of others by which new urban fabric developments will take place. There is also the risk that the differences between cities, as well as within cities themselves, and human pressures on natural resources (water, soil, etc.) will become greater.

On the other hand, urbanization processes in Middle East countries, similar to the other countries in the Global South, are caused by push and pull factors, the structure of socioeconomic, socio-spatial growth, political transformation, and the stage of economic development are always in transitions process (Abu-Lughod and Richard Hay 1977; Rakodi 1997). Population growth has been a major driver for the rapid urban expansion within the countries of the Middle East and for informal housing quarters that are highly vulnerable to disaster (UNDP 2003). It has been witnessed that the proportion of the global urban population living in slums dropped by 20% between 2000 and 2016, although the absolute numbers of those experiencing slum-like conditions increased to more than a billion, meaning that the pressure on cities due to rising demand for services and migration from rural areas shows no signs of abating (Nicklin and Cornwell 2019). Both past and projected population growth (see Table 3.2) are crucial for assessing future vulnerabilities to urban informality and transitions processes.

Urbanization trends have differed significantly. In the Middle East and North Africa, urbanization rates have differed (1950–2000) as have the projections until 2030 (see Table 3.4). Between 1950 and 2000, the most rapid increase in urbanization occurred in Libya, from 18.6% to 87.6% and in Lebanon from 22.7% to 89.7%. A lesser growth rate occurred in Egypt, from 31.9% to 42.7%. By 2030, about 94.6% of the population of Israel will be urban, 93.9% in Lebanon and 92% in Libya. While, in 1950, Egypt with 31.9% had the highest urbanization rate in Northern Africa, by 2030, it would have the lowest at an expected 54.4%. In one way or another, these trends have affected the level and types of urban transitions processes in which the spatial growth dominates the transformation of the urban agglomerations, and they subsequently have major impacts on the natural resources. For example, the rapid informal urban growth of Cairo on previously agricultural land, namely, *Shubra El Kheima* (a suburb of Cairo), grew 25-fold. Similarly, the multiplication rates of Tripoli, Amman, and Rabat grew 10- to 15-fold; Istanbul, Aleppo, Damascus, Beirut, Ankara, Casablanca, Tel-Aviv and Izmir five to ten fold, and Cairo, Tunis, Alexandria and Algiers two to five fold. Among the 30 largest urban agglomerations in 1950, two were in the Mediterranean: Milan at number 14 with 3.633 million and Cairo at number 25 with 2.420 million. In 2000, Cairo was number 20 with 9.462 million, and Istanbul number 22 with 8.953 million. The current number of residents in the Greater Cairo region, at more than 19.44 million, exceeds the total population of Egypt in 1919 (12–13 million) (Cammett et al. 2015).

Of the countries with more diversified economies, Lebanon and Jordan had the highest estimated growth rate of 4.01%, while Egypt, with 1.91%, had the lowest. Common factors influencing growth in this group of countries include the continuing weakness of world economic activity, low export demand, and the low number of non-Arab tourist arrivals. However, intraregional tourism increased in 2002, benefiting the more diversified economies with the capacity to absorb tourism, including, in particular, Jordan, Lebanon, and Syria. It is expected that in 2030 the urbanization rates of the Middle East countries will have increased, and Lebanon and Israel will record the highest urbanization rates of the region (see Table 3.4).

On the other hand, four Asian countries of the Middle East are "at risk", i.e., experiencing moderate to high slum growth rates but also having a moderate incidence of slums that require remedial policies to reverse slum growth. Iraq, Jordan, Lebanon, and Saudi Arabia have witnessed sustained annual growth in the number of slum dwellers (2.5–4.0%). This shows that slum growth is not confined to lower-income countries, but also extends to middle-income and high-income countries that experienced a drop in economic performance in the 1990s. Some of these countries (Lebanon, Iraq, and Jordan) have in recent years also been engulfed in the political turmoil that has exacerbated the refugee crisis and worsened conditions in their cities (UN-Habitat 2006).

While growth in the more diversified economies is generally more uniform than in the Gulf States, the economies and, therefore, the factors influencing growth in that group of countries are more diverse, and, subsequently, so are the transitions. In Egypt and Jordan, the fiscal stance was tight due to strained fiscal positions. Moreover, the increasing requirement of central government for financing has led to some degree of crowding out private borrowing which, combined with reduced domestic credit and liquidity, has decreased private spending. While Lebanon has been negatively affected by high-interest rates, a tight fiscal position, and perceived unsustainable public finances, it has managed to improve confidence and lower interest rate prospects as a result of the successful outcome of the Paris II donor conference in November (ESCWA 2003). Currently, Lebanon is suffering from tremendous transitions in socioeconomic and political conditions within which urban regimes and urban-fabric developments have confronted the country. While Egypt, Lebanon, and Jordan have suffered from sharps drops in national domestic growth production due to Covid-19, subsequently, the urban regimes within these countries have changed their policies to cope with the global crises, and many developments projects have been postponed. Moreover, the explosion 4 August, 2020, in Beirut harbor put further pressure on the state as more than 60,000 households lost their domiciles.

According to (UN 2018), in Arab countries "urban growth rates will remain higher than total population growth rates in the foreseeable future... urban growth has been the result of rural-to-urban migration as well as high fertility and declining rates of mortality." The UN notes that "Many cities are now going through a critical phase of development, marked by dwindling resources, increasing poverty, and serious environmental degradation."

Demographic change has affected the acceleration of urban informality in various ways. Population size, population growth, age structure, and educational, health, and socioeconomic characteristics all influence the increasing demand for informal accommodations, as do migration patterns. The overall impact of these trends on informality and urban-fabric developments have varied from one country or one region to another. Population movements have also at times led to deforestation and other negative environmental consequences. In some Middle East countries, a key concern has been rapid population aging, e.g., in Lebanon, driven by low fertility levels and the impact on health care and other social services. In other Middle Eastern countries, high population growth rates, e.g., in Egypt and Turkey, will see them facing major problems in their urban agglomerations and urban-fabric developments.

3.3.2 Rural–Urban Migration

In general, the countries of the Global South are still urbanizing rapidly under the influence of rural–urban migration, as well as natural population increases (Friedmann 2011). Certainly the nations of the Middle East have been subject to the push and pull of contradictory and competing forces, national, and regional, as well as international, not just in recent years, but for generations (AbuLughod 1971; Drakakis-Smith 2000). Such migration has often been exacerbated by strong pull factors, such as the greater availability of job opportunities and social services in towns or cities, as much as by push factors related to the state's inattention to agriculture and the scarcity of infrastructure, job opportunities, and social services in remote and/or rural areas (Cammett et al. 2015). Conversely, rapid urbanization and the effects of globalization. including the loss of cultural identity, the increase in multinational corporate urbanism (Gharipour 2016), significant transnational immigration, the dualism of economic, the spread of urban informality, and social exclusion, have resulted in changing the form of Middle Eastern cities, and subsequently the formulations of urban-fabric developments, over the last 20 years. Cairo, Beirut, and Amman doubled in their size over two decades, and the corresponding societal transitions of urban regimes have influenced socio-spatial developments.

Research has established the generalization that the urbanization in the countries of the Global South does not follow the historical path of the originally industrialized nations (Roy and AL-Sayyad 2004; Dwyer 1975; Ehteshami 2007; Drakakis-Smith 2000; Fiedler 2014). In Europe and North America—as industrialization and urbanization proceeded in tandem—a large working class was formed in the growing cities (Hall 2014 (fourth edition); Gilbert and Gugler 1992; Peattie 1996). With some important exceptions, urbanization in the Global South has been associated with uninformed development and dependent industrialization in which working-class employment expands only to a limited degree, and national control is compromised by foreign investment. Production is capital intensive, favoring narrow luxury and export markets, and backward linkages to new capital goods industries are reduced by imports under terms of trade, as, in general, labor absorption by industry is low (Arrighi and Saul 1973). The result is a pattern variously described as urban involution (McGee 1982), over-urbanization, pre-urbanization and informal/arbitrary urbanization (Soliman 2004a), suburbanization (Hall 2014 (fourth edition)), tertiarization (Evans and Timberlake 1980), or subaltern urbanization (Roy and AL-Sayyad 2004). A pattern develops whereby the rapidly growing urban population of Middle East cities is increasingly compressed into overcrowded commercial and social services sectors that create dual cities or divided cities, e.g., Cairo, Amman, and Beirut (AbuLughod 1971; AlSayyad 2006; Sims 2010; Fawaz 2004; Khirfan 2017). Hall and Pfeiffer (2000) emphasized one category of urbanization that they call "informal hypergrowth" cities. Expanding outward from these exploding and swollen cities, this phenomenon is not purely restricted to the cities of the south, but through migration "some cities of the developed world are invaded by the Global South", also rendering them ungovernable.

The factors influencing the decision to migrate are varied and complex. Because migration is a selective process affecting individuals with certain economic, social, educational, and demographic characteristics, the relative influence of economic and noneconomic factors may vary not only between nations and regions but also within defined geographic areas and populations (Todaro and Smith 2012). This pattern of urbanization is, to a certain extent, related to various decisions for moving from one area to another within a city, from rural to urban areas, from region to region in the Middle East countries, or from one continent to another as an international shift, and it has been explained by the following five hypotheses.

A central hypothesis is that individuals move (with varying degrees of ease) in response to economic incentives and follow economic opportunities. If location incentives are distorted, so presumably is the growth process. Distorted location incentives may also result in higher than necessary social environmental costs that are mostly offset by the increase in productivity and wealth that usually accompanies urbanization. The growth performance of the Middle East was significantly below that of most other developing countries of the world, both in absolute and per capita terms. The region's feeble economic performance reflects a slowdown in the performance of the oil sector[3], low rates of economic activity throughout the world, and the negative impact on the private sector of heightened political and conflict tensions in the region. As of 2010, the informal sector had come to dominate the region, with nearly 70% of the labor force employed in this sector in a typical MENA country (Mehta and Dastur 2008). In other words, diversification of income sources has been a key factor for the vast majority of households in improving their livelihoods (Sunam 2015), but has accelerated the spread of urban informality and, subsequently, has affected the landscape developments.

Another hypothesis is that people move to urban areas for social and political reasons (Gugler and Flanagan 1978; Abu Lughod 1971). Migration to urban areas can provide an escape from family and cultural constraints, such as restricted land access or a low level of female independence (Todaro and Smith 2012). Migrants may also seek to acquire cash income to contribute to birth wealth as the money economy increasingly penetrates marriage rituals (Gugler and Flanagan 1978). Migration to an urban area may also occur because of an expected increase in social status and standing—the perception that the *"high life"* can be found among the *"bright lights"* myth— migrants have been lured to the city by exaggerated tales of high income and technologically advanced living, especially by returned migrants who "wished to convey to others a positives image of themselves and their experiences". Lebanon and Egypt are among the various countries in the Middle East where a huge number of immigrants decide to move from rural areas for the purpose of "bright light of" the urban areas which changed the urban fabric of cities of Egypt and Lebanon as urban informality came to dominate the transitions.

A third hypothesis is that in the last two decades individuals moved in response the incentives derived from information-and-communication networks. This new wave of immigrants attracts a certain stratum of people to urban areas. It is argued (Castells and Hall 2004 (electronic edition)) that the new global-and-knowledge economy and the emerging informational society have indeed a new spatial form, which develops in a variety of social and geographical contexts: megacities—Cairo among them.[4] Megacities articulate the global economy (Koonings and Kruijt 2009), link up the informational networks, and concentrate the world's power. But they are also the depositories of all these segments of the population who fight to survive, as well as of those groups who want to make visible their dereliction so that they will not die ignored in areas bypassed by communication networks. Because communication is the essence of human activity, all domains of social life are being modified by the pervasive uses of the internet. Castells (2015) argues that a new social form, the network society, is being constituted around the planet, albeit in a diversity of shapes, and with considerable differences in its consequences for people's lives,

---

[3]Oil prices reached their peak in June 2004 at $42 per a barrel, while in September 2008 they approached $140 per barrel and dropped in October 2008 below $65 after the international financial crisis. After the appearance of Covid-19 in late 2019, oil prices further dropped below $40.

[4]The notion of megacities has been popularized by several urban experts in the international arena, most notably by Janice Perlman, founder, and director of the New York-based "Megacities Project." See also Borja and Castells (1997).

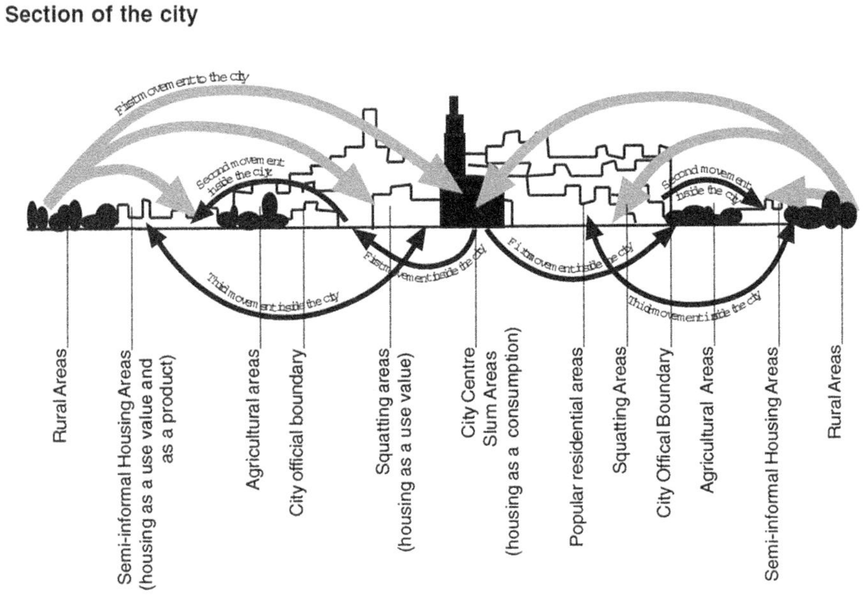

Fig. 3.3 Scheme of how individuals move from rural to urban areas and from one area to another. Source: (Soliman 1996)

depending on history, culture, and institutions. This transformation offers as many opportunities as it raises challenges for future transitions.

A fourth hypothesis is that wars, natural disasters, population displacement, and ethnic conflicts may also lead to an increase in rural–urban migration. Aside from the impact of war on agricultural income through the effects on transport and marketing, war may also push people out of rural areas for sheer safety reasons. In particular, ethnic conflicts increase the danger of living in an area dominated by a persecuted ethnic group because the potential for ethnic cleansing is high in these areas. Urban areas generally have a higher level of ethnic diversity, and thus may be safer havens for persecuted groups. With growing globalization and capitalization, a new international wave of movement has appeared. This movement is accompanied by the acceleration of economic situations.

A final hypothesis is that of internal migration/mobility within the city (see Fig. 3.3). First, the movement is made from rural to urban areas, and, if access to central job markets is paramount, the household can consider renting very small room or a rooftop or staying with a friend when the situation offers centrality and security of tenure but are sometimes too expensive and hold out no hope of eventual ownership. The second mobility is moving from the rooftop to an inner slum area in the old quarter of the city: a location with a poor-quality environment and low rents, or no rent at all, with good access to job opportunities but with no hope of secure tenure. Such illegal dwellers will eventually be forced to evacuate and may move to a squatting area. The third and cheapest housing solution is to squat on publicly owned land, usually on depleted areas in the outskirts and almost always downwind of pollution; negative trade-offs include the very high cost of commuting to work and the government's neglect of infrastructure. The fourth solution, as soon as the settler's income status improves and commonly preferred by the poorest, is to buy a house site in one of the vast semi-informal developments (often on land purchased from a broker or peasant villages) with legal tenure but without official building authorization. Although far from jobs, such sites are secure with a high likelihood for improvement and, after considerable community mobilization and political negotiation, are usually provided with basic municipal services (Soliman 2004a).

Therefore, the population shifts in Middle East countries are linked with various circumstances and dependent upon various incentives. It varies from time to time, as what happened in the past does not necessarily occur nowadays. Also, the rhythm of movement between countries, within one country and within a city, has become faster than before, and it has occurred over a short period of time in which urban informality flourished. The phenomenon of population mobility is also characterized by planned moves rather than arbitrary moves due to the availability of information and knowledge that become available through the network (Castells 2010 2nd edition) and other means of data availability for a certain area.

3.4 Political Mobilization

It is argued (Abu-Lughod 1965) that rapid urbanization in the south countries generates political mobilization and that uprooted masses of rural migrants would overwhelm the carrying capacity of cities leading to unemployment, disappointed expectations, alienation, misery, and vulnerability to the appeal of extremist movements. On the other hand, since the late 1940s, Middle East countries, especially Arab countries, have witnessed political mobilization and transitions due to the political struggle against foreign occupation, and struggle for office. As a result, a social movement has emerged (Singerman and Amar 2006), and huge population shifts from rural to urban areas occurred as people pursue job opportunities, affordable shelter, and improvement in their standard of living.

As political events of the Suez Crisis and the 1960s socialization movements, unfolded in the early 1970s, this jaundiced view was transformed in social studies that challenged characterizations of the urban poor as "marginal masses" (Bayat 2012) and argued for their active participation in community organizations and social movements (Perlman 1976; Castells 2015; Gilbert and Ward 1985). During this era, most Middle East countries experienced the development of industrial activities, a flourishing of trade unions, and massive waves of immigrants from rural to urban areas. Although some of these marginal masses' movements dwindled as their material goals were won, new forms of mobilization in the form of urban informality seemed to take their place. Those who mostly settled in informal areas were on the periphery of cities or in peri-urban areas (Sims 2010; Soliman 2004a; Fawaz 2004; Khirfan 2017) but also joined by a new wave of regional migration into Arab countries.

During the late 1960s and looking ahead to a time when labor absorption would begin to fail, Middle East cities saw a period of renewal from socioeconomic transitions. Nationalism trends dominated this era in which the role of the private sector shrunk, while the public sector flourished. Variations on characteristics of the state is varied between nationalism and liberalism, and party system, by which have resulted in squatter communities themselves being manipulated by the system as much as vice versa: Egypt, Syria, Sudan and Algeria, for instance. Clients' participation might generate real material gains (not simply concessions) in a multiparty system yet expose the poor to unnecessary risks under authoritarian regimes. Thus, changes of urban regimes, dynamics of state, and party have a major influence on socioeconomic transformation and to a large extent on the process of urbanization followed by a huge wave of rural migration to urban areas, as well as the spread of urban informality. It is argued (Soliman 2004b) that, at certain times and under certain circumstance, especially during election campaigns or during political crisis, the state might act in favor of the urban poor, and in another situation, against them.

The war and conflict with Israel of 1967–1973, the Lebanese civil war (1975–1990), Iraqi invasion of Kuwait, and the G.W. Bush invasion of Iraq caused a major disturbance in the built environment of Middle Eastern cities. These periods caused massive waves of people from rural areas to urban areas, and from country to another, and witnessed the influx of masses of refugees to certain countries of the Middle East. Also, changes of demographic structures have occurred. As a result of people mobilization, many peri-urban areas of cities transformed into illegal residential areas accompanied by informal economic activities and basic social amenities.

It seemed that a new era opened in Middle East political history when, early in 2005, in Cairo, Egypt, another urban insurgency appeared in the "new social movements" of community collectives within slum neighborhoods. The Muslim Brotherhood (in Arabic, *al-Ikhwan al-Muslimin, or simply al-Ikhwan*) and more privileged students group (Singerman and Amar 2006) began to cooperate, demanding fair elections, affordable shelter, job opportunities, a decent quality of life, equity, and respect for human rights. After a controversial beginning under the aegis of marginality theory (Mansoob 2002), analyses of political mobilization by diverse urban movements and constituencies are very much in fashion today (Caroline and Yatzimirsky 2014; Parnell and Oldfield 2014).

The same may be said of rival interpretations based on cooptation and the patron-client model. Joan Nelson's survey (Nelson 1979) of cities in Africa, Asia, and Latin America compares participation of the poor in political parties, special-interest groups, and ethnic associations with a fourth type, clientelism, and concludes that "patron-client links are the most universal." According to this theory, the poor are fundamentally satisfied with urban life, too busy making ends meet to protest, and, as time goes on, become astute petitioners in the "demand-making" process (Cornelius 1975). Moreover, Cairo is becoming the "protest city" where many protesters are frequently gathering in front of the *Shura* Council in downtown Cairo to present their demands for socioeconomic and political rights. In response to recent incidents of social unrest in Egypt, many movements have been formed, such as the *6th April* and *Enough* (*Kifaya* in Arabic) movements (Singerman and Amar 2006; Castells 2015), protesting the conditions in which "so many people live."

Although clientelism[5] is usually inferred from cross-sectional studies, some interpretations deal with changing historical conditions. The new Cairo school of urban studies (Singerman and Amar 2006) illustrates that a passive working class emerged under specific circumstances. As urbanization accelerated in cities of the 1970s and 1980s, their skilled working classes that once had included a high proportion of European craftsmen were diluted and depoliticized. Internal rural–urban migrants accustomed to the caudillismo of landed patrons were politically immature, and they found life in the cities generally satisfying given their own recent past. Political participation took the form of deference and patronage under colonization and imperialism. It is remarkable how few riots—even food riots— there have been in the great Middle East cities during a period when the mass of their impoverished and economically marginal inhabitants has multiplied and inflation, as often as not, went uncontrolled. This happened between 1977 and 1986 in Egyptian cities, and it was, to a certain extent, the main cause of the Civil War in Lebanon. In the summer of 2005, Cairo's tens of thousands of protestors brought the world's attention to a set of dynamics and protagonists colliding at the urban crossroads of an assertive, outward-looking Middle East (Singerman and Amar 2006). The Arab Uprisings in 2010/11 caught people, governments, and many academics by surprise (Gause 2011). Many countries in which uprisings took place have experienced the survival of authoritarian rule through repression (e.g., Bahrain), counter-revolution (Egypt), civil war, and the collapse of state structures (Libya, Syria), or processes of reform and 'façade democratization' (Morocco, Jordan) designed to maintain the substance of authoritarian regimes untouched (Malmvig 2014). Recently, after the Beirut explosion on 4 August 2020, Lebanese protestors comprising high and low-income groups demanded a changing the leaders and requested the assistance of the international donors, especially France, to end the corruption in the country.

Bayat (2010), Khirfan (2019), and Fawaz (2004) support these observations with specific reference to Egyptian and Lebanese slum dwellers arguing that political conservatism is the rational posture for these upwardly mobile groups who aspire to a middle-class lifestyle. Similar conclusions (Teti et al. 2018) about political passivity derive from critical analyses of state and party systems that co-opt the urban poor with small concessions in the form of services and subsidies, while elites grab the lion's share of privilege and income. The urban poor became increasingly disaffected with corrupt regimes that maintained power by rewarding a narrow political and economic elite, while excluding much of the population from economic gains, social mobility, and the political forum. In these interpretations, the urban poor are defended as rational actors making the best of austere circumstances rather

[5]Clientelism is defined as the particularistic allocation of state resources aimed at maximizing a political actor's probability of election. This definition encompasses equally what political scientists have traditionally called 'clientelism', 'patronage', 'brokerage politics', and 'pork-barrel politics.'

than gullible subjects of elite manipulation. Although the theory of clientelism has enjoyed its greatest success as a critique of jeremiads about social disintegration, it continues to resonate with pragmatic urban accounts of urban. Most contributors (Stokes et al. 2013) to this debate have defended by turns sober clientelism or hopeful activism; a few have analyzed social conditions that might explain varied instances of both.

In the anguished period of debt and international financial crises of 2008 and structural adjustment since the early 1980s, the Arab Spring, the Covid-19 pandemic, and 2000s, activist interpretations of southern urban politics have gained ground. Accordingly, the Middle Eastern cities have been facing a triple transformation: densification of popular urban tissues, a proliferation of spontaneous housing, and considerable urban informality. Simple and legible urban structures have been replaced by fragmented configurations, with suburban, subaltern, and peri-urban spaces offering the image of a still incomplete patchwork in cities of the Middle East.

3.5 Urban Conflict

A theoretical dialog about the conditions, forms, and outcomes of urban political conflict in developed and developing countries has been initiated (Roy 2009). The aim is ambitious but grows out of an extensive research literature that seems overdue for codification of its regional and case-study variation (Forrest et al. 2017; Roy 2009; Caroline and Yatzimirsky 2014; Singerman and Amar 2006; Khirfan 2019). The theory is intentionally general and a provocation to the venerable empirical tradition that has remained overly descriptive and localized. The argument precedes in two steps. First are matters of distinction, types of collective action, and their institutional contexts in which sociotechnical niches are playing a great role in transitions processes. Next, the discussion turns to the substance of the influence of globalization on urban life in poor countries as it is shaped by the intersection of economy, state, and civil society (Friedmann 2011) by which sociotechnical regimes have played a great role in transitions processes. Collective action, or grassroots action, may take many forms, ranging from noncompliant acts of everyday resistance to sustained movements for reform and revolution (Gilbert and Gugler 1992; Singerman and Amar 2006) to form a strong urban niche for development processes. For our present purposes, three types of collective action, or grassroots action, cover a broad spectrum of popular mobilization in the south cities while they also make key distinctions often masked in the debate over activism versus clientelism. Labor action includes mobilization arising in the sphere of income and employment representing the interests of workers in such concrete actions such as strikes, job actions, and demonstrations protesting unemployment or policies specifically harmful to labor. A high proportion of laborers live in informal residential areas, by their strikes they could affect the national economy, subsequently increasing the rate of price inflation for various commodities. An example is the recent strike in Amman, Jordan, against the increase in the oil price. Collective consumption action includes mobilization by consumers for urban services, action focused on the availability of collective or public goods (e.g., land, housing, transportation, education, health, and urban services, e.g., water, streets, sewers, electrification). It is expressed in actions such as land invasions, squatter protests, and street demonstrations. Political and human-rights transition involves popular mobilization around non-material issues of justice, representation, security, freedom from repression, and democratization because these are expressed in marches, demonstrations, vigils, hunger strikes, and similar acts of conscience, such as the post-explosion demonstrations in Beirut on August 2020.

In the empirical world, of course, these three forms of collective/grassroots action are often compounded. Mass demonstrations protest austerity policies that simultaneously threaten to cause unemployment and restrict collective consumption of subsidized food, land, housing, transportation, and social equity. In Egypt, severe economic collapse has resulted in hunger strikes by street sweepers invoking their human rights to lost jobs (Al-Ahram 2010a, b). All this requires that overt acts be analyzed in their complexity, that distinctions are drawn, and in the end that certain serviceable categories be employed. But this is a general problem of observation and classification that should not prevent careful generalization.

The security situation stabilized in Beirut's suburbs on Friday, 27 May 2004, ending the clashes that started a day before between the Lebanese Army and rioters taking part in a demonstration organized by the General Labor Confederation. While securing main roads leading to the airport and other essential destinations, the army withdrew its troops from the southern suburbs and redeployed its units away from angry crowds in *Hay al-Sellom* and other suburbs, where thousands of mourners gathered to bury their dead while chanting anti-government slogans. Six people, including five civilians, died during the clashes, which also left dozens wounded. Tensions remained high among rioters who took to the streets in several areas including *Bourj al-Barajneh, Hay al-Sellom*, and *Hadath*, setting up roadblocks of burning tires and disrupting traffic movement from one neighborhood to another (Daily Star 2004). The recent clash, between the Lebanese Army and demonstrators, after the Beirut explosion on 4 August 2020 left two persons dead.

A similar threefold distinction refers to theorized influences on collective/grassroots action. A complete understanding of urban political conflict must reckon with the interacting forces of economic structure, state policy, and civil society. Each type of collective action is affected in specifiable ways by distinct configurations of all these institutional forces. Conversely, no causal connection has been proposed between, for example, labor action and the economy to the neglect of state and civil society as decisive, interactive influences—and so forth with respect to each action type. For example, the protesters sought change in Egypt (Castells 2015), the Rebel movement opposed the role of the Muslim Brotherhood in Egypt in 2013, and the Lebanese movement on August 2019 and 2020 sought political change, social justice, a decent life, and an end of corruption.

On the other hand, in pre-capitalist, agricultural societies, labor and natural resources were the key ingredients for economic growth. Since the Industrial Revolution, this changed to energy resources (like steam, oil, coal, electricity, etc.). Throughout these earlier periods, new knowledge has always been applied to the way goods and services are produced in order to invest them more efficiently. Under informationalism, it is how one uses information in production that is different. More specifically, it is the way that information can act on itself. By using the new technologies in unison, one can generate new knowledge and combine knowledge and information in different ways in order to innovate. This ability is the hallmark of economic success today. This method of production has been organized between the state, working class, and capital by which the reproduction of capital has flourished and led into a new economic boom, such as in the Gulf States. The economy has been modified and enhanced through informationalism, which is accelerating the transitions processes.

The *core activities* of the economy have therefore been internationalized, using the new technologies. The new economy, or recycle economy, has been kick-started by major breakthroughs in five areas of technological progress: microelectronics, computing (both hardware and software), telecommunications, optoelectronics (advances in fiber optics), and biogenetic engineering. These core activities include financial markets, high value-added exports (like cars, or computers), multinational corporations (MNCs), science and technology (S&T), and the production of highly skilled persons. All these activities can now operate across the globe at the same time. These things are possible because of combinations of computing and telecommunication technologies and are possible with transitions of the sociotechnical regimes. Various customs, whatever their interest, can purchase various commodities in a short time at the lowest price. On the other hand, a particular customer can also negotiate or compete within the international market to obtain goods and services, but the bottom strata or the marginal strata of the society could not often obtain or reach these goods and services. Globalization can produce different goods and services, but it is still the elite or super-rich (Forrest et al. 2017) who can obtain them. Thus, globalization could create a new duality in urban areas "across the global": some people are capable of obtaining globalized goods and services while others cannot and are marginalized. Is this what Castells (Castells 2010 2nd edition) call the Fourth World? Castells says that the condition of about two-thirds of the world's people has remained the same or deteriorated in the last 10 years because of globalization's devastating impact.

Thus, it seems that the new economy and the transition of sociotechnical regimes combined with the urbanization process have a great impact on low-income groups in various ways. First, it prohibits the urban poor from obtaining goods and services at prices they can afford to pay. Second, it assumes that the new economy will decrease commodities' prices, but this has seldom happened. Third, most of the Middle East countries have witnessed a high level of inflation and its subsequent effect on consumer purchases. Consequently, the demand for various goods and services decreases over time. Finally, the new economy, or new economic reforms, allows for speeding up speculation in the international and national market which affects by one way or another the level of demand for the urban poor and reduces the ability to buy commodities.

Within the next few decades, globalization will create an illiterate society, or at least will exclude certain countries in the South that are not capable of keeping up with the information revolution, unless new technologies are to be distributed equally across the globe. How can this be done? Return to Hernando de Soto's idea (De Soto 2000) of reproduction of capital as he mentions that "the poor inhabitants of developing countries could raise their standards of living if they were able to use the assets they do hold in the local and global markets. And they have managed to amass a considerable amount in assets, some US$ 10 trillion in real estate alone." The question is how to change these assets from dead capital into liquid capital, and how would it be used. Who will pay the legal owners the amount in assets to give the poor title to it? (Gilbert 2012). Recently, Egypt issued a temporary law of reconciliation number 17 of 2019 amended by-law number 1 of 2020 to formalize the illegal housing against certain fees are paid to the government. The implementation of this law started in July 2020 and was to be ended in September 2020. This law will issue a legal building license for people who are seeking reconciliation, but there are many variables that are not yet resolved. Among them is that a building license does not legalize the land tenure: It is only prevents demolishing the illegal buildings and keeps the status quo of the existing spatial situation. The impact of this law is not yet evaluated. It is the same idea of De Soto to use the assets legally in the formal real estate market, as well as to use these as collateral in the official economy. To set up a proper policy for urban informality, a comprehensive strategy should be implemented based on the concept of the ASUST (explained in Chap. 1) in which all variables and challenges of the process of sustainable transitions should be put into force.

The network logic of globalization is innovative and creative. But it is also brutal and socially destructive. Whilst linking up all the innovative and productive centers of the world, it also acts to exclude those societies that are unable to hook into the new technological and information-based economy. These societies become, "structurally irrelevant" to the new global economy. Economic structure refers to characteristic and relatively permanent features of the urban economy, such as the sectoral composition of the labor force, the pattern of industrialization, occupational mobility, formal and informal organization, employment potential, and growth patterns. Key social institutions and state agencies have lost their legitimacy. New forms of social identity are emerging around new things like ethnicity, religion, and community, which are creating new sociotechnical niches to control urban-fabric developments.

In addition, these economic developments are accompanied by other social and spatial changes that have drawn attention to the decline of the nation-state. They are, first, the decline of the social welfare system. This includes the demise of a social safety net including unemployment, housing, health, and other social benefits for the poor and sick. It also affects protected labor markets, minimum-wage legislation, and the system of industrial relations built around negotiation with a strongly organized union movement. This will lead to the continuity of spreading urban informality and subsequent changes in urban-fabric developments to cope with such transitions. Second, there are changes in the labor market centered mainly on the rise of 'flexible labor' and the growth of part-time, self-employed, non-formal and other forms of temporary and outsourced employment. Alas, this will result in the flourishing of the informal economy within cities of the Middle East and will lead to creation of yet more urban informality. Finally, changes occur in class consciousness, which many workers have, based in large factories all making the same product moving along the same conveyor belt. Working-class consciousness has been diluted by a consensus-seeking industrial relations system. This system drew workers into value-based concepts such as being a 'citizen', being a member of a 'nation', and benefiting from the protection of a common state. The gap between various working classes will amplify and marginalize them away from the center or the mainstream, and the mainstream is perceived as the site of power (Bayat 2012). Under the new economy, these forms of national and class identity are falling away, with the social organization of work today becoming highly individualized, and with each individual worker performing very different tasks. The previous social institutions and social processes that formed the bedrock of the nation-state are now in decline. In the process, the nation-state has lost its legitimacy and power.

3.6 Conclusion

During the past few decades, the cities of the Global South in general, and of the Middle East in particular, have witnessed a remarkable and in many ways unprecedented demographic and spatial growth spurt. Despite some slowdown in rates of increase in the past few years as a result of falling wages, contracting social services, and changing demographic trends, contemporary urban areas remain the growth poles of economic progress and the lightning poles of political and social unrest. The previous discussion has indicated two predominant trends associated with urbanization process; first, the Middle East is urbanizing at an ever-increasing rate. It is populated largely by urban people and its socioeconomic organization and movements are increasingly articulated in urban systems. The second trend concerns the growth of socioeconomic development. Put simply, urbanization in the Middle East has not been accompanied by concomitant economic prosperity as it was in Western societies, but rather socio-spatial changes have a great impact on cities. Urbanization that occurs without adequate industrialization, appropriate technology, sufficient formal employment, or secure wages, has condemned large segments of the population to an underclass.

Reducing population growth through reductions in absolute poverty and inequality, particularly for women, along with the expanded provision of family planning and rural health services, all are various attempts to diminish and eventually eliminate rapid urban growth. Clearly, any long-term transition in Middle East employment and urbanization problems must involve a lowering of current high rates of population growth. Even though the size of the labor force for the next two decades is already determined by today's birth rates, the hidden momentum of population growth applies equally to labor-force growth. Moreover, reducing rural population growth with its delayed impact on the urban labor supply provides an essential ingredient in any strategy to combat the severe employment problems that Middle East countries face now and will in future years. These changes will have major influences on sociotechnical regimes to achieve sustainable development.

Looking at the changes of the rate of urbanization, as a sociotechnical landscape, in the last six decades it seems that three-quarters of the region will be urbanized by the year 2030. However, it is expected that urban informality will proceed in parallel unless urgent urban policies will have emerged. With the expected increase of urban informality and its subsequent informal economy, the urban gap will be wider, and the division of cities will be great. To avoid the future urban crisis, the Rio de Janeiro Urban Forum (2010) introduced six dialogues to triumph over the future needs of the planet. These are: (1) taking forward the right to the city, (2) bridging the urban divide, (3) equal access to shelter, (4) cultural diversity in cities, (5) governance and participation, and finally, (6) inclusive, sustainable urbanization. The question is are these dialogues apt for preventing future urban sprawl in the South? It is difficult to argue with any of these dialogues, and even SDGs does not achieve its objectives (Satterthwaite 2014). Doubtless, there are times and places in which activism or clientelism reigns, just as there are social conditions that promote a transition from a situation to another. The real drawback with these formulations—and of many of the summaries of vivid urban ethnographies—is that they tend to privilege description over explanation. Too often the research question becomes whether activism or clientelism, as sociotechnical niches, is the essential state of the masses rather than whether certain conditions are conducive to more or less participation, different forms of conflict and cooperation, changing arrangements of power, and so forth. Too often, polemics with no unambiguous empirical solution substitute for explanatory propositions capable of evaluation. Too often the debate over activism vs. clientelism leads to an impasse.

Going back to Lady Jackson's book (Ward 1976), *The Home of Man* directly relates to the United Nations Conference on Human Settlements—there is in a sense a synthesis of all these concerns, for it is in human settlements that nearly all people play out their individual lives, benefiting or suffering from a world system which can be judged to be in either a crisis of dissolution or the throes of regeneration. More than 44 years after Lady Jackson's warning, very little progress has been made on the planet, and the rate of growth of urbanization is still approaching the highest levels in history.

A central element of the Carnoy's and Castells' (1993) ideas on globalization are their view that globalization has precipitated the decline of the nation-state. This happened mainly because globalization causes all national economies to depend upon the performance of the international financial markets, especially after the Covid-19 pandemic. And these financial markets are globally integrated and have resulted in international crises, such as what happened in 2008. The states in the Middle East and other Southern countries have therefore lost control over monetary policies and interest rates—key levers of national economic policy. This has affected the purchasing power of the urban poor as most commodities became too expensive for them to afford, especially in housing sectors, and urban informality continues to flourish (Soliman 2010). Nowhere is this dilemma more visible or the resulting problems more intractable than in the crowded cities of the Middle East, where projections of urban population growth remain the highest in the world.

An appropriate rural–urban economic balance should be created. A more appropriate balance between rural and urban economic and noneconomic opportunities appears to be indispensable for ameliorating urban and rural unemployment problems and slowing the pace of rural–urban migration. The main thrust of this activity should lie in the integrated development of the rural sector, the spread of small-scale industries throughout the countryside, and the reorientation of economic activity and social investments toward rural areas. A reaction comes in the form of initiatives for the disadvantaged group to achieve their right to the city via grassroots mobilization and informal practices to fill the gap created by socioeconomic exclusion.

Most of the Middle East countries are suffering a slow rate of progress and change, which can be realized without the advocates of the extreme importance of time. The speeding up of development progress and its performance in the right time and at the right place is the main dimension in challenging the rapid growth of urbanization processes to meet the future needs for the next generations. On the other hand, to govern such development progress, new planning innovations are needed to guide and control such development and to take into consideration the rapid socioeconomic, political, and the network of globalization transitions that would accompany such development progress. No doubt, urbanization often contributes, and sometimes detracts from, the quality of people's lives, but one cannot tell by how much. This would require a real shift from the old wisdom to a new systemic innovation through transformations in sociotechnical and socio-ecological systems. As depicted in Fig. 3.3, this transformation is attributed to four elements: state performance, access to urban determinants, realignment of urban society, and the effectiveness of capital. These elements are changeable according to the prevailing circumstances and due to the transformative process of a given setting.

References

Abu Lughod, J. 1971. *Cairo: 1001 Years of the City Victorious*. Princeton: The Princeton University Press.

Abu-Lughod, J. 1965. A Tale of Two Cities: The Origins of Modern Cairo. *Comparative Studies in Society and History* 12 (1): 429–457.

Abu-Lughod, J., and Richard Hay Jr., eds. 1977. *Third World Urbanization*. New York: Methuen.

Al Sayyad, N. 2006. Whose Cairo? In *Cosmopolitan: Politics, Culture, and Urban Space in the New Globalized Middle East*, ed. Singerman and P. Amar, 539–542. Cairo: American Unversity Press in Cairo.

Al-Ahram. 2010a. *Egyptian Newspaper*. April 7th.

———. 2010b. *Egyptian Newspaper*. May 8th.

Arrighi, G., and J.S. Saul. 1973. *Essays on the Political Economy in Africa*. London: Monthly Review Press.

Bayat, A. 2010. *Life as Politics: How Ordinary People Change the Middle East*. Amsterdam: Amsterdam University Press.

———. 2012. Marginality: Curse or Cure. In *Marginality and Exclusion in Egypt*, ed. Ray Bush and Habib Ayeb, 14–27. Cairo: American University Press in Cairo.
Bicknell, J., D. Dodman, and D. Satterthwaite. 2009. *Adapting Cities to Climate Change: Understanding and Addressing the Development Challenges*. London: Earthscan.
Cammett, M., I. Diwan, A. Richards, and J. Waterbury. 2015. *A Political Econmoy of the Middle East*. Boulder: Westview Press.
Carnoy, M., and M. Castells. 1993. *The New Global Economy in the Information Age : Reflections on our*. Pennsylvania, USA: The Pennsylvania State University.
Caroline, M., and S. Yatzimirsky. 2014. *Megacity Slums: Social Exclusion, Space and Urban Policies in Brazil and India*. London: Imperial College Press.
Castells, M. 1997. *The Power of Identity*. Malden: Blackwell Publishers Inc.
———. 2010. *The Rise of the Network Society*. 2nd ed. West Sussex: Wiley.
———. 2015. *Networks of Outrage and Hope: Social Movements in the Internet Age*. Cambridge: Polity Press.
Castells, M., and P. Hall. 2004. *Technopoles of the World: The Making of Twenty First Century Industrial Complexes* (electronic edition). Oxon: Routledge.
Chadwick, G. 1971. *A System View of Planning: Towards a Theory of Urban and Regional Process*. Oxford: Pergamon Press Ldt, Headington Hill Hall.
Clark, D. 2003. *Urban World: Global City*. 2nd ed. London: Routledge.
Coenen, L., P. Benneworth, and B. Truffer. 2012. Toward a Spatial Perspective on Sustainability Transitions. *Research Policy* 41: 968–979.
Cornelius, W. 1975. *Politics and Migrant Poor in Mexico City*. Stanford: Stanford University Press.
Daily Star. 2004. *Lebanese Newspaper*. May 28.
De Soto, H. 2000. *The Mestry of Capital: Why Capitalism Triumphs in the Westand Fails Everywhere Else*. London: Blaack Swan.
Drakakis-Smith, D. 2000. *Third World Cities*. 2nd ed. London: Routledge.
Duffy, H. 1995. *Competitive Cities: Succeeding in the Global Economy*. London: E & FN Spon.
Dwyer, D.J. 1975. *People and Housing in Third World Cities*. London: Longman.
Ehteshami, A. 2007. *Globalization and Geopolitics in the Middle East: Old Games New Rules*. Oxon: Routledge.
ESCWA. 2003. *Survey of Economic and Social Developments in the ESCWA Region*. New York: United Naations.
Evans, P.B., and M. Timberlake. 1980. Dependence, Inequality, and the Growth of the Teritiary: A Comparative Analysis of Less Developed Countries. *American Sociological Review* 45: 531–552.
Fawaz, M. 2004. *Strategizing for Housing: An Investigation of the Production and Regulation of Low-Income Housing in the Suburbs of Beirut*. Massachusetts Institute of Technology, Department of Urban Studies and Planning, PhD.
Fiedler, J. 2014. *Urbanisation, Unlimited: A Thematic Journey*. Cham: Springer.
Forrest, R., S. Yee Koh, and B. Wissink. 2017. *Cities and the Super-Rich: Real Estate, Elite Practices, and Urban Political Economies*. New York: Palgrave Macmillan.
Frantzeskaki, N., V.C. Broto, L. Coenen, and D. Loorbach. 2017. Urban Sustainability Transitions: The Dynamics and Opportunities of Sustainablity Transitions in Cities. In *Urban Sustainability Transitions*, ed. N. Frantzeskaki, V.C. Broto, L. Coenen, and D. Loorbach, 1–19. Oxon: Routledge.
Frantzeskaki, N., M. Bach, K. Holscher, and F. Avelino. 2018a.*Co-CreatingSustainable Urban Futures: A Primer on Applying Transition Management in Cities*. Cham: Springer.
Frantzeskaki, N., M. Bach, and P. Mguni. 2018c. Understanding the Urban Context and Its Challenges. In *Co-Creating Sustainable Urban Futures: A Primer on Applying Transition Management in Cities*, ed. Niki Frantzeskaki et al. Tokyo/Heidelberg/New York/Dordrecht/London: Springer.
Friedmann, J. 2011. *Insurgencies: Essays in Planning Theory*. Oxon: Routledge.
Gause, G. 2011. Why Middle East Studies Missed the Arab Spring? *Foreign Affairs* 90 (4): 81–90.

Gharipour, M. 2016. Urban Landscape: Public Space and Environment in Cities of the Contemporary Middle East. In *Contemporary Urban Landscapes of the Middle East*, ed. Mohammad Gharipour, 1–21. Oxon: Routledge.

Gilbert, A. 2012. De Soto's the Mystery of Capital: Reflection on the Book's Public Impact. *IDPR* 34: 5–16.

Gilbert, A., and L. Gugler. 1992. *Poverty and development urbanization in the Third World*. Oxford: Oxford University Press.

Gilbert, A., and P. Ward. 1985. *Housing, the State and the Poor: Policy and Practice in Three Latin American Cities*. Cambridge, MA: Cambridge University Press.

Goulet, D. 2006. *Development Ethics at Work*. Oxon: Routledge.

Gugler, J., and W. Flanagan. 1978. *Urbanization and Social Change in West Africa*. Cambridge: Cambridge University Press.

Hall, P. 2014. *Cities of Tomorrow: An Intellectual History of Urban Planning and Design Since 1880*. 4th ed. London: Blackwell.

Hall, P., and U. Pfeiffer. 2000. *Urban Future 21: A Global Agenda for Twenty First Century Cities*. Lonodon: Routledge.

Harvey, D. 1985. *The Urbanization of Capital: Studies in the History and Theory of Capitalist Urbanization 2*. Oxford: Basil Blackwell Ltd.

———. 2016. *The Ways of the World*. London: Profile Books Ltd.

IIHS. 2016. *Slum Almanac 2015/2016: Tracking Improvement in the Lives of Slum Dwellers*. New Delhi: Indian Institute for Human Settlements.

Khirfan, Luna (ed.). 2017. *Order and Disorder: Urban Governance and the Making of Middle Eastern Cities*. Montreal: McGill-Queens University Press.

Khirfan, L. 2019. Planning and Unplanning Amman: Between Formal Planning and Non-traditional Agency. In *The Palgrave Handbook of Bottom-Up Urbanism*, ed. M. Arefi and C. Kickert, 201-218. Cham, Switzerland: Springer International Publishing.

Koonings, K., and D. Kruijt. 2009. *Megacities: The Politics of Urban Exclusion and Violence in the Global South*. London: Zed Books.

Malmvig, H. 2014. Free Us from Power: Governmentality, Counter-Conduct, and Simulation in European Democracy and Reform Promotion in the Arab World. *International Political Sociology* 8: 293–310.

Mansoob, M.S. 2002. *Globalization, Marginalization and Development*. London: Routledge.

McGee, T. 1982. Labour Mobility in Fragmented Labour Market, the Role of Circulatory Migration in Rural-Urban Relations in Asia. In *Towards a Political Economy of Urbanization in Third World Countries*, ed. H. Safa, 47–66. Delhi: Oxford University Press.

Mehta, B., and A. Dastur. 2008. *Approaches to Urban Slums: A Multimedia Sourcebook on Adaptive and Proactive Strategies*. Washington, DC: World Bank.

Nelson, J. 1979. *Access to Power: Politics and the Urban Poor in Developing Nations*. Princeton: Princeton University Press.

Nicklin, S, and B. Cornwell. 2019. *A Better World Volume 5: Actions and Commitments to the Sustainable Development Goals*. Towards the Achievement of SDG 11. London: Tudor Rose.

Parnell, S., and S. Oldfield. 2014. *The Routledge Handbook on Cities of the Global South*. Abingdon/Oxon: Routledge.

Peattie, L. 1996. Urban Research in the 1990s. In *Preparing for the Urban Future: Global Pressures and Local Forces*, ed. M. Cohen, B. Ruble, J. Tulchin, and A. Garland, 371–391. Washington, DC: The Woodrow Wilson Center Press.

Perlman. 1976. *The Myth of Marginality: Urban Poverty and Politics in Rio de Janeiro*. Stanford: California University Press.

Piketty, T. 2014. *Capital in the Twenty-First Century*. London: Th e Belknap Press of Harvard University Press.

Pittfor, A. 2019. System Change, Not Climate Change. *Internationa Politices and Society*.https://www.ips-journal.eu/regions/global/article/show/system-change-not-climate-change-3712/.

Potter, R.B., D. Darmame, N. Barhamb, and S. Nortcliff. 2009. "Ever-Growing Amman", Jordan: Urban Expansion, Social Polarisation and Contemporary Urban Planning Issues. *Habitat International* 33: 81–92.

Rakodi, C. 1997. *The Urban Challenge in Africa: Growth and Management of Its Large Cities*. Tokyo: United Nations University Press.

Richardson, L., C. Durose, and B. Perry. 2018. Coproducing Urban Governance. *Politics and Governance* 6 (1): 145–149.

Roy. 2009. The 21st-Century Metropolis: New Geographies of Theory. *Regional Studies* 43 (6): 819–830.

Roy, A., and N. Al-Sayyad. 2004. *Urban informality: Transnational perspectives from the Middle East, Latin America, and South Asia*. Lanham: Lexington Books.

Satterthwaite, D. 2014. Guiding the Goals: Empowering Local Actors. *SAIS Review of International Affairs* 34(2, Summer-Fall): 51–61.

Sims, D. 2010. *Understanding Cairo: The Logic of a City Out of Control*. Cairo: America University Press in Cairo.

Singerman, D., and P. Amar. 2006. *Cairo Cosmopolitan: Politics, Culture, and Urban Space in the New Globalized Middle East*. Cairo: The American University in Cairo Press.

Soliman, A. 1996. Legitimizing Informal Housing: Accommodating Low-Income Groups in Alexandria, Egypt. *Environment and Urbanization Journal* 8 (1): 183–193.

———. 2004a. *A Possible Way Out: Formalizing Housing Informality in Egyptian Cities*. Lanham: American University Press.

———. 2004b. Tilting at Sphinxes: Locating Urban Informality in Egyptian Cities. In *Urban Informality in an Era of Liberalization: A Transnational Perspective*, ed. Ananya Roy and Nezear Al Sayyad, 171–208. Lanham: Lexington Books.

———. 2010. Rethinking Urban Informality and the Planning Process in Egypt. *IDPR Journal* 32 (2): 119–143.

Stokes, S.C., T. Dunning, M. Nazareno, and V. Brusco. 2013. *Brokers, Voters, and Clientelism: The Puzzle of Distributive Politics*. Cambridge, MA: Cambridge University Press.

Stren, R. 2002. Local Governance and the Development of Associational Life: An Exploration. In Democratic Governance and Urban Sustainability, ed. J.S. Tulchin, D.H. Varat, and B.A. Ruble, 9–28. Washington, DC: Woodrow Wilson International Center for Scholars.

Sunam, R. 2015. The Significance of Foreign Labour Migration and Land for Poverty Reduction in Nepal. In *Poverty Reduction Policies and Practices in Developing Asia*, ed. Almas Heshmati, Esfandiar Maasoumi, and Guanghua Wan, 218–234. New Delhi: Springer International Publisher.

Taylor, P.J., R.J. Johnnson, and M.J. Watts. 1995. *Geographies of Global Change: Remapping the World in the Late Twentieth Century*. Oxford: Blackwell.

Teti, A., P. Abbott, and F. Cavatorta. 2018. *The Arab Uprisings in Egypt, Jordan and Tunisia: Social, Political and Economic Transformations*. Cham: Palgrave Macmillan.

Todaro, M.P., and S.C. Smith. 2012. *Economic Development*. 11th ed. New York: Addison-Wesley.

Tulchin, J.S., D.H. Varat, and B.A. Ruble. 2002. *Democratic Governance and Urban Sustainability*. Washington, DC: Woodrow Wilson International Center for Scholars.

UN. 2016a. *Habitat III Issue Papers: 20 – Housing*. New York: UN.

———. 2016b. *The New Urban Agenda*. New York: UN.

———. 2018. *Department of Economic and Social Affairs, Population Division, Online Edition*. Population. New York: United Nations.

———. 2020. *Department of Economic and Social Affairs, Population Division, Online Edition*. New York: United Nations Publications.

UNDP. 2003. *World Economic and Social Survey, Various Issues*. Technical Report. Nairobi: United Nations Publication.

UNESCO. 2019. *The United Nations World Water Development Report 2019: Leaving No One Behind*. World Water Assessment Programme. Paris: The United Nations Educational, Scientific and Cultural Organization.

UN-Habitat. 2003. *The Challenge of Slums: Global Report on Human Settlements*. London: Earthscan Publications Ltd.

UN-Habitat. 2006. *Sluma Upgrading Processes and Challenges in Asia-Pacific*. Paper Presented in Asia-Pacific Ministerial Conference, 13–16 December. New Delhi: Asia-Pacific Ministerial Conference.

———. 2009. *Urban World Bridging the Urban Divide, Why Cities Must Built Equally*. Technical. Nairobi: Un-Habitat.

———. 2010. *Cities and Citizens' Series Bridging the Urban Divide: Sao Paulo a Tale of Two Cities*. Urban Planning. Nairobi: Un-Habitat.

———. 2018a. *DG 11 Synthesis Report 2018: Tracking Progress Towards Inclusive, Safe, Resilient and Sustainable Cities and Human Settlements*. SDG 11 Synthesis Report High Level Political Forum 2018. Nairobi: Un-Habitat.

———. 2018b. *Prosperity for All: Enhancing the Informal Economy Through Participatory Slum Upgrading*. Nairobi: UN-Habitat.

———. 2020. *UN-Habitat COVID-19: Response Plan*. Technical. Nairobi: UN-Habitat.

United Nations. 2018. *Department of Economic and Social Affairs, Population Division*. New York: United Nations.

———. 2019. Department of Economic and Social Affairs, Population Division. *World Urbanization Prospects: The 2018 Revision (ST/ESA/SER.A/420)*. New York: United Nations

Ward, B. 1976. *The Home of Man*. Harmondsworth: Penguin Books Ltd.

Chapter 4
The Paradigm of Urban Informality: Laws, Norms, and Practices

Abstract This chapter examines the paradigm of urban informality. ("Informality" is used to describe a range of behaviors and practices unfolding within cities. It explores the concept of urban informality from a perspective of urban system, law, norms, and practice, in cities in the Global South, especially in the Middle East. It gives insights on certain issues and reasons that influence the formulation of the urban informality. This study seeks a middle ground: between urban and transitional perspectives, between the conceptual and empirical, and between the structural and the practical of the urban informality. The chapter highlights the phenomenon of informal areas and takes a quick look at the evolution of slums and how they have developed in the past few decades to understand the causes of the appearance of urban informality in the cities of the Middle East. It also traces the changes of planning thinking and theories towards urban informality and how they have been developed over time and tries to arrive at a general typology for urban informality. It explores the role of urban informality, as an informal sociotechnical regime, that created "a new pattern of urbanity" or "a site of transitions".

Keywords Urban informality · Slums areas · Diversity and complexity of urban informality · Housing informality taxonomies · Informal economy · Social network · Community-driven process · Symptom of unsustainability

4.1 Introduction

Middle East cities exist in a predominantly 'rural' region, but projections for the coming decades forecast rapid urban transition. Although only 40% of the Middle East's aggregate population is defined as urban, most of the population increase already occurs in cities, and a large proportion is living in what are called slum areas. Some of the chief cities in the region boast huge rates of population growth, but the greatest amount of urban growth is concentrated in the intermediate and small cities.

In 2003, UN-Habitat defined slums as a contiguous settlement that has one or more of the following five conditions: inadequate access to safe water; inadequate access to sanitation and other infrastructure; poor structural quality of housing;

A. M. Soliman, *Urban Informality*, https://doi.org/10.1007/978-3-030-68988-9_4

overcrowding; insecure residential status; and their related phenomena. Slums themselves are the physical manifestation of several overlapping forces (UN-Habitat 2003). On the one hand, they are the manifestation of deep poverty, unrealistic regulatory frameworks, ill-conceived policies, inadequate urban planning policy, weak institutional capacity, social exclusion, and larger macroeconomic factors. Slums are the products of dysfunctional land markets, unresponsive financial systems, and a lack of political willingness. But on the other hand, slums are a manifestation of the ingenuity and resilience with which extremely disadvantaged populations have organized themselves in the face of these very urban challenges (The World Bank 2008; Turner 1991).

Generally, "informality" refers to a category of income-generating, servicing or settlement practices that are relatively unregulated or uncontrolled by the state or formal institutions. Informality is treated both as a way to bypass state regulation and institutions, as well as a sector interacting with formal institutions and markets, with rather blurry lines separating the two systems. Urban informality is defined as any act/activity in life that does not obey the prevailing law or does not embrace the umbrella of formal institutions, and operates outside of the moral of a way of life if that is taken to mean housing informality, economic informality, and informal features of life (social, culture, political, and everyday life). See also the definition of urban informality from transitions perspective in Chap. 1). The agglomeration of urban informality, as a sociotechnical landscape, in Middle East cities began to appear in the 1950s and expand in the 1970s when there was little interest in what was at first a very marginal and not very visible phenomenon, neither from governmental nor academic perspectives. The result is a growing divergence in cities of the Middle East between the old and the new part of the urban agglomeration, a "Great Divide" between "establishing", "struggling", and "emerging" urban spaces. The first involves the new gated communities of the elites of the society, the middle, is the formal city characterized by communities that follow the prevailing laws, and the third involves the peri-urban informal settlements, or urban informality, and it formed by low-and middle-income groups. Capital cities in the Middle East and the other cities within the city, the formal city, the informal city and the old city dot "The Middle East urban fabric" (Silver 2010; Sims 2010; Al Sayyad 2006; Khirfan 2019; Elsheshtawy 2004).

Consequently, this chapter gives credit to the importance of urban informality, as a sociotechnical landscape, in shaping the modern Middle East cities. In doing this, it seeks to undertake a concurrent assessment of the key junctions of the encounters between urban informality and this strategically significant region, notably in the broad realms of socioeconomics, politics, and culture. Therefore, this chapter traces the evolution of slum areas from a historical and current perspectives and how planning thinking towards these areas has changed over time. It is important to note that various scholars (AL-Sayyad 2004; Turner 1968) have admitted that, despite these areas having been established illegally, they contributed effectively in managing the process of socioeconomic and political transitions that occurred in cities of the Global South. Despite the diversity and complexity of housing informality, such areas have managed to facilitate cities' growth, emerging urbanism, and perceptions of space that allow urban informality to function as "a new innovation of collective planning" regardless of government rules. On the other hand, accumulation of capital (De Soto 2000) in the form of housing informality and strong social networks among the citizens of the informal areas have a major influence in consolidating the urban informality in Middle East cities.

The chapter is organized as follows: the second section considers the roots of urban informality. Section 4.3 briefly examines the diversity and complexity of urban informality. Section 4.4 focuses on reflections concerning the links between capital, liberalization, and urban informality. Section 4.5 examines social networks within urban informality. The closing section presents the conclusions.

4.2 Roots of Urban Informality

In the literature, there are various thoughts concerning the transformation of slum areas and the appearance of urban informality in cities. To address this problem, it is essential to examines urban informality transition in line with the call for reconfiguring Western-inflected urban theories on urban informality, overcoming asymmetrical ignorance of non-Western cities, and recalibrating geographies of authoritative knowledge (Edensor and Jayne 2011; Roy 2009; Robinson 2006; Gaonkar 2001; Engles 1970). The dominant normative interpretation of urban informality should be critically re-examined within different cultural contexts from all over the world to broaden the understanding of its relevancy and limits (Dempsey and Jenks 2010). This gains a particular significance in the Middle Eastern context, where rapid population growth and intensive informal urbanization processes have culminated in crucial socio-cultural challenges, parallel to a shift in urban pattern from planned to an unplanned/arbitrary pattern (Al Sayyad 2004; Sims 2010; Soliman 2004a).

Slums are not a new phenomenon in modern societies. They are as older than the Industrial Revolution. Patrick Geddes in his book Cities in Evolution (Geddes 1915) stated that an essential achievement of the "Industrial (i.e., Paleotechnic) Age" is slum areas, which are spreading everywhere in the British cities. He classified inhabitants into middle-class who are living in semi-slums, while wealthy quarters are living in super-slums, and the poor are living in the poorest slums. In his influential book *The City in History*, Lewis Mumford (Mumford 1962) explored the development of urban civilizations in which slums formulated the structure of modern cities and were partially responsible for many social problems seen in Western society. This underlies Geddes and Mumford's thoughts on how to handle the urban transitions of informality in a sustainable way. This was the first time, in the literature review, for a work to look at urban transitions from a wider perspective and to improve the quality of the built environment to create a healthy urban fabric. In the Middle East, in Egypt, slums areas have existed for more than 2000 years, such as evidenced by the site named Mary's Tree in the *El Matarya* area adjacent to the city of Cairo which is said to related to the Holy Trip of the Virgin Mary to Egypt (Soliman 2013). In Lebanon, slums date back to the French Mandate period (1923–1943) (CDR 2016).

In the late 1950s to the early 1980s, many scholars (Abrams 1964; Ward 1976, 1982b; Stokes 1962; Gans 1962, 1971; Seely 1959, 1971; Turner and Fichter 1972; Turner 1976; Gilbert and Ward 1985) questioned the spread of slums areas, or what is called squatting areas, on the periphery of cities in developing countries. This phenomenon spanned the Global South with some references from developed countries (Hart 1973; Rakowski 1994b; Perlman 1976; Dwyer 1975; Ward 2004; Roy 2004; Cirolia 2017; Devlin 2018). The conclusion was that the urban poor have the potential for creating a "new mode of urbanity" (Soliman 2013) or freedom to build (Turner and Fichter 1972) or a dimension of urban reconfiguration (Ismail 2014), which is reflected everyday life (Lefebvre 1971), where people manage to house themselves through self-help housing (Turner 1976) at minimum cost to suit their needs for homes and other social and economic necessities.

The first appearance of the term "urban informality" was in the work of Juan Pablo Sainz in the late 1980s (SoJo 2001). De Soto (2000) defines informality as a "survival strategy"—a safety valve for societal tension. More formally, De Soto defined and elaborated a concept of urban informality with four fundamental characteristics. First, it is sociological in character and is to be analyzed by the interdisciplinary combination of law and economics rather than mainstream economics or sociology. Second, it focuses on economic activities and enterprises, rather than on individuals, households, or neighborhoods; it represents "a way of doing things", rather than a fixed population or territory. Third, it bridges the gap between production and reproduction, dealing with the totality of income-generating and expenditure-saving activities. Finally, it is not dualistic because it does not presuppose that the whole economy is or should be divided into two sectors. It could be said that informality exists when the means are illicit but the ends are licit and valuable to the society.

The urban poor across the Global South have created squatter settlements, or shantytowns, informal areas, arbitrary housing development, illegal residential areas, marginal areas, cancer areas, black holes, black spots, unplanned areas, or spontaneous settlements. This terminology is neither tight nor concise (Gilbert and Gugler 1982; Dwyer 1975), but each country has a local generic name for them (Abrams 1964; Ward 1982a). Many researchers (Abrams 1964; Turner 1976; Juppenlatz 1970; Mangin 1967; Sudra 1976; Payne 1974) point out that squatting is a part of a desperate search for cheap shelter and land in urban areas, with the broad generalization that they are generally illegal (Leeds 1969; Lloyd 1979). Also, unplanned areas not only for the needs for housing but are methods for capital accumulation as a way to secure the future.

Despite these urban transformations of informality and subsequent changes in lifestyle, numerous studies in the Global South report several classification and definitions with varied performances of slums for the adaptation of spatial, economic, social, and political circumstances to suit the requirements of the urban poor (McFarlane and Waibel 2012; Roy and AL-Sayyad 2004; Harris 2018; Marx and

Kelling 2019; Soliman 2010; Roy 2005). Thus, the following part examines five dimensions of slums; spatial, economic, political, social, and densifications to explore the learned lessons, transformations over time, and how they are related or unrelated to the Middle East's cities.

Firstly, from a spatial dimension point of view many scholars elaborated and subdivided slums in the context of spatial development. In classical articles, Stocks (1962), Gans (1962), and Seely (1959, 1971) make a distinction between the "squatters" and the "slum dwellers" of a city. Charles Stocks (1962) divides slums into the "slum of hope" and the "slum of despair". By hope is meant the quality of psychological responses by the inhabitants of the slum who try to improve their status. "Despair" denotes a lack of the intention (Harris 2018) for any attempt to change the status of the settlement. Herbert Gans (1962, 1971) says that there are two types of low-rent neighborhoods: the entry area and the area populated by social rejects. In the entry area, newcomers to the city try to adapt their "nonurban institutions" and cultures to the urban milieu. This is the "urban village." The second type of area is populated by social rejects and, in the legal context, this may be called the "urban jungle."

John Seely (1959, 1971) points out that the slum population is characterized by two major differences; the difference between necessity and opportunity and the difference between permanence and change. For some, the slum represents a set of opportunities and potentialities not available as easily, or at all, elsewhere. For others, the slum represents a set of necessities to which, despite their wants, they have been reduced or despaired. There is underlying harmony between the views of these three researchers, but each has a somewhat different perspective. One can divide the slum-type allocating to the degree of improvement or transition to the existing conditions and to the legality of tenure. The first type (slum of hope, urban village, and opportunities) is where people live in an illegal situation, and slums of this type are always in a dynamic process of transitions. The second type (the slum of despair, the urban jungle, and the slum of necessities) is where people live in squalid conditions, but in a legal situation and a static milieu. The first type is a "squatter" and/or informal residential development, while the other is a "slum dweller," who over time will despair. The classical distinction of slums—slum of hope, urban village, and slums of opportunities—classified slums from a spatial point of view and showed how they have the potential to support survival, transition, and modifications of their living conditions over time. Contrarily, the other form of slums offers no hope for survival, rather the likelihood of despair or demolishment over time.

The second dimension is an economic point of view. In the mid-1950s, Lewis (1954) suggested two models for understanding the employment of the new migration of people; first, "a state", a modern capitalist firm sector, and "trade-service"; and a peasant household firm sector. The former sector is where people work in a formal way and follow institutional and legalization processes. The latter is where people work illegally as the multitude of people crowding the city streets in various petty economic activities, while living in the back streets (Wikan 1980). The former is a "modern "mode of employment, while the latter is the "traditional" mode

(Bromley 1979) or 'Family Mode of Production' (Lipton 1984). With the appearance of the concept of "informal sector" in the early 1970s, Hart's paper (Hart 1973) proposed a new two-sector terminology, dividing the economy into 'informal' (an extension of the concept of 'traditional') and 'formal' (more-or-less analogous to 'modern') sectors and emphasizing the significance of self-employment and small enterprises.

Hart (1973) defined the informal sector through three key characteristics; "substantial overlap between providers of capital and providers of labor"; "prevalence of perfect, or rather... near-perfect, competition", and informal sectors consisting largely of "unorganized,'' unincorporated enterprises. The formal/informal dualism was reified in the International Labor Office (ILO) (ILO 1972) report on Kenya, which alerted international organizations to the importance of the informal economy sector to the national economy (Bromley 1979). The modern subdivision of social classes is seen from an economic point of view, whether these classes are living in squalid conditions, and how they have to serve each other. Since then, a link between the housing informality sector and informal economy sector, whatever their scale (e.g. Hart 1973; ILO 1972; Bromley 1979; Lipton 1984; Moser 1994; Guha-Khasnobis et al. 2006), became interrelated and linked with urban informality. It spreads in dualistic and transnational contexts (Ward 2004), and the growing number of the urban poor are make their way in a transitional phase (d'Alençona, et al. 2018; Bayat 2004) and are already making their own road map on the ground as "a new form of urbanity" (Soliman 2013a).

Politically, many interest groups view informality as a mass response to mindless, pompous bureaucracies and to the manipulations of the economic system. Even though it may officially be illegal, it is not immoral because it breaks no basic moral codes and is a simple necessity for the poor to make a living and satisfy their basic needs. Informality is a structure of action that contains both harmonious (adaptation) and contradictory (resistance) relationships. It is a site of power in relation to external discipline and control of power (Laguerre 1994).

Fourth, the social dimension is a crucial aspect of slum taxonomies. James Scott (1986) classified the social classes into two sectors: subordinate class (informal) and superordinate class (formal). Class resistance includes any act(s) by member(s) of a subordinate class that is or is intended either to mitigate or deny claims (e.g., rents, taxes, prestige) made on that class by superordinate classes (e.g., landlords, large farmers, the state) or to advance its own claims (e.g., work, land, charity, respect) vis-à-vis these superordinate classes. Rakowski (1994a) theorized a fundamental typology between two groups; essentially the structuralists and the legalists. The former comprised the ILO perspective and on subordinate class, while the latter included the Hernando De Soto of microenterprise perspective (De Soto 1989) and Scott's perspectives on superordinate class (Scott 1986).

Lastly, others (Cacioly-Jr 2000; Keivani 2018) classified residential areas from densification perspectives, i.e., the more compact are urban areas, then the higher the number of urban poor and the greater the possibilities of informality. The lower the compactness, the higher the welfare and the lower is the informality. The scale of densification systematically contributes to social sustainability and social justice in the Middle Eastern cities with their sociocultural backgrounds. Relevance and complexity of densification predominate over the compactness of the informality in a city. Researchers on social sustainability face several challenges, one of which is that social sustainability remains the least developed in informal areas in terms of the three pillars of sustainable development of theoretical conceptualization and practical operationalization (Denis 2006; Wheaton 1998; Joint AESOP/ACSP Congress 2013; di Milano 2013; Jenks and Burgess 2000).

On the other hand, urban informality is classified into three categories; dualist, legalist and structuralist, blending the social and the economic spheres (Chen 2006; see also Rakowski 1994a). D'Alençona et al. (2018) argued that the dualistic framework—often accompanied by an assumption of formality as the 'norm' and informality as an anomaly—has persisted (see e.g., Angotti 2013; Rodgers et al. 2012), at least in practice and policy, if not in conceptual terms (Watson 2009) (For further debate see d'Alençona et al. (2018)). For De Soto (1989), informality is the key to survival and success that ignores or deliberately breaks unreasonable official rules and regulations to make a living and to satisfy basic needs. In this view, informality occupies an intermediate position between "formality/legality," when all laws and regulations are complied with, and "criminality", when acts are performed that are clearly against official laws, basic morality, and/or the public interest. From the structuralist point of view, Banks et al. (2020) considered urban informality as "a site of critical analysis", offering a new perspective that draws on and extends political economy approaches. It relied on three trends: state-society interactions and changing state attitudes to informality; the significance of diverse actors' agency; and informality as a strategy for diverse groups. In general, the argument brings together an understanding of informality as an opportunity for accumulation of capital with concerns about inequality to ensure the benefits secured through these informal processes are distributed across diverse interest groups.

These debates can be detected in critiques of legalist approaches (Harriss-White 2019; Meagher 2016; Harris 2018; De Soto 2000), the denial of dualist representations and challenge to the separation of formal and informal spheres (Angotti 2013; Al Sayyad 2004; Sims 2010), or recognition of changing structural dynamics influencing the redefinition of the formal and informal (Mitlin and Walnycki 2019; Bromely 2004; Soliman 2019) and the reconciliation between the state and buildings' violators (Soliman 2021a).

In sum, from the previous discussion, it appears that urban informality from a transition perspective is articulated from four opinions as illustrated in Fig. 4.1. First, between the 1950s and 1960s, some scholars saw these slum areas as illegal, or cancer areas that destroyed the built environment and appealed for bulldozing them and replacing them with new residential districts. This can be considered as *'a symptom of unsustainability'*, e.g., it could not be accommodated by the planned city and created all sorts of urban challenges and negative impacts. On the other hand, urban informality is not a euphemism for 'slum' (Dovey et al. 2020), but rather it differs.

Second, in the 1970s and 1980s thinking about urban informality changed, and many scholars considered urban informality as a response to socioeconomic and

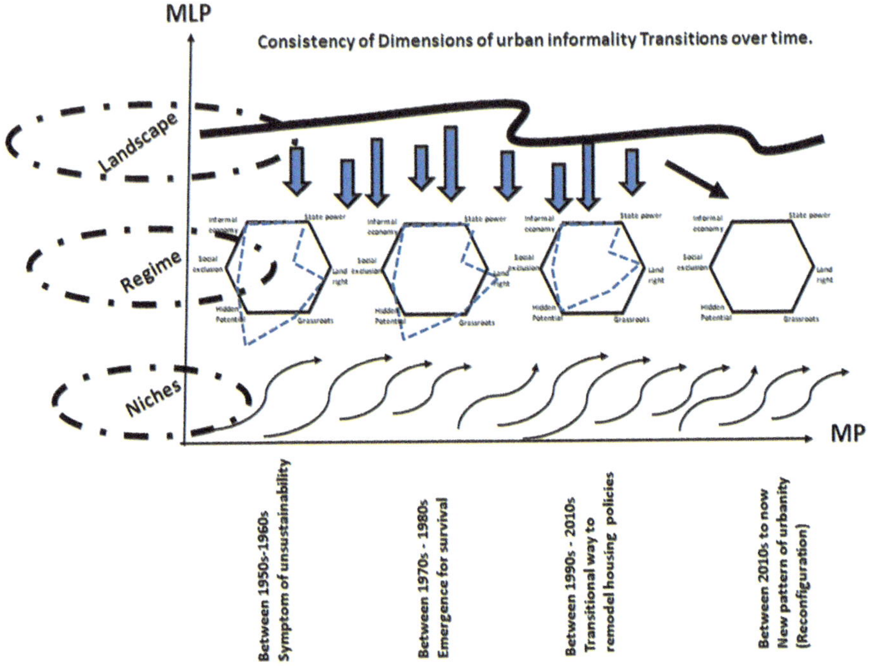

Fig. 4.1 Scheme showing the consistency of the transformation of urban informality as "a site of transitions". (Source: The author)

political transitions that occurred in the Global South and as the emergence of a survival strategy. Third, during the 1990s–2010s, new analysis within international organizations paid much attention to the potentialities of the grassroots to tackle the crisis of housing crises in the Global South and considered urban informality as "a transitional way to remodel housing policies" in the South. Fourth, currently, some scholars from the Global South consider urban informality as "a new pattern of urbanity" (Soliman 2019), "a new way of life" (AL-Sayyad 2004) or "a kind of success story" (Sims 2010). However, these opinions have evolved through various transitions stages due to pressures and tensions between the sociotechnical niche and regime structure over time, and due to numerous changes in the socioeconomic and political transformation of a given environment. The latter include the fluctuation of the power of the state, the evolution of the informal economy, the contribution of the grassroots is determined the level of the land delivery system, the acceleration of the hidden potential of the low-income groups, and the determination of the marginality. Thus, urban informality served as *"a site of transitions"* over time, and the question is how to create consistency and equilibrium among these magnitudes?

4.3 Diversity and Complexity of Urban Informality

The World Bank (2008) and MENA (2005) classified housing informality into two categories: informal and slums. Informal (unauthorized) housing is defined as the housing stock that is not in compliance with current regulations. This includes illegally occupied land (squatters), as well as housing built outside the planned areas (even if the land ownership is legal, and/or housing units constructed outside the construction regulations). Slums, on the other hand, refer to deteriorated living conditions in terms of both their social and physical dimensions (MENA 2005). As illustrated in Table 4.1, the Millennium Development Goals (MDGs) (target 7D) with respect to slum prevention differentiates housing informality into three fundamental forms: slums, informal settlements, and inadequate housing (UN-Habitat 2018b). Others classified slums areas into two categories; unplanned areas and unsafe areas. The latter has four grades contains 12 subtypes (Khalifa 2011).

Recognizing and understanding urban informality is still debatable (d'Alençona et al. 2018; d'Alençon et al. 2016; Roy and AL-Sayyad 2004; Roy 2011; Davis 2006; Sims 2010; Bromely 2004; Banks et al. 2020). This urban informality transformation attracted many scholars to examine the mechanisms of urban informality and its impacts on urban transitions within cities and how people most in need of housing adapted their living environment (Soliman 2004b). The mechanisms of urban informality formulization are well illustrated elsewhere (Soliman 2004b). On the other hand, huge studies were undertaken, but the debate was pursued from a narrow perspective regardless of the influence of the spread of urban informality on the transition of the built environment.

In the late 1960s and early 1970s, the Latin American scene constituted rich ground for studying the development of housing informality and enabled examination of the political economy of informality and the settling basins for rural migrants into the city (Castells 1975; Peattie 1974; AbuLughod 1971; Dwyer 1975; Yeh and Laquian 1979; Roy 2011). Most of the informal settlements have relatively limited urban areas with high population densities. For example, *Manshiet Nasser* in Cairo, *Dahiya Janubiyya* in Beirut, and *Marj Al Hamam* in Amman had around 6000, 4500, 6400 people per hectare, respectively (Hadjitheodorou 1981; UN-Habitat 2011;

Table 4.1 Criteria for defining slums, informal settlements, and inadequate housing

Evaluation criteria	Slums	Informal settlements	Inadequate housing
Access to water	X	X	X
Access to sanitation	X	X	X
Sufficient living area, overcrowding	X		X
Structural quality, durability, and location	X	X	X
Security of tenure	X	X	X
Affordability			X
Accessibility			X
Cultural adequacy			X

Source: UN-Habitat (2018b)

Al-Oun et al. 2010). The housing informality settlements are physically precarious and harbor a complexity of social ills (Mountjoy 1978) referred to as "creeping cancer", "urban sores", "the poorest of the poor" (Juppenlatz 1970), and "growing out of the carapace of the city". In short, the conventional view is that "squatters" experience problems of adjustment to urban "ways of life" living in a "culture of poverty" (Lewis 1966), or a "culture of transition" because the deprived and frustrated residents experience social disorganization. Moreover, a squatter settlement is seen as a breeding ground for radical political behavior. It is a red seedbed in which doctrines of social disorder, conflict, and revolution germinate (Poulantzas 1980).

The housing of low-income groups in the Global South, as well as in the Middle East (Soliman 2012a), has been analyzed in two ways: radical and liberal. Radical analysis within Marxist interpretation has tended to emphasize the production of housing and the way in which it is made available. This is closely linked to, first, the interest in maintaining a labor pool at a minimum housing cost and, second, the domination of capital over the housing process and finally the state's role within housing provision (Burgess 1982; Harms 1982). The liberals focus on housing production as the degree of involvement between community, market, and the state. Focusing on issues of government respectability, derived from this is an interest in regulatory policies that guarantee access to basic resources. It lays emphasis on the four existential features of identity, participation, opportunity, and security (Turner and Fichter 1972; Turner 1976; Payne 1974, 1977). Between the radical and liberal interpretations, there are diverse and conflicting views (labor and capital versus community and market) and amicable alliances between them (the role of the state and the role of affluent groups) (for further debate, see Gilbert and Ward 1985; Abu-Lughod 1969; Peattie 1987). One then comes to see the housing of low-income groups in the Middle East as operating not in isolation but as a part of a complex system of socioeconomic, environmental, and political institutions within which various groups compete and in which the productive, distributive, and consumption processes operate. In other words, from a transitions perspective, the two interpretations have reached consensus on the main driver force of transitions of housing production is a sociotechnical regime in the form of labor, capital, market, the role of the state and affluent groups. On the other hand, the liberal view stressed the role of community, as a sociotechnical niche, which plays an essential role facilitating housing production.

Therefore, one can consider housing as: a commodity or product or as consumption (Burgess 1982; Turner 1976; Soliman 1991); an investment, or as a need; a right or an asset; a conduit for basic services and legal status; a base for employment/working from home; an investment or security for loans; an economic stimulus; and an employment creator (Satterthwaite 2020). Housing informality is a continuity of the outcomes of activities and the functions of capital, state involvement, international organization contribution, and the community concerned. Also, the housing system constitutes the outcomes of the connection and interrelationships between various levels of sociotechnical systems in a given environment. In other words, the housing system is defined here as the outcome of capital or the scale of resources that

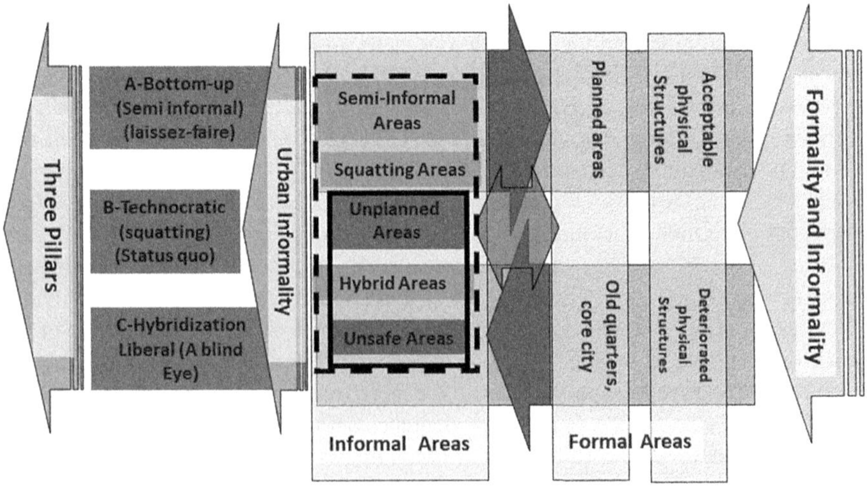

Fig. 4.2 Chart of formal and informal housing, and Taxonomies of informal housing in the Middle East. (Source: The author)

is generated by a number of various interest groups (private, public, and international donors, whatever their legal form and economic status) utilizing the main collective commodities within the market (land, labor, materials, as well as the legislative process that controls the operation of these commodities) in construction, distribution, and control of the housing production, interacting with the development strategy of the government. All of these are formulated into the sociotechnical regime, while housing informality development is a continuous process, or "a site of transition", reflecting innovations by the grassroots, in the form of self-help housing (Yap and De Wandeler 2010; Tunas and Peresthu 2010) or DIY (Do It Yourself), and it is articulated as a sociotechnical niche.

In this context, housing production in the Middle East cities has occurred in the form of a dual system: formal and housing informality. As illustrated in Fig. 4.2, informal or non-conventional housing (Drakakis- Smith 1979; Johnstone 1983) or a subordinate petty commodity (Burgess 1982) is subdivided into three categories: semi-informal, squatter housing, and hybrid or ex-formal (Soliman 2007). The latter might be viewed by scholars as a result of inner conflict (such as in Lebanon (CDR 2016)) or rapid economic transitions (as in the case of Jordan (MENA 2005)) or rapid informal and unplanned growth (as in Egypt).

The diversity, complexity, and diverse nature of housing informality development in the Middle East has resulted in a situation where various taxonomies of housing informality associated with various actors and variety of resources have been created to suit several strata of society. This diversity and complexity of housing informality has occurred on the periphery of cities in the Middle East. It indicates that global restructuring and market liberalization have led to the emergence of new forms of informal development in the last several decades that are often characterized by complex and bizarre intersections of urban and rural restructuring.

As in Hasan's Karachi (2001), informality is no longer the domain of the poor in the Middle Eastern cities but has become a primary avenue to home ownership for the lower-middle and middle classes. Soliman (2004b) ultimately attributes the diversity of housing informality in Egypt and other cities in the Middle East to the great variety of ways through which informal development has taken place and to the role of various actors who sustain it.

In this context, the NUA, coming out of the Habitat III conference in 2016, endorses governments committed to "promoting national, subnational and local housing policies that support the progressive realization of the right to adequate housing for all as a component of the right to an adequate standard of living" (UN 2016). The contravened idea of the NUA is the stress on the power of the state to introduce a strategy build in relation to the extent that actors become empowered and (re-) formulate housing system, taking up roles contributing to a sustainable development to satisfy an adequate standard of living. The main outline is to set a proper framework between a structural regime and a sociotechnical niche to produce a sustainable residential environment. Since the interconnections between the different sectors and different actors in a sustainable way are more evident and relevant in development processes, new urban fabric will be accommodated. In this way, transformation in one sector pushes changes and transformations in other sectors. Thus, the various actors in a sociotechnical regime have to contribute together, formally or informally or both, to support the realization of the right to a better urban future.

To examine all the complexity of the housing informality production, Soliman (2004a, b) created a typology of three main housing informality types: semi-informal, squatting, and hybrid or ex-formal. Each type has subtypes and variants (see Table 4.2).

The first type is sometimes known in the literature the as semi-informal settlement or illegal agricultural subdivision. The second type is often known as squatter housing. The third type involves housing built on public or private land that may have originated as formal housing, but which has been transformed and is now informal or vice versa. This last type is known as ex-formal housing. Indeed, some properties currently in this third category may have passed back and forth between formal and informal status several times. Further, ten subtypes are identified within the three main types, and a total of 15 minor variants are also identified within some of the more complex subtypes.

First, semi-housing informality is not developed through established or state-regulated procedures and does not utilize the recognized institutions of housing and housing finance. Turkey, Lebanon, Jordan, and Egypt are countries in which such phenomena have been recorded (Drakakis-Smith and Fisher 1975; Waterbury 1973; CDR 2016; MENA 2005; Soliman 2004b). Nevertheless, such housing is developed on land for which the owner has legal tenure and a formal occupation permit. Semi-informal settlements are generally found in areas of essentially rural character located on the urban fringe that is interspersed with, surrounded by, or adjacent to undeveloped sites or sites that remain in agricultural use. These settlements often develop in advance of the principal lines of urban growth. Their construction is most

Table 4.2 Taxonomies of informal areas by type and subtype

Typology	Type A Bottom-up (Semi informal) (laissez-faire) on agricultural land		State role	Type B Technocratic (squatting) (Status quo) on desert land		State role	Type C Hybridization (or ex-formal) Liberal (A blind Eye) on public or private land		State role
Subtypes	**A1** On privately owned land		**Laissez-faire**	**B1** On municipal land		**Status quo Blind eye**	**C1 Public housing**	**C1a** By municipalities	**Blind eye**
	A2 On core village land			**B2** On reclaimed land				**C1b** By cooperatives	
	A3 On government land	**A3a** Agrarian reform land		**B3 On Decree land**	**B3a** By assignment to a development company			**C1c** By public-sector companies	
		A3b *Awqaf* land			**B3b** By assignment to the public-sector company			**C1d** By development companies	
		A3c Decree land			**B3c** By assignment to cooperative			**C1e** By armed forces/police	
		A3d Waterbody/lake-edge land			**B3d** On antiquities land		**C2** Dwelling units under rent control		
					B3e On armed forces land		**C3** Dwelling units in ex-permit buildings		
					B3f On land in the public domain		**C4** Dwelling units in a historic city with the confused status		

Source: Soliman (2004b)

noticeable during periods of rapid urban expansion, and they generally appear alongside major urban roads. The state has adapted a status quo attitude towards this category. These settlements may be classified into three subtypes, built on: privately owned land, core village land, and government land. All of these have minor subtypes (see Table 4.2) that are subsequently explained.

(A) On privately owned land (A1): Such housing is generally constructed on illegal subdivisions of agricultural land. Building activity of this nature increases considerably when agricultural areas are incorporated into the municipal boundaries, or as soon as municipalities install basic services to an outlying area. On core village land (A2): The houses may be constructed on traditional, non-cadaster village land already built on before 1950. Peasants in such cases may have replaced their old houses with new ones that occupy a larger area without obtaining building permission. On government or municipalities' agricultural land (A3): Housing here occupies areas that were once legally subdivided by public agencies and that were then sold to private developers for agricultural purposes. Later, the private developers subdivided this land illegally into small plots and sold them to people looking for housing plots. Such activity may be divided into four minor variants. The state has followed laissez-faire policy towards this typology.

Second, squatters are persons who settle on public land without title (hold hand on land or *wada'yad*) or who take unauthorized possession of unoccupied premises. Squatting has meant and continues to mean different things in different cultures and at different times in each culture's existence. In other words, it is a concept that is neither absolute nor static but that evolves over time. Squatting is thus a cultural construct; more specifically, it is a political construct. These settlements may be classified into three subtypes, all of which have minor subtypes (see Table 4.2).

In the Middle East cities, squatter settlements, known as *wada'yad*, have generally developed on state desert lands, or state land, and have often been established outside the formal legal and economic structure of the city. Such settlements can be divided into three subtypes (see Table 4.2). On municipal land (B1): Housing here occupies state desert land controlled (owned) by governorates and their local units. This land is usually located where the governorate has no interest, or where it is too expensive to be publicly developed. Such areas are usually located within municipality boundaries, and land ownership may be in doubt. On reclaimed land (B2): Housing here occupies state desert land that was either sold to investors for reclamation, or was reclaimed by farmers under *wada'yad*, and was subsequently informally subdivided and built upon. The state applied laissez-faire to both previous two types. On decree land (B3): The housing may appear in such areas when ownership of land is in doubt, or when land changes ownership. This type of housing on public-owned land has six minor variants. Blind-eye policy is adapted in this category.

The third type is hybrid that includes residential units in formal areas that have temporarily or permanently acquired degrees of informality. Hybrid housing is that in which the original housing units are built within an institutional framework, but over time it converted informally by one means or another to accommodate two types: formal and informal. As a part of the economic effects of housing informality, low-income groups increase the size of the labor market (by being a supply of labor at low cost) and also increase the market for goods and services (though limited in their ability to buy them). Unlike residential units in informal areas, this type of informality relates to individual dwelling units on a case-by-case basis. In Middle East cities, some units in a formal neighborhood or even an individual building, may be "formal" but others have either been added illegally or been modified over time into illegal configurations. An example is the introduction of a temporary law of reconciliation number 17 of 2019 between the state and illegal housing violators by which the state issued an official building license for illegal buildings against certain fees to be paid to the state. Such transformations are most common in the major metropolitan areas in the Middle East and are of four subtypes (see Table 4.2). Public housing (C1): Public housing in Egypt was initiated around 1950 with the *Medinet el Umaal* (Workers' City) project in the *Imbabah* District of the Greater Cairo Region. Since the early 1960s, the Egyptian government, through a variety of programs, has produced a further supply of between 40,000 and 80,000 housing units per year. There are a number of different types of public housing. While most dates from Egypt's socialist days (1961–1970), others (which sometimes carry equally confusing tenancy) are of more recent origin. Dwelling units under rent control (C2): Vast numbers of residential units in Cairo and Alexandria (as well as in other urban areas in Egypt) and in Lebanon and Jordan are rented and remain under rental contracts that give the tenant near perpetual rights of occupation at fixed or nominal rents. The state applied blind-eye policy to both these two types.

Dwelling units in ex-permit buildings (C3): Starting in the 1970s, due to the Open-Door Policy, the practice of adding floors to existing buildings in formal areas became widespread. At the time, real estate investment, building licensing, and control of construction became quite lax. As a result, there are today a few dwelling units that, although located in formal areas, are irregular in terms of building licenses and may be subject to outstanding fines and/or demolition orders. Most of the dwelling units in these buildings are owned under simple condominium arrangements (*tamlik*), and in most cases, the owners of these units hold assets that are difficult, if not impossible, to transfer/convert—except to a gullible buyer, who then is in a similarly irregular situation.

Units in the historic city with confusing tenure status (C4): The historic cores of both Cairo and Alexandria in Egypt, in Beirut in Lebanon, and in Amman in Jordan were surveyed and cadastered in the 1900–1920 period and were thus included as part of the cities' "formal" areas. However, many of the dwelling units in these areas are subject to rent control, and a few are "ex-permit." Also, many huts have been illegally erected on top of these buildings. Such "roof-toppers" are estimated to number 700,000 and 300,000, respectively, in Greater Cairo and Alexandria. In addition, many residential buildings in cities of Cairo, Alexandria, Beirut, and Amman have disputed ownership due to inheritance or tribe problems and/or religious trust (*awqaf*) involvement, which sometimes extends back for centuries. Many of these buildings have partially or fully fallen into ruin and, although they represent prime real estate, cannot be transferred or otherwise exploited. Occupation of these areas mostly takes the form of renting or occupying public land for a certain period (*El Hekr*) and paying a symbolic rent to governmental or charitable agencies. To these categories, the state has applied a laissez-faire policy.

Several factors lie behind the emergence of the semi-informal, squatter, and hybrid types of housing in the Middle East cities. Among them are land-reform regulations, unfavorable farming conditions, and increasing demand for housing—all of which encourage agricultural landowners to subdivide their land into small lots for sale. Other important factors are the unclear tenure status of desert lands, patterns of agricultural land subdivision based on Islamic inheritance laws, and the construction of mosques within informal settlements to ensure that government agencies do not demolish illegal settlements. The fundamentalist groups in Egypt, Lebanon, and Jordan (MENA 2005) have deployed religion strategically in informal settlements through specific practices that enable them to realize their objectives. The third type of housing informality (hybrid or ex-formal) emphasizes the often-fluid status of housing in the Middle East. Thus, formal housing may be transformed into informal, and vice versa, in response to changing socioeconomic and political conditions. These factors constitute the sociotechnical regime that accelerated the transition and the formulation of the housing informality. On the other hand, a sociotechnical niche fluctuates between the actors and the regime. In general, there are various actors who influence and facilitate the transitions of housing informality development—including landowners, private developers, informal service suppliers (such as brokers and contractors), state agencies, formal-sector institutions (such as banks) and international donors (for further details see Soliman 2004b; Sims 2010). Some actors, such as institutions, make it possible for innovations to cascade and for partnerships and synergies to happen, while other state agencies open windows or lock out or relaxation for some regulation due the current circumstances. However, these changing of attitudes or policies of the state make it possible for innovations to cascade and for partnerships and synergies to happen.

4.4 Social Network and Urban Informality

The community's participation in the housing process was explored in several classic books by John Turner (Turner and Fichter 1972; Turner 1976), Hasan Fathy (1973), and Pierre Bourdieu (1977), all of which elaborated more on social organizational studies. The principle of participation can mean any process by which the users of an environment participate together to shape it (Alexander et al. 1975). Turner (1976) identified the role of social networks in sustaining the production of housing informality in the form of three laws of freedom to build. Those laws are: first, when dwellers control the major decisions and are free to make their own contribution to their own environment; both the process and the environment produced stimulate individual and social wellbeing. Second, the important thing about *housing is not what it is*, but *what it does in people's lives*. The third law is that any deficiencies and imperfections in one's housing are infinitely more tolerable if they are your responsibility rather than somebody else's. Arif Hasan (2001) argued that the capacity and capability of government institutions can never be successfully built without pressure from organized and knowledgeable groups at the grassroots. Social networks, through which stakeholders access the necessary resources or ingredients to acquire housing, result in housing networks. Understanding how social networks contribute to the economic and social fabric of life in developing countries is important (Woolcock and Narayan 2000; Collier 2002), and social cohesion is critical for societies to prosper economically and for development to be sustainable (Adler and Kwon 2002). The interdependent nature of actors (where actors are the individuals within a network) is a key distinguishing characteristic of a social network, and their structures and connectivity and the distribution of power are key components of marketing systems (Layton 2009; Jenkins 2013). Fawaz (2008) stated that, without systematically resorting to the terminology of social networks, the abundant literature that documented the formation of informal settlements during the 1970s and 1980s challenged their condemnation as 'spontaneous' by revealing that these neighborhoods were organized and managed through thick webs of social relationships (Collier 2002; Perlman 1976; Ward 1982a; De Soto 1989).

Social capital, as a reflection to a social network, is the networks of relationships among people who live and work in a particular society, enabling that society to function effectively in various forms of development (Nordstrom and Steier 2015). It is economic, cultural, or social capital in its socially recognized and legitimized form. Piketty (2014) further extended the discussion of capital to include a "patrimonial" form of capital and its creation of terrifying inequalities of wealth and income. There are symbolic, as well as material, dimensions of all three types of capital. Symbolic capital is a 'capital of honor and prestige' (Bourdieu 1977). On the other hand, social networking depends on symbolic capital, which in turn depends on publicity and appreciation; it has to do with prestige, reputation, honor, etc. Material capital is the process of capital accumulation entails investment in material capital equipment, but it also involves investment in human skills and health as well as institutional upgrading. Dimensions capital reflects the collective planning and administration of social production require that not only the means of production but also the distribution of the total product be subject to explicit social control. Thus, social control in the physical environment dominates the right of ways, pattern, and everyday urban life (Lefebvre 1971). Therefore, social networks within informal areas play a crucial role in; first, creating organizations of their own so they can negotiate with government, traders, and NGOs; second, directing assistance through community-driven programs so that they can shape their own destinies; and finally, sustaining their own command of local funds, so that they can eradicate corruption (Woolcock and Narayan 2000). The poor also rely on housing microfinance, community-based finance savings (Landman and Napier 2010), loan groups, and consumer credit for building materials (Ferguson and Smets 2010).

In the field of shaping the environment, Soliman (2019) found that informal settlements operate at least partly outside the official system through three aspects of social networks in squatters' settlements in Egypt. First, the presence of a hierarchy of use, meeting spaces and physical layout are significantly related to social, economic, and climatic considerations. Second, the pattern and the space form were created by the residents themselves without government or professional intervention. The settlers were their own architects, and they formed their settlement according to their needs and requirements. The residents constituted self-reliant communities, where people decide together how to shape their common destinies. All decisions about what to build and how to build it were in the hands of the users (Alexander et al. 1975), which represents the principle of the social networks. Third, the hierarchy of circulation systems within housing informality areas reflected the citizens' cultural and traditional ties. The result was a traditional rural pattern, similar to the citizens' area of origin, with a variety of spatial proportion relating to street widths, building facades, and a hierarchy of space with limited access. On the other hand, the government thinks the above system operates in a certain way, but in practice it operates according to its own social networks, and the government's system fails to engage and therefore fails to work. In the field of delivering goods, it is necessary to distinguish between the responsibility for 'provision', which might be the government's concern, and 'production', which might be done through the social networks. The informal land market has provided housing plots at reasonable prices for most of the population and accelerated land-delivery processes for the urban poor. Contrarily, official programs of the land-delivery system have been consistently ineffective because of their financial incoherence, and laissez-faire policies have led to the near-complete breakdown of the formal land-delivery system. This is manifested in the rapid growth of not only illegal settlements but also in the establishment of informal land markets within most of the large cities.

Scholars (Edensor and Jayne 2011) and practitioners analyzing community-level collective action have become increasingly interested in how relationships based on trust, reciprocal exchange, and social networks, i.e., social capital, affect outcomes. The idea that development plans as such could be directly 'implemented' reflects a very traditional conception of a plan as a spatial blueprint that would steadily be translated into a built form on the ground (Healey 2003). The underlying idea is to create an applicable sustainable environment at the right time and in the right place in which all the stakeholders in the housing process, through a collective process, actively cooperate to manage the process positively for their own benefit, without harming others. The relationships among the three independent systems: state, market, and social network are correlated. The latter reflects strong social ties that formulates local finance customs, as well as acts to create, direct, and sustain cultural identity.

However, the provision of property rights to residents of low-income settlements has become central as the best way is to accelerate their incorporation into the whole housing system. This paradigm has become so significant within multilateral agencies that security of tenure has become the single indicator and is confirmed as a target in the MDGs. The changes in the World Bank policy (Buckley and Kalarickal 2006) since the mid-1990s are motivated by the bank learning much about the composition of a more appropriate policy environment that entails a strong reliance on an active role of the private sector, a recognition of treating land as an important input into the provision of housing services, accounts for a large fraction of total housing costs, and emphasizes the importance of community and non-governmental organization (NGO) involvement. This paradigm has become so significant within multilateral agencies that security of tenure and legalization of land tenure have become the single indicator and confirms a target in the Habitat III protocols (UN-Habitat 2018a; De Soto 2000).

4.5 Reflections on the Linkage Between Capital, Liberalization, and Urban Informality

A fresh view of informal settlements took place in the early 1970s with respect to their built environment (Turner 1976; Turner and Fichter 1972; Mangin 1967; Mangin and Turner 1963; Davis 2006; Bolay et al. 2016; UN-Habitat 2018a) because they represent a solution to the complex problem of rapid urbanization, a housing shortage, and migration. The informal settlements reflect positive elements within the general framework of urban development (e.g., see Turner and Fichter 1972; Ward 1982a; Nelson 1969, 1970; Roy and AL-Sayyad 2004; Sims 2010). Regarding economic context, they make several contributions to national development and aid in the growth of the market. They may also denote the legitimacy of the informal, as an integral part of a unique urban system (Harris 2002; Al Sayyad 2004; Roy 2005). Urban informality is a potential pool of solutions for better urban living (Bolay et al. 2016) and for rethinking the "urban issue" by reconsidering the causes of precarity, their impact on inhabitants, and, more generally, on urban form, now and in the future. It may be said that urban informality constitutes symptoms of unsustainability, but it could upgrade to be integrated with the urban fabric in sustainable way. Back in Egypt, most of popular districts are established in the same way as the formulation of informal residential areas. They have strong social organizations (Laquian 1968; Payne 1977) that provide a degree of support in times of crisis. The structure of social relationships within informal areas was functional in that cooperation among the residents enhanced their survival (Silberstein 1969), mostly through squatter associations (Turner 1967; Martin 1969, 1974; Kendall 1976; Harrison 1979) and local community networks. In other places, squatting areas have political overtones in a very real sense: they are often political radicals, but certainly nationalist (Mangin 1967). Therefore, governments intervened by way of political parties to co-opt or control squatter populations (Ray 1969; Leeds 1972; Castells 1983) or to absorb the anger of the urban poor to alleviate the devastation of the state as happened during the Arab Spring in the Middle East (Fergany 2016). At the same time, committees of various classes were formulated everywhere in villages and Egyptian cities and became the heartbeat of Egyptian society locally rooted and flexibly organized, informal and voluntary (Jamshidi 2014).

Studies by Gilbert and Ward (1985) in three Latin American cities demonstrated that the state has developed highly effective methods of channeling and controlling participation within squatting areas, where these formal channels have served the interests of the state more than those in the communities. Latin America perceptions reveal the social structures and political struggles (such as in Chile) embodied in informal settlements. During the Arab Spring, the residents of urban informality played a great role in getting rid of the *Mubarak* and *Morsi* regimes in the 25th of January Revolution of 2011 and the 30th of June of 2013, subsequently, as well as in early 2019 when they played a great role in changing the political situation in Sudan and Algeria. Finally, the social structures help low-income groups participate in city and material life. They represent personality when dealing with the environment, and the more adaptations are made, the more appropriate the environment, so that this is one of the criteria for evaluating the environmental quality of a given place (Rapoport 1977).

In general, much of this literature approaches housing informality as the study of modernization and urbanization, adopting two contradictory approaches. The first identifies breakdown, anomie, frustration, poverty, and other similar problems among low-income groups and their offspring in the cities. Other studies emphasize the opposite characteristics: family continuity, maintenance of community and community organizations, sustain social networks, and the general ability of the urban poor to adjust to urban life (for further details, see Peattie and Aldrete-Haas 1981; Perlman 1976; Roy and AL-Sayyad 2004) and produce "a new mode of urbanity" or "a new innovation of urbanity", or "a site of transition" to overcome arbitrary urbanization processes (Soliman 2013a). Therefore, the maximization of positive elements and minimization of negative outcomes would open windows of opportunities for a sustainable development.

The first international conference of Habitat in 1976, adopted the ideas of John Turner as an attempt to recover or at least to tackle housing shortages for the growing population of the urban poor in the Global South. Relying on Turner's thoughts (Turner 1968) in his article *"Architecture that works"* for self-help housing that became the main vocabulary for housing the urban poor. This policy emphasized that the most important architectural advantage of the urban poor procedure is that the consequent adaptability of space and structure to changes can be immense. The debate in the early 1980s between Marxist, non-Marxist, and Neutral views concerning the definition of housing was remarkable in re-evaluating self-build (Ward 1982b). The Marxist view (Burgess 1982) considered housing as a commodity changing its category from use value to exchange value and vice versa. While non-Marxist (Turner 1991) thought sees it as a product and looks at low-income housing in terms of use value alone, and the importance of such housing is *"what it does"* rather than *"What it is"*, and the differences between housing as a *"verb"* and as a *"noun"*. The Neutral view has seen housing as a component of the socioeconomic and political process and that housing is a process for participating in the general socio-economic development of a given area (Perlman 1976). Thus, housing informality plays a crucial role in capital accumulation, as well as participating in hastening the economic development of a given area.

However, urban informality in a broad sense is attached to different and unusual kinds of arrangements, networks, activities, and providers, that produce "a new way of life" (Al Sayyad 2004) or "ways of life" living in a "culture of poverty" (Lewis

1966). Many scholars view housing informality, or urban informality, exclusively as a new type of a built environment (Soliman 2013a). The study of urban informality is cross-disciplinary, multidimensional, and strongly correlated with other socioeconomic, political and spatial aspects of the city and society. Thus, what is the difference between housing informality and urban informality? In a simple sense, housing informality reflects the needs for affordable shelter by the bottom strata of the society, while the latter reflects a context of various informal socioeconomic and political activities and a dual system in the political arena. Both influence the quality of the built environment. However, informality as a concept is increasingly recognized as bridging the duality between formal and informal 'sectors' (i.e., economic, spatial, etc.) and processes (i.e., 'a way of life' (Al Sayyad 2004)), *and is defined as a continuum rather than as a condition* (e.g., Jenkins 2013; Roy and AL-Sayyad 2004; Roy 2005; Marrengane and Croese 2021) and defined as the power of capital that guides urban development. It is the urban process under capitalism (Harvey 2016) which is a driver for development and urban transition, however, it is *"a site of transitions"* coping with various changes that occurred within a given environment.

However, urban informality in a broad sense contains various aspects of life, rather than a strategy (d'Alençona et al. 2018), and processes reflecting societal transitions and interacts with the prevailing transitional conditions of a given environment. The role of the state fluctuates with urban informality and varies between overt and covert support (Soliman 2012b). The state acts towards urban informality according to the situation of the sociopolitical arena, not due to the weakness or power of the state as indicated by Kreibich (2012), for example, the introduction of a reconciliation law number 17 of 2019 to issue a legal building license for buildings' violators of the informal areas in Egypt. Studies by Hernando De Soto (2000) look at urban informality as a cornerstone of the national economy because they have real estate assets that could contribute to the national development process. This controversy in housing has opened the path to change the policy toward housing the urban poor. Substantial work took place at the beginning of the new millennium when the United Nations System assigned UN-Habitat the responsibility to persuade the nations to gradually attain the "Cities without Slums" target, also known as "Target 11" in SDGs. One of the three targets of Goal 7 "Ensure Environmental Sustainability," Target 11 is: "By 2020, to have achieved a significant improvement in the lives of at least 100 million slum dwellers" (UN-Habitat 2003). In the very near future, will 100 million slum dwellers' lives have improved by 2020?

In an attempt to draw a taxonomic understanding of informality (Boanada-Fuchs and Vanessa 2018) identified seven dimensions of informality: economic, legal, technical, organizational, political, social, and cultural. Each has many sub taxonomies. This taxonomy is used to tackle 'the informal' by improving communication across sectors, including often excluded voices and various on-the-ground understandings of the same phenomena. It has diverse variables to capture and systematize thematic dimensions and associated meanings of informality, as well as its relation to formality. This taxonomy is a complicated classification to be applied in practice, as there are so many variables overlapping that have to be taken into consideration.

One of the most interesting studies is the working paper that has been produced as a collaboration between N-AERUS and Cities Alliance presented at the Habitat III conference in Quito, as part of the New Urban Agenda (NUA) development process (d'Alençon et al. 2016). Also important is the substantial book (an edited collection) covering urban informality in the Global South entitled *Urban Informality: Transnational perspectives from the Middle East, Latin America and South Asia* (Roy and AL-Sayyad 2004). This offered a very interesting dialogue between Al Sayyad and Roy defined urban informality as the manifestation of informal processes in the urban environment. It explored the main roadmap for challenges of urban informality in the Global South. However, urban informality formulation differs from one country to another, from one city to another, even within a city itself, with various complex pillars (Sims 2010; Soliman 2004b; Davis 2006; Roy and AL-Sayyad 2004).

Urban informality commonly arises in urban areas for a combination of reasons. There are three possibilities: residents are unable or unwilling to conform to regulations; the state may be unable or unwilling to enforce them (Harris 2018); the actors are unable or unwilling to obey the prevailing law in defense of their own interests. The latter is playing a major role in informality transition. In practice, a combination of these is usually in play, at which point it becomes possible to speak about a planning regime that is defined by informality. Harris (2018) defines informality as more than ubiquitous, while Pikner et al. argued (2020) informality is urban commons as a process resulted three modalities; integrating commons, informality dynamics and the constitution of state institutions. It is an inevitable context of planning, a fact of social existence. It is also a fact of social struggle in urban areas to obtain goods and services at the lowest cost as a way of capital accumulation.

Harris (2018) states that the character and importance of the sociopolitical thresholds affect how planning happens. The first threshold is the number of people involved in informal practices, their physical location, and especially their manner of social organization. A second threshold concerns the emergence of embedded social legitimacy and cooperation, a development with practical significance for those involved and, potentially, for planning practices and wider politics. A third threshold is crossed when irregular practices are organized or even led, becoming overt, an unavoidable matter of debate and policy. The final threshold occurs when informality defines the state's whole mode of governance and the form, as well as the practice, of planning. Through the crosscutting of the four thresholds, five modes of informality have emerged: latent, diffuse, embedded, overt, and dominant.

Table 4.3 illustrates three approaches to urban informality that reflect *"a site of transition"* due to different points of views subsequently explained, and which relies on other taxonomies of urban informality explained in Table 4.2. Urban informality transforms overtime and due to various circumstances, that reprise the transitions of socio-spatial, economic, and political situation of a given environment. These taxonomies are grouped into three mains approaches: a bottom-up approach (semi-informal), a technocratic approach (squatting), and a hybridization liberal approach (ex-formal). Devlin (2018) arrived at two categories "informality of needs" and the

Table 4.3 Approaches to urban informality

Informality as:	Sub taxonomies	Main scholars	Strength	Weakness
(Semi informal) Bottom-up Approach	(Desires) Best practice Tactical urbanism Law Overt, and dominant	Devlin (2018), Marx and Kelling (2019), Lydon and Garcia (2015), Silva (2016), Soliman (2004b), Roy and AL-Sayyad (2004), Harris (2018) and Lara-Hernandez et al. (2020)	It promotes new and creative ways in which urban users through informal tactics can spur and foster urban innovation by increasing the competitiveness of cities that contributes to economic development.	This approach relies on a community-driven process in which the role of the state is minimized, and capital plays a great role in the informal land subdivision process which offers housing plots on an affordable basis.
(Squatting) Technocratic approach	(Needs) Cry and demand DIY urbanism Condition Latent, and Diffuse	Devlin (2018), Marx and Kelling (2019), Finn (2014), Iveson (2013), Soliman (2004a), UN-Habitat (2003), ILO (1972), Harris (2018), Lara-Hernandez and Melis (2018), Boanada-Fuchs and Vanessa (2018) and Lara-Hernandez et al. (2020)	It defines informality on the basis of the nature of property rights, planning, infrastructure and level of services. It emerges from the practices of categorization by the state and/or planning derived by the neoliberalism.	This approach provides less usefulness to analyze social processes usually context-specific that identifies the so-called "urban informality". It evidences the blind spot of planners and designers who were trying to tackle urban issues from a top-down approach by missing the everyday activities and the organic social landscape of the streets.

(continued)

Table 4.3 (continued)

Informality as:	Sub taxonomies	Main scholars	Strength	Weakness
Hybridization Liberal approach (ex-formal)	Necessities (Status quo) Everyday authenticity Everyday urbanism Currency (urban commons) Embedded and covert	Devlin (2018), Marx and Kelling (2019), Chase et al. (2008), Soliman (2004a), Canclini (1995), Harris (2018), Boanada-Fuchs and Vanessa (2018) and Pikner et al. (2020)	More vernacular or authentic urbanism informed not by abstract plans but by everyday use and practice These artisans engage in profuse commercial activity that extends across almost all economic activities. They invest the profits from their crafts in the land, animals, housing, and internal fiestas.	The focus on process (theoretically rigorous though it may be), combined with a notion of desirable outcomes that are only roughly sketched, leaves the door open for practitioners to put these ideas to work in a way that lacks any sort of radical edge or concern for the Well-being of disadvantaged communities. The current state of affairs, especially regarding social or political issues.

Source: Adapted from Devlin (2018), Marx and Kelling (2019), Pikner et al. (2020), Lara-Hernandez and Melis (2018), Harris (2018) and Soliman (2004b)

"informality of desires." On the other hand, there is a grey area in between the two categories, namely, informality of the "status quo." Therefore, desires, needs, and the status quo constitute the main aspects of the three approaches. Marx and Kelling (2019) categorized informality in the field of urban studies differently. They identified three main approaches in how Anglophone researchers have studied the phenomenon in the "western world": informality as a condition, as a law, and as a currency. These approaches reflect informality as desires, as status quo, and as needs, respectively, each of which has sub taxonomies. The three approaches are interrelated, linked with each other and are in continuous transition over time. Each has strength and weakness parameters, but all fall under the umbrella of various characteristics and definition of urban informality that are discussed next.

The first approach covers six sub-taxonomies: desires, best practice, tactical urbanism, law, overt, and dominant. This approach relies on the concept of a bottom-up approach in which the grassroots make the main decisions towards their built environment. In fact, while other studies and exhibitions (AL-Sayyad 2004; Soliman 2013a) discuss citizen-driven, community-driven process's bottom-up approaches, there is not even agreement about what to call these practices.

It promotes new and creative ways in which urban users through informal tactics can spur and foster urban innovation by increasing the competitiveness of cities that contributes to economic development. Some scholars have settled on the term informal (Devlin 2011; Mukhija and Loukaitou-Sideris 2014; Ward 2004; Harris 2018), but there are countless other monikers: best practice, Tactical Urbanism (Lydon and Garcia 2015); as a law (Marx and Kelling 2019); and overt and dominant supports (Soliman 1991).

Soliman (2019) argues that such a definition ignores the economic and political dimension of informality, claiming that urban informality is the output of the production, the consumption and the distribution of certain activities and/or commodities in which the capital is the cornerstone of this process. Informality as a law is associated with legal pluralism, which holds that plurality of legal systems coexist, such as state law, religious law, indigenous law, customary laws, and local conventions. In such a plural realm, the law of the state is not necessarily the dominant one, but the community law is. Furthermore, the state might not be able to have the capacity to enforce the law (McAuslan 2003). There is a differentiation between law in text and law in practice, and the latter is applicable on the ground while the former is applied in courts. Thus, it requires more empirical and detailed analysis to understand the context, especially where it assumes a fluidity and coincides with instances of urban informality. Lara-Hernandez and Melis (2018) support this argument with their claim that the cultural dimension plays a decisive role in how people make use of spaces according to customary law.

The second approach covers six sub-taxonomies: needs, cry and demand, DIY urbanism, condition, latent, and diffuse. Devlin (2018) indicated informality as everyday authenticity, informality as best practice, and informality as a (Lefebvrian) cry and demand, DIY (Do-It-Yourself) Urbanism (Finn 2014; Iveson 2013), Open-Source Urbanism (Corsín Jiménez 2014; Parvin 2013), and Guerilla Urbanism (Hou 2010). Informality as a continuous and transformative condition is the most common way of thinking (Banks et al. 2020). Marx and Kelling (2019) criticized the view of the nature of property rights, planning, infrastructure and level of services by arguing that it leads to debates about precision and reliability rather than exploring the reasons about why and how they are applied. But even here, there is divergence concerning the final production of these processes. It defines informality on the basis of the nature of property rights, planning, infrastructure and level of services. It emerges from the practices of categorization by the state and/or planning derived by neoliberalism.

The third approach constitutes six sub-taxonomies: status quo (necessities), everyday authenticity, everyday urbanism (Chase et al. 2008; Silva 2016), currency, embedded, and covert. More vernacular or authentic urbanism is informed not by abstract plans but by everyday use and practice although these artisans engage in profuse commercial activity that extends across almost all economic activities. They invest the profits from their crafts in the land, animals, housing, and internal fiestas. Informality as currency (Marx and Kelling 2019) refers to contexts that might ordinarily be described as 'informal'. However, the research interest is not focused on that informality but on the activities of residents of cities that are essential to the way in which the cities work, and on an understanding of why the residents follow these activities (Guyer 2011; Dovey et al. 2020; d'Alençon et al. 2020). This way of knowing 'informality as currency' exists because it acknowledges that actors in the field consider some activities, people, or spaces as 'informal' and that this consideration has a meaningful social effect—it has a socio-political 'currency' that can be mobilized (Marx and Kelling 2019). Thus, informality as currency approaches the topic with sufficient nuance to understand the social effect of how we (as in, everyone) recognize and interpret our practices and those of others—without buying into these interpretations (Guibrunet 2017).

To sum up, urban informality as desire reflects best practice, tactical urbanism, and usually obedience to the law in certain circumstances. Urban informality as needs mirrors cry and demand and DIY urbanism and does so under certain conditions. Urban informality as status quo reproduces everyday authenticity, everyday urbanism and/or currency. The latter category combines formal and informal status, which might begin formally and over time changes its category into informality and vice versa. The middle category is carried out by the lowest strata of the society, who are considered the bottom of the pyramid of the society, while the former represents higher and middle classes who speculate in real estate to gain high profit by putting affordable land plots on the housing market.

Urban informality is not an unregulated domain but rather is structured through various forms of extra-legal, social, economic, and discursive regulation (Hamdi 2004). Second, it is a way to formulate "*a new space in the city within the city*" and "*a site of transitions*" to squeeze the growing population where various strata of the society find their way to reproduce a reasonable land delivery system to meet their social and economic needs. Third, it is becoming a matter of capital investment in land, and it has a fundamental importance in the urban poor's own rights, for it brings other developmental benefits (for instance, access to services and credit or political voice). In other words, housing informality has become an issue of "Socio-economic Struggle" for meeting the basic human needs that the states in the Middle East since the mid-1970s could not tackle for the vast majority of the population in cities of the Middle East. In this sense, housing informality has become an expression of class power, and in time, has commanded various infrastructures, social services, and legitimacy in urban areas. Fourth, people involved in housing informality have created their distinguished urban pattern that formulates "*a new urbanity*" different than the one produced by the state. Housing informality is a totally "*dynamic space*" as "*a site of transitions*" that better responds to evolving circumstances and contemporary national challenges in a wholly formal and informal private land market, and it is being integrated with the socioeconomic and political circumstances of the country. Urban informality, as "*a site of transitions*", is a continuous process of transitions, changing categories from one state to another, correlating and depending on the urban transitions of a given environment. Finally, some people have the capital and knowledge to benefit from globalization, while others are not yet incapable of being integrated.

Relying on a sociotechnical transitions as evolutionary, interpretive, and conflictual processes (Geels 2020), and reflecting urban informality from three approaches; bottom-up, technocratic, and hybridization as an echo for "*a site of transitions*", a model is formulated to link between this correlation and urban informality in the

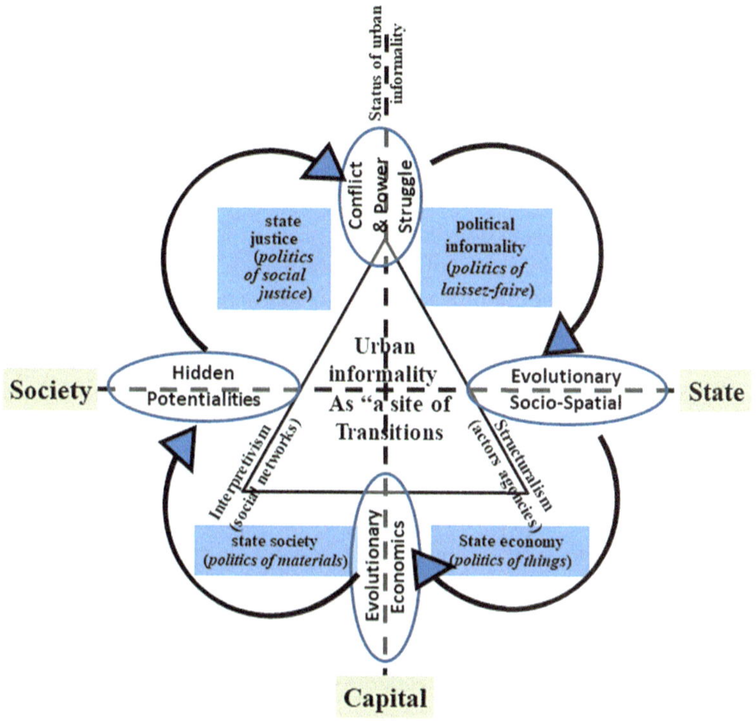

Fig. 4.3 Scheme of the correlation between urban informality and sociotechnical transition. (Source: The author)

Middle East. It is a continuous and commons process. This model tries to highlight the correlation between the three pillars and urban informality and its linkage with four variables: conflict and power struggle, evolutionary socio-spatial transitions, evolutionary economics, and hidden potentialities.

As illustrated in Fig. 4.3, the model is composed of a triangle intersects by two axes, vertical and horizontal. The triangle constitutes three essential foundations: the status of urban informality, structuralism (actors' agencies), and interpretivism (social networks). The horizontal axe reflects the correlation between state and society, while the vertical one represents conflict and power struggle and capital. The interaction between urban informality and the state results in conflict and a power struggle between both in which the state acts informally, as political informality (politics of laissez-faire), towards urban informality to gain evolutionary socio spatial. The linkage between the state and capital takes the form of the state economy (politics of things) to satisfy evolutionary economics. The relationship between capital and society reflects state society (politics of materials) to encourage hidden potentialities of the society, while the latter interacts with urban informality to create a form of justice (politics of social justice). Each of these has certain pathways that depend on the type of urban informality and its connection with other variables (uncertainty or certainty) to satisfy sustainability transitions. Also, an appropriate balance between the three pillars and urban informality will lead to the three elements of the triangle, as already mentioned.

4.6 Conclusion

Despite the enormous discussion that investigates urban informality, there is no agreement on an exact definition of urban informality. Once the critical space of slums as a breeding ground for urban informality that spreads in major urban centers, a huge study was undertaken in various parts of the South, especially in Latin American cities. However, urban informality is growing in a context of increasing poverty, rapid urbanization, and inequity and is becoming the norm (d'Alençona et al. 2018) or the new way of life (Roy and AL-Sayyad 2004) or 'ways of life' (Lewis 1954) or a new mode of urbanity (Soliman 2015a) or "a site of transition" in the Global South and in Middle Eastern cities. The overview of urban informality from a wider-angle shows that this phenomenon has contributed one way or another to tackling the shortage of urban housing, not only for the urban poor but also for the middle and above-middle classes of the society.

Urban informality *exists when the current state does not recognize who owns what, what is going on the market, the capital transition as a driving force for urban development, and the community as a political supporter for its own sake in which all actors play a significant role in the transitional process.*

Urban informality in the socioeconomic and geopolitical frames of analysis, reflecting that the realities of the transformations of those circumstances help shape the urban informality forces now bombarding the Middle East cities. By geopolitics, it understands that "nations or states and the bottom strata of the society", as sociotechnical regime and niche, are engaged in a perpetual struggle for life, the key to which is control over "spaces and places", as sociotechnical landscape, within which urban informality flourishes. Development of these "new spaces and places" within "old spaces" is subject to laws—law in text or law in practice—informal rapid urbanization, rapid population growth, and rural-urban linkages that can be derived from the study of geography and history and successfully applied to informal areas in the Global South.

The persistence of urban informality in Middle East cities, despite the adoption of various regulatory reform measures and targeted programs to curb it (e.g., formalization of informal settlements by installing basic services and lowering standards), continue to be a puzzle for housing policy in Middle East, and are due to a lack of a national urban planning framework, an absence of a transition management, and an absence of a clear land governance policy. Urban informality undoubtedly also has created a major servicing problem and additional expenses for the servicing authorities, which could have been avoided, had more rational means of organizing land for the poor been available. Urban informality as "survival architecture to create permanent facts on the ground" is a danger in terms of the destruction not only of the built environment, but also in terms of the devastation of the social structure of cities in the Middle East and might lead to the destruction of the states in the long term. No doubt, urban informality also leads to a distortion of fertile agricultural land and the urban fabric of cities that cannot be a considerable amount of corruption and favoritism towards clients. It destroys the spatial and agrarian context. Urban informality in all its forms and practices as a habitual national custom has become truly deplorable and utterly condemnable. It has become an alarmingly national phenomenon, an epidemic threatening public order, legal frameworks, and the national pride in its heritage of civil legality, governability, and meanings of urbanity. Informality has become not only "a way of life" in the Middle East's environment, but also a future "urbicide" (Shwayri 2012) for the region's survival and a great threat to urban existence.

On the other hand, urban informality constitutes "a new place within a space" that differs from the one produced by the state, dominating the prevailing urban fabric, generating capital accumulation in which opened a window for overcoming socio-economic struggle, and formulating "a new urbanity" or "a site of transition" relied on a knowledge economy adaptable to the transitions of development process and has the ability to be sustainable if a proper strategy framework applied. In sum, it represents a conflict between formal/informal regimes where, on the one hand, informality is a problem from the formal urban development regime and to control it versus approaches that support informality bottom-up or more DIY that 'fight' the formal approaches by which bring forward a synthesized/integrative an thereby transformative paradigm of "organized informality" or 'formalized informality.' This could be done. as illustrated in detail in Chap. 1 through the application of the ASUST. The ASUST calls for the integration between the three elements (a synopsis of urban informality in the Middle East, ongoing research innovation on governance of sustainable transitions and how to deal with urban informality transition in a sustainable way as the target) and the two dimensions (six access and the mode of tensions) for the determination of production/reproduction, distribution, and consumption of goods and services and the level of social networks, as a sociotechnical landscape, in which the grassroots, as a sociotechnical niche, played a great role in transitions process. This would lock out to opportunities and hidden potentialities for sustainability by understanding the diversity and nature of a sociotechnical niche and by transforming the drivers of urban informality transitions at various socio-spatial levels.

References

Abrams, C. 1964. *Man's Struggle for Shelter in an Urbanizing World*. Cambridge, MA: MIT Press.

Abu Lughod, J. 1971. *Cairo: 1001 Years of the City Victorious*. Princeton: The Princeton University Press.

Abu-Lughod, J. 1969. Varieties of Urban Experience: Contrast, Coexistence and Coalescence in Cairo. In *Middle East Cities*, ed. I.M. Lapidus, 159–187. Stanford: California University Press.

Adler, P., and S. Kwon. 2002. Social Capital Respective or a New Concept. *Academy of Management Review* 27 (1): 17–40.

Al Sayyad, N. 2004. Urban Informality as a "New" Way of Life. In A. Roy and N. Al Sayyad (eds.). *Urban Informality: Transnational Perspectives from the Middle East, Latin America, and South Asia*, 7–30. Lanham: Lexington Books.
———. 2006. Whose Cairo? In D. Singerman and P. Amar (eds.). *Cosmopolitan: Politics, Culture, and Urban Space in the New Globalized Middle East*, 539–542. Cairo: American Unversity Press in Cairo.
Alexander, C., et al. 1975. *The Oregon Experiment*. Mountain View: The Center for Environmental Structure.
Al-Oun, S., M. Al-Homoud, and J. Abdel Jawad. 2010. Fitness Between House Choices versus Market Production in Jordan. *22nd International Housing Research Conference , Istanbul, 4–7 July*.
Angotti, T. 2013. Urban Latin America: Violence, Enclaves and Struggles for Land. *Latin American Perspectives* 40: 5–20.
Banks, N., M. Lombard, and D. Mitlin. 2020. Urban Informality as a Site of Critical Analysis. *The Journal of Development Studies* 56 (2): 223–238.
Bayat, A. 2004. Globalization and the Politics of the Informals in the Global South. In A. Roy and N. AlSayyad (eds.). *Urban Informality: Transnational Perspectives from the Middle East, Latin America, and South Asia*, 79–102. Lanham: Lexington Books.
Boanada-Fuchs, A., and F.B. Vanessa. 2018. Towards a Taxonomic Understanding of Informality. *IDPR* 40 (4): 397–420.
Bolay, Jean-Claude, et al. 2016. *Learning from the Slums for the Development of Emerging Cities*. Cham: Springer.
Bourdieu, P. 1977. *Outline of Theory of Practice*. Cambridge, UK: The University of Cambridge.
Bromely, R. 1979. *The Urban Informal Sector: Critical Perspectives on Employment and Housing Policies*. Headington Hill Hall: Pergamon Press Ltd.
———. 2004. Power, Property, and Poverty: Why De Soto's Mystery of Capital Cannot Be Solved. In A. Roy and N. Al Sayyad (eds.) *Urban Informality: Transnational Perspectives from the Middle East, Latin America, and South Asia*, 271–288. Lanham: Lexington Books.
Buckley, R., and J. Kalarickal, eds. 2006. *Thirty Years of World Bank Shelter Lending What Have We Learned?* Washington, DC: World Bank, The International Bank for Reconstruction and Development.
Burgess, R. 1982. Petty Critique of John Turner's. In P. Ward (ed.). *Self-Help Housing: A Critique*, 55–97. London: Mansell Publishing Co.
Cacioly, Claudio, Jr. 2000. Can Urban Management Deliver the Sustainable City? Guided Densification in Brazil Versus Informal Compactness in Egypt. In Mike Jenks and Rod Burgess (eds.). *Compact Cities: Sustainable Urban Forms for Developing Countries*, 127–140. London: Spon Press.
Canclini, N.G. 1995. *Hybrid Cultures Strategies for Entering and Leaving Modernity*. Minneapolis: University of Minnesota Press.
Castells, M. 1975. Advanced Capitalism, Collective Consumption, and Urban Contradictions: New Sources of Inequality and New Models for Change. In L. Lindberg, R. Alford, C. Crouch, and C. Offe (eds.). *Stress and Contradiction in Modern Capitalism*. Cambridge: Lexington Books.
———. 1983. *The City and Grassroots: A Cross-Cultural Theory of Urban Social Movements*. London: Edward Arnold (Publishers) Ltd.
CDR. 2016. *Habitat III Lebanese National Report*. Final Report. Grand Serail/Beirut/Lebanon: Council for Development and Reconstruction.
Chase, J.L., M. Crawford, and J. Kaliski, eds. 2008. *Everyday Urbanism*. New York: Monacelli Press.
Chen, M. 2006. Rethinking the Informal Economy: Linkages with the Formal Economy and the Formal Regulatory Environment. In R. Kanbur, E. Ostrom, and B. Guba-Khasnobis (eds.). *Linking the Formal and Informal Economy: Concepts and Policies*, 75–92. Oxford: Oxford University Press.
Cirolia, L.R. 2017. Overcoming the Disjunctures: Competing Discourses on Informal Settlements in South Africa. *IDPR* 39 (4): 443–459.

Collier, P. 2002. Social Capital and Poverty: A Microeconomic Perspective. In T. Van Bastelaer and C. Grootaert (eds.). *The Role of Social Capital in Development: An Empirical Assessment*. Cambridge: Cambridge University Press.

Corsín Jiménez, A. 2014. The Right to Infrastructure: A Prototype for Open Source Urbanism. *Environment and Planning D: Society & Space* 32: 342–362.

d'Alençon, P.A., L. Moretto, and H. Smith. 2016. *N-AERUS – Recommendations for the New Urban Agenda*. Working Paper Group, Cities Alliance.

d'Alençon, P.A., H. Smith, E., A. de Andrés, C. Cabrera, J. Fokdal, M. Lombard, A. Mazzolini, E. Michelutti, and Morett. 2020. Interrogating informality: Conceptualisations, practices and policies in the light of the New Urban Agenda. *Habitat International*. https://doi.org/10.1016/j.habitatint.2018.04.007

d'Alençona, P.A., H. Smith, and E.A. de Andres. 2018. Interrogating Informality: Conceptualisations, Practices and Policies in the Light of the New Urban Agenda. *Habitat International* 75: 59–66.

Davis, M. 2006. *Planet of Slums*. London: Verso.

De Soto, H. 1989. *The Other Path: The Invisible Revolution in the Third World*. London: I.B. Taurus.

———. 2000. *The Mestry of Capital: Why Capitalism Triumphs in the Westand Fails Everywhere Else*. London: Blaack Swan.

Dempsey, N., and M. Jenks. 2010. The Future of the Compact City. *Built Environment* 36 (1): 5–8.

Denis, E. 2006. Cairo as Neo-Liberal Capital? From Walled City to Gated Communities. In D. Singerman and P. Amar (eds.). *Cairo Cosmopolitan: Politices, Culture, and Urban Space in the New Globalized Middle East*, 47–71. Cairo: The American University in Cairo Press.

Devlin, R.T. 2011. 'An Area That Governs Itself': Informality, Uncertainty and the Management of Street Vending in New York City. *Planning Theory* 10 (1): 53–65.

———. 2018. Asking 'Third World questions' of First World informality: Using Southern Theory to Parse Needs from Desires in an Analysis of Informal Urbanism of the Global North. *Planning Theory* 17 (4): 568–587.

di Milano, Politecnico. 2013. Cities to be Tamed? Standards and Alternatives in the transformation of the urban South. *The Journal of Urbanism* 1(26). *Cities to be Tamed?* Milan.

Dovey, K., et al. 2020. Viewpoint Informal Settlement Is Not a Euphemism for 'Slum': What's at Stake Beyond the Language? *IDPR*. https://doi.org/10.3828/idpr.2020.14.

Drakakis-Smith, D. 1979. Low-Cost Housing Provision in the Third World: Some Theoretical and Practical Alternatives. In *Housing in Third World Countries*, ed. H.S. Murison and J.P. Lea, 14–22. London: The Macmillan Press Ltd.

Drakakis-Smith, D., and W.B. Fisher. 1975. *Housing in Ankara, Occasional Paper Number 7 (New Series)*. Durham: Department of Geography, University of Durham.

Dwyer, D.J. 1975. *People and Housing in Third World Cities*. London: Longman.

Edensor, E., and M. Jayne, eds. 2011. *Urban Theory Beyond the West: A World of Cities*. London: Taylor & Francis.

Elsheshtawy, Y. (ed.). 2004. *Planning Middle Eastern cities : An Urban Kaleidoscope in a Globalizing world*. London: Routledge

Engles, F. 1970. *The Housing Question*. Moscow: Progress publishers.

Fathy, H. 1973. *Architecture for the Poor: An Experiment in Rural Egypt*. Chicago: The University of Chicago Press.

Fawaz, M. 2008. An Unusual Clique of City-Makers: Social Networks in the Production of a Neighborhood in Beirut (1950–75). *International Journal of Urban and Regional Research* 32 (3): 565–585.

Fergany, N. 2016. *Arab Revolution in the 21st Century? Lessons from Egypt and Tunisia*. Hampshire: Palgrave Macmillan.

Ferguson, B., and P. Smets. 2010. Finance for Incremental Housing; Current Status and Prospects for Expansion. *Habitat International* 34: 288–298.

Finn, D. 2014. DIY Urbanism: Implications for Cities. *Journal of Urbanism* 7 (4): 381–398.

Gans, H. 1962. *The Urban Villagers; Group and Class in the Life of Italian-Americans*. New York: Free Press of Glencoe.

———. 1971. The West End: An Urban Village. In *Internal Structure of the City*, ed. L.S. Bourne, 300–308. Oxford: Oxford University Press.

Gaonkar, D.P. 2001. *Alternative Modernities*. Durham: Duke University Press.

Geddes, P. 1915. *An Introduction of the Town Planning Movement and to the Study for Civics*. London: Williams & Norgate.

Geels, F. 2020. Micro-Foundations of the Multi-Level Perspective on Socio-Technical Transitions: Developing a Multi-Dimensional Model of Agency Through Crossovers Between Social Constructivism, Evolutionary Economics and Neo Institutional Theory. *Technological Forecasting and Social Change* 152. https://doi.org/10.1016/j.techfore.2019.119894.

Gilbert, A., and J. Gugler. 1982. *Cities, Poverty and Development: Urbanisation in the Third World*. Oxford: Oxford University Press.

Gilbert, A., and P. Ward. 1985. *Housing, the State and the Poor: Policy and Practice in Three Latin American Cities*. Cambridge, MA: Cambridge University Press.

Guha-Khasnobis, B., R. Kanbur, and E. Ostrom. 2006. Beyond Formality and Informality. In Ravi Kanbur, Elinor Ostrom, and Basudeb Guha-Khasnobis (eds.). *Linking the Formal and Informal Economy: Concepts and Policies*, 1–18. Oxford: Oxford University Press.

Guibrunet, L. 2017. *The Contribution of the Informal Economy to Urban Sustainability: Case Study of Waste Management in Tepito, Mexico City*. PhD thesis. London: University College London.

Guyer, J.I. 2011. Describing Urban 'No Man's Land' in Africa. *Africa* 81 (3): 474–492.

Hadjitheodorou, N.H. 1981. An Institutional Framework for the Upgrading Process of Manshiet Nasser in Cairo, Egypt. *Ekistics* 48 (286): 71–78.

Hamdi, N. 2004. *Small Change: About the Art of Practice and Limits of Planning in Cities*. London: Earthscan.

Harms, H. 1982. Historical Perspectives on the Practice and Politics of Self-Help Housing. In P. Ward (ed.). *Self-Help Housing: A Critique*, 17–53. London: Mansell.

Harris, R. 2002. A Double Irony: The Originality and Influence of John F.C. Turner. *Habitat International*: 48–56.

———. 2018. Modes of Informal Urban Development: A Global Phenomenon. *Journal of Planning Literature* 33 (3): 267–286.

Harrison, P. 1979. *Inside the Third World*. Harmondsworth: Penguin Books.

Harriss-White, B. 2019. Waste, Social Order, and Physical Disorder in Small-Town India. *The Journal of Development Studies*. https://doi.org/10.1080/00220388.2019.1577386.

Hart, K. 1973. Informal Income Opportunities and Urban Employment in Ghana. *Journal of Modern African Studies* 11 (1): 61–89.

Harvey, D. 2016. *The Ways of the World*. London: Profile Books Ltd.

Hasan, A. 2001. *Working with Communities*. Karachi: City Press.

Healey, P. 2003. Collaborative Planning in Perspective. *Planning Theory* 2 (2): 101–123.

Hou, J. 2010. (Not) Your Everyday Public Space. In J. Hou (ed.). *Insurgent Public Space: Guerilla Urbanism and the Remaking of Contemporary Cities*, 1–44. New York: Routledge.

ILO. 1972. *Employment, Income and Inequality: A Strategy for Increasing Productivity*. Geneva: International Labor Office.

Ismail, S. 2014. The Politics of the Urban Everyday in Cairo: Infrastructures of Oppositional Action. In S. Oldfield and S. Parnell (eds.). *The Routledge Handbook on Cities of the Global South*, 269–280. London: Routledge.

Iveson, K. 2013. Cities Within the City: Do-It-Yourself Urbanism and the Right to the City. *International Journal of Urban and Regional Research* 37 (3): 941–956.

Jamshidi, M. 2014. *The Future of the Arab Spring Civic Entrepreneurship in Politics, Art, and Technology Startups*. Oxford: Elsevier Inc.

Jenkins, P. 2013. *Urbanization, Urbanism and Urbanity in an African City: Home Spaces and House Cultures*. New York: Palgrave Macmillan.

Jenks, M., and R. Burgess. 2000. *Compact Cities: Sustainable Urban Forms for Developing Countries*. London: Spon Press.

Johnstone, M. 1983. Housing Policy and the Urban Poor in Peninsular Malaysia. *Third World Planning Review* 5: 249–272.

Joint AESOP/ACSP Congress. 2013. Planning for Resilient Cities and Regions. In *AESOP/ACSP 5th Joint Congress 2013*. Dublin: University College Dublin.

Juppenlatz, M. 1970. *Cities in Transformation: The Urban Squatter Problem of the Developing World*. Queensland: University of Queensland Press.

Keivani, R. 2018. *The Concept of Densification in the Middle Eastern Cities*. Proposed reasearch. Oxford: Oxford Brookes University.

Kendall, S.H. 1976. The Barangay as a Community in the Philippines. *Ekistics* 42 (242): 16–20.

Khalifa, M.A. 2011. Redefining Slums in Egypt: Unplanned Versus Unsafe Areas. *Habitat International* 35: 40–49.

Khirfan, L. 2019. Planning and Unplanning Amman: Between Formal Planning and Non-traditional Agency, In *The Palgrave Handbook of Bottom-Up Urbanism*, ed. M. Arefi and C. Kickert, 201–218. Cham, Switzerland: Springer International Publishing.

Kreibich, V. 2012. The Mode of Informal Urbanization: Reconciling Social and statutory Regulation in Urban Land Management. In M. Waibel and C. McFarlane (eds.). *Urban Informalities: Reflections on the Formal and Informal*. Surrey: Ashgate.

Laguerre, M.S. 1994. *The Informal City*. New York: St. Martin's Press, Inc.

Landman, K., and M. Napier. 2010. Waiting for a House or Building Your Own? Reconsidering State Provision, Aided and Unaided Self-Help in South Africa. *Habitat International* 34: 299–305.

Laquian, A.A. 1968. *Slums Are for People: The Barrio Magsaysay Pilot Project in the Philippine Urban Community Development*. Manila: Bustamente Press.

Lara-Hernandez, J.A., and A. Melis. 2018. Understanding the Temporary Appropriation in Relationship to Social Sustainability. *Sustainable Cities and Societies*: 1–14.

Lara-Hernandez, et al. 2020. Temporary Appropriation and Urban Informality: Exploring the Subtle Distinction. *Cities* 99. https://doi.org/10.1016/j.cities.2020.102626.

Layton, R.A. 2009. On Economic Growth, Marketing Systems, and the Quality of Life. *Journal of Macro Mark* 29 (4): 349–362.

Leeds, A. 1969. The Significant Variables Determining the Character of squatter Settlements. *America Latina* 12: 44–86.

———.. 1972. *Forms of Squatment Political Organisation: The Politics of Control in Brazil*. M. A. thesis. Austin: University of Texas.

Lefebvre, H. 1971. *Everyday Life in the Modern World*. New York: Harper & Row Publishers.

Lewis, W. Arthur. 1954. Economic Development with Unlimited Supplies of Labor. *Manchester School of Economics and Social Studies* 22 (2): 139–191.

———. 1966. Culture of Poverty. *Scientific American* 215 (4): 19–25.

Lipton, M. 1984. Family, Fungibility and Formality: Rural Advantages of Informal Non-Farm Enterprise versus the Urban-Formal State. In S. Amin (ed.). *Human Resources, Employment and Development, vol. 5: Developing Countries*, 189–242. London: MacMillan, for International Economic Association.

Lloyd, P. 1979. *Slums of Hope? Shanty Towns of the Third World*. Manchester: Manchester University Press.

Lydon, M., and A. Garcia. 2015. *Tactical Urbanism: Short-Term Action for Long-Term Change*. Washington, DC: Island Press.

Mangin. 1967. Latin American Squatter Settlements. A Problem and a Solution. *Latin American Research Review* 2: 65–98.

Mangin, W., and J.F.C. Turner. 1963. Urbansation Case History in Peru. *Architectural Design* 339: 366–374.

Marrengane, N., and S. Croese, eds. 2021. *Reframing the Urban Challenge in Africa: Knowledge Co-production from the South*. Oxon: Routledge.

Martin, R.J. 1969. The Ecology of Squatter Settlement. *Architecture Review* 145 (865): 213–214.

———. 1974. The Architecture of Under-Development, or the Route to Self-Determination in Design. *Architectural Design* 44: 626–643.

Marx, Colin, and Emily Kelling. 2019. Knowing Urban Informalities. *Urban Studies* 56 (3): 494–509.

McAuslan, P. 2003. *Bringing the Law Back In: Essays in Land, Law and Development*. Aldershot: Ashgate.

McFarlane, C., and M. Waibel, eds. 2012. *Urban Informalities: Reflections on the Formal and Informal*. London: Ashgate.

Meagher, K. 2016. Taxing Times: Taxation, Divided Societies and the Informal Economy in Northern Nigeria. *The Journal of Development Studies*. https://doi.org/10.1080/00220388.2016.1262026.

MENA. 2005. *The Macroeconomic and Sectoral Performance of Housing Supply Policies in Selected MENA Countries: A Comparative Analysis*. Dutch Trust Fund for Low Income Housing. Beirout: MENA Region.

Mitlin, D., and A. Walnycki. 2019. Informality as Experimentation: Water Utilities' Strategies for Cost Recovery and Their Consequences for Universal Access. *The Journal of Development Studies*.https://doi.org/10.1080/00220388.2019.157738.

Moser, Carline. 1994. The Informal Sector Debate, Part 1: 1970–1983. In C. Rakowski (ed.). *Contrapunto: The Informal Sector Debate in Latin America*, 11–29. Albany: State University of New York Press.

Mountjoy, A. 1978. *The Third World: Problems and Perspectives*. London: Macmillan.

Mukhija, V., and A. Loukaitou-Sideris. 2014. Introduction. In V. Mukhija and A. Loukaitou-Sideris (eds.). *The Informal American City: Beyond Taco Trucks and Day Labor*, 1–17. Cambridge: MIT Press.

Mumford, Lewis. 1962. *The City in History*. London: Faber.

Nelson, J. 1969. *Migrants, Urban Poverty and Instability in Developing Nations*. Technical. Cambridge, MA: Harvard University, Occasional Papers in International Affairs, No 22.

———. 1970. The Urban Poorp Disruption or Political Integration in Third World Cities? *World Politics* 22 (3): 393–414.

Nordstrom, O.A., and L. Steier. 2015. Social Capital: A Review of its Dimensions and Promise for Future Family Enterprise Research. *International Journal of Entrepreneurial Behavior and Research* 21 (6): 801–813.

Parvin, A. 2013. Architecture (and the Other 99%): Open-Source Architecture and Design Commons. *Architectural Design* 83 (6): 90–95.

Payne, G. 1974. Functions of Informality: Squatter Settlements in Delhi. *Ekistics* 34 (224): 63–66.

———. 1977. *Urban Housing in the Third World*. London: Leonard Hill.

Peattie, L. 1974. The Concept of Marginality as Applied to Squatter Settlements. In *Latin American Urban Research* 4: 101–109, ed. W. Cornelius and F. Trueblood.

———. 1987. Shelter Development and the Poor. In *Shelter, Settlement and Development*, ed. L. Rodwin, 263–280. Boston: Allen & Unwin.

Peattie, L., and J. Aldrete-Haas. 1981. Marginal Settlements in Developing Countries: Researchp Advocacy Of Policy and Evolution of Programs. *Annual Review of Sociology* 7: 157–175.

Perlman. 1976. *The Myth of Marginality: Urban Poverty and Politics in Rio de Janeiro*. Stanford: California University Press.

Piketty, T. 2014. *Capital in the Twenty-First Century*. London: Th e Belknap Press of Harvard University Press.

Pikner, T., Krista Willman, Ari Jokinen. 2020. Urban Commoning as a Vehicle Between Government Institutions and Informality: Collective Gardening Practices in Tampere and Narva. *International Journal of Urban and Regional Research* 44 (4): 711–729.

Poulantzas, N. 1980. *State, Power Socialism*. London: Verso.

Rakowski. 1994a. Introduction: Land and Housing Development in Middle Eastern and North Africa Cities. *International Journal of Urban and regional Research* 12 (1): 1–7.

Rakowski, C.A., ed. 1994b. *Contrapunto: The Informal Sector Debate in Latin America*. Albany: Sunny Press.

Rapoport, A. 1977. *Human Aspects of Urban Form: Towards a Man – Environment Approach to Urban Form and Design*. Oxford: Pergamon Press Ltd., Headington Hill Hall.

Ray, T. 1969. *The Politics of the Barrios of Venezuala*. Berkeley: University of California Press.

Robinson, J. 2006. *Ordinary Cities: Between Modernity and Development*. London: Routledge.
Rodgers, D., J. Beall, and R. Kanbur. 2012. Re-thinking the Latin American City. In J. Beall, R. Kanbur, and D. Rodgers (eds.). *Latin American Urban Development into the 21st Century: Towards a Renewed Perspective on the City*, 3–36. Basingstoke: Palgrave Macmillan.
Roy. 2004. The Gentleman's City: Urban Informality in the Calcutta of New Communism. In A. Roy and N. Al Sayyad (eds.). *Urban informality: Transnational Perspectives from the Middle East, Latin America, and South Asia*, 147–170. Lanham: Lexington Books.
———. 2005. Urban Informality: Toward an Epistemology of Planning. *Journal of the American Planning Associarion* 71 (2): 147–158.
———. 2009. The 21st-Century Metropolis: New Geographies of Theory. *Regional Studies* 43 (6): 819–830.
———. 2011. Slumdog Cities: Rethinking Subaltern Urbanism. *International Journal of Urban and Regional Research* 35 (2): 223–238.
Roy, A., and N. Al-Sayyad. 2004. *Urban informality: Transnational perspectives from the Middle East, Latin America, and South Asia*. Lanham: Lexington Books.
Satterthwaite, D. 2020. Editorial: Getting Housing Back onto the Development Agenda: The Many Roles of Housing and the Many Services It Should Provide Its Inhabitants. *Environment and Urbanization* 32 (1): 3–18. https://doi.org/10.1177/0956247820905212.
Scott, J. 1986. Everyday Form of Peasant Resistance. *Journal of Peasant Studies* 13 (2): 5–35.
Seely, J. 1959. The Slum: Its Nature, Use and Users. *Journal of the American Institute of Planners* 25 (19): 7–11.
———. 1971. The Slum: Its Nature, Use and Users. In *Internal Structure of the City*, ed. L.S. Bourne, 464–474. Oxford: Oxford University Pres.
Shwayri, S.T. 2012. Modern Warfare and the Theorization of the Middle Eastern City. In T. Ednesor and M. Jayne (eds.). *Urban Theory Beyond the West*, 259–270. London: Routledge.
Silberstein, P. 1969. Favela Living: A Personal Solution to Larger Problems. *America Latina* 12 (3): 183–200.
Silva, P. 2016. Tactical Urbanism: Towards an Evolutionary Cities' Approach? *Environment and Planning B: Planning & Design* 43 (6): 1040–1051.
Silver, H. 2010. Divided Cities in the Middle East. *Community and Development* 9: 357–345.
Sims, D. 2010. *Understanding Cairo: The Logic of a City Out of Control*. Cairo: America University Press in Cairo.
SoJo, C. 2001. *Social Development in Latin America: Issues for Public Policy*. Technical report. Washington, DC: World Bank.
Soliman, A.. 1991. *Approaches to Urban Low-Income Housing: Between Theoriticality and Practicality*. The Third International Congress of the Major Metropolises, October 14–19, 1990. Melbourne: The Proceedings of Metropolis'90, 123–131.
———. 2004a. *A Possible Way Out: Formalizing Housing Informality in Egyptian Cities*. Lanham: American University Press.
———. 2004b. Tilting at Sphinxes: Locating Urban Informality in Egyptian Cities. In A. Roy and N. AlSayyad (eds.). *Urban Informality in an Era of Liberalization: A Transnational Perspective*, 171–208. Lanham: Lexington Books.
———. 2007. *Urban Informality in Egyptian Cities: Coping with Diversity*. Fourth Urban Research symposium 2007, Urban Land Use and Land Markets, May. Washington, DC: World Bank.
———. 2010. Rethinking Urban Informality and the Planning Process in Egypt. *IDPR Journal* 32 (2): 119–143.
———. 2012a. Housing and the State in the Middle East. In Susan J. Smith (ed.). *International Encyclopedia of Housing and Home*, 346–354. Amsterdam: Elsevier.
———. 2012b. The Egyptian episode of self-build housing. *Habitat International* 36 (2):226–236
———. 2013a. *Living With/in Ashwaiyyat in Egypt: Is It a New Innovation of Urbanity?* The Flexible City, International Symposium. Oxford: Oxford University, 24–25 October.
———. 2015a. Collective Planning Process: A Driving Seat for Formalising Urban Informality in Egypt. *International Journal of Urban Sustainable Development*, June 1–24.

———. 2019. So It's Always a Chance: Community-Led Solutions to New Urban Expansion. In S. Attia et al. (eds.). *New Cities and Community Extensions in Egypt and the Middle East*, 159–179. Cham: Springer International Publisher
———. 2021a. Informal Cairo: The Making of Urban Fabric. In N. Al Sayyad (ed.). *The Routledge Handbook on Cairo*. New York: Routledge.
Stokes, C. 1962. A Theory of Slums. *Land Economic* 38 (3): 187–197.
Sudra, T. 1976. *Low Income Housing System in Mexico City*. Un-published Ph.D. Dissertation, Cambridge, MA: Massachusetts Institute of Technology.
The World Bank. 2008. *Approaches to Urban Slums*. Washington, DC: The International Bank for Reconstruction and Development/The World Bank.
Tunas, D., and A. Peresthu. 2010. The Self-Help Housing in Indonesia: The Only Option for the Poor? *Habitat International* 34: 315–322.
Turner, J. 1967. Barriers and Channels for Housing Development in Modernising Countries. *Journal of the American Institute of Planners* 33: 167–181.
———. 1968. The Squatter Settlement: Architecture That Works. *Architectural Design* 38: 355–360.
———. 1976. *Housing by People*. London: Marion Boyars Publishers Ltd.
———. 1991. Releasing the Potentials of Self-Help Housing. In D. Ingemann (ed.). *Metropolis in Ascendancy: Housing and Population*, 31–37. Melbourne: Housing and Construction Victoria.
Turner, J., and R. Fichter. 1972. *Freedom to Build: Dwellers Control of the Housing process*. New York: Macmillan.
UN. 2016. *The New Urban Agenda*. New York: UN.
UN-Habitat. 2003. *The challenge of slums: Global report on human settlements 2003*. London: Earthscan Publications Ltd.
———. 2011. *Lebanon Urban profile: Adisk Review Report*. Nariobi: UN-HABITAT.
———. 2018a. *DG 11 Synthesis Report 2018: Tracking Progress Towards Inclusive, Safe, Resilient and Sustainable Cities and Human Settlements*. SDG 11 Synthesis Report High Level Political Forum 2018. Nairobi: Un-Habitat.
———. 2018b. *Prosperity for All: Enhancing the Informal Economy Through Participatory Slum Upgrading*. Nairobi: UN-Habitat.
Ward, B. 1976. *The Home of Man*. Harmondsworth: Penguin Books Ltd.
———. 1982a. Informal Housing: Conventional Wisdoms Reap-Praised. *Built Enviroment* 8 (2): 84–94.
———. 1982b. *Self-Help Housing: A Critique*. London: Alexandrine Press, Mansell Publishing Co.
Ward, P. 2004. Informality of Housing Production at the Urban-Rural Interface: The "Not-so Strange Case" of the Texas Colonias. In A. Roy and N. Al Sayyad (eds.). *Urban Informality: Transnational Perspectives from the Middle East, Latin America, and South Asia*, 243–270. Lanham: Lexington Books.
Waterbury, J. 1973. *Cairo: Third World Metropolis, Part three, Housing and Shelter*. Cairo: American Universities Field Staff Papers, Northeast Africa Series, 189, No. 8 (September).
Watson, V. 2009. Seeing from the South: Refocusing Urban Planning on the Globe's Central Urban Issues. *Urban Studies* 46 (11): 2259–2275.
Wheaton, W.C. 1998. Land Use and Density in Cities with Congestion. *Journal of Urban Economics* 2 (43): 558–572.
Wikan, U. 1980. *Life Among the Poor in Cairo*. London: Tavistock Publications Ltd.
Woolcock, M., and D. Narayan. 2000. Social Capital: Implications for Development Theory, Research, and Policy. *World Bank Research Observer* 15(2): (World Bank Res) 225–249.
Yap, K., and K. De Wandeler. 2010. Self-Help Housing in Bangkok. *Habitat International* 34: 332–341.
Yeh, S., and L.L. Laquian. 1979. *Housing Asia's Millions: Problems, Policies and Prospects for Low-Cost Housing in Southeast Asia*. Ottawa: International Development and Research Centre.

Chapter 5
State Power, Society, Economy, and Urban Informality

Abstract Urban informality, as *a new urbanity transition*, or *"a site of transitions"* formulates the urban fabric of cities in the Middle East. A clear understanding of the state's role in civil society and economy is essential to understand the production, reproduction, distribution, and consumption of urban informality in Middle East cities. This chapter argues that the shifting liaison of the state and civil society has a great impact on the transformation process of urban informality which has altered the urban fabric in the Middle East cities. The relationship between the state, society, and capital is explored to situate the formulation of urban informality. The concept of the state and its role is explored to identify how housing production is related or not related to the political and socioeconomic transitions. It concludes that the transitions of the political and socioeconomic milieu in a given environment have major consequences on urban informality transitions.

Keywords Role of the State · Liberalism · Neoliberalism · Capitalism · Production · Consumption · Distribution

5.1 Introduction

The emergence of urban informality in most Middle Eastern cities was the main output of the tremendous transformation, either formally or informally, of socioeconomic, political, military, and spatial spaces. This acceleration of transitions in the Middle East is largely due to an increase in rural–urban migratory movements, frequently caused by the expulsion of agricultural workers because of the modernization of tasks. It is also a result of the industrialization process, sociopolitical transitions, warfare conflicts, and growth of the informal economy in the metropolitan areas of the Middle East. In almost all countries in the South, the influx into the cities of emigrants from rural zones markedly accentuates the cultural diversity in ethnically diverse countries like Lebanon.

In the last few decades, four important trends have attracted the intention of international and local organizations concerned with urban policies within cities on the planet. The rapid growth of cities continues to accelerate in terms of both their

A. M. Soliman, *Urban Informality*, https://doi.org/10.1007/978-3-030-68988-9_5

population and their physical pressure on the natural environment. The other occurrence, since the mid-1990s and early 2020s in many countries in the South, is the emergence of readjustments in governance and municipal management, at the local and national levels. The third trend is the influence of globalization on capital accumulation, at the national and international levels, and its influence on the urban fabric of human settlements. Finally, the Habitat III conference of 2016 and the NUA as a new strategy to implement SDGs, especially Goal 11 target 7, have introduced urban sustainability innovations to enhance the living conditions of the built environment.

These trends are linked to the Habitat Conference in Vancouver (1976) and match the United Nations Millennium Development Goals (MDGs). The aims are, first, to give the responsibility of governments to prepare spatial strategy plans and adopt human settlement policies to guide socioeconomic development efforts. Second, introduce genuine innovations in institutional form and practice and attempt to respond to at least some of the pressures of urban growth and expansion emerging within the built environment. Third, within the global information era, seek to achieve significant improvements in meeting the requirements in the lives of people in the South and make available to them the best and lowest-priced goods and services. Finally, after the September 11th attacks, impose or introduce new reforms and readjustments in government structure of the South, especially the Middle East countries, to combat terrorism, develop education systems, and regulate arbitrary developments that help to foster terrorist mentalities.

This chapter tries to sketch out the role of the state and how is it changeable towards society over time, and how this has a major effect on the urban informality transitions process. This chapter focuses on two themes, first, the conceptual, theoretical, and policy-relevant linkages among the shifting interaction of the state towards civil society, economy, urban informality, and urban conflict. The second is a clear understanding capable of rationalizing the recent transformation of socioeconomic and political transitions and to enable the varied function of states and its influence on the economy and society. The assumption is that the state's conflict may combine various forms and conflict within the forms themselves, as well as the extent to which each generates insurgency or patronage (such as what happened in Egypt) and may vary with time and circumstance.

The first part of this chapter examines recent trends and future scenarios for the shifting liaison between the state and civil society, with special emphasis on the Middle East. Then, the linkages between state, capital, and society are explored. The transformation of socioeconomics, spatial spaces, the political milieu, and urban informality are examined in the fourth part. The conclusion proposes that social and spatial changes should be understood as useful political tools for categorizing places and spaces in Middle East cities. These places and spaces are spatial boundaries that defend and separate the majority from the minority or separate the formality from the informality and reproduce the existing power relationships.

5.2 The Shifting Liaison of the State and Civil Society

There are various concepts of "state," "regime," and "government", despite these variations, but all consitute a part of sociotechnical regime in sustainable transitions. A state is a political structure with a centralized government that maintains a monopoly on the legitimate use of power within a certain geographical territory (Salmon and Imber 2008). The state is composed of a society and a political and administrative structure ruled in a democratic or in un-democratic way comprised of the various agencies or branches of organizational structure such as the executive, legislature, administration or civil service, judiciary, police, and military. It is a "democratic" state in which no class or group can assure its permanent political predominance (Miliband 1969), such as in the case of Lebanon (Shwayri 2012). The "state" refers to a set of different but interrelated and often overlapping theories about a certain range of political, socioeconomic, and physical transitions.

The definition of the state can be seen as a part of an ideological conflict because different interpretations lead to various theories of a state function and, as a result, validate different political and social strategies. Most political theories have classified the role of the state into two categories (Dahl 1961; Blokland 2016). The first is known as "conservative" or "a partisan instrument" approach that treats capitalism as a given and then concentrates on the function of states in a capitalist society. These theories tend to see the state as a neutral entity separated from society and the economy. The second is Marxist and anarchist theories that view politics as intimately tied into with economic relations and emphasize the nexus between economic power and political power. Hence, the state is a partisan instrument that primarily serves the interests of the elites. Marx and Engles (Miliband 1969) described the state in two ways, one as a "bourgeois democracy" and the other as a "democracy." The former tended to suggest that the state is a "bourgeois democracy" in which an economically dominant class rules through democratic institutions, rather than by way of dictatorship. The latter is based on the claim that they are regimes in which, precisely because of their democratic institutions, no class or group can assure its permanent political predominance. However, the state, whether it is thought to be "bourgeois democratic" or simply "democratic," and societies do have significant similarities not only in economic but also in political terms, and states' roles and power towards their societies differ.

On the other hand, any state, through the organizational structure, expresses a relationship of class domination (and consequently has social bases towards a certain class), assumes an ideology as if in the common interest, and develops and implements policies that respond to the fundamental pact of domination, or certain priorities (Harvey 2009), and neglect the other issues. The state, whatever its structure, should be subjected as it is to a multitude of conflicting pressures from organized groups and interests and cannot show any marked bias towards some and against others: its special role, in fact, is to accommodate and reconcile them all (Miliband 1969). In the Middle East, the state's role varies from one country to another, and dichotomies of "bourgeois democratic" and/or "democratic" exist. In the former, the state acts towards the elite of society to sustain economic growth, while in the latter, the state might act to enhance the welfare of the middle and low-income groups to sustain its political influence on the society. Thus, the state power is subject to the level of socioeconomic transitions.

A regime is a system for the state or the rule of privileged elites (Robinson and Jams 2012), who control the political economy (Piketty 2014) and exercise power over society regardless of the organizational structure of the country. A more recent emergence is that of 'regime theory,' which acknowledges the place of skills in the politics surrounding an issue. but it recognizes other dynamics, including legitimacy and rules of conduct that may not so quickly be reduced or adjusted (Salmon and Imber 2008). Accordingly, many forms of regimes exist: capitalist, democratic, political (Miliband 1969), oppressive, autocratic, and repressive. The last four types of regimes have occurred in the Middle East and Latin American countries. The first three mainly represent developed countries. All forms are linked directly or indirectly with the dynamics of private capital accumulation and inevitably lead to the concentration of wealth in ever fewer hands (Piketty 2014; Harvey 2016). Capitalism can accommodate itself to many different types of regimes including all the ones just cited.

The government consists of agencies that make laws, carry, monitor, and implement them out. The notion of the state is broader, referring to the other agencies as well as the culture of the socity that transcends the actions of governments. It is not surprising that the government and state should often appear as synonymous because it is the government that speaks on the state's behalf. But the fact that the government does intervene in the name of the country and is formally invested within the state does not mean that it effectively controls that power (Miliband 1969). But in the case of Lebanon, the government has the upper hand in controlling the power over society, while the state reflects the ethnic diversity of the country. However, this distinction is not always recognized, and often the two terms are used interchangeably. Each successive government is composed of particular individuals comprising an honored body who dominate political decision making as a reflection of the state's polcy, and they are separated by the organization from the population as a whole. Therefore, in sketching out the role of the state in sustainablity transitions, cultural composition and the economic performance are two components that have major impacts on the power or the weakness of the state towards a society.

Social and political scientists have formulated a variety of theoretical explanations of the functions of the state in modern societies (Amin 1980; Bell 1982; Blokland 2016). These may loosely be categorized into theories that stress the positive and benevolent qualities of the state and those that regard it as domineering, manipulative, and sectional. While the former explanations emphasize consensus, the latter stress conflict. To be more specific, other scholars (Jessop 2010, 2013; Amin 2006) have arrived at four dominant approaches regarding the function of the state to the problem of urban informality transitions—the pluralist perspective (behavioralist), the instrumentalist view, the structuralist view, and the functionalist view derived from the confluence of elitist, neo-Marxist, and Marxist thought.

First, pluralism sees the state as a neutral body that enacts the motivation of whichever groups dominate the electoral and capitalism process. Within the pluralist tradition, Robert Dahl (Dahl 1961; Blokland 2016) developed the theory of the political state as a neutral arena for contending interests or its agencies as merely another set of interest groups. With power competitively arranged in society, state policy is a product of recurrent bargaining. Although pluralism recognizes the existence of inequality, it asserts that all groups have an opportunity to pressure the state. The pluralist approach suggests that the modern democratic state's actions are the result of pressures applied by a variety of organized interests. Dahl (1961) called this kind of state a polyarchy. Pluralism has been challenged on the grounds that the vast majority of people in high leadership positions are members of the wealthy upper class. Critics of pluralism claim that the state serves the interests of the upper class rather than equitably serving the interests of all social groups.

Pluralism sees the state as a set of political institutions standing outside civil society, and it is this position of externality and superiority that enables it to regulate and mediate the conflicts within civil society, using force if necessary. The state should ideally both represent the divisions within society and resolve these differences in the general interest (Bell 1982). Within the same tradition (Owen 2004), pluralism has received a somewhat different and more sophisticated interpretation. Rather than representing collective interests, the pluralist theory claims that the state mediates between sectional interests, compromising the demands of various associations and classes. The theory posits that power in society is dispersed between various interest groups that compete with each other in an open political system or during development transitions. Blokland (2016) argued that the individualization, differentiation, and rationalization constitute a sociotechnical niche that characterize modern societies and are also pre-eminently taken as a point of departure by the political theory of pluralism. In this context, the state has two roles in urban informality processes: a direct/overt role in facilitating, legally, various commodities for housing production for society; and an indirect/covert role, in facilitating, illegally, the spread of urban informality, in certain locations, in certain times to serve the interest of the elite, as well as the interest of the state. Thus, the state acts as a political informality (Goodfellow 2020). The pluralism perspective lasted until the early 1970s and was replaced by "modernization" discourses. Pluralism did not so much challenge the realist orthodoxy as provide a wider perspective on global interactions (Salmon and Imber 2008).

Second, the instrumentalist perspective views the state as an instrument to assure the reproduction of the labor force, eases demands and, in so far as it is a public investment, involves the devaluation of social capital. It counteracts the tendency for the rate of profit to fall (Castells 1977). The state is comprehended as a tool or instrument, its materiality having no relevance of its own, or the state is seen as having no autonomy whatever (Martin 2008). It is simply reducible to state power and the class that manipulates the instrument (Poulantzas 1979). State intervention in the urban informality is subordinated to the interests of the monopolies in two ways. First, social investments only come after the needs of industrial enterprises for direct aid have been satisfied, and, second, there is a constant tendency to make publicly subsidized sectors profitable, so that they can be transferred to the private sector interests. In this view, state intervention in the urban informality is as an agent of private capital, intervening in the conflicts over the built environment between the working class and the appropriators of the significant interests within a given area. The way in which the state acts in the housing process for low-income groups in Middle East countries is not merely for the welfare of the poor but for productive capital (Peattie 1979), which is dominated by the affluent groups. Therefore, state intervention in the housing process is a significant act in favor of the dominant groups to facilitate the interests of the monopolies of a society, and it does little for low-income groups. In this respect, why should the state think of setting up housing policies to help low-income groups? Is it for the benefit of the dominant groups, or is it for the benefit of the current regime to sustain their political influence, or is it for economic and social reasons?

Third, the structuralist view conceives of the role of the state as an essential function of social capital. As such, the state's role as a 'factor of cohesion' never exists outside class 'practices' as some autonomous subject-like entity with a will of its own but was the condensate of a relationship of power between struggling classes (Martin 2008). Therefore, within the structuralist view, the state acts in a way to satisfy both the powerful groups and low-income groups, and it may operate in favor of one particular group against another to sustain its power to enable the supply of cheap labor that upholds the capitalist system. This context was extant in the mid-1950s, during the movement of freedom in certain Middle East countries, where many states provided huge public housing for low-income groups, e.g., in Egypt, Lebanon, Syria, and Algeria. Thus, the state satisfied the needs of the lower class for its own benefit to hold power over society and to gain moral support for its power. The distinction between structuralism and instrumentalism rests on the idea that the latter predicts that the state will invariably protect both the long and short-term interests of the dominant classes. The former anticipates that it will preserve their long-term benefits, while occasionally sacrificing their short-term investments for the sake of the dominant classes' political hegemony (Poulantzas 1979; Martin 2008), as well as avoiding the problems and/or the conflict which might occur from the low-income groups.

Fourth, modern functionalism offers a similar interpretation regarding the exercise of power by the state as being essential for the maintenance of the social system and the realization of its social goals (Mohan 2011). In functionalist theory, therefore, the state embodies the interests of society, and its function has fluctuated between liberalism and neoliberalism. Liberalism (Jessop 2002) is a complex, multifaceted phenomenon. It is a conceptual ensemble with variable patterns of economic, political, social organization, and ideological discourse; it is a strongly contested strategic concept for restructuring market-state relations with many disputes over its scope, application, and limitations in modern societies. Neoliberalism (Jessop 2002) calls for: the liberalization and deregulation of economic transactions, not only within national borders but also—and more importantly—across these borders; the privatization of state-owned enterprises and state-provided services; the use of market proxies in the residual public sector; and the treatment of public welfare spending as a cost of international production, rather than as a source of domestic demand.

Marx's comment that the state is but a committee for managing the ordinary affairs of the whole bourgeoisie continues to inspire contemporary Marxian interpretations. Although some neo-Marxists recognize that state–economy and state–society relationships are much more complicated than classical Marxism suggests, many scholars agree that the state in capitalist societies, as in the most of Middle East, is primarily concerned with capital accumulation (see, e.g., Castells 1997; Ehteshami 2007; Gerges 2015; Cammett et al. 2015). On the other hand, Avelino (Avelino et al. 2016) argued that transition management sometimes adopts a shallow relational ontology of 'the social' and 'the material', without explicitly accounting for 'the political'. But the political has played two crucial roles in determining the level and the output of sustainability transitions, the *'politics of things'* and *'politics of materiality'*. The latter addressed two different foci of material transition politics. First, the politics of materiality is located at the interface between governing regimes and practices on the one hand, and socio-material arrangements on the other hand. Second, the politics of materiality is located at dispersed geographies and everyday social practices. Thus, the dispersed nature of power and agency is interrelated and exchanged between niches and regimes, as institutional power is generally assumed to be centered in the 'regime' and agency for change is primarily attributed to 'niches' in response to landscape developments (Robinson 2006; Robinson and Jams 2012; Avelino et al. 2016; Edensor and Jayne 2011). Therefore, niche–regime interaction is often then conceptualized in a dichotomous manner. While "The politics of things" refers to the way in which things, objects, infrastructures, and physical space remain crucial to political communication in a digital age as well as to the manner in which bodies, objects, and urban space become politicized and digitally remediated (Willems 2019). On the other hand, "the politics of shadow" play a valuable role in identifying the needs of the bottom strata of society and transferee it to the decision-makers or professional. It reflects how various shadow actors are playing an active role in determining the level of integration between niche and regime. It argued (Forrester 2019) that "the politics of shadow" could provide an objective basis for ethics to judge the justice of institutions could also be used to guide the political action and moral decisions of individuals.

Therefore, elitism and structuralism argue that the capitalist state is not defined by the class composition of the state personal but rather by the position occupied by the state in the capitalist mode of production (Poulantzas 1979). However, with the overthrow of capitalism, state power will be seized by proletarians who will govern in the interests of the proletariat. Elite theories of the state have emerged partly as a reaction to the utopian elements in Marxian political thought (Jessop 2013), but the proletariat and elite cannot be separated. The latter has relied on the former for the reproduction of capital, and the former has relied on the latter for survival, but both could not be separated. This represents the *"political of things"* as mentioned previously. However, 'the conceptualization of power and politics', whatever the form of the political institution, and their relationships to questions of urban informality and social justice, require further elaboration, but still the tensions between

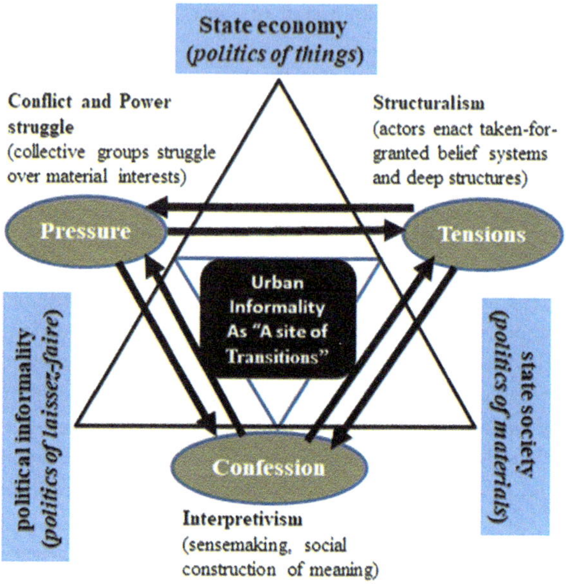

Fig. 5.1 Diagram of the three trends of state towards the society and outputs. (Source: The Author)

democratic governance and the radical steps are deemed necessary for sustainable development (Stirling 2011).

The main problem is that most of Middle East countries suffer from instability of the political situation, and, thus, how to manage the sustainability transition in a rapid political transition. As illustrated in Fig. 5.1, state action in transitions towards a society varies and may folllow three essential trends: state economy (*politics of things*), state society (*politics of materials*), and political informality (*politics of laissez-faire*; Goodfellow 2020). The three trends overlap and are interconnecting over time according to transitions of socioeconomic and political circumstances. The output of these interaction between the state economy and state society are tensions between both, on the other hand, the correlation between state society and state of informality is recognition among them, while the integration of state of informality and state economy causes pressure among both. Therefore, the interrelation of the three trends creates an environment by which the transitions take place. The process of the interaction between the three trends of the state play an extensive part in determining the methods of the state intervention in the transitions process.

However, 'the political' in transitions is so complex that it requires a multitude of analytical lenses in and adjacent to the field of transition studies (Avelino et al. 2016), and this might lead to perspectives through which we can grasp the niches–regime interactions in a more dynamic and fine-grained way. Thus, the state power fluctuates and is dominated by those who exercise power and who are empowered by and with whom. However, urban informality in the Middle East has produced, within prevailing 'state' structures, visions that may provide guidance to transformative efforts, but they are likely to reflect basic assumptions of the status quo. The state also has its own regulatory framework in a given society that provides official legitimacy to certain processes and organizations, in what is called political informality, often just described as 'messy', 'chaotic', or 'fluid'. (Goodfellow 2020). However, as perspectives through which to grasp the niches–regime interactions in a more dynamic and fine-grained way, they are needed (Avelino et al. 2016).

5.3 State, Capital, Society, and Housing

The political economy of development (Cammett et al. 2015) is the outcome of, and interaction between, three domains: the state (state policies, and state structures); the economic (capital accumulation, agents operating, and how the economy behaves over time); and society (social actors, whether groups or individuals). The three pillars in the context of the Middle East are interdependent, interrelated, and interconnected. Each pillar influences and shape the other two; each is, therefore, both cause and effect, both starting point and outcome of housing production and even for urban informality.

As illustrated in Fig. 5.2, the relationship or cooperation between the three pillars—state, society, and economy (capital)—would result in organizing the level of production/reproduction (supply), distribution of commodities (demand), and the level of consumption of various commodities, all of which constitute the modes of informal housing systems by which the transition process occurred. These three pillars may conflict regarding the scope of anarchic market relations, collective decision making, and spontaneous self-organization, as well as the formal/informal and substantive freedoms available to economic, legal, and civil subjects. A shift also occurs from state to market forces (capital/economy) and partnership-based forms of governance (interrelation between state, society, and capital), reflecting the

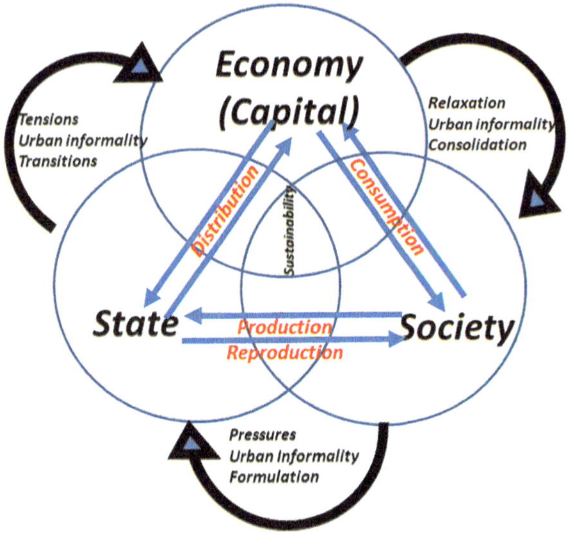

Fig. 5.2 Venn diagram of the relationship or cooperation between three pillars: state, society, and economy, and the modes of informal housing. (Source: The author)

neoliberal belief in the probability, if not inevitability, of state failure/weakness, or power/pressure, and/or the need to involve relevant stakeholders in supply-side policies (community driven process). Thus, the state, as a sociotechnical regime, and the community, as a sociotechnical niche, overlap and may be integrated and correlated. This transformation occurs over time due to various socioeconomic and political circumstances in which the state role alternates.

Ideologically, liberalism claims that economic, political, and social relationships are best organized through formally free choices of the rational actors who seek to advance their own material or ideal interests in an institutional framework that, by accident or design, maximizes the scope for formally free choice or maximizes the speed of transitions. Social and economic inequalities are just only if they result in compensating benefits for everyone, and, in particular, for the least advantaged members of society (Rawls 1971). Economically, it endorses the expansion of the market economy—i.e., spreading the commodity form to all factors of production/reproduction (including labor power, and housing components)—and formally free, monetized exchange to as many social practices as possible. Looking more closely at the dynamics of inheritance, as in the case of Middle East, and then at the global dynamics of wealth, it is already quite clear, however that it is an illusion to think that something about the nature of modern growth or the laws of the market economy ensures that inequality of wealth will decrease and harmonious stability will be achieved (Piketty 2014).

Politically, it implies that collective decision making should involve a constitutional state with limited substantive powers of economic and social intervention, and a commitment to maximizing the formal freedom of actors in the economy and the substantive freedom of legally recognized subjects in the public sphere to accelerate transitions (Jessop 2002). The latter is based in turn on spontaneous freedom of associations collectively or individually, or both, to pursue any social activities that are not forbidden by constitutionally valid laws. In the case of treating housing as consumption, the relationship among the three pillars would outweigh the commodities according to the level of production and distribution or as supply and demand principles. However, the function of the state towards housing in a liberalism viewpoint tries to balance or meet social consumption for all strata of the society. On the other hand, the state role in housing production fluctuates, covertly and overtly, according to socioeconomic transitions and political circumstances and the role of the elite in the development process.

Most of these theories appear to contain at least some valid observations and to fit the realities of at least some societies in the Middle East, and it is challenging to choose between them. However, most political scientists today recognize the centrality of the state in society: The state not only dominates power relationships but increasingly controls economic and social life. The recent financial crisis of 2008, and that caused by the Covid-19 pandemic, have shown that to adjust the global market, any state, especially the most significant seven states in the world, has to enforce a specific regulation to control the sharp drop of the financial markets (Harvey 2016). Therefore, the attempts to shift the wealth from Western countries to another economic agglomeration appear in the atmosphere, in other words, it is logical that it "localized the global and globalize[d] the locality." Therefore, the Middle Eastern countries have an excellent chance to allocate their resources for the yield, for tackling urban informality, and to hold their hand on the stability and security of their societies. This would require reconsidering and specifying the dichotomy between niches and regimes. One way to address this challenge in this niche–regime dichotomy is by unpacking and specifying the notion of 'regime'. Instead of conceiving a regime as a sociotechnical system, one can they consider it as a sociopolitical constellation embedded in economic structures (Swilling et al. 2015), in which the modes of informal housing systems are considered a cornerstone of the economy.

Figure 5.3 illustrates a model that represents the modes of informal housing systems of production, reproduction, consumption, and distribution in Middle Eastern cities, and which identifies the housing types within the cities. Modes of informal housing systems are based on the degree of supply, demand, and construction that are linked to various forms of capital, within which various types of housing production can be identified. To a certain extent, formal housing production is constituted within legal institutions and depends on a large amount of capital that determines the level of supply, demand, and construction. On the other hand, modes of informal housing systems follow the same steps as formal housing but on a smaller scale, with a variation in capital involved in terms of semi- and petty

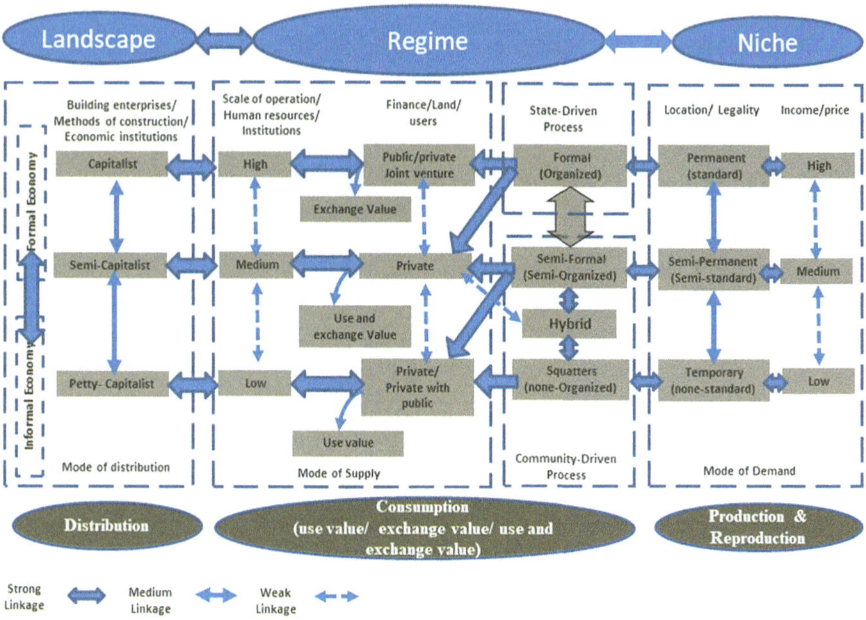

Fig. 5.3 Scheme of the modes of informal housing systems of production, reproduction, consumption, and distribution in Middle Eastern cities. (Source: Adapted from Soliman 1988; Grin et al. 2010)

capitalists (Peattie 1979), legal institutions, and modes of construction, according to the requirements of middle and low-income groups. Semi-informal housing is primarily for use and exchange value, and squatter housing primarily for use value. The differences between formal and informal housing relate specifically to the legality of tenure, the legality of construction, the increased differentiation of the workforce, the more generalized use of waged labor, the increased significance of the means of production, and the purpose of acquiring the housing unit (for further discussion, see (Soliman 1988). The two markets—formal and informal—are integrally linked, and housing activity outside the formal capitalist market cannot be called non-capitalist, as if it were unaffected by the prevailing mode of production. The two housing types, formal and informal (semi-informal, hybrid and squatter housing) are linked to each other in that the formal housing is dependent on a flow of labor from the latter, while the latter obtain their capital and resources from the former. The mode of housing demand constitutes the sociotechnical-niche level in which the level of finance/income determines the type of housing supply. On the other hand, the mode of housing supply constitutes the sociotechnical-regime level in which formal/informal institutions are regulated by the housing supply. The sociotechnical landscape formulates the final production of housing either formally or informally. The integration and interrelation between housing demand and housing supply constitutes the housing market in which sociotechnical niches and regimes interplay. Also, the linkage between the state, society, and market fluctuates according to the circumstances of the development processes and the action of the state towards the society.

Therefore, this model constitutes the middle way of the multi-level perspective (Grin et al. 2010) and composes a new conception of the sociotechnical regime, which is the main facilitator for increasing and regulating housing production, hence the urban informality. All these efforts are especially challenging for local government, which is the level closest, or the main pathway, to the people and most immediately responsible for ensuring a situation conducive to good business and stable civil society, while the sociotechnical niches compose a set of new reforms that facilitate the formalization process of housing production and urban informality. Reforms of public-sector management or private-sector development cannot do what is desired for the national economy until they are adapted to the requirements of urban governance and management and implemented on the municipal level.

Perhaps most surprising is that some progressive social movements, as a sociotechnical niche, have also appealed to themes of security and antiterrorism to bolster their legitimacy and forward their agenda after the 9/11 tragedy and the Arab Spring. Therefore, the role of the Middle Eastern states, especially after the Arab Spring, is being reconsidered and reconfirmed to strengthen its essential functions of facilitating housing markets and correcting market failures, promoting economic and social stability, and ensuring distributional equity. This process has included efforts to expand opportunities for citizen participation, to increase innovation, openness, and cost-effectiveness ("reinventing" and "rightsizing" government), and to rebalance public and private sector roles. The four global trends—urbanization, decentralization, globalization, and government reform—underscore the increasing importance of cities and towns in national development (World Bank 2000). Also, housing production transitions are combined and interrelated with the methods and instruments, whatever their legal statues, for conducting this process, and the way that the state communicates with the society. Regarding the latter, Leftwich (2010) recognized six key institutional characteristics of the state role towards the society to persuade a sustainable development. Therefore, transition management, as a part of planning process, might organize this production in a reasonable and sustainable way.

5.4 The State and Housing

The Middle Eastern states have responded to the emergent needs for housing, and try to encourage the communities in socioeconomic development and to increase housing production and do so via five principal modes: demobilization, participatory, manipulative mode, cost-recovery mode, and transitional socioeconomic and political e.g. Cairo and Amman cases (Ababsa and Abu Hussein 2020) (see Table 5.1). Of course, these are idealized types of responses that do not fit every situation and sometimes exist in combinations or variations.

First, in the 'demobilization mode', the states in the Middle East actively sought to promote community mobilization, from rural to urban areas, and their contribution as mechanisms for sustaining the political system, and the low-income groups were used by politicians to strengthen their own political power and continuance. During the Nasserism era in Egypt, from 1954 to 1970, most of the states of the Middle East followed the paradigm of Nasserism, which reflected socialist ideology and trends, both at the regional and national levels, in attracting the lowest strata of the society towards the new political transformation. Thus, Nasser's ideology was spread throughout the Middle East, including in Syria, Libya, Lebanon, Iraq, Algeria, Yemen, and Palestine, attracting the lowest strata of the society against the external international powers. It was aimed at satisfying the urgent needs of goods and services for the bottom strata of the societies, so that they would be a buffer zone against the external 'imperialism'. It was to be the basis to establish a 'Socialist Society' and to mobilize the working classes from rural to urban areas to accelerate the industrialization and development processes. In the late 1950s, most of the Middle Eastern countries witnessed the introduction of the development of social housing programs for constructing low-cost housing to accommodate middle and low-income groups. In Egypt, the first public housing program was introduced in 1954, and in Lebanon during presidency of Chehab (1958–1964) (Fawaz and Peillen 2003). The construction of workers' housing, sometimes called dormitory towns, was mainly for workers and employees who were attached to major industrial centers established at that time. In Syria, after the establishment of the union with Egypt in 1958, similar policies were applied after 1960 when President Nasser (then President of the United Arab Republic) initiated public housing programs that were assigned

Table 5.1 The modes of the state on housing polices and transitions performances

The state's mode	Housing policies	Transitions performances
Demobilization mode	Socialist ideology Mobilize the working classes from rural to urban areas Social housing programs Shrinking of the involvement of the private sector	Steering for Political performance and certain impacts
Participatory mode	Active but ad hoc intervention Acceleration of the informal housing production Involvement of the private developers New housing programs	Putting socioeconomic transitions in context of other priorities of actors
Manipulative mode	Poorly implemented housing policy Fluctuation in housing policies Flourishing of informal economy Acceptance of a status quo for arbitrary urban growth	Steering innovation dynamic, e.g., preference to solve sustainability challenges by technology innovation
Cost-recovery mode	Involvement of community participation The potential value of the assets of informal housing Facilitating real estate markets Promote public-private partnerships	Steering for sustainable development more broadly over multiple dimensions
Transitional socioeconomic and political mode	Cope with the transitional of political circumstances Shrinking of the role of the state Introduction of social Housing program Establishment of new towns Resettlement of unsafe areas Introduction of a reconciliation law	Steering as focused on systems change and regimes, and steering of radical transformation.

Source: The author

the roles of securing land to set up housing projects, provide utilities, and sell and distribute plots.

Other Middle Eastern countries, such as Libya, Iraq, and Algeria, followed suit and supplied social housing for low-income groups. The socialist movement was in full swing as governments endeavored to become solely responsible for providing housing for middle and low-income groups, while the involvement of the private sector was reduced (Hopkins 1969). In the mid-1960s, housing policy shifted into condominiums in new large-scale land subdivisions such as Madinet Al Awqaf, Maadi, and Heliopolis districts near Cairo, Egypt, in the periphery of Damascus and Aleppo in Syria, and in the periphery of Amman in Jordan (Ababsa and Abu Hussein 2020) taking advantage of the liberal credit terms available under the government-subsidized cooperative housing programs. Libya and Sudan followed the Egyptian model after the announcement of the new union between the three countries in 1958. In Syria, after 1958, the law was changed so that Palestinians were allowed to own one house per person, but they were still not allowed to own farmland (United Nations 2005), and the Syrian government allocated land plots for housing development to low-income groups.

Similarly, Jordan, Yemen, and Lebanon witnessed a transformation in housing policy with the elimination of the role of the private sector in housing supply, causing a sharp decline in the housing market. However, the 1967 war with Israel, the subsequent war of attrition, and the run-up to the 1973 war froze formal housing programs in the Middle Eastern countries as public funds were reserved for the war effort. Large urban centers had to accommodate a significant number of internal and external migrants, especially in Egypt, Syria, Lebanon, and Jordan. Shifting demobilization—from one urban to another urban area—had taken place, with the urban fabric of the cities in confrontation with Israel changing radically, displacement of many residents from warfront cities, and a huge increase in housing production in other cities, but much of it was informal, such as in Cairo, Alexandria, Damascus, Aleppo, Amman, and Beirut. The informal housing development accounted for between 25% and 35% of total housing production. Thus, the political policy was the driving force for housing production and eliminated the private sector from such development, and the urban poor had no choice but to act illegally.

Secondly, the 'participatory mode' was characterized by an elastic housing program through encouragement of the involvement of the private sector in decision making and in housing production for accelerating the development process. During the era 1973–1981, the Middle Eastern countries witnessed a period of active but ad hoc intervention. After the 1973 war, the socioeconomic and political milieu in the Middle East changed dramatically, and the oil prices increased exponentially. The next phase of alteration, 1978–1983 (often referred to as the 'boom years') was based on the large contingent of emigrants coming from Jordan (Kadhim and Rajjal 1988; Ababsa and Abu Hussein 2020) Syria, and Egypt (Soliman 2004a, b), who worked in the oil-rich states of the region, many of whom sent back substantial remittances, and this accelerated the boom in informal housing production. Consequently, Middle Eastern states intervened more positively in the housing market and recognized that the private sectors offered an important means to tackle the housing shortage. Housing policies involved the private developers and embraced three sets of actors—the formal and informal private sectors, and the public sector—where the private housing developers were considered important actors in the housing development in new and old towns. Thus, the private sector gained the privilege of becoming the main developer for housing production. At the same time, the states upgraded sites and services and self-help programs for housing low-income groups that were advocated by the first congress of UN-Habitat in 1976. This was achieved by facilitating the flow of subsidized building materials, forming cooperative housing societies, and allowing a free market within the formal private housing sector. In the late 1970s, the role played by the informal sector as an indispensable part of housing production was recognized in the Middle Eastern countries with at least 75, 55, and 60% of housing construction being undertaken by the uncontrolled private sector in Egypt, Syria, and Jordan, respectively (Kattaa and Al Cheikh Hussein 2009; Potter et al. 2009; Soliman 2004). Adequate shelter for all was the declared objective to be achieved through private and public sectors. The idea was a shift in focus from providing completely subsidized housing to providing other means of housing in terms of goods, services, or both.

Consequently, site and services and upgrading programs, core housing, and wet-core units were developed to house this bottom stratum of the society (see Fig. 5.4). Many projects were introduced in various Middle Eastern cities, e.g., in Egypt—Alexandria, Ismailia, Helwan, Aswan, Cairo; in Syria—Damascus, and Aleppo; in Jordan—Amman; in Lebanon—Saidon; and in Sudan—Khartoum. This was based upon two main principles: first, to allocate large land plots in various locations within the cities to cover various social classes of the population involved; and, second, to leave housing production in the hands of private developers who catered only to a small minority of the privileged who could afford to build their own homes.

Third, the 'manipulative mode' constituted a vaguely formulated or poorly implemented housing policy whereby, at the beginning of the 1980s, the states in the Middle East played an arbitrary role within the housing market and began to impose new restrictions on private developers in terms of building procedures and regulations. Yet by 1986, the new-found orientation of international agencies induced new economic reforms, and the states once again recognized the private sector as the principal housing sector to develop housing construction for middle- and low-income groups (Soliman 2004a, b; Ababsa and Abu Hussein 2020). This era witnessed a great fluctuation in housing policies and the real estate market, where new private developers in the form of money-lending companies became the major supplier of housing units, not for the poor, but for the middle and the upper-middle classes. The informal money-lending companies absorbed the cash flow from the market and controlled the circulation of capital in the real estate market. The urban poor continued to rely on informal methods to shelter themselves. Due to the arbitrary housing policies in this era, many laws were passed and later amended and replaced by new ones, e.g., in Egypt, Syria, Jordan, and Lebanon. The latter witnessed civil war and had many governments within the state where many informal areas had been created on the periphery of the major urban centers. Therefore, the states had accepted a status quo for arbitrary urban growth in seeking to recover from the financial burden that it had inherited from the 1973 war, housing shortages in the 1970s, and the increasing price of oil, and began to speed up the process of implementing new economic policies, making the best use they could of the substantial remittances from people who worked abroad.

Fig. 5.4 Photos illustrating the core housing scheme in Alexandria City, Egypt. (Source: The author)

Fourth, the 'cost-recovery mode' was an attempt to involve community participation programs and the private sector on a cost-recovery basis to increase the level of housing production and for the states' own ends. Lack of resources, in cash, in kind, or both, forced the states to follow the cost-recovery mode to increase housing production for the urban poor. One of the main issues to be tackled was taking account of the invisible resources of middle- and low-income groups and guiding their efforts to participate in housing construction. As De Soto (De Soto 2000) noted, "the potential value of an asset, to effectively 'unlock' $9.3 trillion of what he calls 'dead capital', and so allows us to control it". Therefore, with the beginning of the new millennium, and 9/11, the roles of states were recast to strengthen their essential functions of facilitating real estate markets, correcting market failures, promoting socioeconomic stability, and ensuring distributional equity. The states, especially in Egypt, Lebanon, and Jordan, initiated a new economic policy through privatization programs that paid attention to the urban poor. This process included efforts to expand opportunities for citizen participation; to increase innovation, openness, and cost-effectiveness; and to promote public-private partnerships. All these efforts were challenges for local governments for ensuring an environment conducive to good business and a strong civil society and for alleviating the heavy financial burden to be paid on housing projects. Egypt, Jordan, Lebanon, and Algeria were committed to poverty alleviation and had acquired clear momentum through the explicit adoption of the goal of reducing poverty to 15% by 2015 (United Nations 2005). The states of Egypt, and Jordan explicitly articulated a package of actions and programs designed to empower the poor on a cost-recovery basis. The cost-recovery programs, policies, and strategies were rapidly developed with the participation and consensus of all stakeholders under various national and international housing projects. The important point was the stability of the implementation of the housing policies and its contribution to the development process within the Middle Eastern countries.

In the meantime, after ten years of relative deprivation in real estate market investment (1994–2004), a boom in the private housing development occurred in the Middle Eastern countries, especially in Egypt, Jordan, and Lebanon, and certain Gulf countries. International, regional, and national real estate companies entered the Middle Eastern market; prices rapidly rose, forcing the urban poor to acquire their goods and services in their own way. Urban informality was perceived as potentially dangerous, and states introduced new approaches to attack the belts of poverty on the periphery of urban areas, and to prevent the growth of illegal housing development (Potter et al. 2009; Soliman 2007).

Fifth, the 'transitional socioeconomic and political mode' was an attempt to cope with the transitional state of affairs that occurred after the peaceful revolutions of the Arab Spring in Egypt and other Middle Eastern countries. These changes are in responding to the increasing demand for housing units to absorb the massive anger of the societies, and in coping with the SDGs (UN-Habitat 2018a; Ababsa and Abu Hussein 2020). The role of the state fluctuated, while the housing mechanisms controlled by the private sector in informal housing development began flourishing. As an immediate response of the interim governments in Egypt and Tunisia to modify the current housing delivery system for low-income groups and to meet the rapid demand for housing units, these governments intervened in several ways. First, an inventory of all public housing units located in various Egyptian and Tunisian cities, either completed or non-finished housing units, was made to help distribute them to the applicants most in need of shelter. It is estimated that about one million and 150,000 applications were handled by the responsible authorities in Egypt and Tunisia, respectively, in a short period of less than 15 days after the fall of the Mubarak and Ben Ali regimes. The second method was to speed up the housing production in a short time by distributing of serviced land plots (with an average area ranging between 350–450 m^2) on the outskirts of cities or in the new urban communities in Egypt to private real estate developers at prices to suit low-income groups to build residential blocks up to ten floors in height. This technique was dependent on self-assistance through the formulation of owners' associations between a private developer and those seeking access to housing on a cost-recovery basis. The idea is to formulate cooperation between the public and private sectors, where the former will provide goods in the form of serviced land that the private sector is incapable of doing, while the latter will be enabled to carry out the construction process on an incremental basis at its own expense. The third method was to maintain an inventory of the actual number of closed and vacant housing units in the cities to exploit them as rental units in the real estate market. Up until now, nothing has been done to alter the rent control law. In Egypt, a recent estimation calculated that a figure of ten million housing units are closed-up and/or have no occupants (CAPMAS 2016a) of which 2.6 million were closed-up and 7.4 million were vacant. It was proposed that to overcome this dilemma, some privilege will be allocated to the property owners of vacant units by exempting them from paying properties tax and providing financial support to the property owners ranges between 25,000–30,000 Egyptian pounds (LE)[1] with zero interest. This financial support is to be paid off in monthly installments of 120 LE to the state, out of the actual rental value of a free contract between property owners and tenants. The differences between the actual rent agreed upon with the owner of the property and the reimbursement of the loan will be covered by the tenant over a period of 10–15 years.

Fourth, governments try to temporarily control the unregulated construction on agricultural land and existing buildings, where some individuals have taken advantage of the transitional circumstances that the countries have passed through for personal interest, building on agricultural land the exploitation of which can be accomplished due to the absence of a security presence in Egypt, Lebanon, Jordan and other countries in the region. It is estimated that the illegal conversion of agricultural land to informal residential development in Egypt was about 25–30,000 feddan[2] within a period of 20 days after the outbreak of the youth revolution in January 2011. This is unlike the many roles of the establishment in violation of the conditions to build on the existing buildings in cities of Egypt, estimated at around 15–20,000 residential units in the same period. The interim government is reluctant to implement laws for fear of the negative perception at street level of the responsible authorities in the country. It could be said that, during this critical transitional period, the housing mechanisms are fully controlled by the masses of the people without adhering to the laws governing the construction process and the shrinking role of government in controlling the real estate market. In other words, the role of the state shifted from an enabler to an observer, as it lost control over the built environment. Recently, Egypt issued a temporary reconciliation law number 17 of 2019 to legalize the urban informality. The outcome of this

---

[1]One US dollar is equal 16.0 Egyptian pounds at the time of the writing of this manuscript.

[2]One feddan is equal to 0.42ha.

law is not yet evaluated. It is expected that Egypt and other Middle Eastern countries will face many urban problems because of the indiscriminate construction and the rapid erosion of agricultural lands in over time.

Nine years after the Arab Spring, the political arena of the Middle East has changed dramatically, e.g., an alteration or abolishment of the states of Syria, Libya, Yemen, and Iraq. The appearance of what is called Islamic Phanatic Groups and the spread of military assemblages where and when the main policy for the states is to control its territory and fight terrorism. Despite the unstable political situations of cities in the Middle East, huge luxury projects in the form of gated communities have been constructed, not for the poor, but for the affluent groups.

In Egypt, the state has intervened in real estate markets and housing production as a public developer through three ways: a one-million housing-unit program; the construction of the fourth generation of new cities; and the construction of the new administrative capital on the east side of Cairo. The former was announced after the 30th June revolution of 2013 as an attempt to tackle the shortage of housing for various strata of the society. The fourth generation of new cities started with building twin cities in Upper Egypt, followed by establishing a city with two million inhabitants (1.882 million, (Attia 2019) of New Al Alamein as a model for the new generation of sustainable cities in Egypt, located on the North Coast (Attia 2019), in addition to a new city named Gallala as an international touristic city. A huge new city is currently under construction as a new capital of the country. The aim of the erection of the new administrative capital is to absorb the increasing pressure of the population on Cairo and to attract foreign investments to the country. It is an ambitious program, but the question remained how it could be financed. The main financial resource comes from selling land, housing units, and housing plots at the market prices rate in which land and housing units have rapidly increased and exceeded the affordability of the lowest strata of the society. Land and housing units in the current real estate market became a scarce commodity not only for the bottom strata of the society but also for the middle and above middle classes. In Egypt, a temporary Law No. 17 of reconciliation in building violations was introduced in 2019 to issue a formalize license for the informal buildings. It is expected that the implementation of this law will provide revenue of around 30–40 billion Egyptian pounds. This revenue will be used to finance the social housing program, as well as the needed social amenities. Thus, the Egyptian government changed its role from enabler to a public developer by which the real estate market has rapidly increased.

Due to the Syrian civil war and conflict, many people were forcibly displaced to neighboring countries such as Egypt, Jordan, and Lebanon. These international refugees put further pressure on these countries, which increases the crisis for housing. It is estimated that Egypt, Jordan, and Lebanon hosted around 5, 2.5, 1.5 million Syrian refugees, respectively. However, urban informality became not just useful for accommodating local people, but it also changed its function from merely local to include an international dimension.

In sum, the modes of the state in the housing process represent a crucial instrument for urban transformations operating at the cognitive and normative levels of decision making targeted by transition management (vision, pathways). However, their institutional legacy in terms of actor networks involved and routines developed appears to form a major burden for a stronger engagement with sustainability. Moreover, given their informal and strongly place-based character, huge differences exist in terms of the approach, scope, scale, and liability of the housing process, even though increasing international peer-to-peer exchange and networking may play an important role in the future formation of standards (Table 5.1). The construction of affordable housing by the private sector in the Middle Eastern cities faces two major challenges: availability of affordable serviced land and access to housing finance by both developers and homebuyers. Complementary and integrated government policies to supply affordable serviced land, support mortgage markets, and microfinance and more efficient land registration will be critical in leveraging and encouraging the private sector to invest in affordable housing.

5.5 Political Strain

The Middle East experience has rarely been incorporated into the housing literature within the context of the changing nature of the state and civil society (Roy and AL-Sayyad 2004). On the other hand, the Egyptian, Lebanese, and Jordanian experiences have highlighted how religion and nationalism, as sociotechnical niches, are increasingly invoked in urban struggles for goods and services. Both Latin America and South Asia offer rich traditions of theorizing about the structures and mechanisms of state power (Al Sayyad 2004). Urban informality in the context of wartime and urban crises, e.g., Lebanon and Jordan, point to the crucial role of ethnic politics, nationalism and refugee movements in which enforce pressure on the states (Ababsa and Abu Hussein 2020). It is also true in the case of Israel/Palestine, where the informal land rights of settlers have been so intensely ethnicized that they are now a major hurdle to the prospects of peace (Yifiachel and Yakobi 2004). It is well known that the Israeli state, having prevented the Palestinians, either in the Gaza Strip or the West Bank, from building legally, has left them with no alternative but to expand informally. That same state for a while found itself in the position of having to declare its own citizens "squatters" to justify their forced removal as a part of a political compromise. Thus, the state has a double role, formal and informal, the latter encouraged informality, and so acted via political informality (Goodfellow 2020) to sustain its political influence on Palestinian territory. In the former case, the state acted formally to prevent further urban development, as well as to squeeze the Palestinians into a narrow strip. How the political and urban resources, as a sociotechnical regime, are rendered public or comprehensible, and what are the modalities of a social transaction for an authorized or spatial acknowledgment? How is urban informality produced from day to day? What is the nature of the relationships between urban informality and the city, or between the sociotechnical niche and structure regime? How are the notions of community, or grassroots, and territory to be specified? Are these cases of urban informality indicative of a new urban milieu, or a new innovation, of a new territorial question; or, on the contrary, are they indicative of vernacular societies reflecting local appurtenances?

Intervention in, or management of, urban informality sometimes takes place through large-scale development, sometime through test operations intended for duplication, or sometime through international enabling and participatory approaches. What significance is to be accorded to the principals involved in the reliability of operations and in their transfer to a large territorial scale (or indeed to standardization, sustainability, normalization of the urban informality), principles often defended by the operators or developers? How often is the praise of the inhabitants' competence to develop the city (local entrepreneurial or communal spirit) valorized by the institutional actors, as well as instrumentalized, to justify, or not, the disengagement of the public authorities from the urban informality?

Neither the state nor urban planners, as a part of a sociotechnical regime, can regulate, or manage, the growth of informal areas, or represent the interests of the citizens. Rather, both might facilitate the formalization or transitions of urban informality (Banks et al. 2020), as happened in Egypt though the introduction of a temporary law of reconciliation in 2019 for issuing an official license to building codes' violators. But how does one disengage or cut oneself loose or disjuncture the conflicts that are arise between the state and people who are most in need of shelter? The retreat or absence of the public authorities from the administration of the urban informality, the crisis of confidence and trust between governed and governing, and the deficit of urban policies cannot be concealed (Soliman 2019; Banks et al. 2020). On the other hand, citizens' involvement in defining, or in putting pressure on, the urban context is crucial to stimulate public participation by investigating strategies, as an innovation for transitions, to integrate informal areas into cities' urban fabrics (e.g. Banha project in Egypt (Soliman 2017), and upgrading projects in Amman (Ababsa and Abu Hussein 2020)). In this context, the favored success opportunities are bottom-up decision-making scenarios (involving/initiated through implemented by grassroots) over top-down scenarios (coerced by central governmental authorizes). In other words, the decreasing tensions between the niche and the regime and building bridges of trust between the grassroots and the state in the form of decision makers would facilitate sustainable development. A valuable example of a bottom-up model is "grassroots" movements/organizations that target participatory democracy as a larger system of governance (Soliman 2012a, b; Soliman 2017). Grassroots initiatives give rise to effective community groups and are typically more organized in terms of initiation, structural organization, and mode of operation in comparison to traditional power structures (Soliman 2017; Banks et al. 2020).

Urban researchers have not yet exhausted the analyses of the exercise of power in the city, of the mechanisms for correcting urban inequalities between central and peripheral quarters, or of the modes of shared governance now appearing. To have access to a "minimum right to the city," city-dwellers are more and more frequently organizing themselves on very localized levels. They group themselves according to various modalities and potentialities that extend from traditional communalism to formal associations (secular, religious) and by way of political violence.

The assertion of the inhabitant-actor, as an active sociotechnical niche, although not inevitably creating new identities, nevertheless favors the emergence of new communities of interest and the reconstitution of alliances (private/public, persons of want/officials, individually or collectively, and religious figures/politicians). The organization of the inhabitant as a collective actor in the urban informality is often recognized at the setting up or negotiation of urban projects in the vicinity. How do the state and the municipalities, who make the rules of the urban game, assimilate these forms and non-forms of government which elude them? How are the relationships between state and citizen in different contexts (from the hyper-centralization in Egypt to the attempt at decentralization in Lebanon) to be considered in the light of initiatives at the level of the urban informality and of negotiations and demands? How does popular management in the urban informality by the inhabitants or shared management with new forms of partnership, between the associative sectors, private and public, affect urban politics and the role of institutional actors? Finally, can one evoke a political culture of "urban legality or para-legality"? The conflicts and strategies of the collective actor, as a niche, for access to the city quite often favor a redeployment of the rules of power of the regime. Significant advances concerning the right to the city have been acquired in Latin America, but does the notion of the urban movement have significance in the Middle East?

In the Middle Eastern cities, it appears that petty-commodity production is an ever-existing trait in a capitalist economy and is growing to formulate an informal economy that constitutes a high proportion of the country's national economy. Since the liberal economic reforms of the 1980s, however, political settlements have started to change. The well-connected private-sector elite has gained increasing prominence alongside the "state bourgeoisie." Even if most business associations remain weak vehicles for the organization and representation of private-sector interests, informal channels have enabled well-connected elites to transmit their preferences to rulers, gain preferential access to business opportunities, and enhance their private holdings (Cammett et al. 2015). There has been development of a large informal economy that has little access to state largesse or economic opportunities and typically supports the opposition (usually Islamic). Jordanians of Palestinian origin tend to dominate the private sector and the informal economy in Jordan, while 82% of all entrepreneurs in Egypt worked in the informal sector (Galal 2004). In Lebanon, the informal economy holds a dominant place in the country, especially since 1975 (Fawaz and Peillen 2003).

Due to the presence of the informal economy, the accumulation of capital resulting from this production grew on a large scale and does contribute much to social consumption. In this way, the state is starting to rely on the working class and informal economy that is the root of developing the working capital in various investments. Thus, the governing body, or political informality (Goodfellow 2020), is required to assist, or at least facilitate, the production of housing (formally or informally) and to maintain the status quo of urban informality to house this class and hence guarantee the reproduction of capital and sustain social consumption. Nevertheless, petty commodity production represented in building housing units for the proletariat actually interacts with and depends on capitalist production (Piketty 2014) and is integrated with the formal economy, and neither of the two can be divorced from the country's national economy and vice versa. Therefore, the formal institutions act in collaboration with the lower classes of the society to achieve and sustain social consumption. In other words, sociotechnical regimes are integrated with sociotechnical niches, and they are not separated but interrelated.

Within this context, O'Connor (1973) identifies two fundamental purposes of state expenditure: social expenses, and social capital, as well as debates within transition research on the 'politics of things' (Hendriks 2009) and the 'politics of materiality' (Vellema 2011). The former includes spending on coercive items such as welfare and military costs. They are nonproductive but relate to the state's social-control function. Social capital and politics of materiality are further subdivided into social investment and social consumption. The former increases labor productivity and consists of the infrastructure the state provides for the private sector such as roads, electricity, and water supply. Social consumption lowers labor productivity and consists of investment in housing and education.

Within these perspectives of state expenditure, one will concentrate on social consumption which is a part of capital accumulation in which housing recently became a capital accumulation (exchange value) rather than a commodity for use value. Leilani Farha (Farha 2017) stated that housing "is now a means to secure and accumulate wealth rather than a place to live in, to raise a family and thrive within a community." Housing lost its value as a universal human right. A significant component of social consumption is the collective means of consumption (public transport, schools, parks, health facilities, public housing, etc.), which are increasingly important for securing the supply of labor power and social relationships in capitalist cities (Castells 1977; Castells 1997). The inability of private capital, to provide certain collective means of consumption necessary for its survival and expansion in the urban environment, calls for greater and greater state intervention to the extent that the institution of the state (public and local) has become increasingly responsible for the management of everyday urban life. Urban informality is crucial as a means of consumption that provides for cheap labor, housing, and informal economy itself, making possible the lowering of the social wage assumed by capital or the state in the economies of developing countries (Portes and Walton 1976). Thus, urban informality had already become a significant component of the urban social structure (Castells 1983), as well as the informal economy. Their permanent connection to the political system has to be set up, both for them to survive and for the political system to be able to maintain social control over this popular sector. Therefore, deals, trust networks, and building bridges between the state and the actors of urban informality is essential to arrive at a compromise for sustainable transitions.

In a recent debate (Amin 2011; Blokland 2016), argued that, as power flows away from the nation-state upwards to the global level, power is also flowing downwards to sub-national organizations (local and regional authorities, communities, NGOs and CBOs, etc.). On the other hand, a new basis for community and group identity unleashed by globalization processes presents welcome opportunities for more powers of self-determination and the expansion of democracy, participation and personal and group freedom, as happened during the Arab Spring, and recently during 2019 in Algeria and Sudan. A third point relates to regulations and the question of spatial scale in the prospect of increasing globalization (Burgess et al. 1997). Whatever the case, the current attempts to deal with globalization and environmental and spatial trends within the formula "think global, act locally" will certainly not be sufficient by themselves (Burgess et al. 1997; Hamdi 2004).

Recently, social media play an important role and drew the attention of many commentators and even senior politicians, and it helped in many emerging revolts in Middle Eastern and Latin American countries. It suggests that the new movements cannot and must not underestimate component of the offline, local community-based activism in any democratization of a technological commons (Smith et al. 2017).

A major component of social consumption provided by the state includes public transport, roads, schools, parks, health facilities, public housing, land provision, etc., which are increasingly important for securing the supply of labor power and social relationships in capitalist societies. The inability of private capital to provide certain collective means of consumption, especially housing for low-income groups that are necessary for its own survival and expansion in the urban environment. Thus, state intervention has become increasingly responsible for housing provision and the management of everyday urban life (Lefebvre 1971) and people in urban informality sectors have created "a new way of life" (Roy and AL-Sayyad 2004). It is argued that global restructuring and market liberalization have led to the emergence of new forms of informal development within the Middle Eastern cities in the last several decades (Cammett et al. 2015; Soliman 2019) that are often characterized by complex and bizarre intersections of urban and rural restructuring, and interactions of a state and a society. Urban informality is no longer the domain of the poor in Middle Eastern cities but has become a primary avenue to home ownership for the lower-middle and middle classes (Soliman 2004), and even for upper classes (d'Alençon et al. 2016) and is an expression of class power (Roy 2009) who initiate the pressure on a state power towards informality. Davis, Soliman, and Roy (Davis 2006; Soliman 2008; Roy 2009) provided the first global overview of the diverse religious, ethnic, and political movements competing for the souls of the new urban poor and concluded with a provocative meditations on the "war on terrorism" as an incipient world war between the American empire and the poor of the world's slums.

Therefore, political decisions depend on the nature and interests of a state and its social and economic objectives. In this way, the state reflects the distribution of power within a society in various ways: covert, semi-covert, overt, and semi-overt (Soliman 2004). However, a recent debate (Banks et al. 2020) demonstrates that the political economy has molded the role of the state into four major directions: state-society interactions and changing state (overt); attitudes to informality (semi-overt); the significance of diverse actors' agency within both formality and informality (covert); and informality as strategy for diverse groups (semi-covert). As illustrated in Fig. 5.5, all these diversions and fluctuation of the role of the state are related to social consumption which has resulted in organizing the consumption, production, reproduction, and distribution of human, natural and other resources. The result of these diversions reflects four outputs: socioeconomic well-being; coping with existing situation; exclusion of the bottom of the pyramid of the society; and a conflict between the state and the urban poor. These outputs determine the statute of urban informality in the prevailing urban fabric and the way in which might integrate or be resilience with the content of the urban agglomeration in a sustainable way.

In Egypt and Jordan, states intervention in the housing field have followed the four deviations and mirrored the powerful interests of both the affluent class and the poor, depending on the development transitions process within a given area.

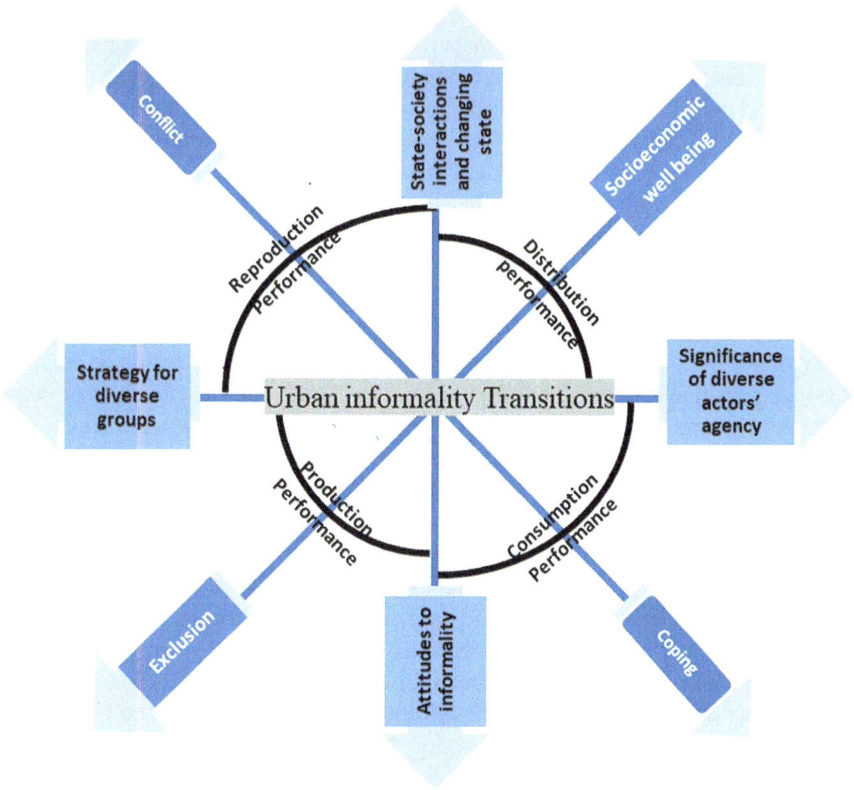

Fig. 5.5 Diagram of diversions and fluctuation of the role of the state towards to social consumption. (Source: The author)

However, social investments and social consumption have generally only been made in Egypt after the needs of industrial enterprises for direct aid have been satisfied. Moreover, there has also been a tendency to ensure that publicly subsidized projects are profitable, so they can eventually be transferred to private-sector interests. In this sense, state interventions in housing have privileged the interests of private capital over the local working class (the lowest strata of the society). An early example of this was state intervention in the housing supply during the New Era movement of the mid-1950s and early 1960s.

Thus, by the early 1970s, much of the attention of the Middle East government's attitude had shifted from "rapid" to "participatory" research and development in which local people maintain significant control over the development process. The involvement of local people in the housing process became a major theme for providing reasonable shelter for low-income groups. These local people have intervened positively in facilitating the land-delivery system for housing development, and they became major contributors in this process (Payne 1999; Ababsa and Abu Hussein 2020), contributing rapidly in the informal economy sector. There are various reasons why a government should become involved in housing. The most common involve humanitarian reasons, functionalism, social control, and human-rights considerations. In relation to housing, human-rights considerations have recently become one of the Middle East states' main concerns, as well as in the broader Global South. Today, states debate centers on who is in need of housing, what constitutes adequate housing, how the government may ensure that those in need obtain housing, and how government resources may best be spent to fulfill housing needs. Also, the sociotechnical regimes has cooperated with sociotechnical niches to achieve the needs of the low-income groups for social consumption.

5.6 Retrospect and Prospect

Housing informality is crucial as a means of consumption that provides low-wage labor, itself making possible the lowering of the social wage assumed by capital or the state in the economies of developing countries (Portes and Walton 1976). Housing informality, as "a site of transitions", has become a major component of the urban social structure, social realm, and political economy. Their permanent connection to the political system has to be set up, both for them to survive and for the political system to be able to maintain social control over this popular sector. As experience and insight has grown, it became evident that local people, who had previously been viewed as passive "subjects," "clients" or "beneficiaries," had much to contribute to the sustainability transitions process. As these approaches were adopted and further modified, the depth and validity of local people's experiences and knowledge became clear (Miltin and Thompson 1995; Roy 2009; Soliman 2017, 2019).

The differences between state, government, and regime are reflected by the instruments, techniques, and socioeconomic circumstances employed. Reflecting a basic spectrum of policy options, including legalization, economic, social and informational instruments, and urban planning has accomplished a gradual shift from an exclusive reliance on state-driven approach towards a growing emphasis on negotiation, communication, and participatory commitments (community-driven process, collaborative and participatory strategic planning). At the same time, planning approaches for steering urban development have undergone a shift from government to governance, including the involvement of the stakeholders, and of interaction formats designed for this, but lack democratic legitimacy in the Middle East and implementation capacity—a dilemma that requires continuous re-calibrations between these processes. This is of equal importance for transition management since its approach essentially aims to add a layer of informal (meta-) governance guided by (cross boundary) systems thinking, not by concerns for representation (Wolfram 2018).

These theories reject the idea that the state will ever wither away and suggest instead that the domination of political life by the oligarchy is both ubiquitous and inevitable. The overthrow of any elite merely results in its replacement by another elite as happened in some of Middle East countries after the Arab Spring (Cammett et al. 2015), e.g., Egypt and Jordan. In more recent versions of the theory, the power of elites is said to derive from their occupation of central positions in society both within and outside the state (Castells 1997). The theory holds that the state seeks to articulate its influence on powerful corporate groups in modern societies. Although the theory acknowledges the dispersal of power, it is not conceptualized in the fluid terms of pluralism but seen instead as flowing through a stable set of compromising relationships between power structures and society. Also, the state acts autonomously seeking to further its own interests. In the corporate society, the state manipulates, co-opts, and compromises sectional social interests, trying to subordinate them to its own, and at the same time sustain the elite's power.

Very different explanations of the state offered by theories that emphasize conflict and domination in politics, economic and social organization, and many states in the Middle East follow a combination of liberalism, as in the case of Lebanon and Jordan, and neoliberalism, as in the Egyptian case. In the mid-1970s, many Middle Eastern countries combined liberalism and neoliberalism concepts in handling the housing problem in which many international housing policies have been applied, such as upgrading, sites and services, and self-help, as an approach to tackling urban informality (Soliman 2012b).

During the last six decades, cities in the Middle East have witnessed a remarkable and, in many ways, unprecedented increase in levels of urban informality. Urban informality is often construed as a land-use problem. Yet it arises and persists due to socioeconomic and culture necessities and material deprivation, physical and social insecurity, and ethnic and class prejudices. As a result, failing housing policies, poor development of social services, changing demographic trends, and contemporary urban informal areas remain the growth points in society and are also potential lightning rods for political and social unrest as happened in Egypt, Libya, Yemen, and Tunisia. In a context characterized by the disengagement of the state and meager local finances, Middle Eastern countries face the contradictory goals of meeting the aspirations for integration in a globalizing economy with the growing difficulties of meeting the needs of the local population and effective management of the urban environment. With the rapid growth of population, governments will continue to face increasing demand for housing and will need to facilitate land-delivery systems paying special attention to the needs of the poor and to the need to increase access to employment and income-earning opportunities. Nowhere is this dilemma more visible or are the resulting problems more intractable than in the crowded cities of the Middle East, where projections of urban informality growth remain the highest in the South.

It seems that the old Marxist idea that assumes that the government's cadre plays the role of ruling bourgeoisie overall social groups, whereby the ruling leadership is represented by the government's body and decision-making leadership, while the other class is represented by the workers (proletariat) who serve the ruling body, is currently not workable. Rather, new transitions have taken place within the civil society, where NGOs and international agencies have played a major role in the relationship between the state and the society. Also, the current changes of the nature of the state and the influence of globalization and a new order of good governance shed light on changing the attitude of the state towards the society by which the state became the main *enabler*, one way or another, for housing production for the bottom strata of the society. In other words, the states in the Middle Eastern countries are preparing, first, to integrate development activities and link housing to larger urban systems of employment and production, and, second, to decentralize resources to support local enterprise and home building.

After the Arab Spring, four main challenges are facing the Middle Eastern governments in tackling housing informality in urban areas. The first is that changes that are taking place within civil society, where NGOs and international agencies have come to play a major role in the relationship between the state and the society. The second is the change in the nature of the state and the influence of globalization requiring good governance in which the state becomes the principal enabler of housing production for the bottom strata of society. The third is that public sector management reform and private sector development cannot proceed effectively unless and until they are adapting to the requirements of good urban governance that can be implemented at the municipal level. The fourth is that a new era for democracy has emerged in various Middle Eastern countries (e.g., the revolts in Egypt and Tunisia) in which civil society, CBOs, and NGOs will play a major role in conducting the sustainable transitions process, especially in housing delivery systems. Perhaps the perpetual challenge that faces the current transitional circumstances in the Middle Eastern countries is the absence of deep understandings of the seriousness of loss of stability, security, a retreat of priorities, and the collapse of the economy. The accomplishment of all targets, enhancing housing delivery systems, and poverty reduction would require the creation of an environment capable of implementing incrementally the priorities of comprehensive sustainable development and to speed the return to a life of natural conditions for the welfare of the societies.

The challenges facing the Middle Eastern governments are the following: to rectify the imbalances between housing demand and supply; to redistribute demographic concentration from the old urban centers; and to close the gap between what the low-income groups can afford to pay and the minimum cost of a legal plot with basic services. Middle Eastern governments need also to integrate development activities and link housing to larger urban systems of employment and production. They must also decentralize resources to support local enterprise and home building at the local and national levels, providing transparency, accountability, responsibility, and trust in their vision for the future. Closer collaboration or partnership between various stakeholders is required to create the required capital for the development process. Urbanization is not exclusively a challenge for cities alone, rather it is a challenge to face chaos in cities as a result of transitional circumstances in the Middle Eastern cities.

A reaction comes in the form of initiatives of the disadvantaged group to achieve their right to the city via grassroots mobilization and informal practices to fill the gap created by socioeconomic and political exclusion. So, is there a possible way out for urban informality transitions? Is it possible to transfer the urban informality from "a new way of life" into "life of the beautiful period" or "a better quality of life?"

References

Ababsa, M., and A.Z. Abu Hussein. 2020. Metropolitan Amman: Comprehensive Climate Plans. In Shagun Mehrotra, Lincoln L. Lewis, Mariana Orloff, and Beth Olberding (eds.). *Volume II of Greater Than Parts: A Metropolitan Opportunity*. Washington, DC: The World Bank.

Al Sayyad, N. 2004. Urban Informality as a "New" Way of Life. In A. Roy and N. Al Sayyad (eds.). *Urban Informality: Transnational Perspectives from the Middle East, Latin America, and South Asia*, pp. 7–30. Lanham: Lexington Books.

Amin, S. 1980. *Class and Nation: Historically and in the Current Crisis*. London: Heinemann.

———. 2006. *A Life Looking Forward: Memoirs of an Independent Marxist*. London: Zed Books Ltd.

———. 2011. *Ending the Crisis of Capitalism or Ending Capitalism?* Oxford: Pambazuka Press.

Attia, Sahar. 2019. Al Alamein New City, a Sustainability Battle to Win. In Attia Sahar et al (eds.). *New Cities and Community Extensions in Egypt and the Middle East: Visions and Challenges*, pp. 1–18. Cham: Springer International Publisher

Avelino, F., J. Grin, B. Pel, and S. Jhagroe. 2016. The Politics of Sustainability Transitions. *Journal of Environmental Policy & Planning* 18 (5): 557–567.

Banks, N., M. Lombard, and D. Mitlin. 2020. Urban Informality as a Site of Critical Analysis. *The Journal of Development Studies* 56 (2): 223–238.

Bell, W. 1982. *Tilting at Windmills: Considerations on the Nature of the State*. London: University College London, Department of Geography, occasional Paper No. 40.

Blokland, H. 2016. *Pluralism, Democracy and Political Knowledge: Robert A. Dahl and His Critics on Modern Politics*. New York: Routledge.

Burgess, R., M. Carmona, and T. Kolstee. 1997. *The Challenge of Sustainable Cities: Neoliberalism and Urban Strategies in Developing Countries*. London: Zed Books Ltd.

Cammett, M., I. Diwan, A. Richards, and J. Waterbury. 2015. *A Political Econmoy of the Middle East*. Boulder: Westview Press.

CAPMAS. 2016a. *General Statistics for Population and Housing: Population Census*. Cairo, Egypt: Central Agency for Public Mobilization and Statistics

Castells, M. 1977. *The Urban Question*. London: Edward Amold (Publishers) Ltd.

———. 1983. *The City and Grassroots: A Cross-Cultural Theory of Urban Social Movements*. London: Edward Arnold (Publishers) Ltd.

———. 1997. *The Power of Identity*. Malden: Blackwell Publishers Inc.

d'Alençon, P.A., L. Moretto, and H. Smith. 2016. *N-AERUS – Recommendations for the New Urban Agenda*. Working Paper Group, Cities Alliance.

Dahl, R.A. 1961. *Who Governs? Democracy and Power in an American City*. Clinton: Yale University.

Davis, M. 2006. *Planet of Slums*. London: Verso.

De Soto, H. 2000. *The Mestry of Capital: Why Capitalism Triumphs in the Westand Fails Everywhere Else*. London: Blaack Swan.

Edensor, E., and M. Jayne, eds. 2011. *Urban Theory Beyond the West: A World of Cities*. London: Taylor & Francis.

Ehteshami, A. 2007. *Globalization and Geopolitics in the Middle East: Old Games New Rules*. Oxon: Routledge.

Farha, L. 2017. *Global-Commodity-Housing-Has-Lost-Its-Currency Universal*. http://citiscope.org/story/2017/-.

Fawaz, M., and I. Peillen. 2003. *Understanding Slums: Case Studies for the Global Report on Human Settlements 2003: Urban Slums Reports: The Case of Beirut, Lebanon*. Nairobi: Un-Habitat.

Forrester, K. 2019. *In the Shadow of Justice: Postwar Liberalism and the Remaking of Political Philosophy*. Oxfordshire: Princeton University Press.

Galal, A. 2004. *Potential Winners and Losers from Business Formalization*. Working Paper Series 95. Cairo: The Egyptian Center for Economic Studies.

Gerges, F.A. 2015. *Contentious Politics in the Middle East: Popular Resistance and Marginalized Activism Beyond the Arab Uprisings*. New York: Palgrave Macmillan.
Goodfellow, T. 2020. Political Informality: Deals, Trust Networks, and the Negotiation of Value in the Urban Realm. *The Journal of Development Studies* 56 (2): 278–294.
Grin, J., J. Rotmans, and J. Schot. 2010. *Transitions to Sustainable Development: New Directions in the Study of Long Term Transformative Change*. New York: Routledge.
Hamdi, N. 2004. *Small Change: About the Art of Practice and Limits of Planning in Cities*. London: Earthscan.
Harvey, D. 2009. *Cosmopolitanism and the Geographies of Freedom*. New York: Columbia University Press.
———. 2016. *The Ways of the World*. London: Profile Books Ltd.
Hendriks, C.M. 2009. Policy Design Without Democracy? Making Democratic Sense of Transition Management. *Policy Sciences* 42 (4): 341–368.
Hopkins, H. 1969. *Egypt, the Crucible-the Unfinished Revolution of the Arab World*. London: Martin Secker & Warburg Limited.
Jessop. 2002. Liberalism, Neoliberalism and Urban Governance: A State Therortical Perspective. *Antipode* 34 (3): 452–472.
Jessop, Bob. 2010. What Follows Neo-liberalism? The Deepening Contradictions of US Domination and the Struggle for a New Global Order. In Bob Jessop, Richard Westra, and Robert Albritton (eds.). *Political Economy and Global Capitalism: The 21st Century, Present and Future*, pp. 67–88. London: Anthem Press.
———. 2013. *The Capitalist State: Marxist Theories and Methods*. Oxford: Martin Robertson & Company Ltd.
Kadhim, A.M., and Y. Rajjal. 1988. City Profile: Amman. *Cities* 5 (4): 318–325.
Kattaa, M., and S. Al Cheikh Hussein. 2009. Women Entrepreneurs Facing the Informality in Rural and Remote Areas in Syria. *European Journal of Social Sciences* 11 (4): 624–642.
Lefebvre, H. 1971. *Everyday Life in the Modern World*. New York: Harper & Row Publishers.
Leftwich, A. 2010. *Redefining Politics. People, Resources and Power*. New York: Routledge.
Martin, J. 2008. *The Poulantzas Reader: Marxism, Law and the State*. London: VERSO.
Miliband, R. 1969. *The State in the Capitalist Society*. New York: Basic Books, Inc., Publishers.
Miltin, D., and J. Thompson. 1995. Participatory Approaches in Urban Areas: Strengthening Civil Society or Reinforcing the Status Quo? *Environment and Urbanization* 7: 231–250.
Mohan, B. 2011. *Development, Poverty of Culture, and Social Policy*. London: Palgrave Macmillan.
O'Connor, J. 1973. *The Fiscal Crisis of the State*. New York: St James Press.
Owen, R. 2004. *State, Power and Politics in the Making of the Modern Middle East*. London: Routledge.
Payne, G. 1999. Introduction. In G. Payne (ed.). *Making Common Ground: Public/Private Partnerships in Land for Housing*, pp. 1–16. London: Intermediate Technology Press Ltd.
Peattie, L. 1979. Housing Policy in Developing Countries: Two Puzzles. *World Development* 79: 1017–1022.
Piketty, T. 2014. *Capital in the Twenty-First Century*. London: Th e Belknap Press of Harvard University Press.
Poulantzas, N. 1979. *State, Power Socialism*. London: Verso.
Portes, A., and J. Walton. 1976. *Urban Latin America: The Political Condition from Above and Below*. Austin: University of Texas Press.
Potter, R.B., D. Darmame, N. Barhamb, and S. Nortcliff. 2009. "Ever-Growing Amman", Jordan: Urban Expansion, Social Polarisation and Contemporary Urban Planning Issues. *Habitat International* 33: 81–92.
Rawls, J. 1971. *A Theory of Justice*. Cambridge, MA: Belknap Press.
Robinson, J. 2006. *Ordinary Cities: Between Modernity and Development*. London: Routledge.
Robinson, Daron Acemoglu, and A. Jams. 2012. *Why Nations Fail: The Origins of Power, Prosperity, and Poverty*. London: Profile Books Ltd.
Roy. 2009. The 21st-Century Metropolis: New Geographies of Theory. *Regional Studies* 43 (6): 819–830.

Roy, A., and N. Al-Sayyad (eds.). 2004. *Urban informality: Transnational perspectives from the Middle East, Latin America, and South Asia*. Lanham: Lexington Books.

Salmon, T.C., and M.F. Imber. 2008. *Issues in International Relations*. Oxford: Routledge.

Shwayri, S.T. 2012. Modern Warfare and the Theorization of the Middle Eastern City. In T. Edensor and M. Jayne (eds.). *Urban Theory Beyond the West*, pp. 259–270. London: Routledge.

Smith, A., M. Fressoli, D. Abrol, E. Arond, and A. Ely. 2017. *Grassroots Innovation Movements*. London: Routledge.

Soliman, A. 1988. Housing Mechanisms in Egypt: A Critique. *Journal of Housing and Environmental Research* 4 (1): 31–50.

———. 2004a. A *Possible Way Out: Formalizing Housing Informality in Egyptian Cities*. Lanham: American University Press

———. 2004b. Tilting at Sphinxes: Locating Urban Informality in Egyptian Cities. In *Urban Informality in an Era of Liberalization: A Transnational Perspective*, ed. Ananya Roy and Nezear Al Sayyad, 171–208. Lanham: Lexington Books.

———. 2007. *Urban Informality in Egyptian Cities: Coping with Diversity*. Fourth Urban Research symposium 2007, Urban Land Use and Land Markets, May. Washington, DC: World Bank.

———. 2008. Diversity of Ethnicity and State Involvement on Urban Informality in Beirut. *Theoretical and Empirical Researches in Urban Management, Year 3, Number 9*, 15–32.

———. 2012a. Housing and the State in the Middle East. In Susan J. Smith (ed.). *International Encyclopedia of Housing and Home*, 346–354. Amsterdam: Elsevier.

———. 2012b. The Egyptian episode of self-build housing. *Habitat International* 36 (2):226–236

———. 2017. Land Readjustment as a Mechanism for New Urban Land Expansion in Egypt: Experimenting Participatory Inclusive Processes. *International Journal of Urban Sustainable Development* 9 (3): 313–332.

———. 2019. So It's Always a Chance: Community-Led Solutions to New Urban Expansion. In S. Attia et al. (eds.). *New Cities and Community Extensions in Egypt and the Middle East*, pp. 159–179. Cham: Springer International Publisher

Stirling, A. 2011. Pluralising Progress: From Integrative Transitions to Transformative Diversity. *Environmental Innovation and Societal Transitions* 1 (1): 82–88.

Swilling, M.J., K. Musephine, and J. Wakeford. 2015. Developmental States and Sustainability Transitions: Prospects of a Just Transition in South Africa. *Journal of Environmental Policy and Planning* 5 (5): 1–23.

UN. 2018. *Department of Economic and Social Affairs, Population Division, Online Edition*. Population. New York: United Nations.

UN-Habitat. 2018a. *DG 11 Synthesis Report 2018: Tracking Progress Towards Inclusive, Safe, Resilient and Sustainable Cities and Human Settlements*. SDG 11 Synthesis Report High Level Political Forum 2018. Nairobi: Un-Habitat.

United Nations. 2005. *Second National Report on the MDGs in the Syrian Arab Republic*. Damascus: United Nations.

Vellema, S. 2011. *Transformation and Sustainability in Agriculture: Connecting Practice with Social Theory*. Wageningen: Wageningen Academic Publishers.

Willems, W. 2019. 'The Politics of Things': Digital Media, Urban Space, and the Materiality of Publics. *Media, Culture & Society* 41 (8): 1192–1209.

Wolfram, M. 2018. Urban Planning and Transition Management: Rationalities, Instruments and Dialectics. In N. Frantzeskaki et al. (eds.). *Co-creating Sustainable Urban Futures: A Primer on Applying Transition Management in Cities*, 103–125. Berlin: Springer International Publisher

World Bank. 2000. *Cities in Transition: World Bank Urban and Local Government Strategy. Urban Planning*. Washington, DC: The International Bank for Reconstruction.

Yifiachel, O., and H. Yakobi. 2004. Control Resistance, And Informality: Urban Ethnocracy in Beer-Sheva, Israel. In A. Roy and N. Al Sayyad (eds.). *Urban Informality: Transnational Perspectives from the Middle East, Latin America, and South Asia*, 209–239. Lanham: Lexington Books.

Chapter 6
Land Rights, Governance, and Urban Informality

Abstract In recent years, intensive research programs on system innovations and transitions have been conducted aiming at the development of urban sustainability globally . The cities in the Middle East suffer from spontaneous, rapid urbanization, peri-urban land, and the conversion of agricultural land into urban uses, all of which became a fertile arena for land-delivery systems for the urban informality. One of the main urban challenges problems that faces the Middle East is the scarcity and insecurity of land to accommodate the growing population. Two themes are explored in this chapter: an investigation of the conversion of agricultural land into urban informality, and understanding awareness at the local, national, and international levels about the importance of secured land through proper land management for altering land markets forces, governance, and sustainability transitions on urban informality in Middle East. It is assumed that the appearance of informal land markets can be attributed to the three pillars: state, capital and social context, and that proper land governance/management concerns the property rights to land which can be defined, exchanged, and transformed. The chapter promotes blends consideration of the three pillars that would arrive at reasonable policies to fulfill the requirements of a practical land-delivery system that satisfies the needs of the growing population, as well as operates in line with the government's policies to achieve sustainability.

Keywords Land rights · Land tenure · Land conversion · Land management · Land governance · Land delivery · Transitions land management

6.1 Introduction

A variety of solutions in the urban context can be provided through science, technology, and innovation in urban planning. These innovations range from high technology-based solutions to retrofitting and other innovative approaches to urban planning, governance, and social context that employ more basic technologies. Each urban setting in the Middle East faces different urban challenges and obstacles and has different planning innovation and technology needs. Globally, such a solution is

A. M. Soliman, *Urban Informality*, https://doi.org/10.1007/978-3-030-68988-9_6

confronted by a triple challenge: environmental degradation, climatic change, and social inequality (Kanger and Schot 2018). These challenges are fundamentally linked to the First Deep Transition, namely, the functioning of societal subsystems such as poverty, duality, and economy (e.g., (Piketty 2014; Moore 2015) or the particular "Western" mode of thinking (e.g., (LaFreniere 2007; Lamba 2010), or Latin American mode of thinking (Roy 2009). The Second-Deep Transition (Kanger and Schot 2018) refers to the creation and expansion of a wide range of sociotechnical systems for the provision of transport, energy, food, housing, healthcare, communications, etc., or tackling social inequality. The Third-Deep Transition reflects the marginality and the segregation of a certain strata of the society in the urban context. One of the main components of societal subsystems is the land-delivery system, as a component of sociotechnical regime, for erecting housing and other socioeconomic amenities.

The Global South's cities are growing at an extraordinary pace. While urbanization is increasing the standard of living for many, this has not been inclusive, and arbitrary urban growth patterns create several spatial challenges impeding sustainable development. These challenges are urban sprawl, urban informality, a lack of infrastructure, depletion of resources, environmental deterioration, and a risk of natural disasters. One of the major challenges of rapid urbanization is lack of appropriate land-delivery systems for housing the urban poor in cities of the South. The land question is central to economic analysis, as well as a cornerstone of regime structure. Its use in an urban context is crucial in shaping how effectively cities function and who gets the principal benefits of urban economy transitions (Stilwell and Jordan 2004), especially concerning land ownership.

Appropriated land is the main component not only for housing production but also for all components of land uses in urban development in cities. The land question is central to socioeconomic development and to urban growth. Article 17 of the Universal Declaration of Human Rights recognizes the rights of everyone to own property, either alone or in association with others, and that no one should be arbitrarily deprived of their property. The failure of the Middle East governments' land and housing policies and the inability of the formal private sector to provide land for housing and other social amenities for the poor have strengthened the appeal of informal land markets and the plea for good land governance (Durand-lasserve 2005; Payne et al. 2009; Soliman 2010). Most of urban informality, housing informality, and other land-use informality in the Middle East has spread on the periphery of the large urban centers. The challenge of land tenure has attracted the attention of researchers, motivating them to research a proper system for securing land in simple, affordable ways (Payne 2002).

Therefore, this chapter explores two main arguments: first, it argues how recent work, such as sustainability transitions on urban planning innovations, can perpetuate agricultural land conversion to urban informality. Second, it raises the awareness on the local, national, and international levels about the importance of secured land through proper management of the alteration of land-markets' forces, governance, and sustainability transitions of urban informality in Middle East. The aim is to create practical and applicable land-delivery systems and to remodel the current planning policies to serve the immediate and future population needs for housing plots in a cooperative way to alleviate poverty, on one hand, and to avoid the obliteration of the built environment, on the other. This chapter consists of five sections. The first examines land conversion into other functions, mainly residential uses. Section 6.3 elaborates on urban land management of urban informality, while land tenure is discussed in the Sect. 6.4. Land markets in informal areas are explored in Sect. 6.5, and the chapter concludes with suggestions for future intervention into accelerating land-delivery systems for the urban poor of the Middle East.

6.2 Land Conversion

In the absence of planning and regulatory frameworks in the Middle Eastern cities, urban informality spread on the fringe of the urban areas and conquered what is called peri-urban areas. These areas face severe environmental, economic, and property-related challenges and gradually lose their role in supplying cities with food, energy, water, housing, social amenities, and ecosystem services. The post-2015 agenda and Habitat III (particularly Habitat III Issue Paper 10 on Urban-Rural Linkages), which translated into the SDG 11a target and subsequently into the NUA, call for support for positive economic, social, and environmental linkage between peri-urban and urban areas to strengthen national development (UN-Habitat 2018a). Rapid migration of people to expanding cities in the Middle East and the resulting population growth (e.g., Egypt annually reports 2.5 million births) make it harder to predict, plan, and build the required housing and the needed social services. Urban informality results in the urbanization of land, mainly on peri-urban areas, at higher rates than urbanization of people. It damages the environment and affects the livelihood of peri-urban areas by converting land into uses that could otherwise be utilized for agriculture, housing, and other activities, e.g., in Sao Paulo, Brazil (Torres et al. 2007) and Cairo, Egypt (Soliman 2012a, c) In the Middle East, one of the most important contexts of the arbitrary urban transition is the peri-urban interface or peri-urban periphery. There is a connection of rural and urban activities, formal and informal institutions, poverty, inequality, environmental degradation (Allen 2014), and formal/informal land-market forces (Soliman 2015) that are the most often closely associated issues. As a result of the speedy urban expansion in the cities of the Middle East, like Cairo, Beirut, and Amman, land has become a primary obstacle for urban development. This results in increasing exclusion from the benefits of urbanization, urban growth, and from urban policy processes. A shift from a 'Chicago School' of urban sociology to the 'Los Angeles School' of postmodern geography has materialized to absorb and promulgate understanding of rapid urban growth (Roy, 2014). This dominant path is driven by the cycle of neoliberal restructuring of peri-urban space that facilitates the co-option of environmental agendas, exclusive urban greening and clean-up, and control responses to the crises of rapid urbanization. Many urban informalities of peri-urban residents in the Middle Eastern cities face the exacerbation of environmental degradation and health

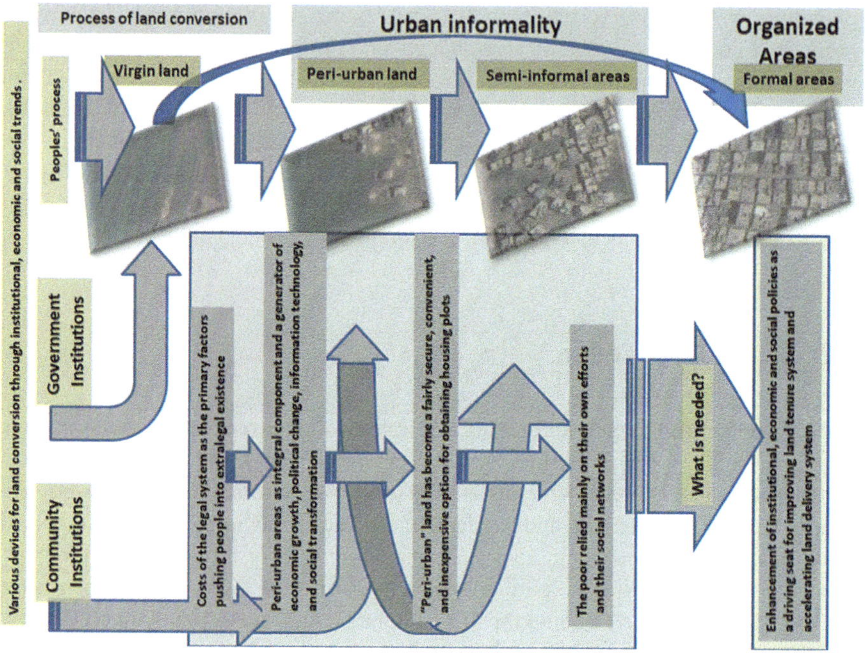

Fig. 6.1 Diagram of the land conversion process in the Middle Eastern cities. (Source: adapted from Soliman 2019)

risks associated with the negative impacts of urbanization. This is accompanied by increasing exclusion from political and development processes that further reinforces the urban agenda of neoliberal restructuring. *It is argued that the peri-urban is a critical frontier* of sustainability and that peri-urban innovation processes, if appropriately recognized and nurtured, may hold the potential to be transformative and to challenge the dominant urban trajectory (Marshall and Dolley 2018). While rural and peri-urban populations benefit from new economic dynamism in manufacturing and services brought by urbanization, they do not always enjoy improvements in quality of life due to adverse environmental consequences of urbanization, as well as lack of tenure security.

The vague framework for land use has led many people in the Middle East to illegally convert agricultural land or peri-urban areas (Colsaet et al. 2018) to other uses. The conversion of peri-urban land (usually agricultural land) to urban informality, for housing or other activities, through a peoples' process (Soliman 2010) became the main mechanism for urban transition in the Middle East. This conversion created land, rights, wealth, and innovation, by which it achieved a realistic land-delivery system, facilitated access to cheap housing plots and other social and economic amenities, and enabled the urban poor to obtain land plots for housing on acceptable terms and conditions (Jenkins 2004; Soliman 2010, 2012a, c).

As illustrated in Fig. 6.1, the processes of land conversion to urban informality are attributed to a peoples' process (Soliman 2019; UN-Habitat 2016) associated with government and community institution who have managed to develop and transfer peri-urban land to urban use. In recent years, agricultural land acquisition, with some sort of tenure security, or "peri-urban" land or land taken in the Middle East cities has become a fairly secure, convenient, and inexpensive option for obtaining housing plots (Jenkins 2004; Soliman 2010; Marshall and Dolley 2018). Despite the fact that this land has been characterized by profound overcrowding and physical compactness and has not been officially documented for land subdivision (Soliman 2017), it has offered a solution for an alternative land-delivery system for the urban poor.

In addition, the loss of agricultural land inflicts major impacts on the Green and Brown Agendas (Duquennois and Newman 2009), and, over time, has changed the patterns of urban morphology in the urban matrix of cities of the Middle East and elsewhere (Rashed et al. 2005). Presently, there are more widely used labels such as: "urban-rural fringe", "urban transition zone," "semi-informal area," (Soliman 1996), "urban informality," (Al Sayyad 2004), "land take" (Colsaet et al. 2018), brown fields (Soliman 2013), "hybrid areas" (Simon 2008), and the "Extended Metropolitan Area" (Lynch 2005), which have, to some extent, become interchangeable with "peri-urban" areas. On the other hand, land take can be defined as "the change in the amount of agricultural, forest and other semi-natural and natural land taken by urban and other artificial land development (Diana Reckien and Karecha 2008). Iaquinta and Drescher (2001) proposed five typologies for peri-urban areas that are embedded in the broader rural–urban dynamic: village, diffuse, chain, in-place, and absorbed peri-urban. The latter is the subject of this study, and it could also be called semi-informal areas (Soliman 1996) or *"hybridity cities within a city"*, e.g. Buluq El Dakror in Cairo, and Sahab in Amman. It constitutes areas adjusted to the official boundary of the city or areas that are located laterally on the main roads or adjacent to various urban infrastructure. Semi-informal areas are often characterized by diverse populations, growth and expansion, heterogeneity of land use, morphological conditions and densities of built-up areas, demographic changes, and complex functional relationships and social structures (DEFID 1999; FAO 2001) and are mainly ruled under traditional, communal or customary tenure arrangements.

"Semi-informal areas" refers to the areas of essentially urban character located on the urban fringe, interspersed with, surrounded by, or adjacent to main roads or infrastructure, or adjacent to undeveloped sites or ones with agricultural uses where urban and rural development processes meet, mix, and interrelate, usually on the periphery of cities. As Fig. 6.1 illustrates, semi-informal areas are transitional zones between urban and rural areas that are undergoing urbanization and are progressively assuming many of the characteristics of urban areas. Semi-informal areas include areas sealed by construction and urban infrastructure, as well as urban green areas and sport and leisure facilities. Semi-informal areas articulate the urban informality in cities of the Middle East. The exact delimitation of commonly agreed-upon urban informality designates land consumption for urban development that can take the form of either lower or higher density, economic activities, social activities, or dispersed development—or all combined.

The processes of land conversion of semi-informal areas or peri-urban areas from a particular use, usually agricultural, to residential or economic uses are linked to numerous variables: public infrastructure policies, land tenure status, fundamental change in sociotechnical systems for agricultural productivity, locational preferences, material mobility, along with the social ecological systems with which these intersect (Soliman 2004b). This system-wide change with a *second deep transition* (Schot 2016) involves shifts in the direction of change across multiple scales of public infrastructure policies, political changes or transitions, social networks changes, land-market forces, and the type and multi-actors involved in the process of the transitions. This system-wide change amounts to what Schot (Schot 2016) called a Second-Deep Transition, involving shifts in the direction of change across multiple sociotechnical systems on multiple scales. Schot and Steinmueller (Schot and Steinmueller 2016) argue that, for such deep and wholesale changes to occur, a new framing of innovation policy is required that goes beyond innovation for growth. As illustrated in Fig. 6.2, the El Rezqa area as a land readjustment project was implemented as a new innovation tool to control the arbitrary development of peri-urban areas on the periphery of Benha City (Soliman 2017).

This innovation of transitions passes through various processes/mechanisms, such as state-driven processes, hybrid-driven processes, and community-driven processes. Each process has a number of actors and may be overlapping with other two processes. The interpolation of sociotechnical regimes reflected on hidden potential actions of local communities, technical assistance, and a cooperation of local authority as a cooperation/participation process among all actors involved in transitions process (Soliman 2017, 2019).

Various bodies participate or cooperate in formulating, accelerating, and encouraging informal processes/mechanisms of agricultural land conversion or semi-informal areas to urban informality in cities of the Middle East (Soliman 2004b). As Fig. 6.3 diagrams, these bodies constitute the regime level in the sociotechnical perspective that are responsible for accelerating and/or suspending land-delivery systems for the urban poor. These include public institutions (facilitators) and governmental agencies (regulators), whatever their forms and organizations; small proprietors (providers) who own larger or smaller parcels on the periphery of the city's boundaries; the private developers (operators) who bought large plots in key locations of the agricultural areas; prospective customers of land supplied by the operators; shadow consultant who acts as a hidden facilitator between the main consultant and the stakeholders, and finally, the informal credit system (the *Gamaiyyat*) (Soliman 2010), which contributes by providing the necessary cash for purchasing land within cities.

Public institutions (facilitators) and governmental agencies (regulators), whatever their form and organization, have frequently encroached on the peripheral agricultural areas by constructing some governmental buildings, such as public-housing complexes, administrative buildings, and/or educational complexes or creating road networks passing by, or linked with, peri-urban areas. Because these are public premises, various services, such as water and electricity supplies and sewerage systems, are then installed. This indirect public-sector intervention, or political

Fig. 6.2 Photos illustrating the process of land conversion of peri-urban areas into residential development in El Rezqa area on the periphery of Benha city, which is located approximately 45 km north of Cairo, Egypt's capital. (Source: The author)

informality (Goodfellow 2020), has changed the land-use pattern from agricultural areas to urban land uses and integrated them into the urban fabric of cities. It has also encouraged private developers to follow the same procedures in changing the use of agricultural areas. Some entities, working in the engineering departments in cities' municipalities, as a part of the sociotechnical regime, indirectly participate in the issuing of documents, stating that the land is unsuitable for cultivation. In all cases, these documents are used in the Middle East cities' courts. Other people in municipalities are illegally facilitating the construction process on land plots in urban informality against lump sums of money.

Small proprietors (providers) who own small parcels on the periphery of cities' boundaries are also involved. Public intervention into agricultural areas has

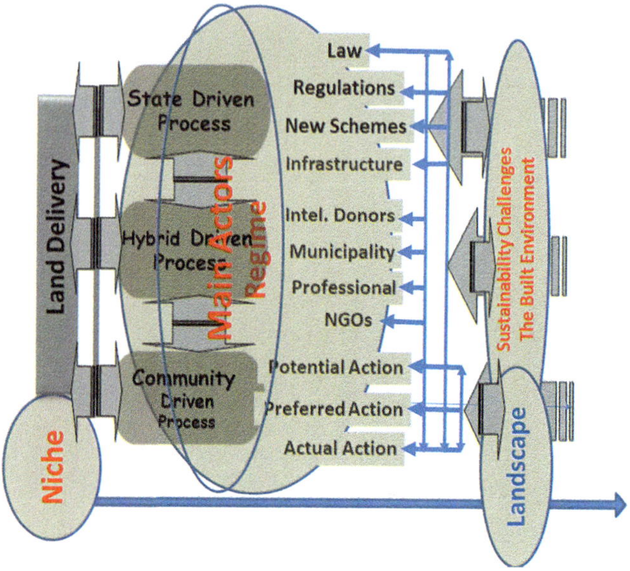

Fig. 6.3 Scheme of the three processes of land-delivery system (state-driven process, hybrid-driven process, and community driven-process) which constitute the regime level in the sociotechnical perspective. (Source: The author)

encouraged them to sell their plots to newcomers and accelerated the processes of peri-urbanization. Unfavorable farming conditions are another factor accelerating the illegal subdivision of agricultural land. In cases where the farmers continue to cultivate their plots, their proximity to urbanizing areas raise problems, such as children and domestic animals from neighboring dwellings spoiling their crops and the shadows cast on crops by adjacent dwellings.

Private developers (operators) who bought large plots of land in key locations of the agricultural areas also play an important role. These private developers slyly responded to state's actions towards enlargement of cities' boundary and acted according to the prevailing environment. In many cases, the private developers who owned large areas built small mosques or *Zouhia* (a tiny mosque) an the border of their lands, and donated to the Waqf authority, in full knowledge that the local authority dare not demolish a place of prayer (Soliman 1999). Later, the authorities agreed to provide services to *Zouhia*, and, therefore, the private developers had moral support from the local authority, as well as basic services installed on their lands.

Prospective customers of land supply are the engines of the process. These are generally a mixture of the low-income and middle-income households most in need of housing sites, and of high-income groups who are looking for land investments for land speculation. They represent around 40–50% of the total population of cities in the Middle East. They always purchase their land on an incremental basis based on a gentle agreement with the private developers, at conditions agreeable to both.

Financial institutions (merchants), whether in private or public sectors or international donors, participate by providing the necessary cash for purchasing land within cities. The roles of the public, international donors, and private commercial developers vary according to the level of interest, risk, and benefit. Each seeks to maximize the rate of return on investment at the lowest level of risk but act jointly to reach the same goal without harming the other. The private-sector land-delivery system operates in two basic ways. The first involves financing land acquisition and providing plots through informal contracting arrangements. This often entails illegal land subdivision or acquiring large parcels on the periphery of urban areas to supply plots on a commercial basis for people most in need. The second involves financing and operating the land-delivery system through informal contracting arrangements. The role of international donors differs depending on which sized locations for rent are available through official contracting arrangements for a long period of time. The financial conventions system (the *Gamaiyyat)*, within local social networks and cultural norms, is introduced among the beneficiaries in a way that suites their financial needs, requirements, and responsibilities through issuing unofficial documents between them to guarantee the rights of the other.

The success and failure of the agricultural land conversion depends upon the number of people who acquire land, the tenure status of the land acquired, the presence or absence of pressure from the landlord, the strengths of groups backing the conversion process, and the sensitivity of the governments to pressure from such groups. Agricultural land conversion has also attracted further intruders and settlers to cities of the Middle East who are seeking land speculation and looking for social services and large land parcels close to cities at affordable prices. Also, "the new urbanity" produced by the agricultural land conversion has created a pattern and space form different than the old parts of cities in the Middle East. The residents themselves without government or professional intervention created this pattern, mainly grid iron pattern. The settlers were their own architects, and they formed their settlements according to their needs and requirements. All decisions about what to build and how to build were in the hands of the users, which represents the principle of people's processes (Soliman 2010, 2019). On the other hand, Cirolia (2017) argues, "interventions to improve the functioning of land markets and address urban poverty should recast the politics of informality and identity as an opportunity for the urban poor to use various forms of capital and association to incrementally and progressively build pathways to urban citizenship." The operational complexity of linkages between the "informal" economy and informal land market are correlated (De Soto 2000; Bromely 2004; Roy 2009). Within this context, it is interesting to consider how the poor acquired land and how the land acquisition is managed by which the subdivision process takes place in a sustainable way.

6.3 Urban Land Management

Urban land governance is a key issue in controlling land use in a given environment, and the state is the only body that possesses the authority tools to create a legal, fiscal, and regulatory framework to land governance. It is argued that land governance can act as an engine for economic growth, shared prosperity and social justice in cities of the Middle East (Nada and Sims 2020; De Soto 2000; GLTN's Arab Land Initiative 2021; Soliman 2021b).In Middle Eastern cities, a lack of accurate information system on land issues (De Soto 2000), the absence of participatory decision-making structures, and a lack of clarity of cadastral information have left the land market in the hands of speculators looking for high profit. Also, the lack of understanding of the mechanisms of land tenure from Islamic conceptions of land and property rights has left land tenure in the cities of the Middle East in doubt (Sait and Lim 2006). In Muslim societies, the range of land- tenure forms contributes to the development of authentic and innovative strategies for enhancing access to land and land rights and allowing for enhanced land-delivery systems (Sait and Lim 2006). It is argued that the field of governance studies concerns the changes in governmental practices and organization from hierarchical to decentralized and horizontal structures (Grin et al. 2010), and facilitates tackling the main challenges that face land issues in the Global South.

There are different views, definitions, and perspectives attributed to land governance by various scholars. Land governance concerns the rules, process, and structure through which decisions are made about access to land and its uses, the manner in which the decisions are implemented and enforced, and the ways that competition for land interests are managed (Palmer et al. 2009). Kauffman (2017) defined governance as "the management of the course of events in a social system". This definition highlights that governance is not merely a process resulting from structure: it contains agency. It involves the purposive efforts by both state and non-state actors to "steer" society toward the pursuit of particular goals (Andonova et al. 2009; Brand 2005; Kjaer 2004). It is argued that public governance is important throughout the policy-making cycle, not only the design but also the implementation and evaluation of inclusive growth policies (OECD 2016). Coase (1960) noted that good institutions, such as strong property rights, reduce transaction costs and consequently support economic development.

Assessing land-governance indicators in cities of the Middle East does not rely on an accurate information system in land issue, participatory decision-making structures, diversity of societal actors (see the previous section), market forces, and land registration that would reflect the reality of the land dynamics of a given environment. For example, 90, 60, and 70% of urban areas in Egypt, Lebanon, and Jordan, respectively, are not officially registered. However, transitions land management identifies forms of interactive and participatory decision making as ways to create societal consensus and/or pressure as a counterbalance to more hierarchical or bottom-up market approaches (March and Olson 1995; Kanie and Biermann 2017; Edelenbos 2005).

Transitions Land Management (TLM) can be thought of as the system of strategic processes and tools, as well as institutions, rules, market forces, urban growth

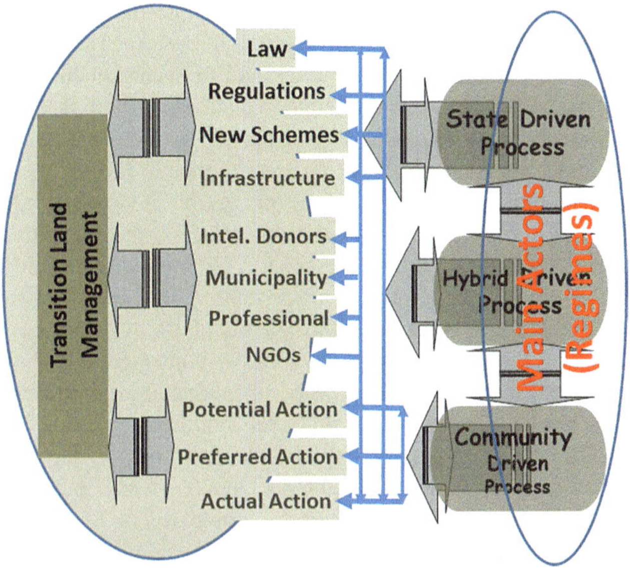

Fig. 6.4 Diagram outlining the three driven processes of transition land management in Middle Eastern cities. (Source: adapted from Soliman 2019)

(biophysical aspects), and interactions for effective policymaking. A central role for land management is the involvement of non-governmental organizations (NGOs) and Community Based Organizations (CBOs) as the main engine to communicate between the state and the community concerned. Kauffman (Kauffman 2017) argues that NGOs and CBOs do this by three devices: constructing network ties with organizations and actors that have the potential to influence targeted populations; working to change these actors' way of thinking about the issues they seek to govern; persuading them to use their own resources (and those available to them through their network ties) to take action; as well as, persuading people who are looking for land plots of the future benefit of capital accumulation to secure their future status.

As illustrated in Fig. 6.4, TLM is a continuous process to regulate/alter urban growth, market forces, institutions, local context, and various rules. It is applied to formal and informal land sectors alike. The former follows the prevailing law, and the latter follows customary rules. TLM is seen as a Strategic Niche Management (SNM), i.e., as a crucial form of policy intervention to enable the creation of robust and influential niches (Frantzeskaki et al. 2017). Niches are seen as "a course of events to protect spaces and places" for experimenting with alternative sociotechnical configurations, interacted and correlated with the institutions or the regime structure (Smith and Raven 2012). The sociotechnical regime is a collection of actions, or a process of transitions, by various actors varying between socio-spatial transitions, markets forces, institutions, local context, and rules. Derived from Geels and Deuten (Geels and Deuten 2006), sociotechnical landscapes are spatial configurations (formal/informal) but conceived of as 'cosmopolitan' networks constituted of 'local initiatives' and 'trans-local' intermediaries that may span various scales. Sociotechnical niche formation is described as a progression of transitions in which intermediaries refine lessons from market forces and people's initiatives and offer transferrable knowledge to institutions, who then reinterpret and apply rules in their local contexts. This supports the consolidation of learnings (learning by doing and doing by learning) and replication of successful practices, thereby increasing the influence of the niche on regime actors to adopt new solutions (Raven et al. 2008). The continuous transition process, therefore, overlaps and is sequenced into phases, beginning with isolated initiatives ('local phase'—formal/informal), to the first exchanges of experiences among initiatives ('inter-local phase' formal/informal), and the increased aggregation of knowledge across initiatives ('trans-local phase' formal/informal), towards the consolidation of a robust landscape that coordinates local projects (formal/informal) and exerts strong influence on the regime (institution—formal/informal) (Geels and Deuten 2006).

Within this context, it is interesting to consider how the poor have acquired land, and how the actors have conducted the land acquisition and subdivision process in cities of the Middle East. The operational complexity of linkages between the "informal" economy and informal land market in the Middle East are interrelated (De Soto 2000; Bromely 2004). The states in the Middle East have institutions and policies in place that allow right holders to easily enforce those rights and exercise them in line with their values and aspirations and in ways that benefit society as a whole. The weakness of the institutional framework in the Middle East's cities has led to a failure to recognize existing rights, the presence of tenure insecurity, curbing of investments in land, increased potential for conflict, and diversion of resources that can be more productively deployed elsewhere to the defense of property claims. Finally, is the issue of the illegality of recognition of existing land rights. Institutional obstacles of urban governance and its effectiveness on the land- delivery system were noticeable in Middle Eastern cities, as well as constraints embedded in the political order and its logic of top-down rule that was enforced in land supply (Dorman 2007), but could act as political informality. It argued that informal politics is ubiquitous, but varies significantly in terms of its interface with formal institutions and the degree to which it is shaped by shared norms and expectations (Goodfellow 2020). Also, the linkage between urban transitions, market forces, institutions, local context, and rules constitute the interaction between niches and regimes, formally or informally, by which sustainable/unsustainable transition is taken place.

The weak linkage between the social institutions and state-society relationships in the Middle East have reinforced disputes over land regulation and transactions (Soliman 2012a, c; 2019). The importance of government policies and regulations that support and facilitate informal mechanisms of land access is essential (Durand-lasserve 2005; Hendriks 2008). Therefore, as long as, the correlation between local and inter-local phases, formal and informal, is dropped, as in the Middle East, the greater the opportunity for the appearance of informal land markets by which urban transitions have been excluded from planning control (e.g., the arbitrary urban expansion of Cairo, Beirut, and Amman). On the other hand, further agricultural land loss in the form of peri-urban areas has increased unemployment and poverty within cities in the Middle East, and mismanagement of vast desert public land and what could be seen as "a well-oiled machine" has resulted in a great loss of assets (Sims 2014), has decreased the quality of life, and has had irreversible effects on the environment (Toth 2009). There is some theoretical or philosophical debate as to whether land itself can be 'owned', but little dispute that there are Islamic rights to 'use' and 'possess' land. In principle, the rights to land are linked to land use, and the person who seizes the land has priority over another who has failed to use it. Following the same logic, only productive land should create wealth (UN-Habitat 2011).

From a market forces point of view, De Soto in his study in Egypt argued that the formalization of land tenure can bridge the gap between the informal and formal economy by converting "dead capital" to "life capital." This message speaks directly to the first and second economy (De Soto 2000). However, Gilbert (Gilbert 2012) argued that, if the message about titling is popular and influential, its application has, so far, not been shown to contribute much good or, fortunately, much harm. But it could be said that formal land title would facilitate obtaining various services. It is argued that one of the root causes for non-inclusion of informal settlements in the provision of services relates to the legal tenure/occupation of the land on which they are located (UNESCO 2019). On the other hand, both De Soto (2000) and Gilbert (2012) dismissed the alteration of urban informality with land value at a low price into urban land value with a high price which created wealth and increased the assets of the urban poor and middle-income groups, regardless who secured formal tenure (Soliman 2010). Gilbert (Gilbert 2012), in his criticism of De Soto's Mystery, claims six main arguments about the relationship of land title and housing production, but he dismissed the importance of urban land governance and people's process of urban informality within the urban transitions which is integrated with informal markets, informal institutions' mechanisms, and the actors who play out these processes.`

From the perspective of socio-spatial growth, land values and uses are determined by physical characteristics such as the location of land and by crucial factors such as social, cultural, economic, and political situations, whether rural, urban, or peri-urban. Extensive research has attributed the occurrence of physical expansion in cities of the Middle East to problems in input markets, especially in regard to price distortions caused by restrictive governments' regulations and laws on land reforms, security of tenure, financial credit, and building regulations (Doebele 1994; Pamuk 2000; Jenkins 2004; Leduka 2004; Gelder 2009; Pritchard 2013; Lall et al. 2009).

TLM is the output of interaction between niches and regimes by which both play a great role in facilitating or unstacking the land-delivery system. Therefore, TLM is a continuous process that is controlled and interlinked, formally or informally, by the state through its institutional and regulatory structure, and it is operated and implemented by people in a society that interacts or is driven by various market forces, either formal or informal, in which the built environment has been shaped. Most of the states in the Middle East, as well as the rest of the Global South, suffer from grasping and ill-structuring systemic failures that have crept into the societal systems.

The branch of land management that pertains to transition land management, along with multilevel institutions, suggest that new, more flexible multi-actor network-based governance is required to address the constellation of sustainability challenges in the urban context of Middle Eastern cities. This is an unambiguous contrast to more traditional, top-down, centralized government that may exclude the meaningful participation of the private sector, civil society, scholars, and other groups. While a range of actors has played central roles, political leadership is a key trigger of land sustainability transition. Most of the states in the Middle East suffer from the absence of management models that engage new actors in creative ways: exploring and exploiting interactions between urban growth, sustainability, and other urban priorities and the growing awareness behind a political body that favors avoiding environmental risk-taking.

From Fig. 6.4, it appears that there are various challenges facing a proper land management system with LTM in cities of the Middle East. There are three driven processes; state, hybrid, and community, which manage and control land-delivery systems and direct land management systems. The state-driven process manages and facilitates land-delivery systems through four procedures: law, regulations, new schemes, and level of infrastructures. The hybrid-driven process facilitates land delivery system through international donors, municipalities, professionals, NGOs, and CBOs. Community-driven process controls and manages land-delivery systems through three actions: potential, preferred, and actual actions. These three driven processes are interrelated and overlap due to the current and immediate circumstances that each environment passes through. Therefore, these three driven processes constitute the basis for formulating sustainable land governance.

However, proper land governance concerns the property rights to land that can be defined, exchanged and transformed and the way in which public oversight over land use, land management, and taxation are exercised. Proper land governance also tolerates: the way the state land and private land are managed, acquired, and disposed of; the nature and quality of land information available to the public; the ease with which it can be accessed or modified; and the way in which disputes are resolved and conflict is managed (Deininger et al. 2012). It encompasses statutory, customary, and religious institutions. It includes state structures such as land agencies, courts, and ministries responsible for land, as well as non-statutory actors such as traditional bodies and informal actors. It covers both the legal and policy framework for land uses, as well as traditional and informal practices that enjoy social legitimacy. Therefore, the challenges of land management, especially for the urban poor, are the outcomes of unclear frameworks for land use, the absence of institutional frameworks, weakness of social institutions, the deterioration of economic situations, and lack of attention to the socio-spatial effects that could affect the city's growth and citizens in the future.

6.4 Land Ownership, Tenure and Title

Most of the literature review (Archer 1999; McAuslan 1985; Fernandes and Varley 1998; Payne 1999, 2002; Nunan and Devas 2004; De Soto 2000; Payne et al. 2009) emphasized the importance of land tenure as a way to increase the supply of land for housing the urban poor at reasonable prices, and securing land tenure that can promote economic activities in the localities. Lately, three 'schools of thought' characterize the debate about land reforms in Africa. On the one hand, there are those who argue that land policies should be rooted in a theory of social capital, especially the African traditional land-tenure system. On the other hand, there are those who are convinced that individualized tenurial systems are more effective and desirable (Obeng-Odoom 2012). The third is that, through historicism and rigorous anthropological studies, both longitudinally and comparatively, there is now abundant proof that common property in land is feasible and property in customary is natural for people's lives, but with a tendency to chaos, crisis, and grave inequalities (Obeng-Odoom 2016).

Legalization of land tenure is, therefore, a critical issue in relation to urban informality because it ultimately implies legality and hence, in some sense, formality. Because the urban informality contains informal residential areas and various informal socioeconomic activities that encompass small- and medium-scale entrepreneurs each needs special treatment for formalization or regularization. But legalization is itself multi-dimensional, and as the land issue is considered as a main component of housing production, formally or informally, the security of land tenure is essential in the formalization process. With this broad generalization, lack of secure rights to land and property has been highlighted as a key factor in explaining the limited success of capital accumulation in the Global South (Doebele 1987; Fernandes and Varley 1998; De Soto 1989), and considered a main impediment for legalization policy. It is also, an essential for poverty reduction, enhancing social value, eradicating social inequality, and refining the local economy.

Formalization of land tenure of the informal housing sector obviously needs a comprehensive set of regulations and policies rather than a piecemeal approach in which one issue is tackled while others are ignored. This realization led the 1976 UN Conference on Human Settlements (Habitat) to adopt seven coherent sets of urban-land recommendations (McAuslan 1985). Following this conference, an edited volume as an output of a seminar entitled "Land for housing the poor: towards positive action in Asian cities" ended up with seven subject areas that require action on land for housing the poor (Angel, et al. 1985). One of the main recommendations is pursuing research and development activities focusing on the land- tenure issue. It was suggested that every society, at every stage of its development, must invent a paradigm of land tenure that optimizes efficiency, while maintaining enough equity to keep the society stable (Doebele 1994). But what type of secure tenure is needed? How has the urban poor managed to secure land? Has their land tenure secured their status and sustained their collateral? How are regulations reflected in land values?

In 2000, the Millennium Development Goals (MDGs) emphasized Goal number 7 to ensure environmental sustainability in which target number 4, to be achieved by the year 2020 is significant improvement in the lives of at least 100 million slum dwellers. The SDGs target 11 explicitly acknowledges the global importance of security of land tenure as key in urban areas for the provision of services but also offers a foundation for access to a basic means of production land (UN-Habitat 2018). The impacts of insecure tenure and its links with poverty, and thus the role of secure tenure in poverty reduction, is a prime concern. According to the adherents of tenure legalization, the provision of property titles establishes the tenure security needed for residents to invest in housing improvement, enabling land markets (The World Bank 2008; Fernandes and Varley 1998) while also empowering housing investment (Cohen 2009), and improving local economies (De Soto 2000). Thus, legalization of land tenure is articulated to balance the social value of land and housing with its economic value and include diverse, incremental, and socially acceptable forms of land tenure and property rights (d'Alençon et al. 2016).

However, article 17 of the Universal Declaration of Human Rights 2 recognizes the rights of everyone to own property either alone or in association with others and that no one should be arbitrarily deprived of their property. In addition, the right to an adequate standard of living and security (Article 25) entails a universal right to adequate shelter (UN-Habitat 2008). Worldwide, urban land cover occupied 0.47% of the total land area of countries, ranging from 0.62% in all developed countries to only 0.37% in developing countries (Angel et al. 2011). On the other hand, the most positive aspect of the changes in the World Bank policy (Buckley and Kalarickal 2006) since the mid-1990s is that the bank has recognized the active role of the private sector, a recognition of the high speed at which market-based housing finance flows into world economies, treating land as an important input in the provision of housing services, which accounts for a large fraction of total housing costs, and emphasizes the importance of community and NGOs involvement.

Subsequently, arguments in land reforms (McAuslan 2013; Harris 2014) gained momentum after the publication of "The Mystery of Capital" by Peruvian economist Hernando De Soto in 2000. The poor, De Soto argues, are often unable to secure formal property rights, such as land titles, to the land on which they live or farm because of poor governance, corruption, and/or overly complex bureaucracies. De Soto's proposition that the formalization of land tenure can bridge the gap between the informal and the formal economy by bringing 'dead capital' to life speaks straight to the first and second economic objectives. He added that property was the "genesis of the rest of the market economy", and that "without property titles, the merging of the first and second economies would be impossible" (De Soto 2000). On the other hand, the conversion of agricultural land that has de jure recognition to housing informality, with land value at low prices, is the changes of rural uses into urban land with de facto recognition that has the privilege of high prices. This conversion created wealth and increased the assets of the urban poor, regardless of the scarcity of securing formal tenure or even having a title, formally or informally (Soliman 2010).

The roots of rules governing the land in the Middle East originated from the Ottoman Empire's regulations during that period (1517–1797). In sum, all lands under the Ottoman Empire was state land and classified into two categories: private and public. The latter constituted four typologies; *Otealak, Kharaj, Iltizam,* and *Waquf,* while the former comprised two types; Sultan (*Rzeq*) and *El Heteta.* All private lands were owned by the sultan or emir (prince) and distributed among princes and relatives to the ruler of the country. Later, the French mandate (1798–1801) sustained the same land typologies along with creating cadastral maps for land in the major cities in the Middle East.

During the era of Mohamed Ali (1805–1845), Egypt and the eastern part of the Mediterranean Sea (Lebanon, Palestine, Syria, and Jordan) were part of the Egyptian empire. Land codes were issued and classified the land into two types: privately owned (*milk*) land under certain conditions; and public domain land called *Miri* (land under the custody of the *Amiri*, or prince). The *eltezam* (commitment) concept was abolished and replaced with a tax system according to Islamic rule. Muslims would pay a tithe called *ushuri* equivalent to one-tenth of the production of their land, while non-Muslims—if allowed to keep their land—would pay a tribute called *kharja.* Military officers were awarded lands called *iqtaa* for their services (with the obligation to cultivate the land).

According to the land classification, a system of taxation applied to every piece of land, either privately owned (*milk*) or part of a domain called "*Miri*" (land under the custody of the *Amiri*, or prince), that included all arable fields. *Miri* holders would enjoy the use of the land after being registered officially and against payment of a fee (*tapu*), on condition that it remained cultivated (Fischhbach 2000). *Milk* land was directly composed of former *ushuri* and *kharaj* lands and also included land for dwellings with appurtenant plots not exceeding one dunum. Two other state's land categories were introduced: *matruka* (such as roads, threshing floors or marketplaces) and *mewat* ('dead' or wasteland such as mountains and rocky places far from villages). *Waqf* land was another class of state's land dedicated to some pious purpose and put under the custody of God, and therefore "protected by the strongest legal and religious sanctions known to Muslim law from seizure by the state or its officers" (Sela 2014). Private land under certain conditions were created upon payment of a fee, being issued a title deed to the continued usufruct of the land, the land being then reclassified as *Miri* (Fischhbach 2000). If land was left uncultivated for three years (and then called *mahlul*), it would revert to the state. The British initiated a land-valuation study, dividing the surveyed area in *hawds* ('basins' of no less than 25 ha) and ascribing a taxation figure to each, based on the estimated productive value of the land (grade of productivity). A land-settlement program ensued, whereby plots would be ascribed to specific owners.

On the other hand, land sharing or *Musha* in Islamic law provides a good alternative to reduce land prices, as well as to accelerate land-delivery systems in urban areas. By systematizing existing insights provided by Richard Schlatter in his book, *Private Property: A History of An Idea* (1951), Obeng-Odoom demonstrated that there is no singular "property rights approach" in political economy but rather

Table 6.1 The typologies of land ownership in the Middle East

	TYPE A *On agricultural land*		*TYPE B* *On desert land*	
SUBTYPES	*A1 On privately owned land*	***A1a*** *Privately registered owned land*	*B1 On Decree land*	***B1a*** *By assignment to a development company*
		A1b *Privately no registered owned land*		***B1b*** *By assignment to a public-sector company*
		A1c *Privately in doubt ownership*		***B1c*** *By assignment to cooperative*
	A2 On State land	***A2a*** *Agrarian reform land*		***B1d*** *On antiquities and cemetery land*
		A2b *Railway land*		***B1e*** *On armed forces land*
		A2c *Irrigation Land (waterbody/lake/canal-edge land)*		***B1f*** *On police forces land*
		A2d *Ministry of agricultural land*	***B2*** *On State domain land*	
	A3 *On municipal land*		***B3*** *On municipal land*	
	A4 *On endowment (waqf or Bait-Elmaal) land*		***B4*** *On reclaimed land*	
	A5 *Herasa land*		***B5*** *On new urban communities' land*	

Source: The author

several property rights approaches not only between heterodox and orthodox political economy but also within heterodox perspectives (Obeng-Odoom 2016).

In the Middle East, two types of land ownership exist: public and private land and cover agricultural and desert land, where the latter is owned by the state, and the former varied in its ownership between the public and private sectors. Typologies of land in cities of the Middle East are classified agricultural land according to five main sub-categories, and desert land categorized into five sub-categories. The ownership of agricultural land is divided into privately owned agricultural land, state-owned agricultural land, municipal land, the endowment (*waqf* or *Bait-Elmaal*) land, and *Herasa* land (see Table 6.1). The first two types are subdivided into five types where privately owned land is subdivided into three sub-categories, with state agricultural land subdivided into five sub-categories. The ownership of desert land owned by the state is divided into five types in which the first type is classified into five sub-categories. All public land is subject to a higher committee of the national land organization who set the roles, procedures, and sale for public land and determine the actual land market price. This division goes back to Ottoman land law, culminating in the Ottoman Land Code of 1858, which was further shaped and distorted by colonial encounters (Sait and Lim 2006) and modified in the period1805–1845 by Mohamed Ali's era in Egypt.

Most of the land in cities of the Middle East originally carry official land registrations. By the time the land tenure, especially agricultural land, are changed ownerships, documentation, characteristics, and lost the accurate information and details which are essential for official registration. On the other hand, private agricultural land ownership has varied and changed over time. It is argued that land tenure involves a complex set of formal and informal land rights, ranging from

Full security					
High security					
Moderate security					
Low security					
No Security					
Rights of holding	**Officially Registered ownership**	**Officially Registered Ownership without tenure**	**De Jure tenure recognition**	**De facto recognition**	**Waad Yaad (Squatting)**
Full Ownership	•	•	-	-	-
Full Ownership without tenure	•	•	-	-	-
Rights of Tenure without ownership	•	•	•	•	-
Strong Land tenure	•	-	•	-	-
Weak Land tenure	•	-	-	•	-

Fig. 6.5 The relationship between land ownership and land tenure in Egypt. (Source: The author)

various rights of use to conditional or full rights to dispose of the land and is called a "continuum of land rights" (UN-Habitat 2015).

Not every owner of a piece of land is a landholder, and not every landholder is a property owner, but to combine ownership and tenure one has to have the right of ownership and tenure that depend on the three rights of awareness, collective, and permanent and the three powers of exploitation, use, and disposition. The Egyptian Civil Law (borrowed from Ottoman land law) determines various ways—in articles no. 870 and 984—to gain full ownership of a property through seven ways (Soliman 2019). Therefore, the right of the ownership in the current Egyptian Civil Law indicates that "the owner of the thing has, in the limits of the law, the right to use, exploit and dispose of his own property".

As Fig. 6.5 illustrates, official full ownership has to be acquired through official registration with the Property Department and has to hold a hand on the property. If the landowner does not hold a hand or possess the property, the official ownership lacks the right to use, exploit and dispose of his own property, and this means that the landowner has lost one of the rights.

The legal right of land ownership is a right to possession which allows the landlord all the authority of using, exploiting, and disposing the land at all times. The ownership of land is characterized by three aspects and measured with three rights. The first is collective land ownership that gives authority to a few landlords all the rights to use their land according to the prevailing law. The main principle of possession is the proclamation and not the prohibition or adherence to others. Second, temporary ownership is rights' awareness and is confined to a landlord so that others are not allowed to share ownership of the land or interfere in the affairs of

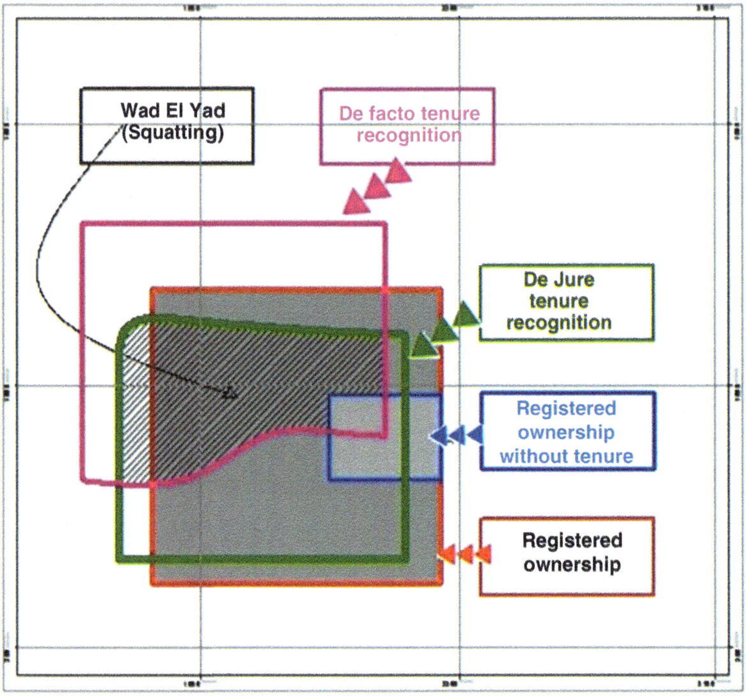

Fig. 6.6 Map demonstrating the process of land subdivision and transformation of ownership of land. (Source: The author)

the property unless it is determined by law or under certain agreements. A third is permanent ownership in which the landlord has the privilege to use, exploit, and dispose of his own property. As long as land remains an enduring theme, does not go away, but the demise of the theme or mortality does not change it by changing a person on the fact that the owner is always right effect.

The right to property authorizes the owner with respect to three powers, i.e., the use, exploitation, and disposition of land. The use and exploitation are interrelated because the use and exploitation are applied to commodities and the difference between them subsists is in the beneficiary of the capital. If the property owner is the user, and if a person holds the land and not the ownership and paid a certain amount of money, s/he is called a beneficiary based on payment of money. The consequences of the distinction that the property owner has full authority over his property, and if not, the owner will comply with the preservation of land in the condition in which it was received. Hence, there is a difference between ownership and tenure: the property owner could own the land and use it however he wishes, while a user has the right to benefit from it , and here it is called possession of a piece of land.

As illustrated in Fig. 6.6, land ownership has changed from formal security of ownership and accurate documentation into a variety of informal documents. These informal documents vary between *Orafi* (unofficial contract), court documents, custom agreement, *Musha* occupation (Arabic for "shared"), and *Hiyazah-land holding*. Most of the urban land tenure in the Middle East currently has no official land registration documents.

Also, there is a difference between land tenure (*Hiyazah*—land holding) and land ownership (*Miluk*—property owned), where the former is the use value, and the latter is the exchange and use value. The contribution of illegal land subdivision to housing informality is not only in relation to the use value but also to the exchange value. There is some theoretical or philosophical debate (Soliman 2004a) as to whether land itself can be 'owned', as all land on earth is owned by God *(El Miluk Lela Alla Whado),* but little dispute that there are Islamic rights to 'use' and 'possess' land. In principle, the rights to land are linked to land use, and the person who uses the land has priority over another who has failed to use it. Following the same logic, only productive land should create wealth (UN-HABITAT 2011), and that logic is in practice through the development of urban informality on the peri-urban areas of the Middle East.

In Western society, a land trust is similar to the endowment land (*waqf*, plural *awqaf*) in Islamic countries. It is a key Islamic institution that has incorporated within its legal sphere vast areas of land within the Muslim world, connected firmly with the religious precept of charity. *Waqf* land has two classifications: *waqf khayri* (charity), and *waqf ahli* (private). The former can occur in three ways; the beneficiaries die, at which point the property is absorbed into the *waqf khayri*; the property deteriorates to the point where the original value is totally dissipated, at which time it is returned to the open market as freehold; or long-term leases on the property are granted to investors with capital. *Waqf ahli* (private family) is transferred from the owner of a property to a religious foundation, but owners continue to receive its revenues personally during their lifetimes. After death, the revenues from the property will be transferred to the descendants, merely arranging for their eventual disbursement to a charitable purpose, if and when the owner dies. Often the owner himself is appointed to administrate his property, although after several generations this administration generally passes into the hands of a professional. After many generations (under the condition of the *waqf*), the *waqf ahli* could be transformed into freehold land, but this often may require a court order.

Due to the differences of land ownership, legal/illegal land subdivision, and for various bodies who govern the land possession in the Middle East, vast lands have fragmented which has left the security of land in doubt. This process of land fragmentation is caused through three occurrences: land reform, inheritance, and illegal land subdivisions.

In the Middle East, land reforms played a crucial role in the fragmentation of land tenure, e.g., land reforms implemented roughly in the period 1952–1996 reduced but did not eliminate the inequalities inherited from the past. Egypt has implemented land reforms since 1952, and most of other Middle East countries have enacted the same reforms. On the eve of the 1952 Revolution, ownership of agricultural land in Egypt was primarily concentrated in a few hands (Osman 2010). The land reform Law No. 178 of 1952, the subsequent laws No. 127 and No. 15 of 1961 and 1963, respectively, limited individual ownership to 50 feddans per individual and 100 feddans for a family. It targeted the property of the upper class of landowners, dubbed "feudalists" by the government, for distribution to the poorest villagers and the landless. According to these laws, the ownership of agricultural land is the right to use the land only for cultivation and not for other uses and gives its owner the power to use, abuse, and dispose of the land for cultivating. It should be emphasized that the system of law to ownership, which respond to the owner is no different in origin according to the right place. Then the basic principle that this system is no different as if the place for this property is also agricultural land or building land or the land is desert. In controlling the individual ownership of agricultural land, either land barren or desert land, every contract carrier of ownership resulting in a violation of these provisions is void.

These land reforms have divided agricultural land into small parcels, and over time these lands became fragmented and comprised the peri-urban areas that later became the site of the informal housing sector in cities of the Middle East. It can be said that the states have collaborated in speeding the process of land fragmentation, as well as changing the land tenure from private into public and vice versa. This leads to the 'web of tenure' in contemporary Middle East societies. It is argued (Sait and Lim 2006) that engagement with Islamic dimensions of land may potentially support land-rights initiatives in Muslim societies and has implications for programs relating to land administration, land registration, urban planning, and environmental sustainability.

Secondly, Middle Eastern societies generally derive their inheritance rules from religious sources for the division of an individual's property upon death, some of which are controversial. Yet, it is argued that the application of these formal inheritance rules pertaining to designated shares must be understood in a broader socio-cultural and economic context and within wider inheritance systems in practice (Sait and Lim 2006). Islamic inheritance laws dictate that female children should inherit a half share if one complete share goes to male children in property, which often results in the land being idle or abandoned when multiple progenies cannot agree over its subdivision or sale. Thus, fragmentation of land has occurred and resulted in the subdivision of land into small parcels. In other cases, when the landlord died and had no children, his property was transferred to the Public Land Property Organization or into the endowment. Thus, the land changes its category from private land in the form of a freehold system into a public title. So, land tenure in cities of the Middle East is transferable from the public to private and from private to public, and this process has a major impact on a land-delivery system and tenure security. Over time, this land is divided by the local authority into small plots and put up for sale at auctions.

Third, illegal land subdivisions play a major role in the fragmentation of land. Virtually all agricultural land on the fringes of the Middle Eastern cities is composed of irrigated, intensely cultivated land holdings. Most of these holdings are very small. Also, most of these holdings are private, that is, under family freehold tenure. In addition, some are land-reform lands (in Egypt, *islah al ziraa*, under the Ministry of Agriculture), again farmed by smallholder tenants. In both of these cases, the rights of tenants are strong, approaching those of freehold tenure. Within an agricultural area, there will be a number of main and secondary irrigation channels (*Turaa*) and drains (*Mussaref*) whose rights-of-way represent the only lands in the public domain. These are normally 10–30 m wide (including associated embankment roads). In mature urbanized informal areas, it is these rights-of-way that have been converted into main streets. Any attempts at the legal conversion of these agricultural lands for urban use confront three main problems. First, there is lack of legality and no up-to-date registration; second, extremely fragmented land holdings in long, narrow strips are often involved; and finally, a fringe informal subdivision of agricultural land into building lots has often developed.

In Middle Eastern countries, the agricultural land subdivision is a product of inheritance. According to Islamic law, all heirs obtain a share of the inheritance, but a son inherits a share double of that of a daughter. Such a process of subdivision is dynamic and occurs piecemeal throughout a locality, increasing disparities in parcel size. However, since each new plot must have access to an irrigation canal and a public road, the result is generally one of linear plots (*Ahwad*) up to 400-m long and 120-m wide. Generation after generation, such plots have been further subdivided into narrower strips separated by small irrigation channels (*Missqa*). Some plots today are as narrow as 15 m. As localities become increasingly urbanized, some canals dry up and are added to the width of contiguous roads. Furthermore, as disputes arise among heirs, courts have often transferred the judgment to engineers with expertise in subdividing land according to Islamic law. Such subdivisions usually occur according to the local measurement. Large agricultural parcels (*Ahwad*) are usually subdivided within the pattern of large irrigation canals and drains, and each *Hoa'ad* (large parcel) must have side reservations for paths and canal cleaning. The pattern of the old irrigation system thus usually defines the main and secondary streets in an informal housing settlement on formerly agricultural land.

The way that the government intervenes in this process (through auction, land privatization, or changing the nature of land tenure) is considered a crucial phenomenon in facilitating the fragmentation of land into smaller plots. Private speculators have contributed in accelerating the process of land fragmentation. With rapid urbanization processes and the acceleration of population growth, private developers find a profitable area to increase their return by purchasing a large area of agricultural land and illegally subdividing it for residential use. In all cases, the Middle Eastern cities tend to become bound by often idle properties whose owners are either unidentified or have no interest in either releasing or developing land for residential use, let alone to low-income groups..

So, what is the way ahead? In absolute terms, the number of slum dwellers in the developing world is actually growing and will continue to rise in the near future. The progress made on the slum target has not been sufficient to offset the growth of informal settlements in the developing world, where the number of urban residents living in slum conditions is now estimated at some 828 million, compared to 657 million in 1990 and 767 million in 2000. It is argued (Satterthwaite 2014) that neither MDGs nor SDGs achieved its objectives in decreasing the number of slums.

Redoubled efforts will be needed to improve the lives of the growing numbers of urban poor in cities and metropolises across the developing world. It is important to locate tenure policy within a wider policy context. Governments must create a legal framework that protects the rights of all citizens, especially the poor, and enables them to secure their land through a systematic way at an acceptable cost. The costs of engaging the legal system are too high, and states have to offer some incentives to encourage people to enter the formal system. Tenure policies are more likely to succeed when they encourage the social and spatial integration of urban areas operating in conjunction with physical planning policies designed for this purpose. Finally, it is imperative that tenure policies recognize the links between tenure status and livelihoods.

6.5 Land Market Forces

The variation of land ownership in the Middle East is widely diverse and many public and private lands are being appropriated and/or squatted upon by people who are unaware of how the land can be illegally acquired. In addition to these broad underlying conditions, the main premises for the acceleration of the informal land market in the Middle East may be attributed to the following variables. First, a desire among owners of agricultural lands to subdivide illegally their land into small plots for sale, and the increasing demand for such plots. Second, the subdivision of inherited agricultural land, causing farm plots to become too small to be cultivated economically. Third, the wish of the owners of large plots who are employed in sectors other than agriculture to avoid the difficulty of finding suitable farming tenants, and who are putting their land up for sale for development instead. Finally, unfavorable farming conditions, such as when crops on lands close to urbanizing areas are damaged by children and domestic activities from neighboring dwellings or are overshadowed by adjacent dwellings.

The scarcity and increasing prices of urban land have accelerated the demand for peri-urban land that became a fertile market for both real estate speculators and people who are looking for residential plots at a reasonable price. Private developers, some people in the official institutions, and speculators are playing a major role in the formulation of the informal land market. The advantages of the informal land market are that the land price is lower than urban land in the city. It does not pass through the complicated procedures for registration, circumventing paying the required taxes, saving time during the purchasing process, relying on an informal financial agreement at a reasonable basis, having some sort of tenure as *de facto* recognition, and finally, being guaranteed (*de jure* recognition from the buyer) during the construction process of a building.

However, both formal and informal land markets operate and collaborate within the same system of a given environment and are mainly related to the formal/informal political-economic system. Therefore, the land market in the Middle East, as well as in other cities in the Global South, is influenced by three major processes

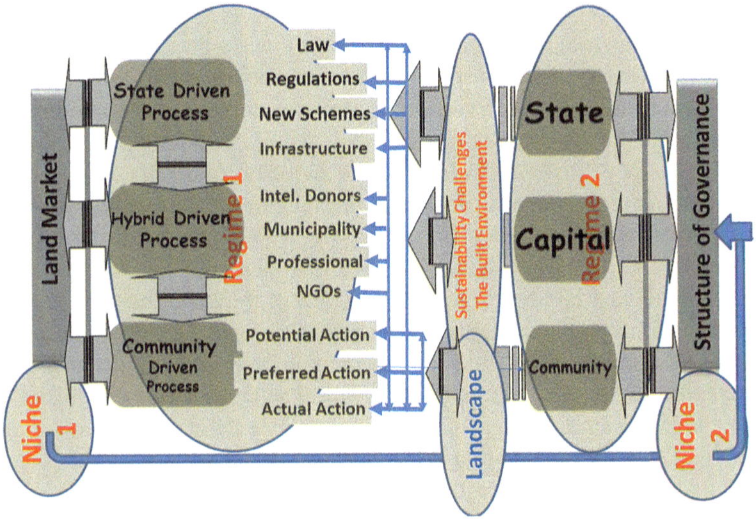

Fig. 6.7 Diagram of the relationship between the land market and the three driven processes. (Source: The author)

of action: state-oriented processes, hybrid processes, and community-driven process. These processes are interlinked and correlated with the three pillars of sociotechnical context, i.e., the state, capital, and the social context. The action and interaction between the three processes and the three pillars influence the formulation of an informal land market and the quality of the built environment and have a major effect on the climate change of given area. It is a complex system process which operates in two directions—horizontal and vertical— and all components of the process interlink and influence each other, resulting in a certain output of commodities at diverse locations and sizes matching the requirements of different strata of citizens. Both processes, horizontal and vertical, have a major impact on the socio-spatial context, as well as the socioeconomic and political atmospheres that are influencing the degree of quality of the built environment being produced, the standards of living, and everyday lives of the inhabitants.

As illustrated in Fig. 6.7, two niches operate in two opposite directions: land markets and a system of governance. Both influence the final output of the sociotechnical landscape being produced and combine to formulate the built environment, guaranteeing sustainability challenges. In between the niches and landscape, two regimes have existed: The first constitutes the three driven process, and the second represents the three main pillars. The duality of the regimes has been reflected in the final output of the market in which the formal/informal market grew.

The land market as the first niche is influenced by the three driven processes: state, hybrid, and community. The state-driven process is linked with prevailing law, regulation and policies, new schemes, and the infrastructure being introduced. The hybrid-driven process is correlated with prevailing law, international donors, local municipality, professional, NGOs, and CBOs. The community-driven process relies on the hidden potential actions of urban informality, as a cooperation/participation process among the urban poor and the way that they formulate the informal urban expansions that have met their requirements and their need of housing plots. The hidden potential actions rely on preferred, effective, and real actions (Soliman 2019).

In relation to the state-driven process, the law is considered the main instrument to organize the relationship between citizens and various aspects of socioeconomic and political activities. It also organizes the relationship between customers and various governmental institutions in purchasing, registering, and using their land in the informal land market. It should be noted that there is a difference between the letter of the law and law in practice or the spirit of the law. The former is used in official documents and institutions, while the latter is used on the ground among the residents. As a key concept, this spawns the twin approaches of transformational versus traditional: the former aims at change designed to ensure social justice in land laws, while the latter aims to continue the overall thrust of the spatial approach to land laws and land administration (McAuslan 2013).

Land and housing market regulations have two bases in economics (Henderson 2009). The first articulated basis involves neighborhood externalities. This is a form of land-use planning that determines density, open-space regulations, the requirements for infrastructure, construction regulations setting minimum quality standards, and health and environmental standards. Standards and regulations are very costly, not just because of their specific restrictions but because of the need to have inspection and permitting processes with rights of appeal. The second basis for regulation is exclusionary zoning. Exclusionary regulations are used to protect the interests of existing residents who want to regulate the flow of people into their community and may therefore not be concerned with improvements for the society as a whole. There are three aspects of exclusion. The first relates to a situation in which a community has special standards of housing, natural amenities, and public services, e.g., gated communities. People may crowd into that community to take advantage of the amenities. The second and more common basis given for exclusionary zoning is stratification of the population by income into different communities, for the purposes of consuming local public goods, and this constitutes the old quarter of a city. The third basis involves the areas outside the official boundary of the city and relates to urban informality which accommodates newcomers or migrants, or both, who are looking for cheap housing plots.

There are generally accepted justifications for government interventions in setting up new schemes in urban land markets (Napier 2009). These are: elimination of market imperfections and failures to increase operating efficiencies, removing externalities so that the social costs of land market outcomes correspond more closely to private costs; and redistribution of a society's scarce resources so that disadvantaged groups can share in society's output.

The impact of the state-driven process can be seen in the planning and programming of large-scale public investment in infrastructure because the state should install services that the community cannot provide for themselves. The provision of basic infrastructure has been taken as the starting point in assessing the demand for investment in housing plots and urban infrastructures, such as water supply, sanitation, electricity, and solid-waste management. The existing infrastructure reflects the assessment that governments did not have the financial resources to provide subsidized shelter and services to growing numbers of urban residents and that most countries did not have the policies in place to provide incentives for the private sector to respond to this demand.

On the one hand, the community-driven process results from the potential actions of the urban poor and their relative impact on housing production. Potential action is defined as the priorities of the participants (government, professionals, and the poor) and their ability, liability, and willingness regarding the production of housing plots being supplied. It constitutes three important themes: preferred, effective, and actual actions. The potential actions, among the urban poor, are in relation to secured land plots, and its relationship to the housing process as use and/or exchange value, and the way it had been created, developed, and invested. The actual action are the means and methods that can be provided by the three collaborating pillars (state, capital, and social context) in the land market to achieve a state of balance between supply and demand for land and ensure the availability of suitable land for various classes in the community. These means could be summarized in three main aspects: site selection, land acquisition, piecemeal tenure occupation, and informal local planning conventions. On the other hand, effective action involves the possibilities that can be provided by cooperating among the three pillars for possession and ownership of land in informal residential areas, which could affect the housing systems. The preferred action is the fundamental option favored by the three pillars' cooperating systems and mechanisms of acquisition and ownership of residential land (Soliman 2019).

It should be said that the duality of the land market in the Middle East is reinforced over time, and not a new phenomenon. The appearance of the informal land market is attributed to the failure of state policy to regulate the land market and housing production. It is also attributed to the increasing demand for shelter, the rapid population growth, and arbitrary urbanization processes.

6.6 Conclusion

One can do worse than to learn from the past to avoid problems in the future. Going back to the mid-1960s, when the states of the Middle East initiated an ambitious plan to enlarge the big cities and adjusted agricultural areas as they did so, they did not take into consideration either the appearance of the duality of the land market or the increasing demand for housing. Therefore, the speculators invaded agricultural land and illegally sub-divided it into small land plots without planning regulation or planning control. This typical model continues to operate in most of the Middle Eastern cities.

As soon as a new scheme or a state-planning project was implemented, new urban informality backed by the state emerged. It is often the power of the state that determines what is informal and what is not. This means that informality is not an unregulated domain but rather is structured through various forms of extra-legal, social, various actors, and discursive regulation. Throughout the growth of cities in the Middle East, as well as other cities in the Global South, the invasion of agricultural areas was started formally by the state and continued informally by the private developers as people in need sought housing plots. Second, transforming peri-urban areas to urban informality is not only for housing but is a way to formulate "a new space in the city within the city" to squeeze the growing population, a process through which various strata of the society find their way to produce a reasonable land delivery system to meet their social and economic needs. Peri-urban or agricultural land conversion became a matter of investment in land, and it has a fundamental importance in the urban poor's own rights because it brings other developmental benefits (for instance, access to services and credit or political voice). In other words, conversion of agricultural land became an issue of "socio-spatial struggle" for meeting the basic human needs that the state could not satisfy for the vast majority of the population in the Middle East. Also, the conversion of agricultural land into other uses accelerates the deterioration of the built environment, as well as leading to a decrease in cultivated areas. In this sense, urban informality became an expression of class power, and, in time, commanded different infrastructures, social services, and legitimacy in urban informality. Third, people involved in urban informality have created their distinctive urban pattern that formulates "a new socio-spatial urbanity" different than the one produced by the state. It is a totally crucial mechanism in a wholly private formal and informal land market that is being integrated with the socioeconomic and political transitions of a country. Therefore, it is the time to understand the philosophy of land markets in urban informality and direct these mechanisms into a sustainable system that is socially, physically, and economically feasible to assure that future urban growth benefits the whole society.

The correlation between the three pillars and the three driven processes influence the transitions of an informal land market, and vice versa. These processes have to be adjusted to guide people to appropriate areas that the state would like to see developed in a sustainable way. However, this would require a proper planning transitions framework, as well as would a political willingness to develop new areas in a sustainable way. This then is the foundation for society-driven urban development policy, or people's processes, as opposed to a state or market-driven policy. From this perspective, the organizations of civil society need to be enabled to adapt and strengthen and guide their own development, engaging with the state and the market where possible and desirable. They would appear to be the most productive means to assist the urban poor because the state and the market are increasingly unable and unwilling to do so.

However, through the various potentialities, constraints, and the facts that the Middle East has been motivated by the political goal of saving and rescuing agricultural land—from the perspective of an overall control desired by the local powers—the passage from "a market-driven policy to"a society-driven urban development policy has become vital and represents an inescapable reversal. A society-driven urban development policy would allow communities to negotiate from a position of much greater power and thereby secure for themselves successive improvements to their neighborhood and explore the economic mechanisms that exist to reevaluate urban externalities. Thus, suggestions that work based on these concepts can make possible a more nuanced and appropriate analytical basis for understanding the perception of Middle East's urban dwellers, as well as mechanisms for consolidating urban informality.

Land plots and the rights to casual, non-customary ownership of land in the space of the public domain should be realized as a starting point. This would lead to transition land management that is relevant and appropriate to the nature and potentials of low-income strata, as well as leading to activating and enlarging the real estate market in informal residential areas and new virgin areas. This transition land management should be based on more flexible, multi-actor network-based, governance, address the constellation of sustainability challenges in the local urban context, and an appropriate legislation process. Among the legislation applied in Egypt is the *Segel Aini* record, which should be opened to informal settlements for the registration of land plots. Also existing in Islamic countries is *musha* (Arabic for shared) lands or common land, mostly relating to rural agricultural land where the custom is 'reallocating land in unequal shares (at regular intervals) to which a customary right of ownership attaches.

However, various actors are involved in setting up a proper framework for reasonable and practical land delivery systems. These would require a correlation between the multi-actor's perspective and alternative transition ideas on experimentation. Based on the ASUST's concept a correlation as an interference of the four policies of; policy strategy credibility, consistency of dimensions, policy processes systemic change, systematic directions of transitions, and instrument mix comprehensiveness, are the main foundation to formulate a model to increase access for reasonable housing plots. These combinations of policies contain various market forces, different land agents, and multi bodies that are controlling land acquisition. This model should take consideration of social innovation, self-organization, multi-actor approaches, experimentation, digital strategies, cooperatives, etc. to ensure that "Leaving No One Behind."

References

Al Sayyad, N. 2004. Urban Informality as a "New" Way of Life. In *Urban Informality: Transnational Perspectives from the Middle East, Latin America, and South Asia*, ed. A. Roy and N. Al Sayyad, 7–30. Lanham: Lexington Books.

Allen, A. 2014. Peri-Urbanisation and the Political Ecology of Differential Sustainability. In *A Routledge Handbook on Cities of the Global South*, ed. S. Oldfield and S. Parnell, 21–36. London/New York: Routledge.

Andonova, L.B., et al. 2009. Transnational Climate Governance. *Global Environmental Politics* 9 (2): 52–73.

Angel, S., R. Archer, S. Tanphiphat, and E. Weglin. 1985. *Land for Housing the Poor*. Singapore: Selected Books.

Angel, S., J. Parent, D. Clivco, and A. Blei. 2011. *Making Room for a Planet of Cities*. Cambridge: Lincoln Institute of Land Policy.

Archer, R. 1999. The Potential of Land Pooling/Readjustment to Provide Land for Low-Cost Housing in Developing Countries. In *Making Common Ground: Public-Private Partnerships in Land for Housing*, ed. G. Payne, 113–133. London: Intermediate Tech.

Brand, U. 2005. Order and Regulation: Global Governance as a Hegemonic Discourse of International Politics? *Review of International Political Economy* 12 (1): 155–176.

Bromely, R. 2004. Power, Property and Poverty: Why De Soto's Mystery of Capital Cannot Be Solved. In *Urban Informality: Transnational Perspectives from the Middle East, Latin America, and South Asia*, ed. A. Roy and N. Al Sayyad, 271–288. Lanham: Lexington Books.

Buckley, R., and J. Kalarickal, eds. 2006. *Thirty Years of World Bank Shelter Lending What Have We Learned?* Washington, DC: World Bank, The International Bank for Reconstruction and Development.

Cirolia, L.R. 2017. Overcoming the Disjunctures: Competing Discourses on Informal Settlements in South Africa. *IDPR* 39 (4): 443–459.

Coase, R.H. 1960. The Problem of Social Cost. *Journal of Law and Economics*: 1–44.

Cohen, M. 2009. Aid, Density, and Urban Form: Anticipating Dakar. In *Urban Land Markets: Improving Land Management for Successful Urbanization*, ed. S.V. Lall, M. Freire, B. Yuen, R. Rajack, and J.-J. Helluin, 385–398. Dordrecht/Heidelberg: Springer.

Colsaet, A., Y. Laurans, and H. Levrel. 2018. What Drives Land Take and Urban Land Expansion? A Systematic Review. *Land Use Policy* 79: 339–349.

d'Alençon, P.A., L. Moretto, and H. Smith. 2016. *N-AERUS – Recommendations for the New Urban Agenda*. Working Paper Group, Cities Alliance.

De Soto, H. 1989. *The Other Path: The Invisible Revolution in the Third World*. London: I.B. Taurus.

———. 2000. *The Mestry of Capital: Why Capitalism Triumphs in the Westand Fails Everywhere Else*. London: Blaack Swan.

DEFID. 1999. *Literature Review on Peri-Urban Natural Resource Conceptualization and Management Approaches*. London: DEFID, project no. R6949.

Deininger, K., H. Selod, and A. Burns. 2012. *Land Governance Assessment Framework, Identifying and Monitoring Good Practice in the Land Sector*. Washington, DC: The world Bank.

Diana Reckien, D., and J. Karecha. 2008. Sprawl in European Cities: The Comparative Background. In *Urban Sprawl in Europe: Landscapes, Land-Use Change and Policy*, ed. Lila Leontidou, Chris Couch, and Gerhard Petschel-Held, 39–67. Oxford: Blackwell Publishing Ltd.

Doebele, W. 1987. The Evolution of Concepts of Urban Land Tenure in Developing Countries. *Habitat International* 11 (1): 7–22.

———. 1994. Urban Land and Macroeconomic Development: Moving from 'Access for Poor' to Urban Productivity. In *Methodology for Land and Housing Market Analysis*, ed. G. Jones and P.M. Ward, 44–54. Cambridge, MA: Lincoln Land Institute of Land Policy.

Dorman, W. J. 2007. *The politices of Neglect: The Egyptian State in Cairo, 1974–198*. London: PhD. thesis, School of Oriental and African Studies, University of London.

Duquennois, A., and P. Newman. 2009. *Linking the Green and Brown Agendas: A Case Study on Cairo, Egypt*. Technical. Nairobi: UN-Habitat, Unpublished case study prepared for the Global Report on Human Settlements.

Durand-lasserve, A. 2005. Land for Housing the Poor in African Cities: Are Neo-Customary Processes an Effective Alternative to Formal Systems? In *Urban Futures, Economic Growth and Poverty Reduction*, ed. N. Hamdi. London: ITDG Publishing.

Edelenbos, J. 2005. Institutional Implications of Interactive Governance: Insights from Dutch Practice. *Governance* 18 (1): 111–134.

FAO. 2001. *Urban and Peri-Urban Agriculture*. Rome: Food and Agriculture Organization of the United Nations.

Fernandes, E., and A. Varley, eds. 1998. *Illegal Cities: Law and Urban Change in Developing Countries*. London: Zed Books Ltd.

Fischhbach, M. 2000. *State, Society, and Land in Jordan*. Boston: Brill.

Frantzeskaki, N., V.C. Broto, L. Coenen, and D. Loorbach. 2017. *Urban Sustainability Transitions*. New York: Routledge.
Geels, F., and J.J. Deuten. 2006. Local and Global Dynamics in Technological Development: A Socio-Cognitive Perspective on Knowledge Flows and Lessons from Reinforced Concrete. *Science and Public Policy* 33: 265–275.
Gelder, Jean-Louis Van. 2009. Legal Tenure Security, Perceived Tenure Security and Housing Improvement in Buenos Aires: An Attempt Towards Integration. *IJURR* 33: 126–146.
Gilbert, A. 2012. De Soto's the Mystery of Capital: Reflection on the Book's Public Impact. *IDPR* 34: 5–16.
GLTN's Arab Land Initiative. 2021. Cairo: Second Arab Land Initiative conference.
Goodfellow, T. 2020. Political Informality: Deals, Trust Networks, and the Negotiation of Value in the Urban Realm. *The Journal of Development Studies* 56 (2): 278–294.
Grin, J., J. Rotmans, and J. Schot. 2010. *Transitions to Sustainable Development: New Directions in the Study of Long Term Transformative Change*. New York: Routledge.
Harris, R. 2014. Urban Land Market: A Southern Exposoure. In *The Routledge Handbook on Cities of the Global South*, ed. S. Parnell and S. Oldfield, 9–20. Oxon: Routledge.
Henderson, J.V. 2009. The Effect of Residential Land Market Regulations on Urban Welfare. In *Urban Land Markets: Improving Land Management for Successful Urbanization*, ed. S.V. Lall et al., 25–50. Dordrecht/Heidelberg: Springer.
Hendriks, B. 2008. The Social and Economic Impacts of Peri-Urban Access to Land and Secure Tenure for the Poor: The Case of Nairobi, Kenya. *IDPR* 30 (1): 27–66.
Iaquinta, D., and A. Drescher. 2001. More than the Spatial Fringe: An Application of the Periurban Typology to Planning and Management of Natural Resources. In *Rural–Urban Encounters: Managing the Environment of the Peri-Urban Interface*. London: Development Planning Unit; University College.
Jenkins, P. 2004. Beyond the Formal/Informal Dichotomy: Access to Land in Maputo, Mozambique. In *Reconsidering Informality Perspectives from Urban Africa*, ed. Karen Tranberg Hansen and Mariken Vaa, 210–226. Madrid: Grafilur Artes Gráficas.
Kanger, L., and J. Schot. 2018. Deep Transitions: Theorizing the Long-Term Patterns of Socio Technical Cahnge. *Environmental Innovation and Societal Transitions*. https://doi.org/10.1016/j.eist.2018.07.006.
Kanie, N., and F. Biermann, eds. 2017. *Governing Through Goals: Sustainable Development Goals as Governance Innovation*. London: The MIT Press.
Kauffman, Craig M. 2017. *Grassroots Global Governance: Local Watershed Management Experiments and the Evolution of Sustainable Development*. Oxford: Oxford University Press.
Kjaer, A.M. 2004. *Governance*. Cambridge: Polity Press.
LaFreniere, G.F. 2007. *The Decline of Nature: Environmental History and the Western Worldview*. Bethesda/Dublin/London: Academic.
Lall, S.V., M. Freire, B. Yuen, R. Rajack, and J.J. Helluin, eds. 2009. *Urban Land Markets Improving Land Management for Successful Urbanization*. London: Springer.
Lamba, H.S. 2010. Understanding the Ideological Roots of Our Global Crises: A Pre-Requisite for Radical Change. *Futures* 42 (10): 1079–1087.
Leduka, R. 2004. The Law and Access to Land for Housing in Maseru, Lesotho. In *Reconsidering Informality Perspectives from Urban Africa*, ed. K.T. Hansen and M. Vaa, 176–192. Spain: Grafilur Artes Gráficas.
Lynch, Kenneth. 2005. *Rural–Urban Interaction in Developing Countries*. London: Routledge.
March, J.G., and J.P. Olson. 1995. *Democratic Governance*. New York: Free Press.
Marshall, F., and J. Dolley. 2018. Transformative Innovation in Peri-Urban Asia. *Research Policy*. https://doi.org/10.1016/j.respol.2018.10.007.
McAuslan, P. 1985. *Urban Land and Shelter for the Poor*. London: Erathscan.
———. 2013. *Land Law Reform in Eastern Africa, Traditional or Transformative? A Critical Review of 50 Years of Land Law Reform in Eastern Africa, 1961–2011*. Oxon: Routledge.
Moore, J.W. 2015. *Capitalism in the Web of Life: Ecology and the Accumulation of Capital*. London: Verso.
Nada, M. and Sims, D. 2020. *Assessment of Land Governance in Egypt*. Cairo: The World Bank

Napier, M. 2009. Making Urban Land Markets Work Better in South African Cities and Towns: Arguing the Basis for Access by the Poor. In *Urban Land Markets Improving Land Management for Successful Urbanization*, ed. S.V. Lall, 71–100. Dordrecht/Heidelberg: Springer.
Nunan, F., and N. Devas. 2004. Accessing Land and Services: Exclusion or Entitlement? In *Urban Governance, Voice and Poverty in the Developing World*, ed. Nick Devas, 164–185. London: Earthscan Publications Ltd.
Obeng-Odoom, F. 2012. Land Reform in Africa: Theory, Practice, and Outcome. *Habitat International* 36 (1): 161–170.
———. 2016. Property in the Commons: Origins and Paradigms. *Review of Radical Political Economics* 48 (1): 9–19.
OECD. 2016. *The Governance of Inclusive Growth*. Paris: OECD Publishing.
Osman, T. 2010. *Egypt on the Brink: From Nasser to Mubarak*. New Haven/London: Yale University Press.
Palmer, D., S. Fricska, and B. Wehrmann. 2009. *Towards Improved Land Governance: Land Tenure Working Paper 11*. Nairobi: (FAO of UN for human settlement programme).
Pamuk, A. 2000. Informal Institutional Arrangements in Credit, Land Markets and Infrastructure Delivery in Trinidad. *IJURR* 24: 379–396.
Payne, G. 1999. Introduction. In *Making Common Ground: Public/Private Partnerships in Land for Housing*, ed. G. Payne, 1–16. London: Intermediate Technology Press Ltd.
———., ed. 2002. *Land, Rights and Innovation: Improving Security for the Urban Poor*. London: ITDG.
Payne, G., A. Durand-Lasserve, and C. Rakodi. 2009. Social and Economic Impacts of Land Titling Programs in Urban and Periurban Areas: A Short Review of the Literature. In *Urban Land Markets Improving Land Management for Successful Urbanization*, ed. S.V. Lall, M. Freire, B. Yuen, R. Rajack, and J.J. Helluin, 133–162. London: Springer.
Piketty, T. 2014. *Capital in the Twenty-First Century*. London: Th e Belknap Press of Harvard University Press.
Pritchard, M. 2013. Land, Power and Peace: Tenure Formalization, Agricultural Reform, and Livelihood Insecurity in Rural Rwanda. *Land Use Policy* 30: 186–196.
Rashed, T., D. Stow, and D. Fugate. 2005. Measuring Temporal Compositions of Urban Morphology Through Spectral Mixture Analysis: Toward a Soft Approach to Change Analysis in Crowded Cities. *International Journal of Remote Sensing* 26: 699–707.
Raven, R.P.J.M., E. Heiskanen, R. Lovio, M. Hodson, and B. Brohmann. 2008. The Contribution of Local Experiments and Negotiation Processes to Field-Level Learning in Emerging (Niche) Technologies: Meta-analysis of 27 New Energy Projects in Europe. *Bulletin of Science, Technology and Society* 28: 464–477.
Roy. 2009. The 21st-Century Metropolis: New Geographies of Theory. *Regional Studies* 43 (6): 819–830.
Sait, S., and H. Lim. 2006. *Land, Law and Islam: Property and Human Rights in the Muslim World*. London: Zed Books Ltd.
Satterthwaite, D. 2014. Guiding the Goals: Empowering Local Actors. *SAIS Review of International Affairs* 34(2, Summer-Fall): 51–61.
Schot, J. 2016. Confronting the Second Deep Transition Through the Historical Imagination. *Technological Culture* 57 (2): 445–456.
Schot, J., and W.E. Steinmueller. 2016. *Framing Innovation Policy for Transformative Change: Innovation Policy 3.0 - Draft Version 2 (Draft Working Paper)*. Brighton: University of Sussex, Science Policy Research Unit (SPRU).
Sela, A. 2014. The PLO at Fifty: A Historical Perspective. *Contemporary Review of the Middle East* 1 (3): 269–333.
Simon, D. 2008. Urban Environments: Issues on the Peri-Urban Fringe. *Annual Review of Environment and Resources* 33: 167–185.
Sims, D. 2014. *Egypt's Desert Dreams: Development or Disasster*. Cairo: The American University in Cairo Press.

Smith, A., and R. Raven. 2012. What Is Protective Space? Reconsidering Niches in Transitions to Sustainability. *Research Policy* 41: 1025–1036.
Soliman, A. 1996. Legitimizing Informal Housing: Accommodating Low-Income Groups in Alexandria, Egypt. *Environment and Urbanization Journal* 8 (1): 183–193.
———. 1999. Public/Private Partnerships in Three Egyptian Cities. In *Making Common Ground: Public/Private Partnerships in Land Provision for Housing the Urban Poor*, ed. G. Payne. London: Intermediate Technology Publications (for DFID).
———. 2004a. *A Possible Way Out: Formalizing Housing Informality in Egyptian Cities*. Lanham: American University Press.
———. 2004b. Tilting at Sphinxes: Locating Urban Informality in Egyptian Cities. In *Urban Informality in an Era of Liberalization: A Transnational Perspective*, ed. Ananya Roy and Nezear AlSayyad, 171–208. Lanham: Lexington Books.
———. 2010. Rethinking Urban Informality and the Planning Process in Egypt. *IDPR Journal* 32 (2): 119–143.
———. 2012a. Housing and the State in the Middle East. In *InternationalEncyclopedia of Housing and Home*, ed. Susan J. Smith, 346–354. Amsterdam: Elsevier.
———. 2021b. *Rethinking Urban Sustainability Transitions on land readjustment in Egypt*, paper presented on 24th February in Second Arab Land Conference, 22–24 February. Cairo, Egypt.
———. 2012c. Building Bridges with the Grassroots: Housing Formalization Process in Egyptian Cities. *Journal of Housing and the Built Environment* 27 (2): 241–260.
———. 2013. *Living With/in Ashwaiyyat in Egypt: Is It a New Innovation of Urbanity?* The Flexible City, International Symposium. Oxford: Oxford University, 24–25 October.
———. 2015. Collective Planning Process: A Driving Seat for Formalising Urban Informality in Egypt. *International Journal of Urban Sustainable Development*, June 1–24.
———. 2017. Land Readjustment as a Mechanism for New Urban Land Expansion in Egypt: Experimenting Participatory Inclusive Processes. *International Journal of Urban Sustainable Development* 9 (3): 313–332.
———. 2019. So It's Always a Chance: Community-Led Solutions to New Urban Expansion. In *New Cities and Community Extensions in Egypt and the Middle East*, ed. S. Attia et al., 159–179. Cham: Springer.
Stilwell, F., and K. Jordan. 2004. The Political Economy of Land: Putting Henry George in His Place. *The Journal of Australian Political Economy* 54 (December): 119–134.
The World Bank. 2008. *Approaches to Urban Slums*. Washington, DC: The International Bank for Reconstruction and Development/The World Bank.
Torres, H., H. Alves, and M. De Oleiera. 2007. São Paulo Peri-Urban Dynamics: Some Social Causes and Environmental Consequences. *Environment & Urbanization* 19 (1) (World Bank): 207–223.
Toth, O. 2009. *Strategic Plan for an Egyptian Village: A Policy Analysis of the Loss of Agricultural Land in Egypt.* MSc thesis. Wagen: The School of Planning of the College of Design, Architecture, Art and Planning.
UNESCO. 2019. *The United Nations World Water Development Report 2019: Leaving No One Behind.* World Water Assessment Programme. Paris: The United Nations Educational, Scientific and Cultural Organization.
UN-Habitat. 2008. *The State of African Cities 2008: A framework for Addressing Urban Challenges in Africa.* Nairobi: UN-HABITAT.
———. 2011. *Islamic Principles and Land: Opportunities for Engagement.* Nairobi: UN-HABITAT.
———. 2015. *UN-Habitat Global Activities Report 2015: Increasing Synergy for Greater National Ownership.* Nairobi: United Nations Human Settlements Programme (UN-Habitat), 31.
———. 2016. *Evaluation of Participatory Review of Egyptian Planning and Related Urban Development Legislation to Support Sustainable Urban Development, April 2016.* Nairobi: UN-Habitat.
———. 2018a. *DG 11 Synthesis Report 2018: Tracking Progress Towards Inclusive, Safe, Resilient and Sustainable Cities and Human Settlements.* SDG 11 Synthesis Report High Level Political Forum 2018. Nairobi: Un-Habitat.

Chapter 7
Social Exclusivity Versus Inclusivity, Marginality, and Urban Informality

Abstract The recent socioeconomic and political transitions and the spread of urban informality that occurred in the Middle East, especially after the Arab Spring, have led into the devastation of both the physical and socioeconomic structures of cities. Marginality, social diversity, differences, social exclusion, and social segregation dominate the urban fabric of cities of the Middle East. On the other hand, social cohesion among the people who are living in urban informality has managed to create a new type of urbanity suitable and affordable for most of the population who are seeking reasonable shelter in the major urban centers. Given the transformations of urban informality, now is the appropriate time to take stock of the significant shifts. Shifts in rethinking on urban informality, official institutions together with professionals, over the past seven decades, necessitate exploring the emerging challenges related to it for future sustainable transition. It is argued that remodeling the correlation of sociotechnical system between the niche, regime, and landscape as urgent for a proper intervention is essential to maintaining the cohesion of urban informality within the urban context in a sustainable way.

Keywords Social Segregation · Urban Socio-Spatial Divide · Marginality · Social Exclusion · Social Cohesion · Everyday Life · The New Urbanity

7.1 Introduction

Throughout the twentieth century, the urban transition in the Middle East was relatively rapid, and the move to urban living continues at an accelerated pace in several countries in the region. One of the most striking recent features of urbanization in the Middle East has been the emergence of urban informality: cities that have surpassed the limits of their immediate outermost periphery, expanding beyond their administrative boundaries. Informal urbanization has led to the emergence of urban informality, i.e., national centers of the informal economy or political power, such as Cairo, Beirut, and Amman. Informal urbanization in such areas is largely unplanned, fueling the continuous growth of informal settlements, the physical manifestation of urban poverty, and social inequality. Because urbanization is one

A. M. Soliman, *Urban Informality*, https://doi.org/10.1007/978-3-030-68988-9_7

of the global megatrends of our time, an unstoppable and irreversible phenomenon (UN-Habitat 2018a), guided urbanization is a crucial task. Currently, these cities and other cities in the region witness great physical and social divides in which the urban poor occupy the worst urban locations and the rich the healthiest ones. There is a great divide between formal and informal areas, even a wider division between various strata of society, as inequality/marginality became the major phenomena of these cities. The physical divergence has reflected on everyday life within different sectors of a city and is reflected in the social structure of the citizens. As a result, marginality, social exclusion, and social segregation are all represented in the Middle Eastern cities' urban fabric.

Thus, this chapter explores the recent social transitions in cities of the Middle East to understand why these cities are not managed or planned to receive the growing immigrants from rural areas and the influx of refugees. It aims to investigate the impact of urban informality on socioeconomic transitions in the urban fabric of these cities and how the recent transitions in the Middle East have changed the socioeconomic structure of its cities. It is argued that, due to the rapid growth of urban informality, the urban fabric of cities of the Middle East has been devastated. Consequently, marginality, social exclusion, and social segregation have become prominent phenomena, and individual cities have been divided into more cities. This has led to the destruction of the traditional cores of cities in the Middle East and created a situation of social conflict among various strata of the societies within the cities, as well as creating a new way of everyday life for various city settlers. One major consequence of "the new socio-spatial urbanity" that emerged recently in cities of the Middle East has been a triple process of, on the one hand, status quos of marginality, and on the other hand, social exclusion and inequity for people who are living in the urban informality. The third is a healthy emergence of social cohesion among people who live in urban informality. The output of this triple process highlights the importance and the need for proper interventions to integrate urban informality into the urban context in a sustainable way to preserve the urban fabric of cities of the Middle East, as well as to alleviate poverty and to create homogeneous societies. More than 50% of the population in the Middle East work in the informal economy sector (Chen and Carré 2020), and a large proportion of them live in urban informality. Thus, if this proportion are capable of being integrated with the rest of the society, it would create a healthy and sustainable environment, as well as it would improve the performance of the national economy.

The chapter examines the related arguments through the four sections that follow: the first considers the recent urban divide that occurred between various parts of the cities of the Middle East; the second highlights marginality aspects as a result of the urban division. Social exclusion is discussed in the third, and social cohesion follows it. A conclusion formulates and emphasizes the crucial need for proper interventions to tackle urban informality and to guide the future urban expansion in a more sustainable way.

7.2 Urban Socio-Spatial Divide

Historically speaking, Muslim cities may have actually been more hospitable to social diversity than European cities or colonial cities. Islamic cities allowed Muslims, Jews, and Christians to interact relatively freely with one another, "creating a symbiotic space" where the sacred spaces of all three communities formed a homogenous urban mosaic that took a synergistic form (Samman 2006). Diversity in city planning was articulated by Jane Jacobs's call for a cityscape based on multi-use, which she argued would promote economic and social diversity (Jacobs 1961). Diversity in city planning would allow multiple social groups exercising their "rights to the city" and "social networks" as aspects of power relations and spaces and places of production, including various ethnic groups and peoples (Lefebvre 1976; 1991) and as a banner of the struggle against neo-liberalism (Sugranyes and Mathivet 2010).

David Harvey (2012) criticized Henri Lefebvre's 1968 essay *The Right to the City*. "That right", Lefebvre asserted, was both a cry and a demand. The cry was a response to the existential pain of a withering crisis of everyday life in the city. The demand was really a command to look that crisis clearly in the eye and to create an alternative urban life. This view is less alienated, more meaningful, and playful but, as always with Lefebvre, conflictual and dialectical, open to becoming, to encounters (both fearful and pleasurable), and to the perpetual pursuit of unknowable novelty. In other words, this is what Lefebvre called "centrality" with the objective being that the "right to the city" had to be respected. Centrality contains various services, such as the transportation terminals and communications networks, or proximity to some highly concentrated activities (such as a financial center, hotel, etc.), as well as residential areas, while suburbs in Western societies, have beneficial, privileged environments and accommodate more affluent groups. As Lefebvre wrote: "The right to the city legitimizes the refusal to be set aside from urban reality by discrimination or segregation."

Lefebvre's idea of 'something different', criticized by Harvey (2012), does not necessarily arise out of a conscious plan, but more simply out of what people do, act, behave, feel, sense, and come to articulate as they seek meaning in their daily lives. Therefore, the right to the city means the constitution or reconstitution of a spatio-temporal unity, of a coming together rather than fragmentation. On the other hand, Lefebvre's idea of cry, demand, and centrality distinguishes between three classes of people: the first where people live in miserable conditions; the second where another stratum of a society live in areas that have a hope for improvement (semi-informal areas); and the third where people have the privileges of the elite of urbanity. Thus, Lefebvre's idea makes clear social divisions, but he does not formulate an idea for the integration between these stratums.

In the Middle Eastern cities, most of the suburbs or hybrid urbanization areas (Bums and Friedmann 1985) are occupied by the lowest strata of the society and characterized by overcrowding, the worst environment, lack of services, and significant distance from job opportunities. These suburbs, as a component of sociotechnical landscape, contain not only informal residential areas but also an informal economy sector (Chen and Carré 2020) that constitutes a high proportion of the national domestic production. Thus, the suburbs have many functions: economic, social, and even political. Scattered areas in Egypt, Lebanon, and Jordan contain suburbs that accommodate many functions, such as the southern parts of Egypt and Lebanon and the west bank in Amman. Recently, there are luxury suburbs appearing on the periphery of cities of the Middle East that contain gated communities, enjoy good environments, dominate good locations, and have the privilege of housing the more affluent elite.

The process of urbanization once viewed simply as an unstoppable phenomenon (UN-Habitat 2018a) is seen by others in a positive way as an inevitable consequence of urban transitions (Frantzeskaki et al. 2017a, b). It is now seen as an integral component and a generator of economic growth, political change, information technology, and social transformation (Frantzeskaki et al. 2017a, b; Lefebvre 1991; Castells 2010, 2nd edition). The latter is emphasized by the Chicago School (Bums and Friedmann 1985; Park 1967), in which a new cultural formation called "urbanity" emerged. A significant theoretical analysis of the interrelationships between the social and the spatial aspects of the urban phenomenon (Castells 2010, 2nd edition) views urbanization as a process that produces spatial structures and forms, supporting the creation of social relationships for the reproduction of capital and social capital, which may represent the conflict between the state and society.

Due to the recent social movement, either in developed or developing countries, a theoretical dialog about the conditions, forms, and outcomes of urban political and societal conflict has been initiated (Roy 2009). The theory is intentionally general and a provocation to the venerable empirical tradition that has remained overly descriptive and localized. The argument proceeds through two steps. First are matters of distinction, types of collective action and their institutional contexts that constitute the niche innovation. Next, the discussion turns to the substance of the influence of globalization on urban life in poor countries, as well as developed countries; as it constitutes the regime, it is shaped by the intersection of economy, state, and civil society. An example is Yellow Jacket movements in France and other European cities, or what is called the European Autumn. In Paris, on 3 December 2018, a group of protesters encircled the Tomb of the Unknown Soldier located under the Arc de Triomphe and sang the French national anthem, protesting rising gasoline taxes and the high cost of living. There is a new era of social movements in the world.

The Cairo School of Urban Studies (Singerman and Amar 2006) became committed to fusing political-economy and ethnographic methods to reveal the new contours and patterns of modern power emerging in the urban frame. It emphasizes the importance of social movements in formulating their urban destiny (Singerman and Amar 2006). But the Alexandria School of Urban Sustainability Transitions (ASUST) argued that it is not a social movement capable of determining urban destiny rather social networks have a major influence in formulating the social structure in urban areas. Nearly ten years of the Arab Spring, it seems that social movements during the revolt went off in the wrong direction, one in which many nations of the Middle East, such as Syria, Libya, Yemen, and Iraq, have been destroyed. The other is that the displacement of people from one location to another has destroyed the social ties and created unhomogenized communities. The social movement should be for the benefit of society rather than its destruction. Thus, the ASUST should shift from relying on social-movement concepts and practices and shift to social coherence concepts and practices to integrate the various stratums of the society and to build the environment sustainably.

But all these interpretations hardly affect the urban poor in the Middle East. The changes that have occurred following the processes of globalization, urbanization, and the Arab Spring have increased the relevance of social exclusion and sharply affect the urban poor when obtaining goods and services for their survival. So, there are losers and winners in such transitions. The development studies literature (de Haan 2007) has been more controversial on the question of whether globalization has increased inequality, whether it has reduced or increased poverty, and for what reasons. The argument is that globalization has positives and negatives results. The latter (Munck 2005) emphasize how globalization deepens inequities and social exclusion, and others (Sachs 2005) see it has a positive benefit, showing how successful economies have been the ones that have integrated the less developed countries into world markets. In absolute terms, the recent period after the crisis of 2008 and the Covid-19 are generally associated with the economic downturn in some parts of the world, the Arab Uprising, globalization and readjustment, and increasing inequalities in many countries, as well as globally.

Major cities in the Middle East have become loci of contested space, with growing numbers of marginalized people, or "structurally irrelevant people" (Castells 1997) or "invisible" urban enclaves (Sibley 1995), or "a deviance situation" (Singerman 2011) or "urban disenfranchised" (Bayat 2004) or broaden "the bottom of the social pyramid" (Soliman 2019) who are now claiming their rights to the city as well. Accordingly, in the Middle East, the urban fabric exhibits a new contour and pattern, as a reflection of the emerging urbanism of urban informality, distinct from ordinary formality. Very often, these places—despite their often large scale—are not marked on cities' maps and are categorized by the majority as 'illegal', 'irregular', or 'marginal' or "outsiders" or "aliens". These places become the signifiers of the socially constructed and demonized image of the 'other'. People who live in urban informality are affected by limited benefits and are often prevented from participating in many activities that normally form bonds between people. These communities are all the result of labor exploitation, poverty, deprivation, ethnonational antagonisms, and socioeconomic exclusions that have pushed people as "skimming the cream off the milk"—occasionally residents of the city in question—to act 'illegally' and claim their right to the city. Even more excluded are whole cities such as *Buluq El Dakror* in Cairo, *El Dahyia El Ganoubyia* in Beirut (Fawaz 2009) and *Al Wehdat* in Amman which are prevented from being integrated within the national and regional administrative nomenclatures.

Thus, urban informality in cities of the Middle East offers a relevant entry point to explore the urban divide and intra-city inequality. In most of the Middle Eastern cities, urban informality has become visible in the periphery of the major urban centers, which the former Prime Minister of the Egyptian cabinet *Attaf Sadekey* interpreted as a "belt of poverty". A belt of poverty surrounds the traditional and legal core of the city with peripheral areas that form a kind of fortification surrounding the cities. All the while, 'gated developments', 'gated communities', 'fortified enclaves', and other forms of privatized public space sit on the fringe areas of Cairo, Beirut, and Amman. There is Cairo and the other Cairos within the city: the formal city, informal city, gated city, and the old city in "the Egyptian urban fabric" (Silver 2010; Sims 2010; AlSayyad 2006).

Thus, cities in the Middle East are divided into two or three or more cities within the city, and this has divided the society and the urban fabric according to the types of housing and the types of urban fabric that divide the social structure into different categories: the rich, middle, lower, and lowest classes of the society. The urban gap between formal, hybrid, and informal areas has widened and separates the poor and the rich as social segregation became visible. Each part of those cities is characterized by its social composition, income level, way of life, level of density, type of services, quality of the environment, and extent of open space. This has impacted social relationships between various strata of society and is reflected in everyday life for each sector of the city. One of the main features of urban informality is that streets lanes and open spaces, if they exist, are occupied by microbuses and tuk tuk in Egypt, or old Mercedes in Lebanon, which connect various parts of the capital cities. Recently, a new Indian method of transportation has come to some cities of the Middle East, called the tuk tuk; the motor tricycles. They circulate within different narrow parts of informal areas as a new cheap method of transportation. These two transportation methods do not exist in luxury areas and are replaced by modern private cars (Uber and Careem). In general, public transportation does not exist in the informal areas, but it links the main arteries of the city leaving the inner areas, informal areas, to be serviced by informal microbuses and tuk tuks. That is one of the main features of informality within urban informality or informality within urban formality.

The conjunction of conditions linked with urban poverty, violence, and ethnic and migrant concentrations are spatially expressed in "invisible" urban enclaves and urban informality. Many areas of such characteristics are extant in Cairo, Beirut, and Amman. These places have become the symbol of the socially consolidated and the symbolized image of the "outsiders", and of cities that have been divided and where urban areas have been sliced into two sectors, formal and informal, or old and modern, or cry, demand, and centrality (Lefebvre 1976), or legal and illegal, or maybe more than two divisions (Sims 2010; AlSayyad 2006), as in the case of the cities in the Middle East. The communities that have been defined in the literature as "spontaneous settlements," "squatter enclaves", or "shantytowns" or "urban informality" are all the result of labor exploitation, colonial legacy, ethnonational antagonism, improper planning polices, and social exclusion. These have pushed people who have immigrated from rural areas or occasionally residents of the city, or both, in question, to act "illegally" and claim their right to the city. Indeed, the cities in the Middle East separate different types of people, and these differences are visible in the spaces and places they occupy and inhabit. Referring to the body of information that deals with segregation, the squatters or disadvantaged groups are considered to have become isolated from the networked society. On the other hand, network urbanization would reunite a special stratum of society, but only a stratum that is capable of integrating or matching or affording the cost of the networked urbanization.

The socio-spatial organization of the Middle East cities is not an organic or natural process reflecting solely economic differences. Rather it is integrated into unequal urban niches that socio-spatially express power relations that currently stem from the rapid increase of information technology. In the Middle Eastern cities, social exclusion is currently seen in three forms. One is the problematic nature of disadvantaged groups on the periphery of big urban centers. Residential consolidation cannot be seen only as enforced by others, rather it is determined by their occupants. Specific groups may prefer to isolate themselves to protect their common sociopolitical entity and collective consciousness. The segregated group can establish communal functions and a sense of security, as well as political organization when necessary, such as *Dayia El Ganobyia* in Beirut ruled by *Hezbollah* (Bayat 2009) in the southern part of Lebanon. These segregated battlefields are the locations, or the worst location we might say, in which struggles for the "right to the city" take place. Hence, understanding the patterns of segregation in housing, economic activities, and everyday life is tightly linked with the analysis of minority–majority power relations in terms of network urbanization.

Another form is social exclusion within different classes of the society in their ability and capability to adopt information technology. Some people have the capital and knowledge to use information technology and others are not capable of being integrated into that rapid transformation. These areas occupy the old urban fabric and popular areas of cities, presuming the privilege of having the various economic activities, reasonable methods of transportation, and governmental institutions. For example, the old parts of the Gulf States, Lebanon, and Jordan have various resources, capital, and a free market to cope with the network society, but they have limited innovation. In other countries in North Africa, such as Egypt and Libya, many areas lack financial resources and having restricted markets through which the populations are not capable of coping with the information technologies, so a clear distinction is seen between those who have and those who have not.

The third form is a segregated private urban fabric established on the periphery of old urban centers that contain the affluent groups and separates them from other features of urbanity. These gated communities constitute a new urban fabric being introduced into Middle Eastern cities and are characterized by a good environment, low density, and wide open spaces. These areas are considered for many people as a secondary home for recreation and prestigious purposes. People have the capital and knowledge to grasp rapid transitions and possess a future vision for a better way of life.

In a context characterized by disengagement of the state and weak local finances, the Middle Eastern cities are or will be faced with management of the contradictory goals of their aspirations to achieve integration in a globalizing economy. The growing difficulties to meet the needs of the local population and the responsibility of managing the urban environment are another urban challenge. However, the largest cities saw deprivations that were all significant, a fact attributed to a pattern of migration in Egypt and Jordan that tended to be directed towards Cairo and Amman rather than to include smaller cities and towns. On the other hand, a low proportion of immigrants to intermediate-sized cities may be attributed to the fact that these cities have good connections with rural areas through various low-cost transportation systems, in addition to the difficulty of obtaining reasonable housing plots within these cities that discouraged the peasants from moving from their community of origin and those who have preferred daily commutes from and back to their homes.

In sum, the lack of balance between informal urbanization, demographic changes, rural-urban immigrants, arbitrary urban growth, and the absence of a national policy for achieving ongoing socio-spatial transitions have influenced the development processes in the Middle Eastern countries. The consequences are migration from rural to urban areas, the decrease in the rate of agriculture production, the growing number of people who move from rural to urban areas, the increasing gap and diversities between various strata of a society, the increasing pressure on various utilities, and rising number of urban unemployed. These lead to multiplying the need for additional urban utilities. The division between "legal and illegal" quarters, rich and poor, and bad and good have become apparent. Even though various different attempts have been made to guide the urban sprawl, arbitrary urban growth has eaten away at the periphery of agriculture areas that accommodate the new migrants, which has resulted in substantial economic losses and the spread of the informal urban centers.

Moreover, the fragmentation of the urban fabric of cities of the Middle East became palpable and reflected the social exclusion of the societies. This has led to the exclusion of certain strata of the society, i.e., the bottom of the pyramid, from having the privilege of the rights of other urban areas and to be looked down upon by the main tissues of the urban fabric. This social segregation became noticeable in the Middle East's cities in which social networks appeared, demanding a better and decent way of life.

7.3 Marginality and Informality

Most cities of the Middle East are witnessing tremendous transitions in their urban fabric. These transitions have come as a shift from the "urban society mode of regulation" to "an arbitrary urban system mode of irregulating." The urban system in cities of the Middle East has changed from a traditional system to a "hybrid urban system; old patterns in new urbanity." The latter is planned by the urban poor and middle-income groups and are marginalized, as a main output of sociotechnical landscape, on the periphery areas of those cities. Marginality, inequality, and social exclusion/segregation have become the dominating feature of the urban fabric of cities of the Middle East. Marginality combined with poverty, informal economy, and social exclusion reflect the ways in which some people, usually seen as an individual, or parts of distinct social groups or categories of people have been excluded or pushed out to the peripheral physical and economic growth or political development. This followed from the rich and above middle-income classes settling in a traditional pattern, formal areas, and gated communities, while the urban poor and the lowest income groups organized themselves and formulated their own urban pattern and lived as deviants (Singerman 2011), separated from the rest of the urban fabric of the city.

However, the urban poor are usually marginalized and excluded from the urban fabric of the city. The term marginality appears to have been used first by Robert Park, one of the key scholars in the Chicago School of urban sociologists, in his "Human Migration and the Marginal Man" (1928) (Perlman 1976). For Park, the marginal man is a migrant seeking a new way of life in the urban setting, so it is a "cultural and spatial type" and is not quite integrated or seeking to integrate into the mainstream of the new society. As soon as the marginal man has the capability to integrate, he will escape the "lockout of marginality", and this very rarely happens. Perlman (2004) claimed that the urban poor residents are not economically and politically marginal, but are excluded and repressed, and that they are not socially and culturally marginal, but stigmatized and excluded from a closed system. On the other hand, the term marginal is used as a "catch-all" phrase to encompass a range of very different situations, such as the dramatics of poverty, illiteracy, less wealth, or bad luck compared with the "average person" (Bush and Ayeb 2012). Ward (2003) argued that the classic marginality theory of the 1960s has shifted to the so-called "new poverty" of today. The classic marginality of the 1960s came in two primary forms: economic and cultural. Recently, marginality spans a wide spectrum of perspectives that include the social, political, economic, ecological and biophysical. Thus, marginality is an involuntary position and condition of an individual or group at the margins of social, political, economic, ecological, and biophysical systems. The marginal person is prevented from access to resources, assets, services, restraining freedom of choice, preventing the development of capabilities, preventing him/her from participation in social milieu and eventually causing destruction in the urban fabric, but above all marginalization means exclusion from a closed system (Perlman 1976).

On the cultural side, migrants to cities of the Middle East were cast as peasants in cities, carrying with them the trappings of a rural and traditional culture, foisting their traits upon the city, and being "marginal" to the mainstream of city life (Ward 2003). They settle on the periphery of cities, or the worst locations within the city, and formed irregular settlements. Or they are settled in or displaced to locations far from job opportunities and blockaded in locations separated from the rest of the city. On the economic side, if people were poor, it was by virtue of their integration, not their exclusion from formal economic activities (Roberts 1978; Perlman 1976). In the architecture of urban informality in Latin America, attention is directed to the multiple linkages between the formal and informal sectors and to the apparent virtuosity of the latter, and even its capacity for growth (Bromley 1978). This critique was often reflected in the imaginative titles of several of the relevant works: *Rationality in the Slum* (Portes 1972); *The Poor are Like Everyone Else, Oscar*, (Safa 1970); *The Myth of Marginality* (Perlman 1976), *Organizing Strangers* (Roberts 1973). *Social Networks and Local Organization in Self-Help Settlements were an Effective Demonstration of Social Mobility and Survival* (Roberts 1973; Lomnitz 1977; Castells 2010, 2nd edition). The poor were not radical (Ray 1969; Moreno 1970; Moser 1998; Eckstein 1988), nor were they excluded, but were, instead, invariably locked into client list networks that they could mobilize to reasonable effect (Leeds 1972; Cornelius 1975; Stokes et al. 2013).

Others (Perlman 2004; Bayat 2012; Roy 2011a, b; Dorman 2012) assert that marginality has now been extended beyond the original status of the urban poor classes, or poor countries, to include also the economically well off. This means those who, nevertheless, are marginalized in the political realm, or in the domain of lifestyle, or in their position as members of a particular gender, ethnicity, religion, or sexual orientation. Social movements were heavily recognized in the Middle East during and after the Arab Spring demanding the basic human needs of essential physical and essential spiritual needs. The latter constitutes two needs: esteem and freedom. Esteem needs cover security, identity, equity, and respect, while freedom needs ask for free choice, participation, power, and responsibility. In Egypt, Jordan, and Lebanon, for instance, many people, including affluent individuals and social groups, may simply not be allowed to participate effectively in the decision-making processes concerning major domains of public life. On the other hand, many people in urban informality are not classified as poor, nor excluded from the existing system, rather they have illegally participated in some way or another in various activities of the society.

7.4 Social Exclusion

A significant theoretical analysis of the interrelations between the social and the spatial aspects of the urban phenomenon originated during the 1970s from the Marxist school of urban sociology. This literature includes *The Urban Question* (Castells 1977) by M. Castells and David Harvey's *Social Justice and the City* (Harvey 1973), both of which refer to the urban scene as an arena of power relations, shaping cities' meaning, places, and spaces. This school views urbanization as a process that produces socio-spatial structures and forms, supporting the recreation of social relationships for the reproduction of capital. Still, the critique of this attitude calls for a wider analysis of the complex and diverse power relationships within the urban fabric (the conflict between the state and society) (Sandercock 1998) (see Chap. 6). Instead, it is rather a conflict for enhancing the living conditions of a given area and seeking a better life, but it is hardly hitting the urban poor who are self enhancement. On the other hand, affluent groups of the elite of the society rely on the working class and the urban poor to produce their capital. Both are important actors in the development processes and reproduction of capital since they produce/distribute commodities and rely on each other.

A theoretical shift appeared in Castells' book *The City and the Grassroots* (Castells 1983), in which he dissociates himself from class analysis as the only vehicle for social change in the urban context. According to Castells, the autonomous role of the state, gender relationships, and ethnic, national and citizens' movements are among other alternative sources of urban transition. The latter may be a danger to the society as happened in Arab countries during the Arab Spring and is lately spreading in French territories. While (Yacobi and Shechter 2005) argue that these social entities are struggling for political self-management and identity recognition, and working to gain standards of collective consumption, they have opened a path to a comprehensive theory of urbanity, that results not only from the actions of the dominant interests but also from the grassroots alternative to them. In cities of the Middle East, both the grassroots and affluent groups are looking for a better way of life, but it is a proportional process for both. Even in countries, such as Lebanon, where ethnicity dominates political life, all of society participate to increase productivity for the benefit of both the elite and the grassroots (as happened in October 2019 and August 2020 in Lebanon).

Adam Smith explained how capitalism worked, and Karl Marx explained why it didn't. Now the social and economic relations of *The Information Age* have been captured by Manuel Castells. A noteworthy aspects of Castells' work (1997) is its scope—its attempt to define the forces reshaping societies, from cultural beliefs to economic practices to political institutions to technological progress. It is a prodigious effort to overcome the poverty of an approach to the information society based on the fragmentation of the social sciences. Castells (2010, 2nd edition) goes a considerable way to helping us make sense of today's global information economy and our place in it. The old trend of urbanization, formal and informal, is moving further beyond the evolution of the new economy. Therefore, with the beginning of the new millennium, the planet witnessed a new trend in urbanization, which is *"network urbanization."* A central element of *network urbanization* is that globalization will have a great impact in shaping and determining the socio-spatial and spatiotemporal transitions within cities, as well as widening the gap between various strata of society. The recent financial crisis, rising food prices, Covid-19 pandemic, and the Arab Uprising have put further burdens on most governments in the Middle East, by which the urban poor have been sharply hit in obtaining goods and services for their survival. This has created a new section of the Middle Eastern societies that is prepared to do things that would threaten national security. The fact is that the states of those societies do not realize that the greater the social exclusion, as a main output of sociotechnical landscape, the greater the danger to national security. Riots, disturbances, and conflicts between the poor and the governments of the Middle East have occurred, e.g., the recent conflict in *Mahalla El Kobra*, the largest industrial and dormitory city for textile production, in Egypt, and the latest strike in Amman, Jordan, and the delay formulating the Lebanese cabinet.

On the one hand, (Carnoy et al. 1993) argue that the new social movements and initiatives that are emerging in response to the global are forming around 'the local'—and centering of power to regions, localities, NGOs, CBOs and so forth. It is a socio-spatial and spatiotemporal transitions. It entails a shift from representation through the nation-state to representation across a diversity of localized constituencies and institutional forms—be they the revitalization of devolved power in regional governments and municipalities or through the increased activities of NGOs and CBOs. Some of these provide services previously offered by the nation-state (such as Muslim Brothers in Egypt and Jordan). Citizens in the Middle East now identify with these new institutions—shifting power, responsibility, and legitimacy away from the central state. Legitimacy through decentralization and citizen participation in NGOs and CBOs in the Middle East seems to be the new frontier of the state in the twenty-first century. Moreover, globalization occurs at the expense of the Middle East's status, but it concedes that the stabilizing role of the states is ongoing, after the Arab Spring, and the benefits of global capital are gained by the elite while the urban poor have *"the rest of the cake"* and *"cream off the milk"* to contribute in such transformations. In most of the Middle Eastern cities, it has become possible to recognize the urban informality in the periphery of major urban centers and the formulation of the *"Belt of Poverty" or "Belt of Misery"* by which socio-spatial and spatiotemporal transitions have flourished.

On the other hand, Lefebvre's concept of the "right to the city" shifted the discussion from class analysis to wider aspects of power relations, including various ethnic groups and people forming both the urban population and human society at the global level (Lefebvre 1996). There is a complex system that integrates social, political, economic, and cultural dynamics of urban environments. The concept of the "right to the city" has been explored and ways in which many urban dwellers are excluded from the advantages of city life, using the framework to explore links among poverty, inequality, slum formation and economic growth. Others have argued (Landy and Saglio-Yatzimirsky 2014) for "spatial justice" in the form of places in a city (e.g., the right to land, right to shelter, etc.) rather than rights to various social and economic amenities in a city (e.g., the right to education, right to health services, etc.), but the right to the city is more inclusive to city life. The changes that have occurred following the process of globalization have increased the relevance of this approach. Major cities in the Middle East have become loci of contested space, with growing numbers of marginalized people. Castells (1997) defined them as "structurally irrelevant people" who are now claiming their rights to the city as well.

It is possible to generalize about urban spaces across various differing regional, national, and historical contexts. The conjunction of conditions linking with urban poverty, violence, ethnic and migrants' concentration is spatially expressed in "invisible" urban enclaves. Very often, these places—despite their scale—are not marked on the cities' urban fabric and are categorized by the majority as "illicit". Recently, the agglomeration of these spaces became visible on the periphery of cities, and even more became noticeable in the cities' urban fabric. The communities that have been defined in the literature as "spontaneous settlements," "squatter enclaves" or "shantytowns" are all the result of the labor exploitation, colonial legacy, ethnonational antagonism, and social exclusion that have pushed people—occasionally residents of the city in question—to act "illicitly" and claim their right to the city.

In addition, Castells (1997) writes that an educated labor force—a critical ingredient of the informational economy—also requires 'good health, decent housing, psychological stability, [and] cultural fulfillment'—the key qualities of the traditional welfare state. Castells acknowledges this by writing that 'welfare states, minus their bureaucratic undercutting and wasteful civil service perks', could be sources of productivity rather than budgetary burdens. The provision of these benefits is dependent on the restructuring pathways chosen by nation states in adapting to the dictates of globalization—the neo-liberal 'cost-lowering' model, or the more social democratic 'productivity enhancing model' (Castells 1997). On the other hand, Carnoy et al. (1993) conceded that different state models lead to different social policies and institutional arrangements that have a differential impact on the quality of life. They note that the important role of state policies 'in shaping the engagement with globalization but restricted within the context of possibility'. Castells notes the important role of state policies 'in shaping the engagement with globalization but restricted within the context of possibility' or restricted within political of informality (Goodfellow 2020). Indeed, Middle Eastern cities separate different types of people, and these differences are visible in the spaces they occupy and inhabit. Referring to the body of information that deals with segregation, squatters or disadvantaged groups are considered a component of the network society. On the other hand, network urbanization would rejoin a special stratum of society that is capable of integrating or matching the network urbanization.

Social segregation is evident within Middle Eastern cities' residential areas with distinct areas being occupied by high-income, middle-income, and low-income groups. Social exclusion became the main phenomenon for the grassroots in which it describes every kind of deprivation—whether or not relational features are important in its genesis. Thus, the further a residential area is from a high-quality environment, in general, the poorer the quality of services, while the higher the density and the higher the level of air pollution (Soliman 1996). The result is a growing divergence in Middle Eastern cities: between the new, the old, and the haphazard part of the urban agglomeration, a "Great Divide" between "establishing", "struggling" and "emerging" urban spaces. The former space comprise the new gated communities of the elites of the society; the middle, is the formal city and symbolizes the community that follows the prevailing laws; while the latter is *Ashwaiyyat*, or urban informality, or the peri-urban informal settlements and mirrors the marginal groups (see Fig. 7.1). Cairo has many cities within one city: for example, Islamic Cairo, Old Cairo, Khidwai Cairo, Modern Cairo, Dormitory Cairo (Nasr City), Cairo *of Baron Edouard Empain*, and informal Cairo.

On the other hand, various economic regimes in the Middle East are adopting social policies and institutional arrangements that have a differential impact on the quality of life. Middle Eastern cities increasingly separate people: this is visible in the spaces they occupy and inhabit, such as the social divides of *Imbaba* and *El Zamalek* areas in Cairo, the social segregation in West and East Beirut (Soliman 2008), and the differentiation between the Eastern and Western sectors of Amman, where the former is a district for the poor, while the latter is occupied by the elite (Potter et al. 2009). Traditional physical planning decisively reinforces processes of

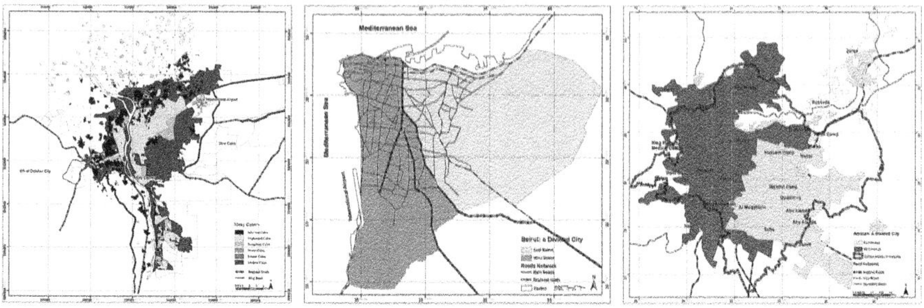

Fig. 7.1 Maps illustrating the many Cairos, Beiruts, and Ammans. (Source: The author)

social exclusion. Disadvantaged groups are set apart from the formal economy society.

Urban informality in Middle Eastern cities originate in part from the informal economic society, social exclusion, and diversity that have emerged through two forms. One is the problematic nature of disadvantaged groups on the periphery or in the most derelict quarters of large urban centers. The segregated group is able to establish communal functions and a sense of security, as well as a political organization when necessary, such as the Muslim Brothers movements and Islamic groups in Egypt and Jordan, and *Hezbollah* in Lebanon (Bayat 2009). Another form, namely social exclusion within urban informality, also occurs from their respective abilities and capacities to take advantage of network urbanization. Social media has played a huge role in connecting many of the people who would eventually join the protests which boasted 35 million participants on 30 June 2013 in various Egyptian cities' squares, and were instrumental, so much so that a new expression has entered the Egyptian lexicon. A central element of network urbanization is that globalization will have a great impact in shaping and determining the socio-spatial transition within cities, as well as widening the gap between different strata of society (Castells 1997), and it would increase the economic divergence in cities. These segregated battlefields are the locations, or the worst locations, in which struggles for the right to the city take place, such as the *Ezbet El Haghana* and *Bulaq El Daqrur* areas in Cairo, and *Dayia El Ganobyia* in Beirut.

7.5 Social Cohesion and Urban Informality

Social cohesion as a reflection of community participation has played a major role in formulating urban informality. Community participation in housing processes was explored in several classical books of John Turner (Turner and Fichter 1972; Turner 1976) and Hasan Fathy (Fathy 1973) while organizational studies were elaborated in Pierre Bourdieu's work (1977). The principle of participation can mean any process by which the users of an environment participate together to shape it (Alexander et al. 1975; Hasan 2001). Turner (1976) identified the role of social networks in sustaining the production of informal housing in the form of three laws of freedom to build, as a reflection of the community-driven process (see Chap. 4).

In cities of the Middle East, activism (Bayat 2000) might exist in urban informality in various types: trade unionism, community activism, social Islamism, NGOs, CBOs, and quiet encroachment, as well as activism expressed in urban mass protests to any kind of human activity—individual or collective, institutional or informal, ranging from survival strategies and resistance to more sustained forms of collective action and social movements, such as *Maspero* Youth Union in Cairo (Delgado 2015). Activism usually exists in large informal settlements to ensure strong solidarity among the residents, as happened in the *Maspero* triangle in Cairo, the recent protest due to increased oil prices in Amman, and the increased rate of exchange of US$ against the Lebanese pound. It could also exist in the famous squares in the Global South, and Global North, e.g., in the former in *Tahrir* square in Cairo during the January and June revolts of 2011 and 2013. The latter happened recently on 1 December 2018, which is called *Black Saturday*, when *Yellow Jackets* moved into L'Arc de Triomphe square in Paris to protest the rise of oil prices in which around 180 were injured and one person was killed. Various types of activism might lead to strong social cohesion, active social networks, strong social solidarity, and vice versa. Also, it could be said that arbitrary social movements can result in violence and crime as happened in Paris.

Understanding how social networks contribute to the economic and social fabric of life in Middle Eastern countries is important (Woolcock and Narayan 2000; Collier 2002), and social cohesion is critical for societies to prosper economically and for development to be sustainable (Adler and Kwon 2002). The interdependent nature of actors (where actors are the individuals or groups within a network) is a key distinguishing characteristic of a social network, and their structures and connectivity and the distribution of power are key components of marketing systems (Layton 2009; Jackson 2010). Fawaz, in Lebanon, stated that without systematically resorting to the terminology of social networks, the abundant literature that documented the formation of informal settlements during the 1970s and 1980s challenged their condemnation as 'spontaneous' by revealing that these neighborhoods were organized (Fawaz 2008) and managed through thick webs of social relations (Collier 1976; Perlman 1976; Ward 1983; De Soto 1989). It is argued (Fawaz 2008; Fawaz, Krijnen and El Samad 2018) that various social agents generate new insights into how (informal land) markets work and the practices of developers in this type of neighborhood, as well as the yet unstudied mechanisms of informal housing production in the Lebanese context.

On the other hand, social networks is economic, cultural, or social capital in its socially recognized and legitimized form. Residents living in urban informality in cities of the Middle East are characterized by various types of capital, as well as material dimensions, which are symbolic, efforts, and cash. Hamdi (2004) and UN-Habitat (2018b) argued that the guru of urban participatory development is looking for ways of connecting people, organizations, and events, i.e., of seeing a strategic opportunity. It means acting practically (locally, nationally, and globally) and thinking strategically, and acting strategically (locally, nationally, and globally) and thinking practically. It is not doing either/or but doing both, and doing it reflectively, practically, collectively, and progressively. This kind of capital has resulted in housing types and forms, as many residents living in urban informality like to imitate their neighbor in designing and building his/her house like the other. Swept equity represents capital of human efforts. Capital of efforts, in a special issue of *Habitat International* (2010, 34:3), presented the view that the poor relied mainly on their own efforts and their social networks (Tunas and Peresthu 2010) and social networks played a major role in negotiating better deals with landowners for the lease or purchase of land (Yap and De Wandeler 2010). Capital in cash, in cities of the Middle East, means that the urban poor, and middle-income classes rely upon an informal credit system called *Ghameyhia* that exists through cooperation between all stakeholders including the household, an intermediary, the landowner and an informal contractor (Soliman 2010). This collective monetary system operates in such a way that each stakeholder's needs in the agricultural land development process are met. Under this system, a certain amount of money from each group member is collected on a monthly basis, and the total amount collected money is paid once for each member each year on an alternating basis. The poor also rely on housing microfinance, community-based finance savings and loan groups, and consumer credit for building materials (Ferguson and Smets 2010). Therefore, informal credit institutions (the *Gamaiyyat* in English cooperatives) that provide the necessary funding for goods and services also participate in the production of informal housing.

Recently, capacity development as a system perspective became widely used in the development process (Baser and Morgan 2008). There are numerous definitions of capacity development (Kuhl 2009; Wignaraja 2009), each reflecting a particular bias or orientation. Some describe capacity development as an approach or process, e.g., toward reduction of poverty, while others see it as a development objective, e.g., targeting the development of individual or organizational capacity. The stronger the capacity development, the higher the degree of social cohesion and social participation in development processes. Capacity development exists in urban informality of the Middle East's cities, including various 'approaches, strategies and methodologies', and seeks to improve performance on various political, social, economic and environmental levels. However, the treatment of housing as a market value/demand market movement is advocated as capacity development to illustrate how to organize, use, produce, and accumulate/distribute resources, whatever these are, at the right time and in the right place, which would increase housing values, accelerate the housing production process, and integrate housing production with development programs on incremental phases for the benefit of the urban poor. This phenomenon is illustrated in various informal areas of Cairo, Beirut, and Amman (Gerges 2015).

As Engels (1970) put it: "The 'working people' remain the collective owners of the houses, factories, and instruments of labor, and will hardly permit their use . . . by individuals or associations without compensation for the cost". However, the former is often viewed as negative modification resulting in increased fear and isolation, while the latter is a collective response in which individuals act jointly to undertake joint activities that they could not accomplish on their own or from the many-faceted obstructions to collective action (Barker and Linden 1985; Breman 2020). The stakeholders as members of a "community" can be part of productive, non-capitalist social processes, which form the basis for collective consciousness and a cooperative ethic (Bayat 2010). The collective planning and administration of social production is well illustrated in urban informality in the Middle East, such as in Cairo and Beirut. Collective planning or conventional planning (Soliman 2013a, b) or collective action requires that not only the means of production but also the distribution and consumption of the final product be subject to explicit social control. Thus, social control in the physical environment dominates the right of ways or right of acquisition and/or both, privacy, behavior, identity, sense of belonging, pattern, and everyday sort of urban life. The collective planning or collective action or process requires trust, transparency, accountability, and responsibility among all the stakeholders for a certain activity to be achieved. It is argued (Breman 2019) that the neoliberal mode of capitalism is a regime of unregulated informality that not only embraces the economy at large but also extends to politics and governance. However, when the provision of property rights to residents of low-income settlements becomes a central theme for a social network, the best way is to accelerate their incorporation into the complete housing system.

This paradigm has become so significant within multilateral agencies that housing and land issues are mentioned in the MDGs in relation to one part of Goal 7. Perhaps the most positive aspect of the changes in the World Bank policy (Buckley and Kalarickal 2006) since the mid-1990s is that the bank has learned much about the composition of a more appropriate policy environment that entails a strong reliance on an active role of the private sector, a recognition of the high speed at which market-based housing finance flows in the world. This treats land as an important input into the provision of housing services, accounts for a large fraction of total economic development, and emphasizes the importance of collective action. This paradigm has become so significant within multilateral agencies that not only has securing of land become the single indicator, but also market mechanisms that facilitate land delivery systems became crucial. Some writers (Roy and AL-Sayyad 2004; Soliman 2007) have argued that formal and informal markets are inseparable and that each impinges on the other. The former depends on the latter and vice versa in housing components and in the distribution of housing commodities within the market, as well as each relying on each other in economic activities. Both depend on the level of social coherence and the degree of social networks, for example, see the Banha project in Egypt (Soliman 2017; UN-Habitat 2016).

It is argued (Soliman 2019) that the people living in urban informality in Egypt have hidden potential actions as a cooperation/participation process and as a social network in the urban informality to deliver residential land on new urban expansion for the citizens. Potential actions of the urban poor and their relative impact on housing production are defined as the priorities of the stakeholders (government, professionals, and the poor) and their ability, liability, and willingness towards the production of the housing being supplied. The potential actions are threefold: preferred, effective, and actual actions. The latter is the means and methods that can be provided by the stakeholders in the land market to achieve a state of balance between supply and demand for land and ensure the availability of suitable land for various classes in the community. Effective action involves the possibilities that can be provided by cooperation among the stakeholders for possession and ownership of land in informal residential areas, which can affect the housing systems. The preferred action is the fundamental option favored by the stakeholders cooperating systems and mechanisms of acquisition and ownership of residential land. Scholars and practitioners (Soliman 2019; UN-Habitat 2016; Breman 2020) analyzing community-level collective action or hidden action have become increasingly interested in how relationships work based on trust, reciprocal exchange, and social networks—social capital affect outcomes. The idea that development plans as such could be directly 'implemented' reflected a very traditional conception of a plan as a spatial blueprint, which would steadily be translated into a built form on the ground (Healey 2003). The underlying idea is to create an applicable sustainable environment at the right time and in the right place in which all the stakeholders in the housing process, through a collective process, are actively cooperating to manage the process positively for their own benefits without harming others.

Therefore, community collective process through social networks within informal areas plays a crucial role in communicating with other stakeholders in various ways. First, organizations of their own so they can negotiate with government, traders, and NGOs to formalize their situation in a sustainable way. Second, direct assistance through community-driven programs so they can shape their own sustainable destinies; and finally, local ownership of funds, so they can end corruption (Woolcock and Narayan 2000). The relationships between the three independent systems or three pillars, state, market, and social network, are correlated. The latter strongly reflects social ties that would formulate local finance customs, as well as sustain cultural identity.

In El Rezqa area of Banha (Qalyoubia), located 45 kilometers away from Cairo, Egypt, the land readjustment project went a step further and piloted a participatory method where landowners were convened to participate and cooperate for the success of the project. All stakeholders reported that a major achievement of the project was the fact that it facilitated networking and cooperation between landlords, academics, technical team, local governance, and central governance (UN-Habitat 2016; Soliman 2017). For the first time in Egypt, Participatory and Inclusive Land Readjustment projects (PILaR) enable cities to significantly increase the supply of serviced land at the urban fringe (UN-Habitat 2015).

7.6 Conclusion

During the last seven decades, cities in the Middle East have witnessed a remarkable and, in many ways, unprecedented increase in levels of urban informality. Urban informality is often construed as land-use-related and economic problems. Yet it arises and persists due to socioeconomic and cultural necessities, cry and demand, and material deprivation, physical and social insecurity, social exclusion, and class prejudices. As a result, falling housing policies, poor development of social services, and changing demographic trends, contemporary urban informal areas remain the growth points in society and are also the potential lightning rods for political and social unrest as happened in Egypt, Libya, Syria, Yemen, and Tunisia. In a context characterized by the disengagement of the state and meager local finances, Middle Eastern countries face the contradictory goals of meeting the aspirations for integration in a globalizing economy and the growing difficulties of meeting the needs of the local population and effective management of the urban environment. This is particularly pronounced in the largest cities, as well as being common in intermediate and small cities. With the rapid growth of population, rural-urban migration, the influx of refugees, flourishing of informal economy, and increasing demand for housing, governments will continue to face the problems that have only accelerated due to the spread of urban informality. It is not only the problem of housing, but it is rather also the destruction of the urban fabric, and everyday urban life that is caused by urban informality in cities of the Middle East. As there is strong link between informal economy and urban informality, the question is how urban informality is linked to formal regulations and the state. It is time to look at the people in the societies and to find a way out to close or at least narrow the gap between the rich and the poor, paying special attention to the needs of the poor, facilitating and enabling social justice among various strata of the societies. At the same time, one must bear in mind that a new era for democracy has emerged in various Middle Eastern countries (the revolt in Egypt, Tunisia, and the current people movements in Algeria and Sudan) in which civil society, CBOs, and NGOs will play a major role in conducting the urban transitions process, especially in housing delivery systems.

The four main challenges face Middle Eastern governments in tackling urban informality in cities. The first is the urban divide in the major cities. The current division is accelerating the social divergence between various classes in the city which is reflected in increasing social segregation, social exclusion, and inequity. It is also leading to the destruction of the urban fabric of a city, and devastation of the national economy in which is created various cities within a city, each having a certain class of the society. The second is the change in the behavior of people who are living in urban informality because they feel marginalized within the city. This has led to changing urban everyday life and the way residents act towards the urban tissues of the city, and towards the rest of its citizens. These transitions of attitude might result in increasing violence and crime in various sectors of a city. The third challenge is increasing the social and economic segregations among the residents within a city. This will lead to accelerating the pressure on informal economy and urban informality as the cheapest location to find affordable houses, away from organized areas to escape from paying taxes, as well as opening the door to accommodate people who are involved with violence and crime. Finally, the challenge is the aspect involving public-sector management reform and private-sector development, which cannot proceed effectively unless and until they have cooperated with citizens in urban informality and to make the best use of the advantages of social cohesion, in a sustainable way, among the people living in urban informality.

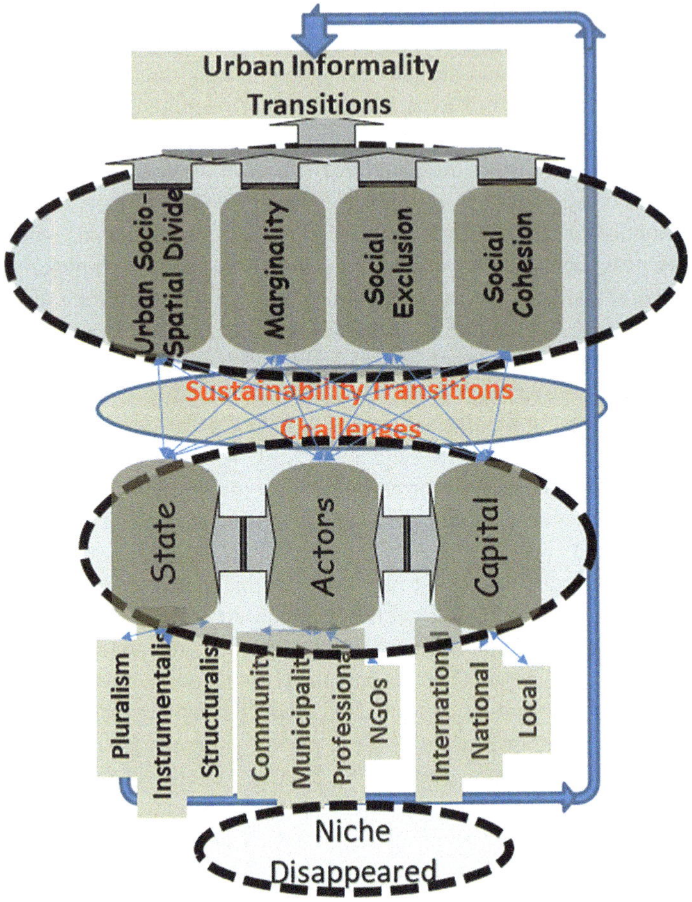

Fig. 7.2 Scheme of the various forces that have led to marginality and social exclusion in Middle Eastern cities. (Source: The Author)

Figure 7.2 shows the various forces that have led to marginality and social exclusion of urban informality. The sociotechnical niche has nearly disappeared, while the sociotechnical landscape in the form of socio-spatial divide, social exclusion, marginality, and social cohesion dominate the diversity the landscape of the society. On the one hand, the sociotechnical regime is shrinking the benefits of the upper classes and the elite for accelerating the economic development of the country. On the other hand, the lowest classes of the society are forced to conquer additional areas under the umbrella of informality in which they expand the current urban fabric full off urban illness.

Given transformations of urban informality including forms of informal economy, now is the appropriate time to take stock of the significant shifts, i.e., shifts in rethinking urban informality over the past seven decades and the exploration of emerging challenges related to it leading to future sustainable transition. A great proportion of those in marginality are capable of improving their statues, a shift from being poor, or living in the shadow of poverty, into improved economic status. This would require changing the state's action from being biased towards a certain stratum of a society into an equilibrium towards all classes of the society, building on three social conditions: social justice, equity, equality, and social safety. These apparatuses crosscutting the sociotechnical regime and sociotechnical niche and overlap and interlink with the sociotechnical landscape. Thus, it is a continuous and communal process, as many times it merges with the other two. Other variables have arisen, such as legalizing the informal economy, labor markets, labor rights and legalization that are essential for sustainable transitions.

MDGs and the Habitat Agenda, including the notable fact that some targets now, including equality in urban areas, have reliable baselines to work with, while many others are new and come as a response to the challenges and opportunities that urban areas face today in search of sustainable development outcomes. If subalterns' classes are excluded from transitions process and wealth creation, this is a systematic consequence of unsustainable process, and so an equitable sociotechnical system should take place. Therefore, a balance between the needs of the bottom of the pyramid, as a sociotechnical niche, and the state's policy, as a sociotechnical regime, is needed to lessen the spread on urban informality. There is a need to invest in social equity in cities to eradicate socioeconomic inequalities, social exclusion, extreme poverty, high unemployment, poor environmental conditions, and the drivers of climate change (UN-Habitat 2018c).

References

Adler, P., and S. Kwon. 2002. Social Capital Respective or a New Concept. *Academy of Management Review* 27 (1): 17–40.

Al Sayyad, N. 2006. Whose Cairo? In *Cosmopolitan: Politics, Culture, and Urban Space in the New Globalized Middle East*, ed. D. Singerman and P. Amar, 539–542. Cairo: American Unversity Press in Cairo.

Alexander, C., et al. 1975. *The Oregon Experiment*. Mountain View: The Center for Environmental Structure.

Barker, I., and R. Linden. 1985. *Community Crime Prevention*. Ottawa: Ministry of Solicitor General Canada.

Baser, H., and P. Morgan. 2008. *Capacity, Change and Performance. European Centre for Development Policy Management*. Dordrecht: HE Maastricht.

Bayat, A. 2000. From "Dangerous Classes" to "Quiet Rebels" Politics of Urban Subaltern in the Global South. *International Sociology* 15 (3): 533–557.

———. 2004. Globalization and the Politics of the Informals in the Global South. In *Urban Informality: Transnational Perspectives from the Middle East, Latin America, and South Asia*, ed. Ananya Roy and Nezar AlSayyad, 79–102. Lanham: Lexington Books.

———. 2009. Middle Eastern Megacities: Social Exclusion, Popular Movements and the Quiet Encroachment of the Urban Poor. In *Megacities: The Politics of Urban Exclusion and Violence in the Global South*, ed. Kees Koonings and Dirk Kruijt, 82–106. London: Zed Books Ltd.

———. 2010. *Life as Politics: How Ordinary People Change the Middle East*. Amsterdam: Amsterdam University Press.

———. 2012. Marginality: Curse or Cure. In *Marginality and Exclusion in Egypt*, ed. Ray Bush and Habib Ayeb, 14–27. Cairo: American University Press in Cairo.

Bourdieu, P. 1977. *Outline of Theory of Practice*. Cambridge, UK: The University of Cambridge.

Breman, J. 2019. *Capitalism, Inequality and Labour in India*. Cambridge, UK: Cambridge University Press.

———. 2020. Informality: The Bane of the Laboring Poor Under Globalized Capitalism. In *The Informal Economy Revisited: Examining the Past, Envisioning the Future*, ed. M. Chen and F. Carré, 31–37. Oxon: Routledge.

Bromley, R. 1978. The Urban Informal Sector: Why Is It Worth Discussing? *World Development* 6: 1033–1039.

Buckley, R., and J. Kalarickal, eds. 2006. *Thirty Years of World Bank Shelter Lending What Have We Learned?* Washington, DC: World Bank, The International Bank for Reconstruction and Development.

Bums, L.S., and J. Friedmann. 1985. *The Art of Planning: Selected Essays of Harvey S. Pedoff.* New York: Plenum Press.

Bush, R., and H. Ayeb. 2012. Introduction: Marginality and Exclusion in Egypt and the Middle East. In *Marginality and Exclusion in Egypt*, ed. R. Bush and H. Ayeb, 3–13. Cairo: American University in Cairo Press.

Carnoy, M., M. Castells, S.S. Cohen, and F.H. Cardoso. 1993. *The New Global Economy in the Information Age: Reflections on Our Changing World*. Pennsylvania: The Pennsylvania State University.

Castells, M. 1977. *The Urban Question*. London: Edward Amold (Publishers) Ltd.

———. 1983. *The City and Grassroots: A Cross-Cultural Theory of Urban Social Movements*. London: Edward Arnold (Publishers) Ltd.

———. 1997. *The Power of Identity*. Malden: Blackwell Publishers Inc.

———. 2010. *The Rise of the Network Society*. 2nd ed. West Sussex: Wiley.

Chen, Marty, and Françoise Carré. 2020. *The Informal Economy revisited: Examining the Past, Envisioning the Future*. Oxon: Routledge.

Collier, D. 1976. *Squatters and Oligarchs*. Baltimore: Johns Hopkins University Press.

Collier, P. 2002. Social Capital and Poverty: A Microeconomic Perspective. In *The Role of Social Capital in Development: An Empirical Assessment*, ed. T. Van Bastelaer and C. Grootaert. Cambridge: Cambridge University Press.

Cornelius, W. 1975. *Politics and Migrant Poor in Mexico City*. Stanford: Stanford University Press.

de Haan, Arjan. 2007. *Reclaiming Social Policy: Globalization, Social Exclusion, and New Poverty Reduction Strategies*. New York: Palgrave Macmillan.

De Soto, H. 1989. *The Other Path: The Invisible Revolution in the Third World*. London: I.B. Taurus.

Delgado, M.C. 2015. Contentious Copts: The Emergence, Success, and Decline of the Maspero Youth Movement in Egypt. In *Contentious Politics in the Middle East: Popular Resistance and Marginalized Activism: beyond the Arab Uprisings*, ed. Fawaz A. Gerges, 259–280. New York: Palgrave Macmillan.

Dorman, W.J. 2012. Exclusion and Informality: The Praetorian Politics of Land Management in Cairo, Egypt. *International Journal of Urban and Regional Research*. https://doi.org/10.1111/j.1468-2427.2012.01202.x1-27.

Eckstein, Susan. 1988. *The Poverty of Revolution: The State and the Urban Poor in Mexico. (Revised edition 1988.)*. Princeton: Princeton University Press.

Engles, F. 1970. *The Housing Question*. Moscow: Progress publishers.

Fathy, H. 1973. *Architecture for the Poor: An Experiment in Rural Egypt*. Chicago: The University of Chicago Press.

Fawaz, M. 2008. An Unusual Clique of City-Makers: Social Networks in the Production of a Neighborhood in Beirut (1950–75). *International Journal of Urban and Regional Research* 32 (3): 565–585.

———. 2009. Neoliberal Urbanity and the Right to the City: A View from Beirut's Periphery. *Development and Change* 40 (5): 827–852.

Fawaz, M., M. Krijnen, and D. El Samad. 2018. A Property Framework for Understanding for Understanding Gentrification. *City* 22 (3): 358–374.

Ferguson, B., and P. Smets. 2010. Finance for Incremental Housing; Current Status and Prospects for Expansion. *Habitat International* 34: 288–298.

Frantzeskaki, N., V.C. Broto, L. Coenen, and D. Loorbach (eds.). 2017a. UrbanSustainability Transitions: The Dynamics and Opportunities of Sustainability Transitions in Cities. In *Urban Sustainability Transitions*, ed. N. Frantzeskaki, V.C. Broto, L. Coenen, and D. Loorbach, 1–19. Oxon: Routledge.

———. 2017b. *Urban Sustainability Transitions*. New York: Routledge.

Gerges, F.A. 2015. *Contentious Politics in the Middle East: Popular Resistance and Marginalized Activism Beyond the Arab Uprisings*. New York: Palgrave Macmillan.

Goodfellow, T. 2020. Political Informality: Deals, Trust Networks, and the Negotiation of Value in the Urban Realm. *The Journal of Development Studies* 56 (2): 278–294.

Hamdi, N. 2004. *Small Change: About the Art of Practice and Limits of Planning in Cities*. London: Earthscan.

Harvey, D. 1973. *Social Justice and the City*. Oxford: Blackwell Publishers.

———. 2012. *Rebel Cities: From the Right to the City to the Urban Revolution*. London: Verso.

Hasan, A. 2001. *Working with Communities*. Karachi: City Press.

Healey, P. 2003. Collaborative Planning in Perspective. *Planning Theory* 2 (2): 101–123.

Jackson, L. 2010. *Influences on Financial Decision Making Among the Rural Poor in Bangladesh*. Unpublished Ph.D. thesis. Sydney: School of Marketing, The University of Western Sydney.

Jacobs, J. 1961. *The Death and Life of Great American Cities*. New York: Vintage Books, a division of Random House Inc.

Kuhl, S. 2009. Capacity Development as the Model for Development Aid Organizations. *Development and Change* 40: 551–577.

Landy, F., and M.C. Saglio-Yatzimirsky. 2014. Conclusion: Spatial Justice, Exclusion and Urban Policies in Brazil and India. In *Megacity Slums: Social Exclusion, Space and Urban Policies in Brazil and India*, ed. Marie-Caroline Saglio-Yatzimirsky and Frédéric Landy, 409–426. Covent Garden/London: Imperial College Press.

Layton, R.A. 2009. On Economic Growth, Marketing Systems, and the Quality of Life. *Journal of Macro Mark* 29 (4): 349–362.

Leeds, A.. 1972. *Forms of Squatment Political Organisation: The Politics of Control in Brazil*. M. A. thesis. Austin: University of Texas.

Lefebvre, H.. 1968. *The Right to the City*. The Anarchist Library.

———. 1976. *The Survival of Captalism: Reproduction of Relation of Production*. New York: Allison and Busby.

———. 1991. *The Production of Space*. Maiden: Blackwell Publishing.

———. 1996. *Writings on Cities*, trans. and ed. E. Kofman and E. Lebas. Oxford: Basil Blackwell.

Lomnitz, Larissa. 1977. *Como sobreviven los marginados. Mexico DF: Siglo XXI. English Translation Networks and Marginality: Life in a Mexican Shantytown*. New York: Academic.

Moreno, Jose Antonio. 1970. *Barrios in Arms; Revolution in Santo Domingo*. Pittsburg: University of Pittsburg Press.

Moser, Carline. 1998. The Asset Vulnerability Framework: Reassessing Urban Poverty Reduction Strategies. *World Development* 26 (1): 1–19.

Munck, R. 2005. *Globalization and Social Exclusion. A Transformationalist Perspective*. Bloomfield: Kumarian Press.

Park, R. 1967. The City: Suggestions for Investigation of Human Behavior in the Urban Environment. In *The City*, ed. R.E. Park, E.W. Burgess, and R.D. McKenzie, 1–46. Chicago: The University of Chicago Press.

Perlman. 1976. *The Myth of Marginality: Urban Poverty and Politics in Rio de Janeiro*. Stanford: California University Press.

Perlman, Janice. 2004. Marginality: From Myth to Reality in the Favelas of Rio de Janeiro, 1969–2002. In *Urban Informality: Transnational Perspectives from the Middle East, Latin America, and South Asia*, ed. Ananya Roy and Al Sayyad, 105–146. Lanham: Lexington Books.
Portes, A. 1972. Rationality in the Slums: An Essay in Interpretive Sociology. *Comparative Studies in Society and History* 14 (3).
Potter, R.B., D. Darmame, N. Barhamb, and S. Nortcliff. 2009. "Ever-Growing Amman", Jordan: Urban Expansion, Social Polarisation and Contemporary Urban Planning Issues. *Habitat International* 33: 81–92.
Ray, T. 1969. *The Politics of the Barrios of Venezuala.* Berkeley: University of California Press.
Roberts, B. 1973. *Organizing Strangers: Poor Families in Guatemala City.* Austin: University of Texas Press.
———. 1978. *Cities of Peasants: The Political Economy of Urbanization in the Third World.* Beverly Hills: Sage Publication.
Roy. 2009. The 21st-Century Metropolis: New Geographies of Theory. *Regional Studies* 43 (6): 819–830.
Roy, A., and N. Al-Sayyad. 2004. *Urban informality: Transnational perspectives from the Middle East, Latin America, and South Asia.* Lanham: Lexington Books.
Roy, A. 2011a. Postcolonial Urbanism: Speed, Hysteria, Mass Dreams. In *Worlding Cities : Asian Experiments and the Art of Being Global*, ed. A. Royand A. Ong, 307–335. Chichester: Blackwell Publishing Limited.
———. 2011b. Slumdog Cities: Rethinking Subaltern Urbanism. *International Journal of Urban and Regional Research* 35 (2):223–238.
Sachs, J. 2005. *The End of Poverty. How We Can Make It Happen in Our Lifetime.* London: Penguin Books.
Safa, Helen. 1970. The Poor Are Like Everyone Else, Oscar. *Psychology Today* 4 (4): 26–32.
Samman, K. 2006. *Cities of God and Nationalism: Mecca, Jerusalem, and Rome as Contested World.* Paradigm: Boulder.
Sandercock, L. 1998. *Toward Cosmopolis.* New York: Wiley.
Sibley, D. 1995. *Geographies of Exclusion: Society and Difference in the West.* London: Routledge.
Silver, H. 2010. Divided Cities in the Middle East. *Community and Development* 9: 357–345.
Sims, D. 2010. *Understanding Cairo: The Logic of a City Out of Control.* Cairo: America University Press in Cairo.
Singerman, D. 2011. The Siege of Imbaba, Egypt's Internal "Other", and the Criminalization of Politics. In *Cairo Contested: Governance, Urban Space, and Global Modernity*, ed. Diane Singerman, 111–144. Cairo: The American University in Cairo Press.
Singerman, D., and P. Amar. 2006. *Cairo Cosmopolitan: Politics, Culture, and Urban Space in the New Globalized Middle East.* Cairo: The American University in Cairo Press.
Soliman, A. 1996. Legitimizing Informal Housing: Accommodating Low-Income Groups in Alexandria, Egypt. *Environment and Urbanization Journal* 8 (1): 183–193.
———. 2007. *Urban Informality in Egyptian Cities: Coping with Diversity.* Fourth Urban Research symposium 2007, Urban Land Use and Land Markets, May. Washington, DC: World Bank.
———. 2008. Diversity of Ethnicity and State Involvement on Urban Informality in Beirut. *Theoretical and Empirical Researches in Urban Management, Year 3, Number 9*: 15–32.
———. 2010. Rethinking Urban Informality and the Planning Process in Egypt. *IDPR Journal* 32 (2): 119–143.
———. 2017. Land Readjustment as a Mechanism for New Urban Land Expansion in Egypt: Experimenting Participatory Inclusive Processes. *International Journal of Urban Sustainable Development* 9 (3): 313–332.
———. 2019. So It's Always a Chance: Community-Led Solutions to New Urban Expansion. In *New Cities and Community Extensions in Egypt and the Middle East*, ed. S. Attia et al., 159–179. Cham: Springer.

Soliman, M. 2013a. *Living With/in Ashwaiyyat in Egypt: Is It a New Innovation of Urbanity? The Flexible City, International Symposium*. Oxford: Oxford University, 24–25 October.

———. 2013b. *The poor in search of shelter: An examination of squatter settlements in Alexandria, Egypt*. Deutschland, Germany: Lap Lambert Academic Publishing.

Stokes, S.C., T. Dunning, M. Nazareno, and V. Brusco. 2013. *Brokers, Voters, and Clientelism: The Puzzle of Distributive Politics*. Cambridge, MA: Cambridge University Press.

Sugranyes, A., and C. Mathivet. 2010. *Cities for All:Proposals and Experiences Towards the Right to the City*. Santiago: Habitat International Coalition (HIC).

Tunas, D., and A. Peresthu. 2010. The Self-Help Housing in Indonesia: The Only Option for the Poor? *Habitat International* 34: 315–322.

Turner, J. 1976. *Housing by People*. London: Marion Boyars Publishers Ltd.

Turner, J., and R. Fichter. 1972. *Freedom to Build: Dwellers Control of the Housing process*. New York: Macmillan.

UN-Habitat. 2015. *UN-Habitat Global Activities Report 2015: Increasing Synergy for Greater National Ownership*. Nairobi: United Nations Human Settlements Programme (UN-Habitat), 31.

———. 2016c. *Evaluation of Participatory Review of Egyptian Planning and Related Urban Development Legislation to Support Sustainable Urban Development, April 2016*. Nairobi: UN-Habitat.

———. 2018a. DG 11 Synthesis Report 2018: Tracking Progress Towards Inclusive, Safe, Resilient and Sustainable Cities and Human Settlements. *SDG11 Synthesis Report High Level Political Forum 2018*. Nairobi: Un-Habitat.

———. 2018b. *Prosperity for All: Enhancing the Informal Economy Through Participatory Slum Upgrading*. Nairobi: UN-Habitat.

———. 2018c. *Addressing the most Vulnerable First -pro-poor climate action in informal settlements*. Thematic Guide, Nairobi: UN-Habitat.

Ward, P. 2003. *From the Marginality of the 1960s, to the "New Poverty" of Today: A LARR Research Forum*. Workshop. Austin: University of Texas at Austin.

Wignaraja, K., ed. 2009. *Capacity Development*. Nariobi: A UNDP Primer. Available at: www.undp.org/capacity

Woolcock, M., and D. Narayan. 2000. Social Capital: Implications for Development Theory, Research, and Policy. *World Bank Research Observer* 15(2): (World Bank Res) 225–249.

Yacobi, H., and R. Shechter. 2005. Rethinking Cities in the Middle East: Political Economy, Planning, and the Lived Space. *The Journal of Architecture*: 499–515.

Yap, K., and K. De Wandeler. 2010. Self-Help Housing in Bangkok. *Habitat International* 34: 332–341.

Part III
Cross-Regional Scene

Chapter 8
The Puzzle of Urban Sustainability Transitions in Urban Informality in Egypt

Abstract The central question of this chapter is: How can urban transitions perspectives assist understanding of the integration of urban informality in the urban context? This question is addressed first by introducing the current forces that influence 'urban informality transitions', and the recent and current socioeconomic and political transitions that bring urban informality to transitions studies. Then the Egyptian urban fabric as a site for urban transitions studies is considered, followed by the main section of the chapter in which four aspects of sociotechnical transitions that influenced the formulation of urban informality are considered. Thus, the sprawl of urban informality, urban transitions, political economy and social capital, and exclusion transitions are highlighted. In reviewing these aspects, of course, others are to some extent neglected. But these four transitions are selected specifically to draw out contributions of urban informality transitions to understanding informal housing and urban change. It concludes that developing urban sustainable transition and management as a visionary proactive approach is seen as a response to lagging, or misfit approaches for public engagement and strategic urban planning to integrate or at least to eliminate informality within the urban context or to convert urban informality into sustainability transitions.

Keywords Informal urbanization · Urban sprawls · *Ashwaiyyat* · Urban transitions · Semi-informal areas · Squatters areas · Hybrid areas

8.1 Introduction

The total area of Egypt is around one million km^2, in which more than 100 million inhabitants live on only 6% percent of the land, mainly along the Nile valley and Suez Canal zone, with the rest being desert. The national average density is 100 people per km^2, but 1660 people per km^2 in the agglomeration's areas including internal water bodies, a very high density by international standards. In Egypt, urban informalization is as old as modern urbanization and urban development, which proceeded in Cairo more than 100 years ago. Janet Abu-Laughed (Abu Lughod 1971), whose main research (Abu-Lughod 1971) covered the period 1957–1961, did

A. M. Soliman, *Urban Informality*, https://doi.org/10.1007/978-3-030-68988-9_8

not mention the history of informal settlements formulation at all, nor did David Sims (Sims 2010).

The urban informality, or what is called *Ashwaiyyat,* has long infested Egyptian cities. This phenomenon set in more than 2000 years ago during the *Holy Journey* of *The Virgin Mary* with her son *Jesus Christ* to Egypt, where she settled in the *El Matarya* area of a small hamlet currently called *Mary's Tree* (Soliman 2013a, b). In the modern era, evidence indicates that the appearance of modern urban informality in Cairo dates to the 1900s, when squatters' areas (*Ezbet El Saiaada, Ezbet El Lemon, Ezbet Abu Toyalhia* and *Tall El Hosen*) were found in the *El Matarya* area northeast of the city of Cairo (GOPP 2012). These squatters' areas accommodated the workers who came from Upper Egypt to perform construction work in the *Heliopolis Oasis,* which was built by the *Belgian Baron Edouard Empain* between 1906 and 1929 (Ilbert 1984). The agglomeration of *Ashwaiyyat* in *El Matarya* as small peri-urban areas appeared in the 1910s and expanded in the 1950s. They were called Ezabs or hamlets, such as *Abu Toyalhia* and *El Lemon*, and there was little interest in what was at first a very marginal and not very visible phenomenon, neither from the governmental or academic perspective. In the1930s, the King Farouk of Egypt granted to the borders' guards to settle in a location 4.5 km from the Cairo city center. This area is called *Ezbet El Hagganiha* and later became one of the largest squatters' areas in Cairo (Soliman 1999).

In the mid-1990s, the state of Egypt, as a way to tackle terrorism in the country, launched an upgrading program to develop some scattered informal areas in Cairo. The main reason was a political intervention to eliminate terrorists who used informal areas as hiding places to avoid arrest. In 2008, the Egyptian government set up a new program to intervene in unsafe informal residential areas and a new typology for informal areas was adopted: unplanned areas and unsafe areas (ISDF 2016; Khalifa 2011). Unsafe areas were classified into six categories that attracted the state's intention as the most dangerous areas, which each needed special treatments and action. The growth performance of urban informality significantly exceeded that of most other development processes, both in terms of size and numbers. The Middle East region's feeble economic performance in 2002 reflected a slowdown in the performance of the oil sector,[1] low rates of economic activity, the international Covid −19 pandemic, and the negative impact on the private sector of heightened political tensions in the region.

During the period 2011–2014 and after the two revolts of 25th of January 2011 and 30th of June 2013, Egypt witnessed the loss of 150,000–200,000 feddan[2] of precious fertile agricultural land to illegal urban sprawl. Combined with this arbitrary urban sprawl, a multitude of urban challenges has emerged: the increase of Egypt's population to nearly 95 million people (CAPMAS 2016a) (in 2020 more

[1]The price of oil reached a new peak in June 2004 of US $42/barrel, while in September 2008, it approached $140/barrel and dropped below $65/barrel in October 2008 during the international economic crisis, and below $40/barrel during and after the Covid-19 in March 2020.

[2]One feddan equals 0.42 hectare.

than 100 million); deterioration of the economic situation; decrease of foreign reserve (from US$36 billion to under US$14 billion in 2013); rising unemployment (to more than 20% nationwide); deprivation; lawlessness; lack of security; disintegration of social conduct; deterioration of human behavior; and rapid expansion of the informal economy. What has happened to the Egyptian built environment? Chaos and urban disorder became the common phenomena of urban life within the Egyptian urban fabric. At the same time, currently, Egypt is rapidly transitioning in all aspects of everyday life, urban development, and politics.

Many scholars have examined various aspects of urban informality in Egypt: a physical point of view, socioeconomic aspects of change, and political transitions (Sims 2010; Soliman 2004a, b, c; Singerman and Amar 2006; Al Sayyad 2006; Kipper and Fischer 2009; Dorman 2013). Informal-housing typologies have been examined and redefined (ISDF 2016; Khalifa 2011; Soliman 2004a, b, c). Various housing policies have been applied in Egypt, but have not met the objectives (Payne 1999; Soliman 1999; Denis 2012). It is not the intention here to review urban informality, or what is called *Ashwaiyyat*[3], in Egypt, rather it is to highlight urban informality transitions in the last seven decades, and how, by whom, when, why, they have emerged.

This chapter examines the urban informality transitions in Egypt that occurred in the last seven decades and their influence on the evolution of urban informality. The study adopts the ethnographic method, in which ethnographic observation is used to review the literature, as well as the experimental background, while parallel ethnographic interviews relied on personal and official documents, as well as informal interviews with various stakeholders.

The conceptual framework of this chapter is the outcome of literature reviews for urban informality and the author's experience and practice in such urban contexts. This scheme was initiated by the author who was involved as an energetic researcher and planner in several areas of the urban informality (Soliman 2004a, b, c), and a participatory land-readjustment project (Soliman 2017), as well as in many strategic urban plans for selected Egyptian cities (GOPP 2012). Having gained experience from previous work, the author tries to support the adaptation and application of urban informality transition in the investigated areas.

The context analysis is based on information obtained from various, complementary data-collection methods: first, literature reviews of reports, books, articles, and data related to urban informality and previous activities of NGOs, CBOs, and local associations; second, informal discussion with selected experts; third, participation in and supervising an upgrading of the urban informality graduation project in Alexandria City organized by the Architecture Department of Pharos University the 2017–2018 academic year; fourth, participatory observation and consultation for a strategic urban plan for *El Matarya* district in Cairo and a land-readjustment project

[3]*Ashwaiyyat* the plural for *ashwaiyya* (literally meaning 'half hazard') is the term used in public to refer to the informal settlements in Egypt; it refers to illegal arbitrary housing development on the periphery of urban centers.

in Benha city (Soliman 2017); fifth, various workshops as a participant for setting up housing policy in Egypt in which urban informality was a primary theme (UN-Habitat and Egypt 2015); sixth, a postgraduate course in governance of sustainability transitions at the Architecture Department, Faculty of Engineering at Alexandria University in the 2017–2018 academic year; and finally, various workshops held by GIZ in Cairo for redefining urban informality in Egypt. All semi-structured discussions were recorded, reviewed, and transcribed. Through this rich information, the idea of handling urban informality from a sustainability perspective was initiated and discussed with experts. Also, the editors of the book series 'Theory and Practice of Urban Sustainability Transitions' (https://www.springer.com/series/13408) have encouraged the author to link urban informality to the sustainability transitions and transition governance.

The spread of urban informality and the consequences of socioeconomic and political transitions had a major influence on the urban informality transitions in which the state played covert and overt roles. It also indicates that the divergence that occurred within a city in which the gap has been widening between the physical and social atmosphere, a gap between poor and rich, informal and formal, social exclusion and inclusion, created the main features of the Egyptian urban fabric. The central question for this chapter is: How can urban transitions perspectives assist understandings of urban informality sustainable transition? The chapter suggests that market and governance reforms, as well as the contribution of the grassroots, could create a common ground for facilitating, integrating, or enabling the transitions of urban informality within the built environment in a sustainable way.

This chapter consists of five sections. The first section is the introduction; it summarizes the chapter contents. The second section gives an overview of the sprawling of urban informality in Egypt. The third section investigates political economy and socio-spatial transitions of urban informality through two lenses: political economy transitions and social capital and exclusion transitions. The fourth section discusses urban transition in urban informality. The final section addresses the necessity to integrate the people innovations, state, and capital in urban informality transitions to be sustainable.

8.2 Sprawl of Urban Informality in Egypt

Most of the inhabited areas of Egypt cover only 6% of its surface, mainly in the Nile Valley and Suez Canal Zone, with the rest being desert and small oases. The population of Egypt has increased by a factor of more than eight since the census of 1907. The population in 1907 was 11.2 million, while the population in the census of 2016 was 94.8 million (CAPMAS 2016a). In 1952 the total population of Egypt was around 18 million, which then doubled in 24 years to reach around 36 million in 1976, and it increased by six fold to exceed more than 100 million in 2018. Today, the country has around 2.5 million births each year (CAPMAS 2016a), of which 62.4% are in urban areas (UN-HABITAT 2008). Egypt comprises North Africa's largest population (UN-HABITAT 2008). The national average population density is 100 per km^2, but 1660 per km^2 in the agglomeration's areas including internal water bodies where it is also very high density. It is expected that, if the annual growth rate in population continues at 1.8%, 2.05%, or 2.56%, the total population in 2050 will reach 158.000, 176.000, 219.000 million, respectively.

Between 1982 and 2004, an estimated 1.2 million feddan of agricultural land was lost to housing informality at an estimated annual, daily, and hourly rate of 54,545; 149.4; and 6.22 feddan, respectively. It is expected that Egypt will lose around 66,300 and 138,000 feddan of the best fertile areas surrounding Egyptian cities and villages, respectively, by the year 2027, while during the 1.5 years after the 25 January revolt of 2011, Egypt has lost 100,000–120,000 feddan to urban informality. Today, estimates indicate that Egypt has lost around 30% of the fertile agricultural land to urban informality since the January Revolt. Studies on Egyptian informal housing have enumerated that there are as many as 8.5 million informal units housing at least 17 million Egyptians in urban areas, of which 4.7 million units are on agricultural land within or outside municipal boundaries, 0.6 million are on government-owned desert land within municipal boundaries, and 3.2 million units (10.2 million Egyptians) are outside administrative village boundaries. Due to the rapid arbitrary urbanization processes within the country, divergence in cities, and even within the city itself, have become noticeable. It is assumed that the informal economy, in parallel with housing informality, has expanded, and the informal market or shadow economy represents around 60% of the national economy (Denis 2012). But whatever assumptions are adopted, the informal economy in Egypt is judged to be substantial, comprising 37–93% of the overall economy, depending on the definition of informality and proxy variables applied (Subrahmanyam 2016). This occurs in a global context where estimates indicate that over 70% of the workforce in developing countries and around 4.3 billion worldwide rely on the informal economy to survive (UNDP 2009).

Currently, about half of Egypt's urban population is concentrated in two major urban centers: Cairo and Alexandria, while the remaining is scattered in 331 small and intermediate urban cities along the Nile River Valley and Delta. Egyptian cities are increasingly becoming complex systems of social, economic, and ecological transition. Applying the concept of *what if?* to the Egyptian environment, one could find the following. Using the moderate growth rate of population of 1.9%, the expected total population will reach around 176.0 million in 2050. This will require approximately doubling the current urban and rural agglomeration to meet the 2050 housing demand and various socio-economic amenities. As applied to the urban informality, if the current trends continue, without any proper intervention, the extent of urban informality will double by 2050. Similarly, if the current housing policy and planning trends continue, it is expected that more than 50% of the future urban and rural agglomeration will expand informally to adjacent agricultural land on the periphery of urban areas, having major effects on natural resources and urban development. Thus, the rapid population growth, without proper planning, is considered a national security threat in which all sorts of development are entangled in

Table 8.1 The total population and size of slum areas in Egypt (2016)

Type of Slums areas	Size in Feddan 0	Percent of total slums area to total agglomeration	Number of slums area	Percent of slum population to total population of Egypt	Estimated size of population (millions) in 2016
Unplanned areas	156,300	10.94	870	39.58	37.52
Unsafe areas	4900	0.34	351	1.14	1.08
Total slum's area	161,200	11.28	1221	40.72	38.60
Total formal area	1,267,300.70			59.28	56.20
Total size of current built up area	1,428,500.70				94.8

Source: (CAPMAS 2016b; ISDF 2016)

this growth. This is a warning for the current government to safeguard the future generation.

As indicated in (Table 8.1), in 2016, there were 38.60 million inhabitants in informal housing areas representing 40.72% of the Egyptian population, of which 39.58% were living in unplanned areas, and 1.14% were living in unsafe areas (UN-Habitat and Egypt 2015). A recent study estimated that informal housing varies 40–60% of the urban housing stock of 27.1 million physical housing units. This yields a population of around 48.3 million who live in informal areas in the year 2020. The total area of urban informality was at least 161.2 thousand feddan which constitutes over 11.28% of the total built-up area in Egypt in about 2016 (CAPMAS 2016b; ISDF 2016), distributed over 1221 areas. On the other hand, President Abd Al-Fatah El Sisi declared more recently that more than 50% of urban and rural agglomerations in Egypt are informal (Soliman 2019). The main problem is that there is no accurate information to indicate the exact number of people who live in informal areas in Egypt. But, relying on the figures given by the ISDF and CAPMAS (CAPMAS 2016b) that the total population of informal settlements constituted 38.6% of the total population of the country, which was about 94.8 million in 2016, it can be said that around 38.6 million were living in informal areas at that time, of 37.52 and 1.08 million were in unplanned and in unsafe areas, respectively (see Table 8.1).

Walking around Cairo's and Alexandria's streets, one can see the difference in the overall aesthetics and planning when comparing older urban patterns and buildings to modern ones. This difference is visible between the organized or formal sectors of the cities and unorganized or informal areas surrounding these cities. The former constitutes the planned and official sectors in which planning control and restrictive regulations, to a certain extent, have existed. These areas have the privilege of having a good environment and reasonable services. The latter is characterized by arbitrary physical growth, or urban informality, and contains or accommodates a high proportion of the informal economy, a sector of the society with its own special way of life. These areas are considered the worst locations in the city and characterized by overcrowding, densification, lack of infrastructure, environmental degradation, and social ills. So, there is a great diversity in the physical structure of Egyptian cities, between old and new, between rich and poor, between good and bad environment, between availability and scarcity of infrastructure, between various features of everyday life. This result in many cities within one city (Al Sayyad 2006) and great diversity in socioeconomic and political life in various sectors of one city. Nowadays, Cairo's urban informality houses more than 11 million inhabitants that represent two-thirds of the population of the city and more than a third of its footprint (Sims 2010).

Informality at the household level triggered the Arab Spring, which had been a 'ticking time bomb' that must be carefully handled by policymakers, particularly in transition countries such as Egypt. Growth in informal arrangements and practices is one reason why Egypt continues to struggle with rising poverty, high unemployment, and sluggish growth—issues that have impeded its smooth recovery from the Arab Spring. Government action on informality, therefore, needs to be carefully considered and evidence-based if Egypt is to enjoy a peaceful transition to democracy characterized by rapid, inclusive growth and poverty reduction. Uncertain factors, such as natural disasters, climate change, energy crises, scarcity of water, political instability, financial crises, food security and terrorist attacks, and urban informality, all play an important role in threatening urban development (Spaans and Waterhout 2017; Soliman 2019).

The combination of rapid population growth, the spread of urban informality on the fringe of urban centers, especially on agricultural land, and the shift from rural to urban areas' concentrations continue to put further pressure on infrastructure, housing, and social services (Sims 2010; Kipper 2009). With the increasing demand for housing plots/units by low-income groups and the difficulty of obtaining cheap shelter within the big urban centers, informality land markets have become, *and will continue to be*, the main feature of urban development in the Egyptian urban fabric (Soliman 2010). These changes have occurred in several aspects of the Egyptian urban fabric: changing the spatial configuration of the physical structure of the cities; causing the disappearance of agricultural land; exacerbating inappropriate or absent good urban governance and chaos in everyday life; changing the attitude and transitions of everyday life and behavior of the citizens; magnifying social exclusion; increasing poverty, the spreading of the informal economy, environmental degradation, and a tremendous lack of urban services. These rapid socioeconomic transitions create a heavy burden on the Egyptian governments and have caused several economic, social, and environmental problems. One is the rapid increase in prices of various commodities, especially housing, and the sharp drop of the value of local currency, especially after the devaluation of November 2016. The complexity of linkages between the "informal" economy and informal land market, the rapid population growth, deterioration of rural areas, immigration from rural to urban areas, insecurity of land tenure, lack of basic services, and environmental

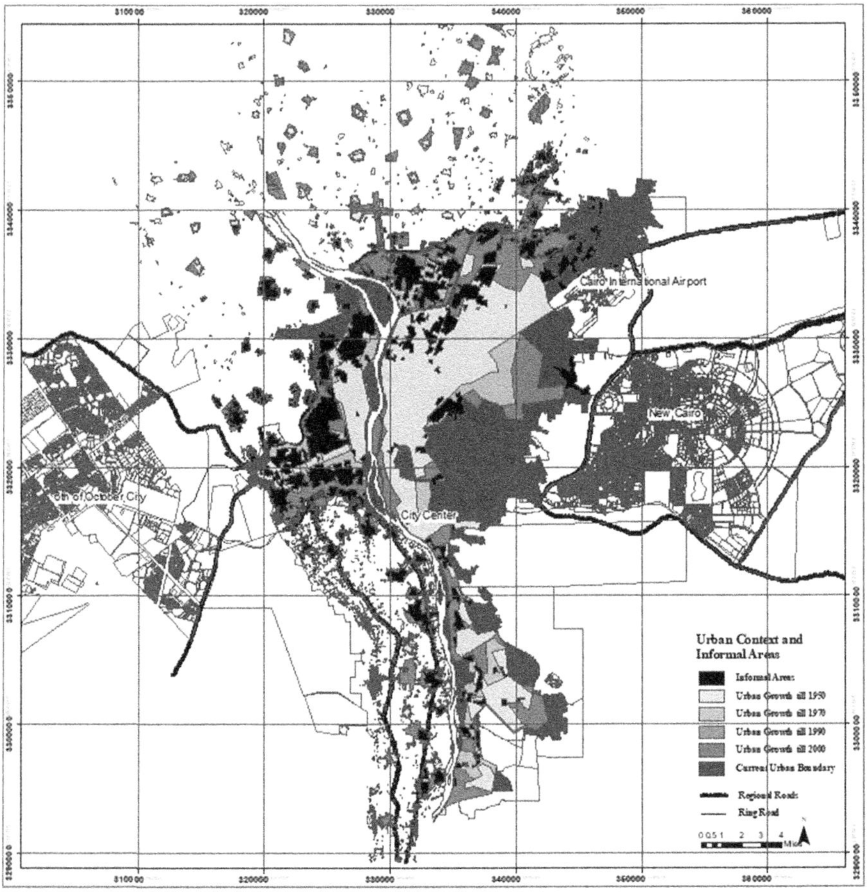

Fig. 8.1 Map of informal housing areas in the Greater Cairo Region. (Source: The author and adapted from (ISDF 2016) and modified by the author in 2020)

degradation are the main urban challenges that face the Egyptian government in the coming years. Among these challenges is the spread of urban informality and the booming of the informal economy, dramatically worsening the Egyptian scene. As illustrated in Figs. 8.1 and 8.2, informal areas compose a belt of poverty or a belt of misery, or a black spot surrounding the western sector of the Greater Cairo region and the southern part of Alexandria City. This is not a pessimistic view, rather it is an attempt to see the reality from a wide perspective to gather explanations for such rapid arbitrary transitions processes and to try to glimpse the future picture of all these transitions on what might come to be.

This brief background sheds light on questioning the future of Egyptian urbanism and transitions, and how to engage the spread of urban informality within urban context. This rapid increase in population occurs in a context of housing, jobs and basic human needs, demands that are rising too rapidly. It is the time to understand

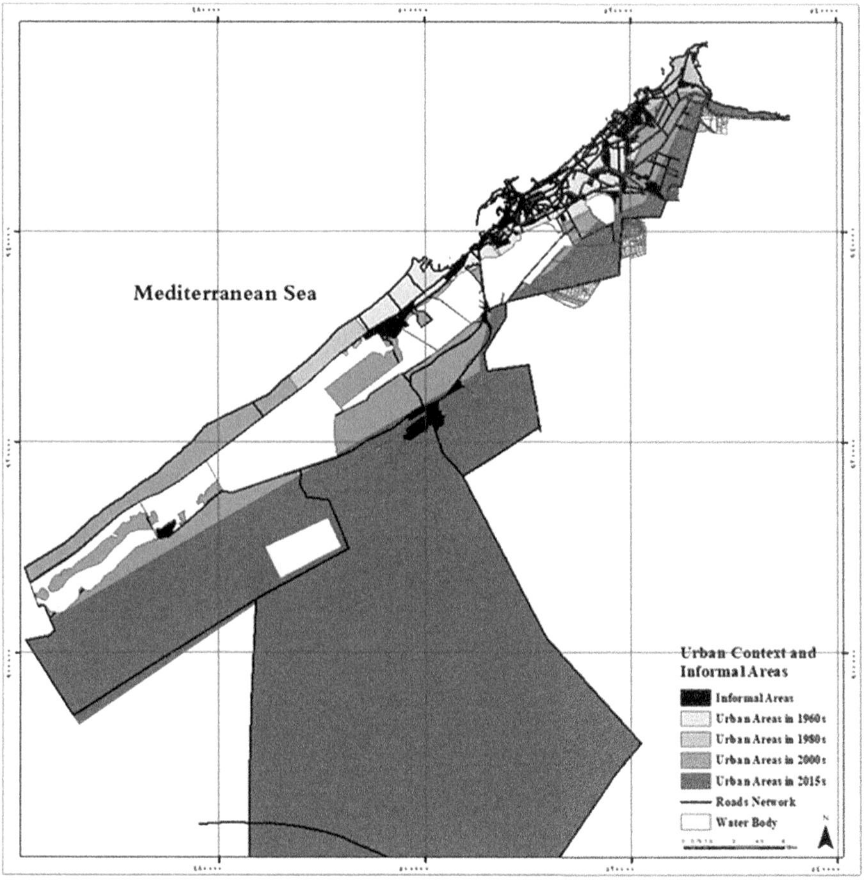

Fig. 8.2 Map of informal housing areas in Alexandria City. (Source: The author and adapted from (ISDF 2016) and modified by the author in 2020)

how the urban poor formulate urban informality in which they secure their land tenure, and how to integrate urban informality within the urban context. If the urban patterns of low-income groups are remodeled through the integration in the urban context, and the informal process of cooperation among the urban poor were formalized, it would enhance the Egyptian built environment, sustain sustainability, and meet the official planning processes' aspirations.

8.3 Political Economy and Socio-Spatial Transitions in Urban Informality

The following section examines the socioeconomic and political transitions that Egypt has passed through and traces these transitions and various policies on the process and the formulation of urban informality during the last seven decades. By transitions is meant purposive, sustained sociotechnical interventions in existing systems or 'regimes' (Geels 2010) designed to remove them or to transform them—in this case, leading to a new, 'sustainable' urban setting. Elsewhere, the Multilevel Perspective (MLP) combines ideas from science and technology studies, as well as policy and management, evolutionary economics, and the sociology of innovation in a heuristic schema for understanding sociotechnical change (Geels et al. 2016a). The aim is to discover how these transitions have affected, positively or negatively, the formulation of urban informality and the urban system as a whole. Therefore, we shall consider sociotechnical changes in the form of political draining, socio- spatial structure, and economic transfer or circular economy (Kirchherr et al. 2017) and their impact on urban informality in Egypt. As stated in the introductory chapter, our main concern is exploring new perspectives in urban sustainability transitions on urban informality in the Middle Eastern countries. The following two subsections illustrate political economy transitions, and finally, social capital and exclusion.

8.3.1 Political Economy Transitions

The rise of urbanization and interest in cities is occurring contemporaneously with attempts to respond to changes in urban informality and the environmental sustainability trends. Due to the transformation of political institutions in 1952, the changing of the social structure of the country from a kingdom into a republic, and the changing from feudalism system into nationalization system, changing from nationalism system into capitalism system, and changing from capitalization system into a chaos system have led into dramatic transitions of the society and the everyday lives of the country's citizens.

Indeed, as illustrated in (Table 8.2) urban informality processes have changed during the socioeconomic and political transitions and passed through five eras. Among these were Nasserism (1952–1970), a period of an extensive intervention; Sadatism (1970–1981), a period of active but ad-hoc intervention; Mubarakism (1981–2011), a period of fluctuating intervention; the post Arab Uprising (2011–2014), a period of arbitrary intervention as the state was absent and retreated from interventionist policies in which housing markets fluctuated constantly; and the current intervention (2014–2020), a period of exhaustive and comprehensive intervention.

Table 8.2 Illustrates socioeconomic, political and reforms transitions during the last seventy years

	Niche		Regime		Landscape
	Economic Transitions	Social Transitions	Reform Transitions	Political Transitions	Output & Policy
1952–191970	Changing feudalism into nationalization	Uprising of the lower class	Agrarian land Reform Law Nos. 178,127, 15 of 1952, 1961 and 1963 respectively limited individual ownerships of land to 50 faddan.	52 revolution, changing from kingdom to republic	Conversion of some villages into cities
Nasserist (producer)	Flourishing Industrialization and drop in cotton crop	Incentives for working classes	The introduction of the operational unit called the Hiyazah (land holding or land tenure).	Military role 52-54 to presidency	Transfer land ownership from private to public
	Shrinking the role of the private sector in economic activities	Encouraging mobility from rural to urban areas	Introduction of the Nationalization Program in 1961	Nationalization of Suez Canal and the Crisis of Suez 1956	Industrial worker housing at Helwan
	Drop in foreign reserve	Changing landless into landowners	Law Nos. 199, 55, 168, 7, 78, of 1952, 1958, 1961, 1965 respectively for rent control by imposing great reduction	A battel to build the High Dam	Madinat El Oqawf (conversion of agricultural land into residential use)
	Drop in EP value	Departure of foreigners	Allowing the conversion of agricultural land into other uses	Non-Alliance organization	Mass public housing program
	Decrease of foreign investment	Social inclusion and social rights		Union and breaking of union with Syria	Naser City (as dormitory city) for low and middle classes
	Shrinking the role of the private sector in real estate's sector	Free education		Yemen Crisis	Resettlement program & introduction of core-housong units (Alexandria)
	Shut down Egyptian Monetary Stock Market	Reduction in transportation fees		Establish the Arab Socialist Union Party to rule the country	Drop of the private housing sector
	Appearance of informal economy	Increasing the welfare of the society		1967 War of Defeat	Absence of peri-urban areas

(continued)

Table 8.2 (continued)

	Niche		Regime		Landscape
	Economic Transitions	Social Transitions	Reform Transitions	Political Transitions	Output & Policy
		Appearance of new middle classes			
		Steady urbanization process			
		Displaced people			
1970–1981	Spreading of informal economy sector	Rapid urbanization process	Relaxation of Regulation	War of attrition	2.5 million persons immigrated from Suez Canal cities
	Sharp drop in foreign currency	Increasing the poverty level	Some decrees for the benefit of the private sector	All efforts directed to recover from war of defeat	The Status Quo of the urban sprawl
	All funds allocated to military purposes ????	Decreasing number of the middle classes	Nos. 49, 136 of 1974, 1977 respectively for rent control	Changing policy towards private sector as a main engine for development	Appearance of squatters' areas
	Flourishing of informal economy sector	Lack of funds for social and health amenities	Incentives for the private sector by exempting from paying taxes for 10 years	Shrinking housing production	?????
	Encouraging the role of private sector	Rapid urbanization growth	Peace agreement with Israel	Open Door policy towards a free market approach	Involvement of private sector, NGOs and contractors' companies in housing production
	Free market approach	Allocate resources for people to organize their own house building	Thus, in the mid-1970s, within the framework of the Open-Door Policy, Egypt was forced to balance the imperatives of capital accumulation against the needs of those with few personal resources.	Use of house building to fuel economy and industry.	Introducing site and services and upgrading schemes
		People relied on their efforts and their social network		New towns policy	Redevelopment of Suez Canal cities
		Building organizations that facilitate central initiatives.		Encourage the role of private developers Law	

1981–2011	Going back to Arab League	Building organizations that facilitate central initiatives.	Issuing ISDF for tackling informal housing problem	Access to use of and satisfaction with goods and services.	Housing microfinance, community-based finance savings and loan groups, and consumer credit for building materials
	Strengthening the relation with USA	Magnetize low income to sustain the stability of the regime.	Issuing Law 4 of 1996 for freeing land from the renters to the owners.	Decentralized resources to support local enterprise and home building.	
	Finalized Camp David agreement	Flourishing of the Muslim Brothers and prejudice			
	Privatizing trends	Rapid agricultural land conversion to other uses			
2011–2020	Encourage liberal economy	Knock down the Muslim Brothers and radical Muslims	New constitution after two revolts	Deflation of Egyptian pound in November 2016	Political support for recovering the unsafe areas
	Privilege for foreign investment.	Attract the mass of population to support the regime	Liberalization of the economy	New Economic reforms	Mass housing production for low-income groups
	Integrate development activities and link housing to larger urban systems of employment and production.	Equilibrium of the various strata of the society.	Recover of Tourism sector	Shrinking subsidies for low-income groups	Encourage the private sector in luxury housing production
		Rapid urbanization process	25th January revolution and the role of Muslim brothers	More incentives for the private sector	Construction of 4th generation of new towns (14)
		Rapid population growth (2.5 million persons annually)	30th June revolution and beginning of economic reform	Housing as a component of socioeconomic process and that housing is a process by which to participate in the general socioeconomic development	

It is important to point out that before 1952 there was no housing policy in the country. The only publicly constructed housing project before then was Workers' City in *Imbabah*, an area close to Cairo, with about 1000 dwelling units. Scattered examples of urban informality and urban transition in Egypt are then presented in order to illustrate how urban and transitions perspectives can contribute in different ways to our understanding of this phenomenon. During this era, dominant political groups influenced the housing process and the formulation of housing policies. Later, under certain circumstances, the state may take steps in favor of the poor, while in others it may act against the poor in favor of dominant political/capitalist groups. In a third situation, the state may act for both affluent and poor people. Political support, in this sense, may be crucial and vary between a overt, covert, and double role of support or political informality.

Nasserism (1952–1970) was a period of extensive interventions where political decisions depended on the six principles of the revolution of 1952; among them was a social justice for all strata of the society. In this way, the state reflected the redistribution of power within the society. State intervention in the housing field followed the interests of the politically powerful, the affluent group and the poor, depending on the development process and various circumstances experienced within the country. However, social investments had generally only been made in Egypt after the needs of industrial enterprises, 1954–1967, for direct support had been satisfied. Because there had also been a tendency to ensure that publicly subsidized projects were profitable, they could eventually be transferred to private-sector interests. In this sense, state interventions in housing had privileged the interests of private capital over the local working class (the lowest strata of the society), as well as satisfied the social needs of the working class.

An early example of this was state intervention in the housing supply during the New Era movement towards socialization and nationalization of the mid-1950s and early 1960s. The state introduced a huge public housing program to serve the increasing demand for housing the urban poor. In 1958, a dormitory city, close to Cairo, called Nasser City, was established to fulfill the housing needs for middle-and-below and middle-income groups. People who worked in the construction of this project were allowed to settle temporarily in the *Manshiet Nasser* area close to Nasser City. By that time, this area had developed into an informal area and become the largest squatter area in Cairo. Many scattered public housing developments were constructed in various cities of the country. Today, most of these projects have deteriorated and reflect the features of informality or hybrid housing, and many additions have been added to the original housing units, or what is called extending housing (Tipple 1999, 2000), which has changed its typology from formal housing into informal housing.

Also, in the early 1960s, the state adopted resettlement program for public benefit projects, such as *El Sideen* area in Alexandria City (see Fig. 8.3). This project, in association with the World Bank, introduced the core housing program. It was the first project in Egypt to adopt the international housing policy that applied the approach of self-help housing (for details see (Soliman 2012a, b, c). It was hoped at the time that the residents would carry out the rest of the construction process

Fig. 8.3 Photos of core housing units in El Siadeen area in Alexandria City. (Source: The author)

according to the planned scheme but that seldom happened. Another concept, following the same program, was applied in a project in *Nadi El Said* area in Alexandria, where the rest of the project had been shifted to be completed by the residents. Today, both areas have evolved from a formal area to the worst informal area in the city and are classified as unsafe areas of the fourth degree according to the taxonomy of ISDF (ISDF 2016).

In the early-1960s, housing policy shifted to condominiums in new large-scale land subdivisions, such as *Madinet Al Awqaf, Maadi*, and *Heliopolis* districts near Cairo, and in the periphery of other cities of Egypt, taking advantage of the liberal credit terms available under the government-subsidized cooperative housing programs. Also, conversion of small villages into small cities in the Nile Delta in which huge housing developments on agricultural land were erected. The introduction of the Agrarian Land Reform Law Nos. 178, 127, 15 of 1952, 1961 and 1963 respectively witnessed a transformation in housing policy with the elimination of the role of the private sector in housing supply, causing a sharp decline in housing production. However, the 1967 war with Israel, the subsequent war of attrition, and the run-up to the 1973 war froze formal housing programs in Egypt and other cities in Middle Eastern countries, as public funds were reserved for the war effort. Large urban centers had to accommodate a significant number of internal and external migrants, especially in Egypt, Syria, and Jordan. Shifting demobilization—from one urban area to another—had taken place, with the urban fabric of the cities during confrontation with Israel changing radically, displacement of many residents from war front cities, and a huge increase in housing production in other cities, but much of it was informal, such as in in Cairo, Alexandria, Amman, and Beirut. The informal housing development accounted for 25–35% of the total housing production.

There are various reasons why a government should become involved in housing production. The most common involve humanitarian reasons, functionalism, social control, political control, and human-rights considerations. In relation to housing during the Nasserism period, political-control considerations had become one of the Egyptian government's main concerns, and the state played a diverse role. Similar to the beginning of this era, the state played an overt role to attract the bottom strata of society to support the new regime, and later, with the economic difficulties, the state played a covert role and enforced subdivision of agricultural land for housing development to satisfy the affluent groups, especially the professionals, the intellectuals, and skilled persons. It could be said that the state has had various roles in housing production, and its roles have changed according to the socioeconomic and political transitions. At the same time, the state initiated the appearance of urban informality by allowing people who were coming from Upper Egypt to settle in the *Manshiet Nasser* area, as the migrants had followed the same procedures. It seems market-based promotion of informality and resource efficiency remains the great hope for the lower classes of cities. Thus, it is not a matter of power/powerlessness of the state; rather, it is based on the various circumstances that the country passes through.

Second, with the beginning of the 1970s, the political arena changed and reached its peak after the triumph of 1973. Thus, in the mid-1970s, within the introduction of the Open-Door Policy, a framework to redistribute population outside the Nile Valley, and the redevelopment of Suez Canal cities was familiarized. Egypt was forced to balance the imperatives of capital accumulation against the needs of those with few personal resources. Accordingly, innovations of housing policy for low-income groups were introduced and implemented in various cities. Among these policies were upgrading programs, site and services schemes, and core and wet housing programs. Most of these programs were implemented in the first generation of the new-towns scheme, except that the sites and services and upgrading projects were implemented in Ismailia city (Davidson and Payne 1985). A master plan for the GCR was set up to divert population growth away from the arable land in the west (Giza), north (*Qalyobia)*, and south of the GCR towards the desert fringes of the new towns to the east and the southeast, relying on a nuclear approach (Egypt and UNDP 2007). Through such an approach, the two governorates Cairo and Giza released, over one decade, some 300 km^2 of public land in the GCR expansion areas (Egypt and UNDP 2007). On the other hand, due to the economic liberty movement, rapid informal urbanization occurred in all parts of the Egyptian cities, and urban informality flourished. Many international organizations, such as the Overseas Development Administration (now DEFID) (UK), the World Bank, UN-Habitat, GTZ (GTZ 2004), and UNDP had become involved in implementing new policies for housing the urban poor, such as basic infrastructure, self-help housing, sites, and services, and upgrading programs (Soliman 2012a, b, c). Urban informality in the GCR constituted around 38.1% of the total housing production in 1976 (Sims 2010), and nearly 30% nationally. This era was a very rich experimental period in housing the urban poor; for example, one of the schemes involved unfinished space units, in which finishing task of the unit was left to the residents or following the concept of an incremental process. It assumed that 70–80% of the new housing stock produced in the GCR 1970–2000 was informal (Egypt and UNDP 2007).

Housing policies involved the private developers and embraced three sets of actors, the formal and informal private sectors, and the public sector, where the private housing developers were considered an important performer in the housing development in new and old towns. Thus, the private sector had gained the privilege of becoming the main supplier for housing production. At the same time, for low-income groups, the states adopted the self-help approach that was advocated by the first conference of UN-HABITAT in 1976. This was achieved by facilitating the flow of subsidized building materials, forming cooperative housing societies, and allowing a free market within the formal private housing sector. In the late 1970s, the role played by the informal sector as an indispensable part of housing production was recognized in Egypt as at least 60% of housing construction was undertaken by the uncontrolled private sector (Soliman 2004a, b, c).

This policy led directly to changes in housing policy, leaving the supply of luxury housing to the private developers, while reserving a role for the state in the provision of housing for the lower classes. Since then, for many reasons, the state has intervened more widely to increase housing production across social strata. One reason was to maintain control over the urban poor. Another was to use housing policy to enhance economic development in order to cope with the increasing needs of the population and to match the goals of international development assistance. A third reason was to overcome the problem of international debt (this was around US$ 45 billion in 1976 and was nearly US$ 28 billion in 1986). Therefore, the states accepted a status quo for arbitrary urban growth in seeking to recover from the financial burden that it had inherited from the 1973 war, housing shortages in the

1970s, and the increasing price of oil, and had to speed up the process of implementing new economic policies, making the best use they could of the substantial remittances from people who worked abroad.

Third, in 1989, former Prime Minister *Attaf Sadekey* compared informal housing areas with cancers, belonging neither to the city nor the village, and as a reflection of rural urbanization. In 1991, to address this problem, the Egyptian government established a wide range of Economic Reform Structure Programs, according to which the private sector was engaged in a policy of market-based remedies. At the beginning of the 1990s, the terrorist movement in Egypt further encouraged politicians to address the problem of informal housing areas, where much of the unrest was concentrated. Thus, in May 1993, after the terrorism of *Imbabah*, President Hosni Mubarak announced a national program to upgrade informal housing in Egypt, to which 106 million Egyptian pounds were allocated.

Thus, the state and donors influenced by the World Bank adopted an "enabling markets" approach to "housing". The enabling markets approach, through the upgrading program, encouraged reforms of various aspects (land titling, property rights, infrastructure, housing finance, housing institutions) of the housing sector, and embraced land issues within a housing framework. This program passed through three phases. The first phase of 1993–2004 brought an increased official presence in such areas. While major improvements have taken place in the form of providing basic infrastructure, such informal settlements still face various problems from the lack of proper planning and legal tenure status. On the whole, central Cairo districts lost over one million inhabitants during the period 1986–1996. Yet, the 1996 census for the GCR showed that this strong inner-city population decline translated into suburban growth to the north and south, contributing to the extension of informal settlements. One of most important laws was Number 96; it was issued in 1996 to resolve the relationship between tenants and landlords of agricultural land, when landlords has been encouraged to convert the land into housing informality. The second phase, 2004–2008, aimed at belting informal areas (*Tahzeem El Ashwaiyyat*) to limit the growth of informal settlements, and installing basic urban services for about 352 informal areas in Cairo and Alexandria. This program did not achieve its objectives: rather it provided the private landlords serviced land at no cost, while informal areas did not upgrade as expected. The upgrading program had modified and involved the grassroots, CBOs, and NGOs as main contributors for development, so participatory upgrading programs emerged. Accordingly, two pilot projects in *Manshiet Nasser* and *Boulaq El Dkrour* in cooperation with The German Technical Cooperation (GTZ) were implemented (GTZ 2004; Runkel 2009).

The third phase is a concept of public–private partnership in housing production. It is aimed at attracting private developers to develop low-income urban areas and "lockout" the latent value of the land for the benefit of the developers to cover the cost recovery of the project. One might ask several questions about this policy, however. Have such private/public partnerships offered a suitable environment for housing development? Can this approach accelerate housing production? And what type of partnership best serves middle- and low-income groups? The track record of such public/private partnerships has not been outstanding. Many of their schemes did not develop further than the design stage. Others changed their objective mid-stream, often ending up as typical government-constructed housing. Such failed partnerships have resulted in a situation of confrontation rather than collaboration. In addition, construction standards, which assumed the existence of predetermined socioeconomic groups, had resulted in mismatches and conflicts that have had a direct and negative effect on environmental quality. Two main groups had benefited from the partnership policy: holders of commercial and industrial capital, and the bureaucratic and technocratic leadership within both the private and government sectors. The poor and those who live on fixed incomes—i.e., middle-income groups—have been hardest hit.

In late 2006, the state announced a social housing program, within the scheme of the Mubarak Youth Housing Project, in which "build your house" or Do It Yourself (DIY) or "*Ibni Beitak*" using self-building techniques was introduced. The scheme has been implemented in "50 sites, distributed over 13 new towns with 90,000 land plots, each with a size of 150 m^2. However, middle-income households with the ability to afford higher down-payments and secure larger loans represent the best target group for the *Ibni Beitak* option (see Fig. 8.4). Another innovative step offered to the owners of the *Ibni Beitak* scheme was to enable those who are willing to continue the construction process to qualify for their household to access the National Housing Subsidy. Despite the innovative approaches adopted by the government and the significant efforts to enhance the subsidy mechanisms to reach the targeted groups, there are major concerns over the sustainability of such efforts (for further details, see (Soliman 2012a, b, c)).

8.3.2 *Institutionalizing Informal Areas*

An important step toward institutionalizing the upgrading of informal settlements in Egypt took place in 2008 through the creation of the Informal Settlements Development Fund (ISDF) headed by the Prime Minister of Local Development (now headed by the Prime Ministry). The ISDF obtained LE 500 million in a revolving fund to be utilized for upgrading the informal areas on a cost-recovery basis. The aims were twofold: upgrading informal areas which have the potential for development and making use of part of the land after upgrading to cover the total expenses of the improvement activities. The ISDF redefined slums areas into two different typologies. The first is unsafe areas with four different grades: exposed to threats to life; a risk to health risk; inappropriate housing conditions; and insecurity of land tenure. The second typology constitute the unplanned areas, which represent the vast majority of informal settlements in Egypt (see Table 8.1). A total of 351 unsafe areas have been identified making up an urgent phase of intervention to be tackled in 2021.

Subsequently, the huge social housing projects of Asmhart phases 1, 2, 3, and 4, Bashair El Kheair 1, 2, and, 3, Thia Misr phases 1, 2, and 3, and other projects have been constructed for the purpose of housing people displaced from unsafe areas to the new social housing. However, such a massive program is required an effective

Fig. 8.4 Photos of the *Ibni Beitak* development in Sadat City. (Source: (Soliman 2012a, b, c)

communication strategy to raise awareness of the public and to offer flexible alternatives for relocating those families living in dangerous areas. This program does not include self-help as an approach for upgrading the unplanned areas. It focuses attention on economic exchange relationships to achieve the cost recovery basis in which housing has been treated as a market value/demand market movement. On the other hand, most of the informal housing in the unplanned areas has been left without any intervention or treatment, so, it is expected that, in the coming few years, the program will continue and be enlarged.

Thus, the state assumed a new double role—widening its control over society and providing funds to improve informal settlements. This is not to over emphasize the role of government and the certainty of success of such tools. So many actors are involved, interests inevitably conflict, and struggles ensue with unpredictable outcomes, and ‘system innovations are characterized by emergent and nonlinear dynamics’ (Geels et al. 2016b; O’Brien 2015). Today, government debate centers on who is in need of housing, what constitutes adequate housing, how the government may ensure that those in need obtain housing, and how government resources may best be exploited to fulfill housing needs. However, these policy orientations have not always been paramount. In Egypt, dominant political groups have influenced the housing process and the formulation of housing policies. As a result, in certain circumstances, the state may take steps in favor of the poor, while in others it may act against the poor in favor of dominant political/capitalist groups. Political support, in this sense, may be crucial.

The shortage of affordable housing for the urban poor was one of the main causes of the recent revolt where Egyptians rose up in January 2011 to achieve the noble objectives of ‘*bread, freedom, dignity, and social justice*’. One of the demands was housing as a social right in which low-income groups asked for affordable housing. To absorb the people’s anger, the multiple transitional governments responded to this huge demand for housing in three ways: turning a blind eye to illegal construction on existing residential buildings; continuing the *laissez-faire* attitudes towards illegal residential development on agricultural land; and finally, by introducing the National Social Housing Programme (NSHP) to build one million housing units within the next five years to meet the immediate demand. Therefore, after the revolt of 2011, and the introduction of NSHP, more than one million people applied for housing to the Ministry of Housing and Urban Development. None of them obtained a housing unit.

The subsequent governments built a dream in the sand for the One Million NSHP before the revolt of June 2013, in which, after the collapse of the Muslims Brothers’ role, a series of steps was taken to implement the project. Some modifications had been made to the NSHP that relied on two principles. They were: first, allocating plots of land of 300–400 m^2 in the new towns to private developers who are able to finish the construction process within two to three years; and, second, continuing the ‘*Ibni Beitak*’ scheme in new towns. This plan recalls Mubarak’s scheme, the Youth National Housing Program (YNHP), in size and ambition. Neither of the principles were actually applied, but a resettlement program for unsafe areas is currently taking place. *Asmart* 1, 2, and 3 projects and the *Tahya Misr* projects constructed completely furnished housing units. There is no exact data to be analyzed. However, an analysis of the YNHP revealed that the government achieved barely 42% of the total target units. Comparing the NSHP to what has already been implemented on the ground, the picture is unclear.

In late 2013, UN-Habitat in cooperation with the GOPP at the Ministry of Housing launched Participatory and Inclusive Land Readjustment (PILaR) project. Two aspects were emphasized: the first to examine LRP as a tool to alleviate the arbitrary development of new urban land expansion in Egypt: the second to explore PILaR as a mechanism to shorten the gap in current Egyptian planning policy, its regulations, its requirements from one side and the existing reality of people’s

Fig 8.5 Photos illustrating the negotiation processes with the landlords (top two) and the final urban pattern of the PILaR project (bottom). (Source: The author)

interests, needs, and demands from other side. As illustrated in Fig. 8.5, a pilot project based on PILaR was applied in Benha city and implemented on the ground, relying on 'technical enablement' of the Technical Team (TT)[4] (UN-Habitat 2016; Soliman 2017). The TT initiated a set of basic principles and choices to get a preliminary agreement before assigning any new reparcel ling of the plots to reach a consensus among the landowners. Throughout the entire process with the landowners, the TT initiated a set of basic principles that prioritized the following five values: transparency, equity, trust, credibility, and efficiency. Thus, these principles

[4]The Technical Team (TT) consisted of UN-HABITAT's members, Mrs. Rania Hedeya, Dr. M. Nada, Mrs. Salma Mousallem, and headed by a private consultant, who is the author of this book. Mrs. Mousallem has played an important role, as a shadow consultant, in communicating with female landowners, as well as helping to set up the three scenarios. All opinions in the current presentation are the sole responsibility of the author, and they do not imply any opinion whatsoever on the part of either the GOPP or UN-HABITAT.

lead to practicable transition management for the project (for further details, see Soliman 2017).

For the first time in Egyptian urban history, this project introduced PILaR as a tool that enables cities to significantly increase the supply of serviced land at the urban fringe through orderly and negotiated processes of LR (UN-Habitat 2015). This project illustrates the challenge for stakeholders and urban practitioners alike to take up process-content thinking and to reflect on how to navigate societal complexity, while mobilizing transformative societal potential towards action for sustainability. As such, an integral part of complex change processes in urban areas was learning about one's new pathways for action, and the interrelated challenges of the city. All stakeholders were involved in the project in which transition management was applied as a response, or even as a 'reaction', to an emerging urban problem that occurred on the spot.

Recently, congruent the first role of President El Sisi, the state shows more intention to intervene in unsafe areas, as a primary concern to increase the standard of living and justify social justice for the bottom strata of the society, as well as to gain political support from a vast majority of people and to prevent any disaster that might happen within these areas—as happened in 2008 in the *El Dweka* area. Thus, the Ministry of Urban Renewal and Informal Settlements (MURIS), introduced in 2014 under decree no. 1252, will be responsible for upgrading slums areas and setting up a future housing policy to alleviate the future spread of informality. A year and a half later, MURIS was amended and its responsibility given to the ISDF. During this period, nothing was done on the ground. Since the middle of 2015, the state changed its housing policy and pursues a road map in four directions: abolishing unsafe areas by 2021; a huge land subdivision program of vacant land in new towns and satellite cities; intervention in the housing market as a developer; and the construction of 14 new towns as a fourth generation of the new town scheme. The current government announced that, by the end of 2021, the country will declare itself as free from slums, but, alas, it will rid of unsafe slums, not from unplanned areas.

Recently, the state introduced a temporary reconciliation law number 7 of 2019, and its amended law number 1 of 2020 to issue a certificate, as de facto tenure recognition between the state and buildings' violators, against certain fees to be paid to the government. The main objective is to put an end of the informal areas and to legalize their status. It is estimated that there are around two million buildings/ housing units that are in violation by which the state will collect around 40 billion Egyptian pounds to formalize the informal buildings/ housing units. This law was in effect from July to end of October 2020, after which all illegal buildings/housing units will be treated as in violation and restricted regulation will be imposed before the Egyptian courts. Because this law has just been applied, no evaluation of its impact can yet be carried out.

Two important new cities are currently under construction, the New Capital Administrative City (NCAC) and Al Alamein New City (ANC). The ANC is located on the north-western coast and extends 90 km west of Alexandria City; it is expected to accommodate 1.6 million people by 2052 (Attia 2019). In the NCAC, there will be exactly 21 residential districts, 25 “dedicated districts”, 663 hospitals and clinics, mosques and churches, and 1.1 million homes housing at least five million residents. It will cost 330.0 billion Egyptian pounds. The currently nameless city (NCAC) would span 700 km^2 (almost as big as Singapore), host a park doubles the size of New York’s Central Park, and a theme park four times as big as Disneyland—all to be completed within five to seven years (The Guardian 2015). Sims (Sims 2019) pointed out that although the NCAC will be twice the size of New York City, most of the housing units will be affordable only to the top 10–15% of Egypt’s households and will consume a great deal of Egypt’s natural resources. Thus, it seems that the NCAC would not absorb the bottom strata of the expected population, rather the elite of the society. In November 2016, the state deflated the local currency by which all prices of the commodities, including housing commodities, increased threefold, so, how will middle-income and low-income groups satisfy their needs for housing? This is a major challenge that faces the state. Currently, the government announced that the New Capital Administrative CITY (NCAC), 45 km east of Cairo, will be in operation at the end of 2021. All civil servants who will work in the NCAC will be housed in the new city of Badar with a distance of 40 km from the NCAC. Gated communities and luxury housing are constructed by the private developers to serve the elite of the society while the other strata of the community have no place to live in the NCAC. Thus, most of the housing units will be affordable only for the top 10-20% of Egypt’s households and will consume a lot of Egypt’s natural resources. Thus, the NCAC will be functioning for the elite of the society while the other classes will serve them, eliminating the bottom strata of the Egyptian social pyramid to move to the city, creating a new form of social segregation in the NCAC. It seems that the planning of the NCAC is not for everyone, but rather is exclusive to certain categories of society, mainly those who can afford it (Elmouelhi 2019).

From the housing production point of view, assuming that each 400 m^2 land plot will contain 16 housing units with an average height of four floors and four units per floor, a residential area of 25 km^2 is required to contain one million housing units. In addition, 50% of the area will be needed for roads, services, and open spaces, bringing the total area needed to meet the requirements of the NSHP to 37.5 km^2. This area constitutes 60 and 40% of the total sizes of the new cities of Sadat and Tenth of Ramadan, respectively. It also represents eleven and nearly 50% of the total serviced built-up area (349.166 km^2) and the total housing production (571,770 housing units) of 17 new towns, respectively, in the year 1990. Moreover, the total cost of land and infrastructure installation under the NSHP will be around US$5 billion, while the total cost of housing construction will be around US$16.66 billion. The total cost of the NSHP will be around US$21 billion. However, an annual estimated sum of US$4.2 billion will be needed to implement the NSHP. This constitutes around three-and-a-half times the total subsidy for NHP housing, which was dispersed over six years and not ever completed. Moreover, it also constitutes the loan amount that the Egyptian government is currently negotiating with the International Monetary Fund to fund the economic reform program. However, the role of the state has changed from an enabler, supporter, or provider to a developer and a contractor in the housing market. What has happened is a sharp increase in land prices, increasing the cost of the housing unit, the scarcity of low-income housing plots, and a decline in the real estate market, formally and informally. For example, a m^2 of housing in the NCAC cost around 30,000–40,000 Egyptian pounds or may be more. The state's road map took three forms. First were the outright economic reforms where the government abolishes fuel, electricity, and water subsidies, franchised other operations, and contracted for other services. Second was the introduction of certain free-market mechanisms, allowing the private sector to put housing units on the market without restriction and according to supply and demand. Third was a program of pragmatic public–private partnership, according to which initial agreements with private management agents might eventually lead to more extensive commitments. As part of the latter process, the government allowed private enterprises to become involved in land development within Egyptian new towns, frequently through partnerships with local governments. This kind of partnership allowed private enterprises to construct gated communities with huge luxury housing projects in newly developed satellite towns. The final one was the introduction of a temporary reconciliation law as an attempt to legalize the informal housing sector.

On the other hand, figures indicate that more than 16.477 million housing units are empty in Egypt and/or under the categories of unused units (CAPMAS 2016a) of which 9.193 million housing units are in urban areas and the remainder in rural areas (see Table 8.3). More luxury housing units are currently under construction in gated communities. The state is currently taking the role of the private developers by erecting more luxury housing units and through sale of luxury housing plots in various new towns. So, there is a problem of housing low-income groups, and at the same time there is a surplus of housing of than 16.4 Million housing units. It is *"a puzzle of housing"* in Egypt.

8.3.3 *Social Capital and Exclusion Transitions*

Significant theoretical analysis of the interrelationships between the social and the spatial aspects of the urban phenomenon refers to the urban as an arena of power relations, shaping cities' meaning and space (Castells 2000). This school views urbanization as a process that produces spatial structures and forms, creates spaces and places, supporting the creation of social relationships for the reproduction of capital that may represent the conflict between the state and society. But it hardly affects the urban poor. The issue of identity in a modern urban context (i.e., 'otherness and boundaries') echoes a long-standing debate concerning the status of 'communities' and spaces within Middle Eastern cities (Miller 2006). This differentiation is questioned by many academics in Egypt, in which people who immigrated and become urbanized to Cairo are classified into two categories; people from Lower Egypt, and people from Upper Egypt, the former as *fallahin*, or peasants, and the latter as *Saiaada*.

Table 8.3 The total number of closed, unused, and empty housing units in Egypt (2016)

total H. units	Total housing units	Total housing	work	Housing & Work	Closed living outside	Closed has alternative residence	Empty completed units	Empty unfinished & unoccupied	unit requires renovation	unit under Demolition by Decree	Unoccupied or closed establishment	other	Total Housing only	closed & unoccupied
Urban	21,643,979	9,998,107	2,398,519	54,184	772,907	1,934,391	2,749,802	2,030,497	204,992	44,593	1,059,452	396,535	11,645,872	9,193,169
Rural	21,329,905	12,307,491	1,610,205	127,960	386,322	958,087	1,916,742	2,300,799	410,462	32,819	650,109	628,909	9,022,414	7,284,249
Total	42,973,884	22,305,598	4,008,724	182,144	1,159,229	2,892,478	4,666,544	4,331,296	615,454	77,412	1,709,561	1,025,444	20,668,286	16,477,418

Source: (CAPMAS 2016a)

With the rapid urbanization wave of the 1950s and 1960s, the migrants who come to join the low-income urban groups have always been perceived as different and potentially dangerous to the social order. Two views were reached. The first was crystallized by the elite and educated people. The old urban elite has always considered rural migrants as threats to the 'refined civilization' of the city. Others described the inhabitants of the mid-1950s along popular peripheries of Cairo as *fallahin*, or peasants (Abu Lughod 1971), and also considered poor migrants a threat to urban stability. By the second half of the 20th century, the early pejorative perception that the dominant Cairo elite had of the traditional popular urban districts had been transferred to the new urban informal settlements (*Al Ashwaiyyat*). On the other hand, people who immigrated from Upper Egypt, called Sa'idi,[5] are often perceived in the public discourse as being more 'particular' than the other Egyptian regional migrants and less prone to integration. They are characterized by ruralization and famous with a collective Sa'idi grouping and settings, e.g., *Ezbet El Saidha* in *El Matarya* area in Cairo. Several academic works and institutional reports did not hesitate to stress the anarchic "ruralization" of Cairo's peripheries, describing them as places of serious sociocultural problems or black spots characterized by their "disruption of social organization" (Egypt Human Development Report 1996). The debate around the ruralization of the urban periphery raises the issue of the migrants' cultural and social integration within the city.

The second view was the politicians' view that saw immigrants as the main human power for industrialization, modernization, and clientelism. With the nationalization and industrialization movements of the 1950s and 1960s, the state encouraged and facilitated the mobility of people from rural areas into urban areas to participate in the industrialization process. Thus, the state treated the immigrants not as proletariats but rather as participants in the development process and accommodated them in remote locations in the city, which indirectly marginalized them from the rest of the society. With the urbanization wave of the 1960s, many peri-urban areas have been converted, formally and informally, to informal residential areas whose belts of informality surrounded the major urban centers. Today, spatial segregation is visible between *Boulq El Dakror* and *El Mohandseen* area, and between *Roushdy* and *Hager El Nawatayah* in Cairo and Alexandria, respectively (Soliman 1996). Improper socioeconomic and spatial transitions are increasingly separating people: this is visible in the spaces they occupy and inhabit, such as the social divides of *Imbaha* and *El Zamalek* areas in Cairo.

This era witnessed tremendous transitions in the Egyptian urban fabric in which huge agricultural areas were formally converted into residential uses, as well as many villages transformed into small cities, e.g. Kotor and El Santa. Thus, the role of the state was to satisfy the vast majority of the population to support the new regime and accelerate the industrialization process. Subsequently, mass production of public housing was 'popularized', mostly located on the periphery of the urban centers

[5]Sa'idi as a single and *Saiaada* as plural is the Egyptian term for someone coming from Upper Egypt

mainly on agricultural areas. Also, resettlement programs have taken place on inaccessible and marginalized agricultural land without basic infrastructure, e.g., the *El Siadeen* area in Alexandria City. Thus, the state encouraged, indirectly, the formulation of urban informality to serve its goals for attracting the support of low-income groups of the ongoing economic and political transitions. On the other hand, low-income groups are moving to these places looking for cheap shelter and might move from one area to another, but these areas are usually the worst locations within a city and over time become marginalized. Therefore, marginalization is determined by people for economic reasons and, by the state, for sociopolitical reasons. At the same time, the state is also responsible for creating marginality by pushing people away from the formal market, or from the formal urban context. In the resettlement program, the Egyptian authority located housing projects for low-income groups on the periphery of the urban contexts which created a new type of slums, as well as marginalizing people from enjoying the availability of services in central areas. Such provision of housing and social welfare was necessary to build popularity among the urban workers and a middle stratum at a time when the state was struggling against both the colonial powers and old internal ruling classes. Thus, the state acted as the moving force of economic and social development on behalf of the populace (Bayat 2009) and acted as the maestro for directing the formalization of marginalization.

Prior to the advent of the victory of 1973 and the political-economic restructuring of the mid-1970s, the Egyptian state was largely dominated by pro-Western rentier states. Many public-sector workers, people displaced from Suez Canal cities', and rural laborers, as well as educated, once well-to-do members of the middle class (government employees and college students), have been pushed into the urban informality on the periphery of cities. In the meantime, the state has gradually been retreating from the social responsibilities that characterized its early populist development and recognized the private sector as the main engine for accelerating the development process. Many social provisions have been withdrawn, and the low-income groups largely have to rely on themselves to survive, i.e., to house themselves informally. In Egypt, for instance, as late as 1977, the Egyptian state introduced sites and services and upgrading programs as an approach to tackle the scarcity of affordable housing for the urban poor. This era witnessed 'deteriorating social conditions', in which informality spread widely in all Egyptian cities. Due to the economic difficulties, the state adopted a *laissez-faire* policy towards urban informality which encouraged further informal-housing developments. The informal urbanization has caused several urban and social problems in the Egyptian urban fabric. Although certain social indicators, such as life expectancy and infant mortality, have improved, unemployment, poverty, and income gaps reportedly increased in the 1990s (Rahman and Westley 2001).

Therefore, due to economic difficulties, and to the dynamics of informal housing processes, the state adopted self-help approaches with international donor demonstration sites and service schemes in Egypt. The largest, most visible, and most successful sites and services and upgrading projects in Egypt were launched in Ismailia in 1978 and 1979 by the Governorate of Ismailia. Although these Ismailia projects became well known within Egyptian professional circles and abroad, such an approach was not followed to any significant scale elsewhere in Egypt (Soliman 1988). In addition, in spite of considerable project preparation efforts made by the World Bank to launch a large sites and services programme under Cairo and Giza governorates in Greater Cairo in 1984-85, these never saw the light of day.

Thus, the state tried to tackle the economic difficulties by increasing the subsidized prices for essential commodities (e.g., sugar and oil), but a social movement 18-19 January 1977 prevented this increase. This movement was carried out by six types of activism expressed in urban mass protests, trade unionism, community activism, political Islamism, non-governmental organizations (NGOs), and quiet encroachment. At the same time, changing housing systems and stronger demand for services was increasing the desirability of suburban living. As a result, lower-income groups moved increasingly away from crowded central locations towards less desirable parts of the city and used a different building style. Thus, the state has followed the *status-quo* policy towards the bottom strata of the society. Marginalization became noticeable throughout the Egyptian cities, and a division between poor and rich, between old and new, also became visible.

The urban riots of the 1980s, e.g., March 1986, were an early expression of discontent with some aspects of neoliberal policies in Egypt, as the state tried to reduce their deficits through austerity policies, such as cuts in consumer subsidies. Although it is difficult to determine the precise profile of the participants, the urban middle and lower classes were among the main actors. In parallel with these riots, Islamism—in particular, 'social Islam' or political Islam—articulates the concerns and struggles of the underprivileged urban Egyptians. The Islamist movements, especially the Muslim Brotherhood, contribute to social welfare and social competition (Bayat 2009) by providing services, such as healthcare, education, and financial aid; at the same time, they offer involvement in community development and a social network, most of which are carried out through local, non-governmental mosques. Political Islam encouraged the establishment of informal residential areas some of which some are characterized by Islamic songs inscribed on the facades of the buildings. The spread of the Islamists all over the country was reflected through stickers displayed on Egyptians' streets with the slogan *"Islam is the Solution"*. Also, women's dress had changed: scarf, veil, muffler, and yashmak came to dominate the attire of girls and women. Thus, everyday life has changed, as Islamic features dominate life and even Islamic dresses dominating shops, so the beautiful period of Egypt has 'gone with the wind'. These transitions of life in the Egyptian scenery are well illustrated in the old Egyptian black-and-white movies (Mofeed and Elgendy 2016) in which everyday life in urban informality dominate the movies in the period between of the 1990s to 2000s. On the other hand, sexual harassment has usually existed in the formal areas, while, in informal areas, it is very rare because such areas usually host homogeneous populations with strong territoriality feelings, mutual trust, and social control where external actors or outsiders can be easily identified. This phenomenon of community cohesion, namely 'collective efficacy' (Sampson, Raudenbush and Earls 1997), can prevent anti-social activities for the common good (Friedrich, Hillier and Chiaradia 2009).

The state, in order to outmaneuver the Islamists and regain legitimacy, is often forced to implement social policies in favor of the poor, and at the same time turn a "blind eye policy" towards illegal subdivision of agricultural land. One typical association, the *Ansar al-Muhammadiya* association in the poor community of *Imbaba,* radical Islamists such as *al-Gama'a al-Islamiyya* and *al-Jihad*, through their use of mosques, were also involved in urban community work. Rural and urban guerrillas concentrated their strategy on armed attacks, targeting the state officials' police and tourism (for further details, see (Badawy 1999). In 1996, the state adopted an upgrading policy in the back streets of Cairo city to control and record people who are the roots of terrorism and as a way to attack them on the ground. Thus, the state intervened heavily in some areas that accommodated such groups, especially *Imbaba*, through upgrading the program[6] as a tool to attack, or to prevent, terrorism.

The world financial crisis of 2008 raised food prices and oil prices, while revolutionary movements and the Covid-19 pandemic have added to the burden on the state. This sharply affects the urban poor in obtaining goods and services for their survival. This has created a new section of Egyptian society that is prepared to do things that could threaten national security. The state does not realize that the greater the social exclusion, the greater the danger to national security. However, urban informality has increasingly used the privilege of a *laissez-faire* policy of the state and has consolidated to form large urban agglomerations that would be impossible to remove or demolish. All these communities are the result of the socioeconomic exclusion that pushed people—occasionally residents of the city in question —to act 'illegally' and claim their right to the city. These places have become the signifiers of the socially constructed and demonized image of the 'outsider' and are categorized by the majority as 'illegal'. Even more, these areas excluded a whole region within a city, such as the western strip and southern part in Greater Cairo Region and Alexandria, respectively, from being integrated within the society. Therefore, riots, disturbances, and conflicts between the poor, low-income, middle-income, and the state were the initial spark for the explosive January Revolt of 2011 in Egypt.

The resignation of the Egyptian president Hosni Mubarak on 11 February after the white revolution of Egyptian youth on 25 January was to protest against poverty, low wages, rampant unemployment, government corruption, and the autocratic governance of the regime. The revolt called for social, economic, and political freedom and justice. The common demands of these popular revolts are the eradication of poverty, social justice, democracy, freedom of expression opinions, a decent life for every person in the society, and the justice of the housing delivery systems as primary elements of human rights.

[6]Western Munira of Imbaba, the stronghold of the Islamists, has been allocated more funding for its development than any other district in North Giza, to the east of Cairo. Between 1992/93 and 1995/96, some 372.5 million Egyptian pounds were spent on constructing, upgrading, and burnishing this area (Al-Ahram 1996).

After one year of the rule of the Islamist ex-President Mohamed Morsi, there is only one aspect of Egyptian politics that all Egyptians seem to agree upon: The only people who have an accurate understanding of the current situation are those who are utterly against the MB policy. The economic crisis mushroomed with a sharp decline in foreign reserves, large budget deficits, spiraling inflation, petrol, gas and bread shortages, lack of productivity, loss of exports, increased imports and a sharp decline in tourism. The Rebel movement was announced by Egyptian youth in April 2013 and collected around 22 million signatures from different strata of the society demanding the resignation of the Islamist president Mohamed Morsi, and to retain the January 25th revolution which was kidnapped without ransom. The Rebel movement asked the Egyptians to gather in the Egyptian squares all over Egyptian cities and villages on 30 June to demand early presidential elections and to meet their requirements of freedom, dignity, social justice, and bread. On Sunday, 30 June, 2013, more than 35 million people all over Egyptian territory peacefully gathered asking for their right to live a decent life and demanding that the Islamist president Mohamed Morsi leave the cabinet. June 30 was midnight for Morsi's Cinderella story, and the MB were surprised with the largest population movement that had ever happened in the history of Egypt. On Monday, 1 July 2013, Egypt's military delivered an ultimatum to Islamist President Mohamed Morsi, stipulating he had 48 hours to satisfy the public's demands or else, or it would impose its own "road map." The communiqué, which was interpreted by some members of the Muslim Brotherhood as a military coup, came on the heels of massive anti-government protests over the weekend that brought the country to a standstill. The streets were occupied again with at least 30 million people on Tuesday, people who were largely asking the military to back up their demand for freedom from the MB. Morsi rejected the military's timeframe in a statement on Monday, saying he had not been consulted and that the ultimatum could "cause confusion in the complex national environment." He continued saying that, if he left the cabinet, the whole country would descend into violence. The underlying meaning was that he remain as the elected president, or violence would spread throughout Egypt. The statement read further: "The presidency confirms that it is going forward on its previously plotted path to promote comprehensive national reconciliation ... regardless of any statements that deepen divisions between citizens."

On Wednesday, 3 July 2013, according to the demands of, and backed up by the majority of the Egyptian population, General Abdel Fattah Al-Sisi, the head of Egypt's armed forces, announced the removal of Egyptian President Mohamed Morsi, suspension of the constitution, and early presidential elections which brought Egypt's latest political crisis to its end. What actually happened in Egypt on 3 July 2013 was a response to the popular demand of tens of millions of Egyptians who took to the streets to demand early presidential elections? The roadmap, presented in the armed forces' statement and which outlines early presidential elections and parliamentary elections, is currently being implemented. It is not a coup d'état; it is a real revolution against corruption of the MB. The massive crowds in the street welcomed the military's intervention deliriously, while all awaited the potential response of enraged Muslim Brotherhood supporters. It is the end of the MB in

Egypt and maybe in the Middle East. At last, Egyptians can breathe clear air without being accused of not being Muslims. Islamic religion asks every Muslim to live in peace and obey the rules of *Qur'an* and *Souna* of Mohamed the Prophet, and that is we what we followed, not what the MB believes.

After the Arab Uprising in 2011, informality in all aspects of life was the main feature of the Egyptian scene. The poorest of the society were hardest hit. A new constitution was issued and opened a new era for a new Egypt. In 2014, the new state, headed by President El Sisi, realized that the poor constitute the majority of the society, are the main engine for development, and are the cornerstone for social and economic stability. Thus, the bottom of the pyramid of the society gained great attention, and a resettlement program was introduced to relocate people living in unsafe areas to new respectful residential areas, as well as a decent-life initiative for the urban poor, announced by President El Sisi. The question is how the majority of low-income groups are viewed by the society as a whole. Looking at various resettlement locations that accommodated the urban poor, it appears that the new locations are on the periphery of the urban centers and fenced off from the main road networks that have created new marginalization areas. Also, all data about everybody living in these projects are recorded and under control. Even though these people have the privilege of owning a house, they have lost many things: local jobs, homogeneity of social networks, accessibility to the city center, traditional ties, and many other social connections. Thus, they found themselves excluded from a modern way of life but are instead living in a new modern prison.

8.4 Urban Transition on Urban Informality

The ASUST perspective as multi and continuous processes conceives of transition as an interference of six trends. These are Policy Strategy Credibility, Consistency of Dimensions, Policy Processes (Coherence of processes), Systematic change (four perspectives), Systematic directions of Transitions, and Instrument mix Comprehensiveness. The first three trends are mainly related to the state's policy development process, while the fourth, fifth and sixth trends are highlighted subsequently. As Fig. 8.6 illustrates, the three trends of the ASUST are linked with the three pillars and the mechanisms of urban informality.

In the systematic change of urban informality, trends echo pressures of the urban informality formulation that constitutes four perspectives: socio-spatial, socioeconomic, socio-ecological, and sociotechnical. The latter constitutes the interaction between niche, regime, and landscape which are formulating the process of urban informality transitions. On the niche level, various uncertainty and certainty aspects emerge in the urban informality transitions. At the regime level, there are interactions between the challenges of socio-spatial, socioeconomic, socio-ecological, and social capital in which the transitions have merged in transformations during the last seven decades. The final outputs of these interactions and actions are the quality and the degree of a sociotechnical landscape being produced. It also reflects the degree and

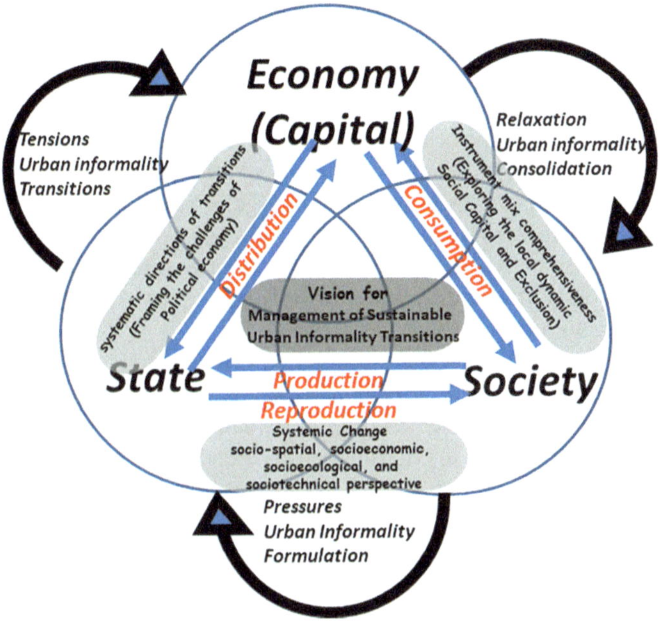

Fig. 8.6 Diagram illustrating the relationship between three pillars, the three stages of the ASUST, the mechanisms of urban informality, and sustainability transitions in Egypt. (Source: The author)

the level of finial production, or the quality of the end of production. It is not as simple as that: It is a complicated process because there are many forces and actors that participate in the final production outputs. Therefore, the participatory action adapted in the formulation of urban informality transitions constitutes a strategic agenda and builds on the transition-management approach as an overarching framework to consider how transitional change towards a sustainable future can be enabled.

Built on a transformative approach, the transitions procedure consists of the following four steps, which are followed in a transformative way: (i) urban informality definition; (ii) urban informality structuring; (iii) data gathering; and (iv) data assessment and analysis, all of which as explained in the Chap. 4. There is a limited academic literature analyzing interventions and/or possible strategies to enable transformative change in developing countries with a focus on co-creation, knowledge coproduction/reproduction, and sustainability transitions (Binz and Truffer 2008; Poustie et al. 2016).

In the systematic directions of transitions, trends mirror the tensions of urban informality transitions between state and political economy. This is the merger of two elements: stakeholders' trends and urban regimes' tasks. The former represents dimensions: continuous growth of cities, emergence of economic reforms, innovations in institutional reform, and innovations in reforms and constitutions. The latter represents six trends: future strategy participatory plans, sustain natural resources and climatic change, genuine innovations in institutional form and practice, achievement of improvement to meet people's requirements, and engagement of youth participation and people movements and demands. However, to reduce these tensions, the participation of grassroots of citizens, CBOs, and NGOs come to realize and share the co-created socioeconomic transition agenda that are enabling changes in a sustainable way. Sustainability-transition studies propose that problems persisting over time require a fundamental change in handling the problem solving (niches), ways of organizing (regimes), and the ways of doing (landscape) to allow for sustainable solutions to be considered and dominate ((Frantzeskaki and De Haan 2009; Grin et al. 2010). Also, learning by doing is a vital element to avoid misinterpretation of urban transitions.

At comprehensive instrumental mix tendency replicates the relaxation of urban informality consolidation in which the housing system is linked between political economy and the society which has needs for housing. It composes two trends: transitions management and the housing-system economy. Urban informality transitions, as a sociotechnical landscape, can occur through complex interactions of the housing-system economy of the production, coproduction/reproduction, consumption, and distribution of goods and services, interrelated with political economy, circular economy, state, and society. To achieve this purpose, total control of the expansion of Egyptian cities is needed, especially through controlled land development, as the main component of housing informality. This can be accomplished by enhancing a networked society, as an active urban niche, stimulated by increased digitalization, informatization, through the great acceleration in resource use, through rescaling, but also by changing the power of state and grassroots, and by transforming the drivers of change at various socioeconomic-spatial levels. The question is how to set up strategic co-creation processes for change-oriented interventions in complex mechanisms of urban informality as a part of adaptive societal (sub-) systems.

As illustrated in Fig. 8.7, the transformative approach is a comprehensive component for the relationship between the challenges of the political economy, socio-spatial, capital, and circular economy in which transformation of urban informality occurs. These transitions depend on the ASUST perspective of systematic change (four perspectives), systematic directions of transitions, and comprehensive instrumental mix . The systematic transformative approach relies on four perspectives: socio-spatial, socioeconomic, socioecological, and sociotechnical, in which the future urban informality transitions emerge on the three levels of niche, regime, and landscape.

First, the development of city-wide visions that go beyond each project is embedded in the city-regional context. These transitions occurred according to three variables: co-evolution multilevel, multi-phases and co-design and learning changes in sociotechnical systems. These variables interact and interrelate with complex systems of a sociotechnical regime (e.g., planning regulation, land-delivery system, capital forces, institutions' management, political transitions), a sociotechnical landscape perspective (e.g., population growth, population mobility, housing production, social ties, 'radical' change in terms of scope of change

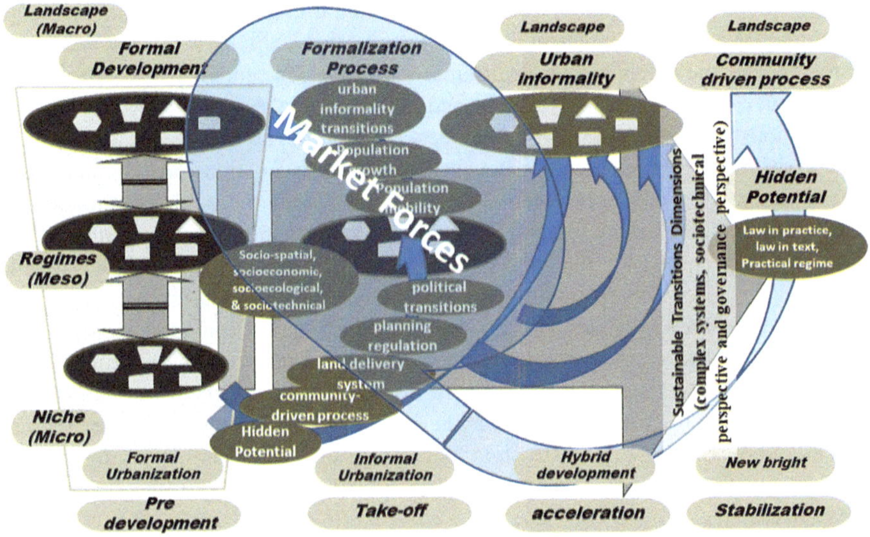

Fig. 8.7 Diagram of the management of sustainability transitions on urban informality's taxonomic in Egypt. (Source: The author and adapted from (Koehler et al. 2017; Grin et al. 2010; Geels and Schot 2007; Frantzeskaki et al. 2017a, b)

traditions, norms etc.), and a sociotechnical niche perspectives (e.g., community-driven process, hidden potentialities, a time period that witnesses transitions).

Thus, as illustrated in Fig. 8.7, the model affects sociotechnical transitions and the MLP through the interaction and correlation among the three levels of niche-innovations, regime, and landscape (Grin et al. 2010; Geels and Schot 2007), while the Multiphases perspective (MP) draws four phases of transition: predevelopment, takeoff, acceleration, and stabilization (Grin, J., et al. 2010). It continues to proceed in two directions, vertically, and horizontally in which interactions occur between the MLP and the MP through various pathways. These pathways highlight socio-cognitive aspects of the regime's rules and their changes that potentially shape socio-technical transitions. This model provides X-curve transition phases and pathways (Loorbach and Oxenaar 2018) through interactions between various levels of the MLP and the MP to create an enabling milieu, sustainable urban management: new residential areas, upgrading of existing informal areas, enhancement of social amenities, guiding new informal development on new bright sites (virgin areas, etc.) for the acceleration of desired transitions.

Second, the final landscape of this process has four outputs: formal development, formalization of informal areas, upgrading of existing urban informality, and, finally, encouraging community-driven processes in newly established areas. The transitions process occurs on three vertical levels: micro, meso and macro, and four horizontal levels. In each phase of the transition, vertically and/or horizontally different strategies and instruments (pathways) can be used to create a sustainable urban fabric for the uptake of natural dynamic transitions of urban informality. The learning concept is the output of gaining pieces of knowledge generated from the several phases of the progress and the mechanisms of different pathways. The micro phase denotes the guided development on virgin sites and relies upon the modification of people innovations in urban informality. It reflects the land development that is organized and controlled by a cooperation between residents and various stakeholders. The meso phase represents governments' intervention, market mechanisms, and the role of NGOs and CBOs in facilitating the integration of urban informality within urban contexts and enhancing the land- delivery system for low and middle-income groups. The macro phase imitates the final output, or the final production, of the previous two levels in which a sustainable urban informality transition is introduced. An integrated and cross-sectoral approach (horizontal and vertical coordination) is developed. Various models developed in this field aim to explain how transitions unfold and how to govern them (Koehler et al. 2017), and these challenges require system innovation, i.e., deep-structural changes of the sociotechnical configurations underlying the respective sectors ((Markard, Raven and Truffer 2012; Van den 2010).

Third, new instruments of urban governance, administration, and management, including increased local responsibilities and social networks through the accumulation of processes promoting 'lock-in' and path dependency (Dixon, Tim et al. 2014) and the willingness of the regime towards transitions are needed. Financing and investing with lasting effects are encouraged to make the best use of concentration of resources and funding on selected target areas. Capitalizing on knowledge, exchanging experience and know-how (benchmarking, networking), or learning by doing or doing by learning are executed. Monitoring progress (ex-ante, mid-term and ex-post evaluations, and indicators) is reached by introducing the concept of learning by doing and doing by learning.

Finally, this model has the advantages that it comes with the dynamic of the duality of economic and social exclusion in Egypt. Another advantage is that it is a flexible process that meets changes beyond the formal/informal and regulatory framework, including also behavioral, cultural, and practical changes caused by rapid transitive informality. The third characteristic it that it is an adaptable model to cope with outer and inner forces (economic, social, and political) that might affect the emergence of societies. Fourth, it copes with natural disasters, local conflicts, poverty degradation, spreading of informal urbanization, fluctuation of the market, etc. Fifth, it is an elastic model that works with a diversity of actors, to be able to challenge incumbent ideas and interests and to creatively adapt methods and tools to different contexts requiring specific skills and capacities. Last, it is flexible with various transitions and allows us to lock-in/out the privileging circumstances.

However, to perform the described process of urban informality transitions, there is a need for a transition-management framework that would generate a governance framework for proper intervention to alter urban informality in a sustainable way. Transition management as a transformative approach means to support governance capacities of cities, as well as the participation of all the stakeholders on the transition process. Transition management seeks to provide impulses for systemic change by creating spaces for developing innovative ideas, through the niche level, practices, through a sociotechnical landscape, and for empowering actors to develop transition experiments (Frantzeskaki et al. 2018b; H€olscher, K. 2018), through the initiatives of the grassroots. It is argued that the transition-management framework (Frantzeskaki et al. 2018a) has shown analytical strengths in different types of applications: theoretical applications position; and advocate transition management. This is as a new governance framework for transformative change, operational applications adopting transition management process tools (e.g., the transition arena, transition experiments) in setting up and realizing participatory processes for scenario development and strategy formulation. It is heuristic applications employ transition management as a descriptive and/or diagnostic lens to understand and explain the dynamics of on-going governance processes (Frantzeskaki et al. 2018a).

As summarized in Table 8.4, therefore, the state has to set up an arena of transitions management through four levels: strategic, tactical, operational, and reflexive (Loorbach 2010). The strategic level should specify problem structuring, vision development, strategic discussions, long-term goal formulation, collective goal, norm setting, and long-term anticipation. The tactical level should achieve goals in a specific context, steering activities, negotiation, collaboration, agenda-setting, and coalition forming etc. The operational level should handle short-term and long-term everyday decisions, and specific projects and experiments. The reflexive should look at learning processes, monitoring, assessment, and evaluation. A transition experiment takes a "societal challenge as a starting point for learning aimed at contributing to a transition" (Van den 2010), rather than a specific solution as is done for many innovation projects.

Table 8.4 Illustrates the arena of transitions of urban informality into the urban context

	Process competences	Substance competences
Strategic	Networking skills	Systems thinking
	Communication skills	Creativity and imagination
	Decisiveness	Problem structuring skills
	Determination	General knowledge
	Leadership	Large network
	Vision	Abstract thinking
Tactical	Negotiation skills	Strategic thinking
	Communication and consensus building	Analytic ability
	Thinking in terms of co-production	Specific knowledge
	Open to new combinations	Innovative ideas
	Coalition building skills	Capacity development
Operational	Short-term, everyday decisions, specific projects and experiments	Real-life societal context
Reflexive	Learning, monitoring, assessment, evaluation	Learning by doing and doing by learning

Source: Adapted from (Loorbach 2010; Van den 2010)

The quality of the built environment depends on the balance, correlation, interlink and actions between state, society and capital, that result in production, coproduction/reproduction, distribution, and consumption of various commodities and resources. Combined with Multilevel perspectives at three structural levels for urban informality (Geels 2010; Geels and Schot 2007; Grin, Rotmans and Schot 2010) of niches, regime, and landscape perspective, this would result in a framework for transition management of sustainability transitions in urban informality (TMSTUI). This framework combines transition-management processes and sustainability-transitions processes.

As evident in Table 8.4, developing transition management as a visionary pro-active approach is seen as a response to lagging, or misfit approaches to public engagement and strategic urban planning (Frantzeskaki et al. 2018b) to integrate, or at least to fit, urban informality into the urban context. The gap between the government policy and the capacity of community members should be narrowed. This means to identify what is of critical importance and what is required for people concerned, as well as what the government can provide and what people can afford. To fill this gap, the participatory inclusive approach is needed. Also, the government should act as an agent, rather than as a speculator or a developer, for the avail of local citizens and for the compatibility and sustainability. Moreover, it should amplify the political goals of rescuing agricultural land and encourage the sustainability of urban development in the remote desert of Egypt. To address this knowledge gap, Fig. 8.6 shows how applying transition management as a transdisciplinary framework organizes a co-creation of first, transformative narratives in the form of visions and second strategies in the form of transition pathways. Transition pathways are the interaction and correlation between political economy and social capital on one side (regime), and between both and urban informality (niches) on the other. The interaction and action between niche and regime are the final vision of a landscape for sustainability transition on urban informality.

8.5 Conclusion

There are indications that middle-income and low-income groups were hurt by the economic liberalization programs of the 1990s, and especially by their acceleration after the revolt of 2011, the devaluation of the local currency in November 2016, and the Covid-19 pandemic. The transformation of economic reforms from the mid-1970s and subsequent transformations and various sociopolitical transitions that occurred between 1980s and 2010s created a difficult situation not only for the low-income groups, but also badly affected the middle-income in which urban informality in the form of illegal agricultural land subdivisions and the acceleration of informal economy have exploded. In the late 1990s and early 2000s, it was mostly small merchants and industrialists, often in the informal sector, who benefited from the pro-market reforms, as well as the small but expanding skilled components of the formal/informal private-sector labor market. Most of the informal economy sector flourished in urban informality locations that enlarged the phenomenon of the disturbance of the urban context of the major urban centers. Urban informality later managed to spawn organizations and social networks and provided the mental, social, and physical space to develop new ideas, common language, and ambitions, as well as new joint projects to provide most of the urban population with goods and services.

Deteriorating socioeconomic conditions combined with high aspirations increased discontent among the middle- and low-income groups, resulting in their gradual withdrawal of support for authoritarian regimes. Given policy lock-in and the threat of political backlash from a key constituency in regime coalitions, it was difficult for governments to eliminate urban informality. As civil servants and employees of state-owned enterprises, the middle and low-income were effectively incorporated into the system, and thus their influence on policy formulation and ability to play the role of an "autonomous actor" have been undercut. Moreover, a sharp spike in income inequality occurred in Egypt—while levels of inequality have varied across the country's society that witnessed mass protests, the country does not exhibit particularly high levels of income inequality.

Urban informality is the outcome of socioeconomic, socio-spatial, and political transitions that occurred in the urban context, established by various actors, institutions, capital, and the role of state as the cornerstone of this transition process. State, capital, and private institutions constitute the three pillars and vary in function according to the status of socioeconomic, political, and socio-spatial transformations. The three pillars are responsible for the circular economy of production, reproduction, distribution, and consumption for commodities and various services, each of which has sub-levels according to the type of the urban informality.

With the focus on the insistent structural, economic, and cultural constraints underlying shift towards more sustainable lifestyles for urban informality, however, the need for sustainable urban transitions and management are essential. Urban sustainability transitions associated with good management contribute critical models for how transition management relates to urban planning and insights into actors and roles in transition management and urban sustainability transitions, as well as validating the democratic legitimacy of transition management in cities (H €olscher 2018). Urban-sustainability transitions also promote long-term fundamental change in existing sociotechnical systems, and transition management provides a tool for understanding how governance, cultural systems, infrastructures and socio-spatial practices co-evolve, creating lock-ins and sweeping adaptive changes (Kemp, Rotmans and Loorbach 2007). Over the years, transition management has been criticized (Frantzeskaki et al. 2018a) around the question of the extent to which transitions can actually be managed, and debated as serving a new 'dogma' for incrementalism and how to deal with complexity and develop proper interventions.

The consensus of the NUA is to put urban areas at the center of achieving sustainable development for the future generations as a key objective for Habitat III of leaving no one behind. This sustainable development reflects transitions theory that postulates successful systems comprising three main dimensions: complex systems, sociotechnical perspectives, and governance perspectives. These dimensions are in continuous processes of adaption and reformulation and are related and linked to the concept of transitions, transitions level, and level of intervention. This inherent complexity requires thinking about urban informality in Egypt as never being finished production and facing continuous change.

Thus, transition management opened dialectic and governance innovation spaces for urban planning in a proactive manner, to seize and to create opportunities for more visionary urban planning. However, transition management, an offshoot of the transitions approach, seeks to provide just this: a new approach to co-create transformative action to deal with and/or steer clear of persistent problems of unsustainability (Tidball et al. 2016).

Therefore, people innovations represent a sociotechnical niche on the micro level in which urgent intervention is required to understand the current mechanisms and to integrate urban informality within the urban context, in sustainable ways. This would require the recognition of the hidden potentialities of the citizens to interact and correlate between political economy, circular economy, official institutions, and social dynamics of citizens that constitute a sociotechnical regime at the meso level. The output of the interlinkage and correlation between the regime and the niche is a sustainable landscape to adjust the dynamic of urban informality and to make it sustainable in the urban context. The main policy for management transitions is to transform urban informality into a main driving force for sustainable development. Urban informality represents not only informal housing production but also an economic method for liquid capital and social-capital accumulation. Thus, urban informality in Egypt does not represent a need for housing so much as it is a matter of collateral investment to create wealth and to secure residents' future from the fluctuation of the economy within and outside the country.

In general, urban informality in Egypt is the output of the domination of capital and its production/reproduction, the size of production, the level of consumption societal technical, and the means of distribution. As citizens have experienced social bargains and come to expect real social mobility as a result of state economic and welfare policies, the inability to advance socioeconomically may have been especially frustrating. In this respect, it may have been the inability of the new "liberalized" order to create sufficient numbers of jobs to absorb the large cohorts of educated youth that generated much of the social frustration that ultimately led to the uprising in 2011. Last, not least, the implementation of the law of reconciliation will has major impacts on the level of housing production for low-income groups. The main question is: Has the state of Egypt the capacity and capability to meet the requirements of housing for the rapid increase of population?

References

Abu Lughod, J. 1971. *Cairo: 1001 Years of the City Victorious*. Princeton: The Princeton University Press.

Al Sayyad, N. 2006. Whose Cairo? In *Cosmopolitan: Politics, Culture, and Urban Space in the New Globalized Middle East*, ed. D. Singerman and P. Amar, 539–542. Cairo: American Unversity Press in Cairo.

Attia, Sahar. 2019. Al Alamein New City, a Sustainability Battle to Win. In *New Cities and Community Extensions in Egypt and the Middle East: Visions and Challenges*, ed. Attia Sahar et al., 1–18. Cham: Springer.

Badawy, M.A.S. 1999. *Islamic Associations in Cairo*. Unpublished MA thesis. Cairo: American University in Cairo.

Bayat, A. 2009. Middle Eastern Megacities: Social Exclusion, Popular Movements and the Quiet Encroachment of the Urban Poor. In *Megacities: The Politics of Urban Exclusion and Violence in the Global South*, ed. Kees Koonings and Dirk Kruijt, 82–106. London: Zed Books Ltd.

Binz, C., and B. Truffer. 2008. Leapfrogging in Infrastructure-Identifying Transition Trajectories Towards Decentralized Urban Water Management Systems in China. *DRUID Conference Paper*. Copenhagen: DRUID, 1–26.

CAPMAS. 2016a. *General Statistics for Population and Housing: Population Census*. Cairo: Central Agency for Public Mobilization and Statistics.

———. 2016b. *A Study for Development and Upgrade Informal Areas in Egypt (in Arabic)*. Cairo: CAPMAS.

Castells, M. 2000. Materials for an Exploratory Theory of the Network Society. *British Journal of Sociology* 51 (1): 5–24.

Davidson, F., and G. Payne. 1985. *Urban Project Manual: A Guide to Preparing Upgrading and New Development Projects Accessible to Low Income Groups*. Liverpool: Liverpool University Press.

Denis, E. 2012. The Commodification of the Ashwa'ryyat: Urban Land, Housing Market Unification, and de Soto's Interventions in Egypt. In *Popular Housing and Urban Land Tenure in the Middle East*, ed. M. Ababsa, M. Dupret, and E. Denis, 227–258. Cairo: American University.

Dorman, W.J. 2013. Exclusion and Informality: The Praetorian Politics of Land Management in Cairo (Egypt). *International Journal of Urban and Regional Research.*

Egypt, Arab Republic of, and UNDP. 2007. *Project 00056471*. New York: Untied Nation Development Program.

Egypt Human Development Report. 1996. *Egypt Human Development Report*. Cairo: UNDP.

Elmouelhi, H. 2019. New administrative Capital - Cairo: Power, Urban Development and Social Injustice – the Official Egyptian Model of Neoliberalism. In *Neoliberale Urbanisierung Stadtentwicklungsprozesse in derarabischen Welt*, ed. Al-Hamarneh, A., Margraff, J., Scharfenort, N., Bielefeld, Germany: Transcript Verlag.

Frantzeskaki, N., and H. de Haan. 2009. Transitions: Two Steps from Theory to Policy. *Futures* 41 (9): 593–606.

Frantzeskaki, N., V.C. Broto, L. Coenen, and D. Loorbach. 2017a. Urban Sustainability Transitions: The Dynamics and Opportunities of Sustainablity Transitions in Cities. In *Urban Sustainability Transitions*, ed. N. Frantzeskaki, V.C. Broto, L. Coenen, and D. Loorbach, 1–19. Oxon: Routledge.

———. 2017b. *Urban Sustainability Transitions*. New York: Routledge.

Frantzeskaki, N., M. Bach, K. Holscher, and F. Avelino. 2018a. *Co-Creating Sustainable Urban Futures: A Primer on Applying Transition Management in Cities*. Cham: Springer.

———. 2018b. Introducing Sustainability Transitions Thinking in Urban Contexts. In *Co-creating Sustainable Urban Future: A Primer on Applying Transition Management in Cities*, ed. Matthew Bach, Katharina Holscher, and Flor Avelino Niki Frantzeskaki, 63–79. Cham: Springer.

Geels, F. 2010. Ontologies, Socio-Technical Transitions (to Sustainability), and the Multilevel Perspective. *Research Policy* 39: 495–510.

Geels, Frank W., and Johan Schot. 2007. Typology of Sociotechnical Transition Pathways. *Research Policy* 36: 399–417.

Geels, F.W., F. Kern, G. Fuchs, N. Hinderer, K.G. Nele, J. Mylan, M. Neukirch, and S. Wassermann. 2016a. The Enactment of Socio-Technical Transition Pathways: A Reformulated Typology and a Comparative Multi-Level Analysis of the German and UK Low-Carbon Electricity Transitions (1990–2014). *Research Policy* 45: 896–913.

Geels, F.W., F. Berkhout, and D.P. van Vuuren. 2016b. Bridging Analytical Approaches for low-Carbon Transitions. *Nature Climate Change* 6: 576–583.

GOPP. 2012. *The Detailed Plan of El Matariya Area in Cairo City*. Cairo: The General Organization for Physical Planning (GOPP).

Grin, J., J. Rotmans, and J. Schot. 2010. *Transitions to Sustainable Development: New Directions in the Study of Long Term Transformative Change*. New York: Routledge.

GTZ. 2004. *New Practice of Participatory Local Development in Egypt's Urban Areas. Policy Paper, Ministry of Planning, Egypt*. Cairo: German Technical and Financial Cooperation.

H€olscher, K. 2018. So What? Transition Management as a Transformative Approach to Support Governance Capacities in Cities. In *Co-creating Sustainable Urban Futures: A Primer on Applying Transition Management in Cities*, ed. N. Frantzeskaki, K. H€olscher, M. Bach, and F. Avelino, 375–396. Cham: Springer.

Ilbert, R. 1984. Heliopolis: Colonial Enterprise and Town Planning Success? *The Expanding Metropolis Coping with the Urban Growth of Cairo*. Cairo: Proceeding of seminar nine in the series architecture transformations in the Islamic World, November 11–15, 36–42.

ISDF. 2016. *Slums Areas in Egypt*. Cairo: ISDF.

Khalifa, M.A. 2011. Redefining Slums in Egypt: Unplanned Versus Unsafe Areas. *Habitat International* 35: 40–49.

Kipper, R. 2009. Cairo: A Broader View. In *Cairo's Informal Areas: Between Urban Challenges and Hidden Potentials*, ed. Regina Kipper and Marton Fischer, 13–15. Lisbon: Norprint SA.

Kipper, R., and M. Fischer. 2009. *Cairo's Informal Areas: Between Urban Challenges and Hidden Potentials*. Cairo: GTZ.

Kirchherr, J., D. Reike, and M. Hekkert. 2017. Conceptualizing the Circular Economy: An Analysis of 114 Definition. *Resources, Conservation & Recycling* 127: 221–232.

Koehler, J., F. Geels, F. Kern, E.. Onsongo, and A. J. Wieczorek. 2017. *A Research Agenda for the Sustainability Transitions Research Network*. Technical. Amsterdam: STRN Working Group, STRN.

Loorbach, D.. 2007. *Transition Management New Mode of Governance for Sustainable Development*. Ph.D. thesis. Grifthoek/Utrecht: International Books.

———. 2010. Transition Management for Sustainable Development: A Prescriptive Complexity-Based Governance Framework. *Governance: An International Journal of Policy Administration and Institutions* 23 (1): 161–183.

Loorbach, D., and S. Oxenaar. 2018. *Counting on Nature: Transitions to a Natural Capital Positive Economy by Creating an Enabling Environment for Natural Capital Approaches*. Rotterdam: Dutch Research Institute for Transitions (DRIFT), Erasmus University.

Miller, C. 2006. Upper Egyptian Regionally Based Communities in Cairo Traditional or Modern Forms of Urbanization? In *Cosmopolitan Cairo: Politics, Culture, and urban Space in the New Global Middle East*, ed. D. Singerman and P. Amar, 375–397. Cairo: The American University Press in Cairo.

Mofeed, R., and N. Elgendy. 2016. Re-Presenting the Egyptian Informal Urbanism in Films Between 1994–2014. *International Journal of Humanities and Social Sciences* 10 (6): 2044–2050.

O'Brien, K. 2015. Political Agency: The Key to Tackling Climate Change. *Science* 350: 1170–1171.

Payne, G. 1999. Introduction. In *Making Common Ground: Public/Private Partnerships in Land for Housing*, ed. G. Payne, 1–16. London: Intermediate Technology Press Ltd.

Poustie, M., N. Frantzeskaki, and R. Brown. 2016. A Transition Scenario for Leapfrogging to a Sustainable Urban Water Future in Port Vila, Vanuatu. *Technological Forecasting and Social Change* 105: 129–139.

Rahman, A., and J. Westley. 2001. The Challenge of Ending Rural Poverty. *Development Policy Review, Volume* 19 (4): 553–562.

Runkel, C. 2009. *The Role of Urban Land Titling in Slum Improvement the Case of Cairo*. Berlin: Institut für Stadtund Regionalplanung, Technische Universität Berlin, Diploma Thesis.

Sims, D. 2010. *Understanding Cairo: The Logic of a City Out of Control*. Cairo: America University Press in Cairo.

———. 2019. *The Distorting Luxury of Infinite Space: Egypt's New Capital, New Cities, and New Industrial Areas*. Cairo: American university Press in Cairo.

Singerman, D., and P. Amar. 2006. *Cairo Cosmopolitan: Politics, Culture, and Urban Space in the New Globalized Middle East*. Cairo: The American University in Cairo Press.

Soliman, A. 1988. Housing Mechanisms in Egypt: A Critique. *Journal of Housing and Environmental* 4 (1): 31–50.

———. 1996. Legitimizing Informal Housing: Accommodating Low-Income Groups in Alexandria, Egypt. *Environment and Urbanization Journal* 8 (1): 183–193.

———. 1999. Public/Private Partnerships in Three Egyptian Cities. In *Making Common Ground: Public/Private Partnerships in Land Provision for Housing the Urban Poor*, ed. G. Payne. London: Intermediate Technology Publications (for DFID).

———. 2004a. *A Possible Way Out: Formalizing Housing Informality in Egyptian Cities*. Lanham: American University Press.

———. 2004b. Tilting at Sphinxes: Locating Urban Informality in Egyptian Cities. In *Urban Informality in an Era of Liberalization: A Transnational Perspective*, ed. Ananya Roy and Nezear AlSayyad, 171–208. Lanham: Lexington Books.

———. 2004c. Regional Planning Scenarios in South Lebanon: The Challenge of Rural-Urban Interactions in the Era of Liberalization and Globalization. *Habitat International Journal* 27 (1): 385–408.

———.. 2010. Rethinking Urban Informality and the Planning Process in Egypt. *IDPR Journal* 32 (2): 119–143.

———. 2012a. Housing and the State in the Middle East. In *International Encyclopedia of Housing and Home*, ed. Susan J. Smith, 346–354. Amsterdam: Elsevier.

———. 2012b. The Egyptian Episode of Self-Build Housing. *Habitat International Journal* 36 (2): 226–236.

———. 2012c. Building Bridges with the Grassroots: Housing Formalization Process in Egyptian Cities. *Journal of Housing and the Built Environment* 27 (2): 241–260.

———. 2013a. *Living With/in Ashwaiyyat in Egypt: Is It a New Innovation of Urbanity?* The Flexible City, International Symposium. Oxford: Oxford University, 24–25 October.

———. 2013b. *The Poor in Search of Shelter: An Examination of Squatter Settlements in Alexandria, Egypt*. Deutschland: Lap Lambert Academic Publishing.

———. 2017. Land Readjustment as a Mechanism for New Urban Land Expansion in Egypt: Experimenting Participatory Inclusive Processes. *International Journal of Urban Sustainable Development* 9 (3): 313–332.

———. 2019. So It's Always a Chance: Community-Led Solutions to New Urban Expansion. In *New Cities and Community Extensions in Egypt and the Middle East*, ed. S. Attia et al., 159–179. Cham: Springer.

Spaans, M., and B. Waterhout. 2017. Building Up Resilience in Cities Worldwide: Rotterdam as Participant in the 100 Resilient Cities Programme. *Cities* 61: 109–116.

Subrahmanyam, G. 2016. *Addressing Informality in Egypt*. Tunis: African Development Bank, AFDB, Working Paper.

The Guardian. 2015. *A New Cairo: Egypt Plans £30bn Purpose-Built Capital in Desert*. London: The Guardian.

Tidball, K.G., N. Frantzeskaki, and T. Elmqvist. 2016. Traps! An Introduction to Expanding Thinking on Persistent Maladaptive States in Pursuit of Resilience. *Sustainability Science* 11: 861.

Tipple, G. 1999. User-Initiated Transformations of Government-Built Housing Stocks: Lessons from Developing Countries. *Journal of Urban Technology* 6 (3): 17–35.

———. 2000. *Extending Themselves: User-Initiated Transformations of Government-Built Housing in Developing Countries*. Liverpool: Liverpool University Press.

UNDP. 2009. *Assessment of Capacity Building Needs of NGOs in Lebanon*. Washington, DC: UNDP.

UN-Habitat. 2008. *The State of African Cities 2008: A framework for Addressing Urban Challenges in Africa*. Nairobi: UN-HABITAT.

———. 2015. *UN-Habitat Global Activities Report 2015: Increasing Synergy for Greater National Ownership*. Nairobi: United Nations Human Settlements Programme (UN-Habitat), 31.

———. 2016. *Evaluation of Participatory Review of Egyptian Planning and Related Urban Development Legislation to Support Sustainable Urban Development, April 2016*. Nairobi: UN-Habitat.

UN-Habitat and Arab Republic of Egypt. 2015. *Egypt Housing Profile*. Cairo: Arab Republic of Egypt.

Chapter 9
Pockets of Urban Informality in Lebanon

Abstract Lebanon is the smallest country in the Middle East but has major influences in drawing, or at least transforming, the socioeconomic and political milieu in the region. Through time, Lebanon passed through several eras of rulers until its independence in 1943 when the urban expansion transitions of those rulers were inherited by the newly independent state. As a result of the tremendous socioeconomic, demographic, and political transitions of Lebanon, arbitrary urban growth dominated the Lebanese scene. What happens when the space of the unformed urban expansion overlaps with the space of Lebanese cities? Unplanned urban growth and urban informality became the main features of urban development transitions by which multifaceted forms of marginalization were set into motion. This is due to the peculiar sectarian character of the Lebanese sociopolitical economy and the formulation of refugee camps. Urban informality in Lebanon is not a physical problem alone. Rather it constitutes a sociopolitical economy and sectarian challenge for the Lebanese government and international agents. Therefore, its treatment differs from that of other countries in the Global South.

Keywords Civil War · Sectarianism · Misery Pockets · Urban Informality Transitions · Religious Sects · Informal Housing Typologies · Refugees

9.1 Introduction

Images of war, mass destruction, genocide, terror, and disaster have long held a prominent place in modern Lebanon, but their psychological or cultural impact remains uncertain. What we do know is that a range of claims, assertions, and discourses have been circulating about Lebanese catastrophes and shocks at least since the 18th century. In the modern era, Lebanon passed has through tremendous socioeconomic, socio-spatial, and political transitions in which the urban structure of Lebanese cities has been formulated (Soliman 2008).

This chapter links together three social-science perspectives in Lebanon—political transitions, sectarianism changes, and socio-spatial transitions—that have had major influences on urban informality, sustainable development transitions, human

A. M. Soliman, *Urban Informality*, https://doi.org/10.1007/978-3-030-68988-9_9

security, and sustainable peace. The chapter connects these four concepts within the research paradigm of 'sustainability transition'. This research paradigm focuses on a large-scale and short- and long-term transformative change of the dominant socio-economic and political-intensive development path by addressing the causes of local and regional changes on the urban fabric of Lebanese cities and the formulation of urban informality.

This chapter examines urban informality within Lebanese territory with a focus on Beirut city, in order to understand the extent to which the religions/ethnicities, along with the political and the demise of pre-war safety net mechanisms, have affected the urban development that is remapping the country. It emphasizes two arguments: (1) the importance of the diversity of state power, ethnicity structure, and international organizations on urban informality; and (2) how these diversities encourage the Lebanese and non-Lebanese when acquiring land plots for housing at reasonable prices. The diversity of government, together with international agents , involvement with such areas and their effect on shaping the urban fabric within Lebanese cities are clearly noticeable. Urbanization sprawl and the creation of urban informality in Lebanon in various periods are briefly reviewed to help explain the various causes for the appearance and the transformation of urban informality in Lebanon.

The post-Syrian crises has amplified the growth of urban informality within the country in which conflict between local politicians and international organizations occurs regarding the legality of Syrian refugees (Nassara and Stel 2019). In order to examine these issues, the chapter examines the processes of formulations and mechanisms of informal areas within Beirut city, as a typical transformation model for urban transitions in the Lebanese territories. It is assumed that the diversity of state power, the influx of refugees, and ethnical structure due to several attributes could be adjusted in favor of upgrading and integrating informal areas within the rest of the urban fabric of the cities by which a homogeneous built environment would be molded.

Due to the multidimensional nature of the Lebanese territories, transformative mixed methods (Mertens 2012) are used as a philosophical framework that focuses on ethics in terms of cultural responsiveness, recognizing those dimensions of diversity that are associated with power differences, building trusting relationships, and developing mixed methods that are conducive to socio-spatial and spatiotemporal changes. The transformative paradigm is a metaphysical framework that "directly engages the complexity encountered by researchers and evaluators in culturally diverse communities when their work is focused on increasing social justice" (Mertens 2009). The transformative paradigm also focuses on the strengths that reside in communities that experience discrimination and oppression on the basis of their cultural values and experiences (Cram and Mertens 2015). This method concentrates on a wide range of appropriate tools, methods, and techniques including literature review, qualitative meta-analysis, qualitative content analysis, on-site empirical investigations, document collection from official administrative units, and multivariate statistical analysis. For data collection, the primary literature analysis was accomplished via secondary data such as Google's maps sources, GIS maps available for Beirut city, national census, and local authority. On-site empirical investigation relied on field survey carried out in 2004 by the students of the faculty of architecture at Beirut Arab University supervised by the author. Updated information on urban informality was gathered during the author's visit to Lebanon in early 2012 and 2014. Recently, on Tuesday, 4 August 2020, Beirut shook with a great explosion in its harbor (The Guardian 2020) which caused a collapse of sociopolitical structure of the country.

The chapter is organized into six sections. The second section sheds light on urbanization sprawl and the creation of urban informality. It is followed by an overview of urban informality transitions, the fourth part highlights political-economy transitions, and the fifth examines social capital and marginality. The last section concludes that urban informality is the most significant phenomenon shaping the built environment within Lebanese cities, especially Beirut city.

9.2 Urbanization Sprawl and the Creation of Urban Informality

Starting from the 18th century, Lebanon was part of Bilad El Sham (Syria, Lebanon, and Jordan), which had been conquered by several Islamic rulers. In 1832, the expansive Egyptian ruler Muhammad Ali sent his troops to take over the Bilad El Sham area, and Beirut became the provisional capital for his administration, while the Mountain of Lebanon was under the control of Druzes and Catholic Maronites. By 1840, however, global politics turned against the Egyptian adventure in Bilad El Sham, and, on 11 September 1840, an Anglo-Turkish fleet heavily bombarded Beirut at the start of a campaign to force Muhammad Ali's forces out of Bilad El Sham. In the wake of the Egyptian withdrawal, Beirut's administrative centrality ironically increased because the Ottomans began to use the city as a center for their own renewed control of the region (Traboulsi 2007).

Lebanon is in the eastern sector of the Mediterranean basin, surrounded by Syria in the north and east, while Palestine constitutes the southern edge of Lebanese territory. The country has a 200-km coastline, running north-east–south-west along the eastern shores of the Mediterranean Sea. Lebanon, with an area of 10,452 km^2, is a highly urbanized country with more than 87% of its population living in urban areas and 64% living in large urban agglomerations (Beirut and its suburbs, Tripoli, Saida, Zahle, and Tyre). Lebanon is divided into six administrative governorates (Mohafazats), namely, Beirut, Mount Lebanon, North Lebanon, South Lebanon, Nabatiyeh, and Beqaa. Beirut and Mount Lebanon accommodate around 50% of the population (UN-HABITAT 2011). A primary characteristic of the five million inhabitants involves their belonging to 18 different Christian and Muslim denominations (Harris 2009).

Moreover, the great asset of Lebanon, its human capital and their entrepreneurial spirit, led to the new state having a laissez-faire economic policy, which helped the

Lebanese economy grow dramatically (Gaspard 2004). Soon Beirut became the center for regional business in banking services, a unique tourism spot, and a busy harbor which helped spur growth in its accommodation facilities, intellectual freedom, and its exotic, extravagant nightlife. Lebanon was the Switzerland of the Middle East during the 1960s. The population of Lebanon is rich in its ethnic and religious diversity. Even in this heterogeneity, common social and moral values underlie the societal structure. This is enhanced in the settlement patterns in cities, especially Beirut, which reinforces the sectarian divisions. The different social groups generally dwell in the urban district dominated by the sect to which they are affiliated. These districts or "urban villages" are found in the suburbs of Beirut and constitute its urban fabric. Unfortunately, Beirut also has become a depressed world of rural immigrants such as Palestinian and Syrian refugees disconnected from the growing wealth and development of the city. A "belt of misery[1]" began to encircle the old core of the city, with the poor and unemployed living in shantytowns on the city's outskirts.

The geography and history of Beirut outlines the seven development stages the city has passed through, awarding it important and special traits as the Lebanese capital city (see Table 9.1 and Fig. 9.1). The first phase began under the Ottoman Empire (1840–1920) when Beirut became a cornerstone of the regional networks. During this period, Muhammad Ali declared Beirut as a provisional capital for the *Bilad El Sham* area. After the struggle against Ottoman Rule, Amir Faysal took over Beirut on 3 October 1918, proclaiming its allegiance to the concept of Arab sovereignty and ending 400 years of Ottoman rule (Dumper and Stanley 2007). After an Anglo-Turkish agreement, Beirut accommodated 50,000 people and became the main center of Lebanese financial and political life. With an increase in openness to the global economy, many quarters started forming outside the city center.

The second phase was the one under the French mandate (1920–1943). On 1 September 1920, the French high commissioner *Henri Gouraud* proclaimed the establishment of the State of Greater Lebanon (Rabil 2011). In 1929, the Beirut Airport was built, and in 1932 the Saint George Hotel was opened as a shining symbol of the French Mandate's interest in infrastructural promotion. By the end of the 19th century, the city's population had increased to 100,000. During this period, the population of Beirut doubled. The Sunni, Maronite, and Greek Orthodox populations all grew by more than 50%, while the Shi'i and Druze communities in the city tripled. Into this mix was added 28,000 Armenians and Syriacs, most fleeing from Turkey (Traboulsi 2007).

The third phase was the post-independence of the country (1943–1975). On 22 November 1943, Lebanon gained its independence and became a member of

[1]Misery Belt or Misery Pockets is a term commonly used by the media and the Lebanese political figures to designate the informal settlements that developed around Beirut's city center in the southern suburbs. The term refers to the miserable conditions characterizing the low-income housing of Beirut suburbs ranging from environmental degradation, low hygienic conditions, over-crowding, and other dreadful living conditions.

Table 9.1 Data on socioeconomic, political and reforms transitions during the last 70 years

Muli-Levels/ Phases of Transitions	Niche		Regime		Landscape
	Economic Transitions	Social Transitions	Reform Transitions	Political Transitions	Output & Policy
The Independence Years (1942–1975)	The establishment of Palestinian's camps after 1948 war constituted "Misery Pockets" on the periphery of Beirut. As of 1954, a decree required land subdivisions to secure the approval of municipal authorities. Flourishing industrialization and drop in silk skills. General Fouad Chehab (1958–64), who established the Central Bank of Lebanon and the Ministry of Planning. This historical commitment to the "free market" was translated in the reluctance of successive Lebanese governments to intervene in the delivery of social services, including the provision of housing (Sadik 1996).	The waves of rural migration to Beirut generally followed religious lines. The 1950s corresponded to the migration of Muslim Shi'ites, at that time the poorest religious group in Lebanon, to Beirut (Faour, 1981). Rapid demographic growth and continuous rural-urban migration. After the 1967 war, large migration flows from southern Lebanon to the suburbs of Beirut further escalated the demand for housing in the area. For Cairo Treaties, see Salibi (1976) and el-Khazen (2000).	A 1954 decree required that all lot subdivisions be approved by the Municipality through an application process to be submitted to the Ministry of Interior. Master Plan for the suburbs of Beirut, Decree 16 948 issued on 23 July 1964. Many of the current institutional set-ups were created.	According to the 1943 National Pact between sectarian leaders, the president would be a Maronite; the prime minister a Sunni; and the parliamentary speaker a Shi'a. During the era of President Camille Chamoun in the 1958 civil war and its influence grew markedly in the 1960s. The 1967 Arab-Israeli war, and especially the 1969 Cairo treaty which gave the Palestinian Liberation Organization a de facto permit to initiate resistance activities against Israel from South Lebanon, created large migration flows from southern Lebanon to the suburbs of Beirut and further escalated the demand for housing in the area. For Cairo Treaties, see (Salibi	In 1899, after cholera crisis, a new Quarantine area near Nahr (river) Beirut was constructed. Zoning regulations for Beirut city with minimum land subdivision of 2000 M2. All transaction land sales should register in the formal notary registries (kateb'adel). In 1952 there were six Palestinians camps. In the early 1960s, once again, Ecochard set up a plan for Greater Beirut. Palestinians displaced to Lebanon after the defeat of 1967 war by the help of UNRWA. By 1975 informal communities had been growing as a peculiar social case in the cultural texture of Beirut. In 1972, Israel marched into South Lebanon and

(continued)

Table 9.1 (continued)

Muli-Levels/ Phases of Transitions	Niche		Regime		Landscape
	Economic Transitions	Social Transitions	Reform Transitions	Political Transitions	Output & Policy
				1988) and (El-Khazen 2000). Nationalization of Suez Canal and the Crisis of Suez 1956 1967 War of Defeat	the invasion included 150 southern villages.
The War Years (1975–1990)	The destruction of the BCD and its hotels; the interruption of activities of the various ports, the international airport, the Casino and the National Hippodrome. The destruction of an already modest industrial sector, and its replacement with low cost, low technology, medium size manufacturing activity. The reduction of the number of private enterprises from 7000 before the war to 3000. The unemployment rate is said to represent 20% of the active population in the 75–90 period. The fall in the level of remittances of the	Rapid urbanization process. Increasing the poverty level. Decreasing number of middle class individuals. Lack of funds for social and health amenities. Rapid urbanization growth Allocate resources for people to organize their own house building. With Israel's two wars (1982–1986) many people from southern country had moved towards Beirut and settled illegally on the periphery of Greater Beirut. People relied on their efforts and their social networks.	Relaxation of Regulation. Some decrees for the benefit of the private sector. Nos. 49, 136 of 1974, 1977 respectively for rent control Incentives for the private sector by exemption from paying taxes for 10 years. Peace agreement with Israel. Continued conflicts between various Lebanese politicians. The emigration of qualified labor (technicians and managers) to the Gulf countries, Europe, the US and Africa.	In 1976, when Amal supported the Syrian intervention, the National Movement and the PLO easily routed it from areas it controlled in Beirut. In 1982 the new Islamist organization was named Hizballah, "the Party of God." In the mid-1980s between rival factions of the Lebanese Forces, after Elie Hobayka had signed the Syrian- sponsored Tripartite Agreement and was overrun by Samir Geagea. This was also the case toward the end of the war in 1988—during the inter-Shi'i wars between Amal and	The continuing civil war of 1975 led to a radical change in the demographic fabric of Beirut's overall urban structure. The status quo of the urban sprawl. The spreading of urban informality changed the land uses within the city and created what is called the "Misery Belt" around Beirut. Appearance of squatter areas. Vertical expansion as well as horizontal 'spill-over' around informal areas. Involvement of private sector, NGOs and

	Lebanese immigrants from $200 million a month in 1982 to $50 million in 1987. Flourishing and spreading of informal economy sector.			Hizballah in Beirut and south Lebanon.	contractors' companies in housing production.
Post War Reconstruction (1990–2005)	After sixteen years of war, Taiif Accord and the political reforms were approved. The Taiif Agreement brought many more stakeholders and categories of actors, old and new, into the political game. The murder of Rafiq Hariri on February 14, 2005. The internal upheavals and the unstable political situation veiling the country since 2005 worsened the physical, social, and economic fabrics.	Building organizations that facilitate central initiatives. Magnetize low income to sustain the stability of the regime. Flourishing Muslim Brothers and prejudice. Sectarian institutions exert pressure on individuals to adopt sectarian identities. The political leader or Zaim is playing a major role in shaping the built environment. Inequality of income, hunger and war, and discriminatory practices in housing markets lead to a disproportionate concentration of ethnic minorities in certain urban areas.	The plan of housing and restoration of dislodged groups as it appears in the Plan 2000 for Reconstruction and Development which has several obstacles. The introduction of the Reconstruction and Development plan in 2000. A state-of-the-art central business district (Solidere) as the epitome for post-war economic development. Rapid agricultural land conversion to other uses. The government made practical attempts to intervene in the domain of housing through the project of Beirut's southern suburb (Elissar).	The Taiif Accord that ended the war, and subsequently abided by its precepts concerning disarmament of all militias, handing over their arsenal to the army and the state in 1992. Lebanon's problematic relations with Syria, and Hizballah's confrontation with Israel are two dimensions in the crisis inaugurated in August 2004 as well as the repercussions for Lebanon up to mid-2008. Decentralized resources to support local enterprise and home building.	Housing microfinance, community-based finance savings and loan groups, and consumer credit for building materials. 45,000 households (of the total number of those dislodged) occupying residences on illegal basis and 120,000 households living under grievous conditions in places unfit for habitation. The project of Beirut's southern suburb (Elissar) to fulfill the requirements of the urban poor.

(continued)

Table 9.1 (continued)

Muli-Levels/ Phases of Transitions	Niche		Regime		Landscape
	Economic Transitions	Social Transitions	Reform Transitions	Political Transitions	Output & Policy
Post Cedar Revolution (2005 to present)	The Lebanese people participated in the largest protests since the assassination of Hariri in 2005 and the withdrawal of Syrian occupying troops. Militant clashes in 2008. Israel attacks against Hizballah in 2006. Toward the Paris III conference in January 2007 adopted the program and approved a package of aid worth $7.4 billion as grants and loans to help revive the Lebanese economy. Direct and indirect losses to the tourism industry amounted to $2 billion. Dramatic decline in the activities of Beirut's stock exchange.	These circumstances have meant the absence of work opportunities which caused massive emigration among Lebanese youth. Two years prior to the July 2006 war, income-related indicators, mainly in employment and economic dependency, worsened from 43 to 52 percent for the entire population. In 2007, reform program to stimulate growth, create employment, reduce poverty, and maintain social and political stability is issued. Many Lebanese, primarily the entrepreneurial and professional people, particularly among the Christians, seek a better future elsewhere.	Dysfunctional land and housing sectors disproportionally affect the poor. Real GDP growth, which achieved a respectable rate of 5–6 percent in the first half of 2006, ended the year on the negative side with a decline of 5 percent. Far more serious is the growth of the gross public debt, which totaled $39.8 billion at the end of June 2007. Consumer and investor conditions deteriorated, deterring tourists, encouraging the departure of skilled workers, and hindering the implementation of much needed structural reforms.	At the height of Lebanon's Cedar Revolution in 2005 to end Syrian occupation. In May 2005, with Syria's enforced military withdrawal and the first free elections since 1972, produced a parliamentary balance close to the probable numerical weight of major political forces. In the mid-2005 parliamentary elections and the formation of a new cabinet under Prime Minister Fuad Siniora, who was toppled by Hizballah. New president, Michel Suleiman, was elected on May 25 but as of the end of June 2008 there was still no government of national unity.	Massive movements of Syrians to Lebanon took place in which more than 1.5 million are currently residing in Lebanon. In June 2008, the "protest tents" put up by the opposition forces under the leadership of Hizballah in the center of Beirut. Hizballah affiliated military officer keeps his post as chief security officer at the Rafiq al-Hariri International Airport in Beirut. Since then many conflicts have raised between Hizballah and other sects in Lebanon, in the 2018 Prime Minster Hariri managed to form a government cabinet.

The Urban Growth of Beirut

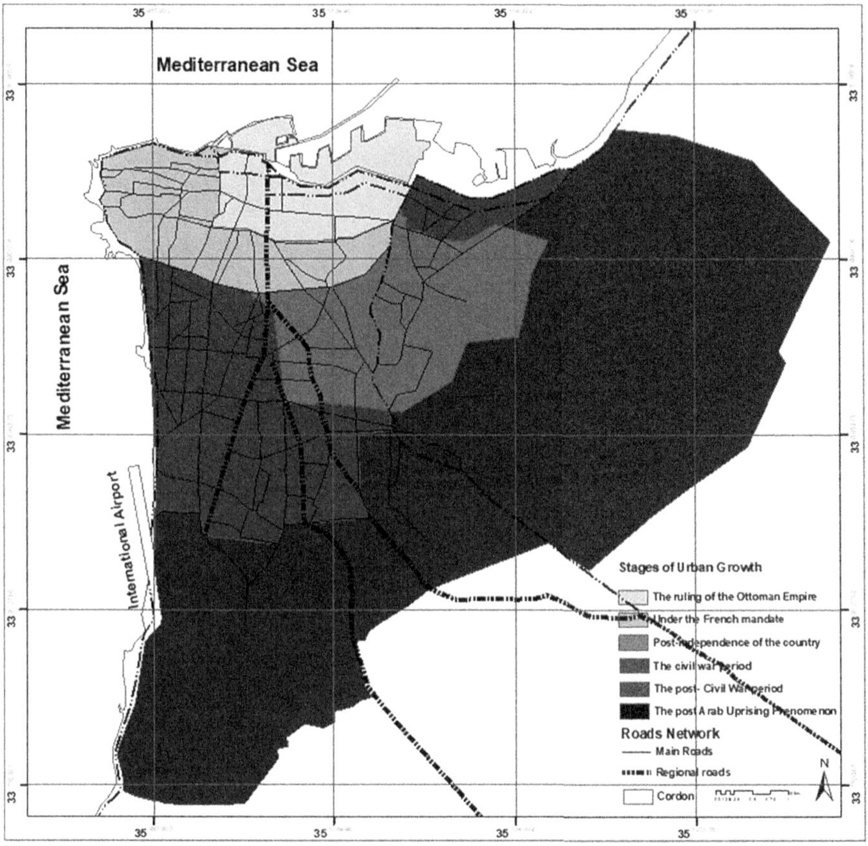

Fig. 9.1 Urban growth of Beirut. (Source: The author and adapted from The Government of Lebanon and the UN 2017)

the Arab League. Beirut became the capital of a free, independent Lebanon, after 20 years under the rule of the French Mandate, during which many of the current institutional instruments were created. The post-independence phase was also marked by rapid population growth and continuous rural–urban migration (Bourgey 1985). During this period, Beirut was somehow demolished to give birth to a gentrified center solely focused on trade and commerce. The old city center started to shift west (from El Gemmaza to Ras Beirut), and a new civic center was created. The third phase consisted of the urbanization sprawl and the creation of the 'Misery Belts' around the city center coupled with rapid demographic growth and massive rural-to-urban migration. The establishment of Palestinian camps after the 1948 war constituted Misery Pockets on the periphery of Beirut.

The fourth phase was the civil war period (1975–1990) that entailed displacements within the city center of different religious populations moving about inside the city. Hence the 40% faction of the Muslim communities in the Eastern suburbs of

Fig. 9.2 Photos of destroyed buildings occupied by squatters in Beirut (photos were taken between 1995–1997). (Source: The author)

Beirut decreased to a mere 5%. As for the Christian faction, it decreased from between 30–40 percent to 5% in West Beirut and the southern suburbs of Beirut. During this era, militias took control over the area and created what is known as the "Green Line", a delineation between East Beirut occupied by Christians and West Beirut, occupied by Muslims (see Fig. 9.1). What later came to be called Beirut's 'misery belt' seemed to follow an axis drawn around the capital along a line starting in the eastern part of city, from the former Armenian camp of Medawar in Quarantine, proceeding towards the Palestinian camps of Tell El-Zaatar, Jisr El-Basha, and Dbayeh and stretching towards the western part of Beirut to include the camps of Shatila, Mar Elias, and Burj El-Barajneh (Salibi, K. 1976).

Due to the conflict between various sects during the civil war, the city witnessed a tremendous displacement of people from south to north and from west to east and vice versa that changed the sociopolitical structure and urban pattern of the city. Also, during Israel's two wars (1982–1986), many people from the south of the country moved towards Beirut and settled illegally on the periphery of Greater Beirut. *Hezb Allah* was founded during the two wars with Israel as part of an Iranian effort to aggregate a variety of militant Lebanese Shia groups into a unified organization. *Hezb Allah* acts as a proxy for Iran in the ongoing Iran–Israel proxy conflict. The creation of *Hezb Allah* had contributed in changing urban pattern in Lebanon which led into changing of sociopolitical structure of the country. Other people squatted in housing units in residential building that had been destroyed by the civil war (see Fig. 9.2). These transitions of the physical and sociopolitical structure of the city resulted in the spread of urban informality, changed the land uses within the city, and created what is called the "Misery Belt" around Beirut. Urban expansion occurring in the form of informal areas developed around major cities and towns throughout the civil war. These informal areas are inhabited by Lebanese peoples who were displaced from their areas of origin or by refugees, mainly Palestinian refugees and smaller refugee communities (such as the Kurdish). Refugee camps housing Palestinian refugees in Lebanon, of which 12 remain today, have been witnessing both vertical expansion and horizontal 'spill-over' around some of them, resulting in the development of informal areas known as Adjacent Areas.

The fifth phase is the post-war period (1991–2005), when large urban reconstruction and building activities took place. The main theme of this epoch was that economic growth is at the heart of establishing and maintaining peace. Hence the focus on post-war recovery and large-scale infrastructure projects (highways, electricity etc.) and on regenerating a state-of-the-art central business district (Solidere) as the epitome of post-war economic development. A notable change in poverty indicators was seen between 1995 and 2004 with a decrease in both the households living in extreme poverty (from 7% to 5%) and those living in relative poverty (from 28% to 18%) (UNDP 2009) (Gaspard 2004). Most poor urban pockets are in the vicinities of major cities in the suburbs of Beirut, Tripoli, and Saida, all of which house the highest number of the Lebanese poor. However, the internal upheavals and the unstable political situation prevailing in the country since 2005 have worsened the physical, social, and economic fabrics, contributing to widening the gap between the rich and the poor, between rural and urban areas, between West Beirut and East Beirut, between Lebanese and non-Lebanese, and within given areas as well (Soliman 2004c). The assassination of Rafik Hariri in 2005 was the crossroad of a political division in which *Hezb Allah* dominated the political atmosphere.

The sixth phase is the post-Arab Uprising Phenomenon (2011 to October 2019). Dysfunctional land and housing sectors disproportionally affect the poor. The beginning of peaceful protest in Syria in April 2011 soon morphed into open revolt during which the first 5000 Syrian refugees entered Lebanon (Janmyr 2018). Since then, massive movements of Syrians into Lebanon have taken place resulting in more than 1.5 million refugees currently residing in Lebanese territory. This number of refugees constitutes a third of Lebanon's population (The Government of Lebanon and the UN 2017). The context of the large-scale Syrian influx to Lebanon opened a political debate between the Lebanese government and the United Nations High Commissioner for Refugees (UNHCR). Lebanon insists that it is not a country of asylum and flatly rejects ratification of the 1951 Convention on the Status of Refugees and its 1967 Protocol. The conflict between UNHCR and the Lebanese government revolves around whether the Syrians are refugees as defined in Lebanese law or are considered as displaced people, who will eventually return to their home. There is fear among Lebanese politicians that the Syrians, like the Palestinians in 1948, will remain in Lebanon. In addition to de facto sealing of Lebanon's borders, the Lebanese policy had devastating effects on the well-being of the country's Syrian refugees, leaving many Syrians living 'illegally' in the country and under extremely harsh and marginalized conditions.

The final phase is the Lebanese movement and Beirut's explosion (post October 2019–4 August 2020). People's movements on 17 October 2019 against the political institutions and corruption demanded social justice, a decent life, the reduction of poverty, and housing rights. Since then, the local currency sharply dropped from 1500 Lebanese Lire (LL) to more than 8000 LL against a USD and a sharp increase in the prices of various commodities, including housing, occurred. The Lebanese

Fig. 9.3 Photos of Beirut's explosion that destroyed a third of its buildings. (Source: The Guardian 2020)

blame *Hezb Allah* of this deflation and for responsibility for all the corruption in the country. After more than nine months, Beirut on Wednesday, 4 August, 2020 suffered a man-made earthquake in the form of a great explosion in its harbor when more than 2750 tons of ammonium nitrate, a compound often used to make fertilizer and bombs, suddenly combusted, perhaps because of a fire that had started nearby (see Fig. 9.3). On Saturday, 8 August, clashes between demonstrators and security forces broke out near Lebanon's Parliament at a protest fueled by the vast public anger over the death and destruction caused by a huge explosion in Beirut's port. This explosion left more than a third of Beirut's buildings destroyed and damaged, more than 200 dead, and the blast injured 5000 people and pushed at least 300,000 persons from their homes. These demonstrations demand the resignation of the president, the head of the parliament, the cabinet prime minister, and all persons in the regime who are responsible for the economic decay and corruption. Also, the Lebanese are demanding a new political institution without ethnic or religious prejudice and are seeking to terminate the power of *Hezb Allah* and expel it from the political life.

As *Hezb Allah* has complete power in determining the system of the political institution, economic and urban development, and more, it controls Beirut's airport and harbor and the official and non-official borders of Lebanese territory. Also, because any socioeconomic development projects or schemes must be approved, controlled, and monitored by *Hezb Allah*, Lebanon needs a careful and sensitive transition to sustainable development. With the current economic difficulties that Lebanon is passing through, the question is how the Lebanese government would accommodate 300,000 homeless? This might lead to the creation of additional slums to those already within the city of Beirut. As of now, Lebanon is facing a nightmare for the coming years.

In proportion to both its geographical and population size, Lebanon hosts the highest number of refugees in the world. As a consequence, the situation for Syrian refugees in Lebanon has become legally more complex and piecemeal, with different systems of law and policies being applied on both a global and a local level. In such blurred circumstances, UNHCR classifies Syrian refugees into four categories: unregistered refugees; registered refugees; those who are not registered but are recorded as refugees; and those who are registered refugees but may at some point have renewed their residency in Lebanon under the sponsorship system (for further details, see (Janmyr 2018; The Government of Lebanon and the UN 2017). Until now, Syrian refugees have been leaving the country in continuous political debate among various sects of the society on one hand, and with UNHCR and international organization on the other. At the same time, UNHCR approved an unprecedented increase in its Lebanon budget— from 49 million USD in 2012 to 362 million USD in 2013. Although the recent changes and revolutions taking place in the Arab countries have brought real estate investment and prices down slightly, the market in Lebanon is still guaranteed to make high profit and still involves laissez-faire economic policy.

Figures on the population in Lebanon are based primarily on estimates. In 1997, it is estimated that the total population of the country was 3.111 million people, with an annual growth rate of 2.7%. In 2001, Lebanon's population was estimated at 4.3 million people. In 2005, the Lebanese population reached 4.8 million people, with 32.5% or 1.3 million people living in Beirut and its suburbs alone (The Ministry of Social Affairs 1996). In 2010, according to the World Bank, Lebanon's population was estimated at around 4,223,000 inhabitants and expected to reach 5.41 million people in 2050 (United Nations 2018). Population has been increasing in the last 30 years at an estimated growth rate of 1.5% (1980–2009). The increase in the population and growth rate has been calculated, over the past three decades, as evolving from 2,784,713 in 1980 to 4,223,553 in 2009, which is a 1.5-fold increase in 30-year period. Prior to the Syrian Crisis, Lebanon had a population of 270,000 Palestinian refugees among its total refugees and migrants. Since the onset of the Syrian Crisis in 2011, Lebanon's population has further increased as it has become host to 1.5 million refugees from Syria, around a third of the total residents of Lebanon.

The average population density on all Lebanese territory is 2999 per km^2 (The Ministry of Environment 2003). Beirut, the capital and largest city of Lebanon, is situated on a peninsula that projects slightly westward into the Mediterranean. Beirut is contained by the Mountains of Lebanon that rise to the east, with an area of roughly 67 km^2. Some sites located outside the municipal boundary are commonly associated with the city. Along the coast line in the southern part of the city, are located the largest informal areas *Dahiya Janubiyya1 and Ouzai*.

The physical features of Lebanese territory played a significant role in settlement patterns. At the geographic level, Lebanon constitutes a rich territory combining coastal and mountainous characteristics. Port cities still constitute essential components of the country's demography and territory. Mountains, on the other hand, have historically hosted large numbers of settlements that have, in the past, created a dynamic and prosperous economy. As an agricultural hinterland, small and medium towns have developed across Lebanon as points of social and economic exchange operating not only within the country but also as 'relay' agglomerations linking the coast to the deep Syrian hinterland.

In the last 60 years, rates of urbanization have increased drastically. Urbanized areas went from 221 km^2 in 1963 to 465 km^2 in 1994 and to 741 km^2 in 2005. Urbanization is expected to reach 884 km^2 in 2030 (UN-HABITAT 2011). This growth has mainly developed around large cities, especially Beirut. However, in recent decades, secondary cities have become increasingly urbanized, with their populations mirroring this growth. At the beginning of the 1960s, Lebanon had a population of only 2 million people and 260 km^2 of urbanized areas; in 1998, these areas increased to 600 km^2 and accommodated around 4 million (DAR - IAURIF 2005). In 2030, assuming a ratio of 170 m^2 of urban area per inhabitant, there will be a total urbanized area of 884 km^2 for a population of 5.2 million people, which represents a growth of 284 km^2, or a half of the current urban agglomeration, within 30 years. The increase in secondary residences, an important factor for the economy of villages, would reach about 11% of the housing stock in 2030 (DAR - IAURIF 2005).

Urbanization processes in Lebanon are categorized by four patterns: concentrated in and around main coastal cities; between secondary cities and peri-urban areas; the formation of informal areas on the peripheries of cities; and the spread of refugees' camps. The pattern of urban expansion in and around main cities has mostly taken place along the Mediterranean coast, stretching 200 km from the north to the south, and around major cities where most of the industrial and commercial centers are located. Developments stretching along the coast include large-scale reclamation projects (public and private leisure projects, dozens of marinas for leisure boats and fisheries), residential development projects, and waste-water treatment plants (Dar-IAURIF 2005). While most of these developments occur with limited consideration of the environmental impact, violations of the public maritime domain are also significant.

Urban expansion occurring between secondary cities and towns has not been occurring only in major cities and peri-urban areas around them (such as Beirut, Tripoli, Saida, and Tyr), but also between secondary cities and towns (such as Zahle – Chtoura; Beirut to Bauchreyeh, Sin el Fil, Fanar, Zalka). Cities have been growing both vertically, with the incessant erection of high-rise residential buildings

Fig. 9.4 Photos illustrating urban informality in Lebanese territory: (top) informal areas in Mountain of Lebanon; (middle right and left) informal areas in Kafrshouba Village in the liberated part of south Lebanon; and (bottom) El Ouzai area (sea resort) south of Beirut. (Source: The author)

and towers, and horizontally, encompassing surrounding peri-urban areas and domains.

Urban expansion occurring in the form of informal areas developed around major cities since the civil war (see Fig. 9.4). These informal areas have converted land use patterns from leisure areas and/or agricultural land to illegal housing units and are inhabited by Lebanese populations who had been displaced from their villages in the liberated part of south Lebanon, or by refugees, mainly Palestinian refugees and smaller refugee communities (such as the Kurdish). The private developers have a prodigious role in illegal land subdivisions, and they cooperated in one way or another to consolidate urban informality.

The most recent development pattern is the spread of refugee camps. These camps harken back to WWI when the former Armenian camp of Medawar in Quarantine was established. After the 1948 war, the Palestinian camps of Shatila and Sabra were erected according to the agreement between the government and UNHCR, and further camps were established after the defeat in the 1967 war. These are Tell El-Zaatar, Jisr El-Basha, and Dbayeh, stretching towards the western part of Beirut to include the camps of Shatila, Mar Elias, and Burj El-Barajneh. Recently, Syrian refugees' camps are arbitrarily spreading in various urban and rural areas of Lebanese territory. This type of expansion has deteriorated not only the urban fabric of cities but has transformed the social structure, in religions and sects, of the territory.

However, the pattern of urbanization in Lebanese territory is, to a certain extent, related to various decisions for moving from rural to urban areas, and it is subjected to four variables. A central variable is that individuals move (with varying degrees of ease) in response to economic incentives and follow economic opportunities. This is predominantly exemplified by rural–urban migration since economic opportunities are scarce in Lebanese rural areas. If location incentives are distorted, so presumably is the growth process. Another variable is that people move to urban areas for social, educational, and political reasons. Migration to urban areas can provide an escape from family and cultural constraints, such as restricted land access or a low level of female independence. It is argued (Castells 2010) that the new global economy and the emerging informational society have indeed spawned a new spatial form that develops in a variety of social and geographical contexts, i.e., megacities—Cairo, Amman, and Beirut among them. This move is accompanied by the acceleration of changing economic situations, e.g., the expansion of East Beirut and the Mount of Lebanon, which squeezes the informal areas in between.

A third variable is that migration to an urban area may also occur because of an expected increase in social status and standing—the perception that the "high life" can be found among the "bright lights" myth—migrants have been lured to the city by exaggerated tales of high income and technologically advanced living, especially by returning migrants who "wished to convey to others a positive and prestigious image of themselves and their experiences."

A fourth variable is that the increase in Palestinian military activity during the early 1970s split the people of Lebanon between those who supported their cause (mainly leftists) and those who opposed their presence in the country (mainly Christians). Thus, between 1970 and 1975, Beirut was boiling with ideas, ideologies, strikes, demonstrations, attacks, violence, and a sense of foreboding. Due to these transitions in Lebanon life, many people have migrated from north to south, west to east, and vice-versa, transitions that changed the demographic structure and the spatial pattern of the urban agglomeration.

A final variable is that the wars and ethnic conflicts that occurred in Lebanon may also have led to increased rural–urban migration. Lebanese society by then was composed of denominational sects of the Muslim and Christian religions. There were frequent tensions between the religions, as well as between the sects (Katkhouda 1998). Quickly, new place names, unknown in the prewar days, began to be used and were adopted by the general public and the media. Therefore, East Beirut was considered predominantly Christian, a pro-West zone, while West Beirut was considered predominately Muslim, pro-Arab, and friendly to the Palestinians. Aside from the impact of war on agricultural income through effects on transport and marketing, war may also push people out of rural areas, especially the southern part of Lebanon, or even from one part to another in urban areas, simply for safety reasons (Soliman 2004c). This was exemplified by the population shifts that occurred from East Beirut to West Beirut and the rapid informal development of *Dahiya Janubiyya*[2].

Therefore, the population shifts and the pattern of urbanization processes in Lebanese territory are linked with various circumstances and have depended on various incentives. Ethnic conflicts, as happened during the civil war (1975–1990), in particular, increase the danger of living in an area dominated by a persecuted ethnic group because the potential for ethnic cleansing is high in informal areas. Urban areas generally have a higher level of ethnic diversity and thus may be safe havens for persecuted groups. In Lebanon, 44% of the total population lives in Greater Beirut (United Nations 2017). In 1975, nearly half of the country's population of 3 million lived in the city; by 1992, its population had dropped to an estimated 300,000 (Dumper and Stanley 2007). Rafik Hariri was prime minister and the architect of the *Taif Accord* who rebuilt Beirut, especially downtown, through his company *Solidere*. In a decade, Hariri rebuilt downtown Beirut, bringing back Lebanese Muslims and Christians to meet in the heart of Beirut, bringing back Arab and international investors, and bringing back Arab and international tourists. Hariri was assassinated on 14 February 2005, and he will always be remembered as a prominent pillar and architect of modern postwar Beirut.

Throughout these turbulences, Lebanon struggled to maintain stability but succeeded in avoiding collapse. The notoriously unstable country of the Levant paradoxically created a certain center of gravity and steadiness. Although not unaffected by the domino effect of the Arab Uprisings, neither the series of revolutions in the region nor the armed conflict next door in Syria could bring about the total breakdown of Lebanon (Wählisch and Felsch 2017). On the other hand, the Syrian crises have changed the physical and demographic structure of Lebanese territory in which the balance between various sects was devastated and the urban

---

[2]Dahiya Janubiyya is divided by the north–south axis of the airport road which creates two different parts. The eastern part combines old dense villages (*Burj Al Barajneh, Mrayjeh, Harit Hrayk, Ghobeyri, Shiyyah*) and peripheral illegal sectors (*Amrusseyeh, Hay Al Sillom*), while the western part contains major illegal sectors (*Jonah, Uzai*) and legal low-density urbanization (*Bir-Hassan, Ramlit Al-Baydah*), together with relatively large non-urbanized areas. Then there is *Uzai*, a loner, and the largest illegal settlement that has been created along the axis linking Beirut and Saida. This area used to be a sea resort for Greater Beirut before the civil war. On the other hand, the southern suburb is not a homogenous space. It has been divided into several territories managed by different players. *Hezb Allah* is hence not its sole constituency or authority. A good number of residents are neither close to *Hezb Allah* nor to the Amal movement, the other major political party active in the area. The southern suburb is inhabited by one-third of the population of Greater Beirut, almost 0.5 million, and occupies an area similar to that of municipal Beirut (16 and 17.6 km^2, respectively).

fabric deteriorated. Also, Lebanon's problematic relations with Syria, and *Hezb Allah's* confrontation with Israel are two dimensions in the crisis inaugurated in August 2004, as well as the repercussions for Lebanon up to mid-2008.

9.3 Urban Informality Transitions

Lebanese territory has been the object of several attempts to regulate arbitrary urban growth. It was only after 1958, following weeks of civil conflict (Gendzier 1999), that most Lebanese public planning agencies were created, and the public sector began to intervene in the direct regulation of urban and social spaces (Tabet 2001). The 1958 conflict had brought to power President Fuid Shehab, a former army commander. Unlike his predecessor, Shehab believed in the necessity of establishing modern institutions and adopting national development plans. In the few years that followed his take-over, several public agencies were created to regulate urban spaces, such as the Directorate General of Urbanism and the Higher Council for Urbanism (Fawaz 2004).

Furthermore, urban and building regulations were adopted in 1961 and became mandatory in all areas of the country; zoning and planning regulations were gradually developed for cities throughout Lebanon. An elaborate network of highways was designed in 1964 for Beirut and its vicinity, and the first five-year national plan was approved (Ghorayeb 1998, Fawaz 2004). Shehab routinely appealed to international consultants, such as IRFED, to develop a comprehensive assessment of socioeconomic conditions in the country to craft a national master plan (IRFED 1963); Constantinos Doxiadis was tasked with developing a comprehensive analysis of the housing sector in 1959 (Rowe and Sarkis 1998), as was Michel Ecochard, who designed two successive master plans for Beirut and its suburbs in 1943 and 1964 (Ghorayeb 1998). One of the outputs of this plan was the first public housing in Saida, which was three stories in height with two housing units per floor (Spectrum 1997). In 1970, there were 485,000 homes in Lebanon, approximately 10% of which were summer second homes, mainly located in Mountain of Lebanon. This number adequately met the demand (Gemayel 1992). However, this process of institutional building slowed as subsequent regimes lost control of the city, particularly by the 1970s, when militia groups challenged the state even before its semi-breakdown during the civil war (O'Ballance 1998).

To free the closed rented housing units, the government issued a new law towards liberalization intended to dramatically escalate the drift towards a seemingly liberalized housing market in January 2013 (Saksouk-Sasso and Bekdache 2015). It provided a rent-to-buy clause as the main condition to end rent control, while asserting a priority to enable tenants to stay in their neighborhoods. In the new law, the compensation fee was dropped in favor of a gradual increase of rent values to reach market prices, which were also linked to property prices and hence unaffordable for the majority of city dwellers. This caused forms of violence between landlords and tenants and produced conflict and risked escalating the situation.

Despite all these attempts to regulate urban growth during the last few decades, urban informality has accelerated for those seeking affordable housing not only for the Lebanese, but also for the massive numbers of refugees from the surrounding countries. Urban informality exists in Lebanese territory and is characterized by internationally enforced urbanization and the appearance of an informal economy due to the failure of the state to satisfy the basic needs of the disadvantage groups of the society. Urban informality seems to be a land-use problem, and it is thus often managed through illegal attempts of the lower and middle strata of the society to restore their "space" or instill "order" to the urban scene. It could be said that urban informality is segregated space organized in response to certain circumstances and which does not observe the prevailing law or system. It arises and persists due to economic necessity and material deprivation and physical and social insecurity, along with ethnic and class prejudice. If so, how does social exclusion create, or at least encourage, the spread of the informal sector within Lebanon?

The transitions of urban informality in Lebanon are due to various circumstances that have confronted the country in the last few decades, and they are linked with the transformation of socioeconomic, military, and political transitions. Ethnic and class prejudice plays a great role in the informality phenomenon in Lebanese urban centers where several key attributes have emerged. To begin, international migration supported by foreign and Arabic powers vying for control of the Mediterranean have used Lebanon as a battlefield for their economic and political conflicts. During the last few decades, a large number of refugees (both Palestinian and non-Palestinian) moved into and settled in the periphery of the Lebanese urban centers, especially in Beirut and Tripoli, which created informal residential areas. Also, extensive rural-urban migration fueled by two Israeli invasions (1978–1982) led to the transformation of most of the open spaces and sea resorts of the southern suburbs of Beirut into large informal areas.

Second, in the early 1970s, the injection of petrodollars into the Lebanese economy accelerated the economic growth within the country that encouraged huge waves of immigrants (insiders and outsiders) to move into the major urban centers, reinforcing inequalities and marginalizing the lowest income groups, both Lebanese and non-Lebanese.

Third, urban informality is created due to internal and external military conflicts that have confronted the country in the last few decades. Also, external military conflicts (the two Israeli invasions) have led to displacement of people from the south of Lebanon to more secure places within the major Lebanese urban centers. This trend of displacement has resulted in population shifts by which Muslims moved from Eastern Beirut to Western Beirut, and Christians moved in the opposite direction. This reorganization of the population was in line with ethnicity and dominated by militarization, religious ideologies, and the maintenance of political structures that govern through patronage, division, and economic oppression.

Fourth, postwar, Lebanon reached the *Taiif* agreement that promoted a new status quo in the territory and deliberately ignored the roots of the civil war, setting about restoring Beirut's old financial role at the expense of the impoverished and marginalized rural and suburban populations. Few political efforts were invested in the return of internally displaced communities living on the outskirts of the Lebanese cities (Halabi 2004).

Fifth, informality is now manifest in new forms and new geographies, both at the rural–urban interface and in terms of development that may serve as a main path for various political wings within the city. The variable geometry of the new world economy and the intensification of the migratory phenomenon, both rural–urban and international, have led to a new population category divided between the rural, urban, and metropolitan settings: a drifting population that moves with the economic tides and according to the permissiveness of institutions, seeking out its survival with variable temporal and spatial features according to circumstances in the Lebanese territories.

Finally, the foundation of urban informality is created for the purpose of defending and controlling ethnic diversity within the society by which informal growth has taken place according to social exclusion and religion. This is the true source of social tension: growing ethnic diversity in Beirut that has not absorbed that diversity and continues to speak of 'immigrants', when, increasingly, it is a matter of nationals of non-European ethnic origin. Urban informality has allowed informal processes to spread not only among the lowest strata of the society, but among what were once seen as the formal lower and middle classes, including the various ethnic groups within the city of Beirut, accentuating the pattern of urban ethnic segregation. Although Beirut accounts for only 40% of the population, it has 42% of the ethnic minority population, which is concentrated particularly in certain districts (Lebanese National Report 1996). In the late 1970s, as a successive wave of displaced Shiite refugees, backed by *Hezb Allah,* fled the chronically embattled villages in southern Lebanon, Dahiya Janubiyya (southern suburb) quickly acquired the label of Beirut's "Misery Belt" or "Belt of Poverty": a ghetto seething with feelings of neglect and abandonment and, hence, accessible to political dissent, mobilization, and violence. Also, Lebanon went deeply into debt to finance reconstruction, and with a debt-to-GDP ratio of 149%, it is today the world's third-most-indebted country (Tierney 2016).

However, there is a significant risk of further deterioration and more violence given the uncertainty inherent in such a tense atmosphere that led the recently elected president of Lebanon, Michel Suleiman, to lament that the "disagreements between the Lebanese have reached the suicide level," which could only serve "the interests of the Israeli enemy. Today, Lebanon's politicians sit in a paralyzed parliament—which has failed to elect a president in more than two years (or pass a budget in over a decade)—where they line their pockets through a system of sectarian patronage (Tierney 2016).

As mapped in Fig. 9.5, the metropolitan of Beirut contains 28 informal residential areas with a total population of 515,000 (including the number of foreign refugees), which accounts for 25% of the city's population. Since 1980, Lebanon has lost to unregulated urbanization some 7% of its cultivated land and 15% of its irrigated land. Most of this development has occurred in the suburbs of Beirut and other

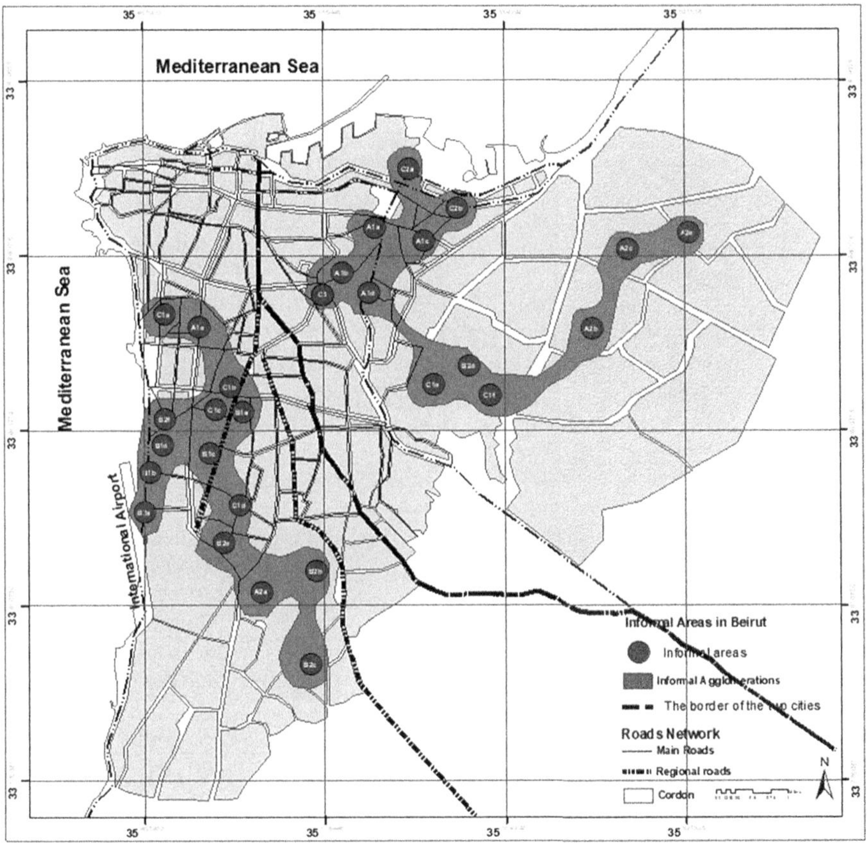

Fig. 9.5 Informal areas of Beirut City. (Source: The author (Soliman 2008))

coastal cities (Masri 1999; Shwayri 2008). As summarized in Table 9.2, a typology of three main informal housing types in Beirut has emerged (Soliman 2004a, b): semi-informal (bottom-up approach), squatting (technocratic approach), and hybrid (hybridization) (as explained in Chap. 4).

Illegal land subdivision (semi-informal areas—bottom-up approach) (A): These areas housed the various waves of rural-to-urban migrants and displacements arriving in Beirut and its suburbs due to the country's industrialization and urbanization processes, coming especially from South Lebanon and the *Beka'a Valley,* where poverty and insecurity (in the case of the South) gradually encouraged a migratory movement. It develops on agricultural land to which the owner has legal tenure and a formal occupation permit. Such areas develop in essentially rural locations on the urban fringe, and they tend to be interspersed with, surrounded by or adjacent to, undeveloped or agricultural sites. They often develop in advance of the principal

Table 9.2 Informal housing typologies in Beirut

Main Typologies	A- Illegal Land Subdivision (Semi-informal areas) Bottom-up approach		B- Illegally Occupied Land (Squatting areas) Technocratic approach		B- Foreign Refugees Camps (Hybrid areas) Hybridization	
Sub typology	**Sub typology**	Popul.	**Sub typology**	Popul.	**Sub typology**	Popul.
	A1- Inside city boundary	23,000	**B1- Inside city limits**	55,000	**C1- Palestinians**	200,000
	A1a- Karam Elziatoon	4500	B1a- Harsh Tabet	10,000	C1a- Mar Elias	
	A1b- Hai Elsyriaan	6000	B1b- Hai Elzahria	5000	C1b- Shatilla	
	A1c- El Nabhia	5000	B1c- Harsh Elkateel	20,000	C1c- Sabra	
	A1d- Harsh Rahal	3000	B1d- Ber Hassan	20,000	C1d- Bourj El Barjneh	
	A1e- Wata Almousautbeh	4500			C1e- Jisr El Pasha	
					C1f -Tal Elzatar	
					C1h- Hay El-Gharby	
					C1i- Said Ghawash	
					C j- Daouq	
	A2-Outside city boundary	12,6000	**B2-Outside city boundary**	96,000	C2- M. Leb. (Armenians)	12,000
	A2a- Hayy el Sellom	12,0000000	B2a- Ouzai	60,000	C2a- Karantina	
	A2b- El zaterihia	2500	B2b- El likely	6000	C2b- Sanjak	
	A2c- El rowaisat	2000	B2c- El amrosahia	5000		
	A2e- Hai El hean	1500	B2d- Dakwanha	5000		
			B2e- El Ramal	5000		
			B2f- El Jounhia	15000	C3-Syriac	3000
	Total Population	14,90000		151000		21,5000

Source: Adopted from (Soliman 2008) and updated in 2019. The typology is the author's typology, while the other data gathered from different sources: mainly (Fawyaz and Bian 2003; Masri 1999; Fawaz 2004), and a field survey carried out by fourth-year students at the faculty of Architectural Engineering, BAU, under supervision of the author, in the period of November–December 2003. It is important to note that the city boundary is defined in this study within a radius of 5 km from the city center (*Sahat El Njemehia*). Details of everyday life in Palestinian camps are illustrated in (Halabi 2004; Martin 2015)

lines of urban growth and are most noticeable around the most rapidly growing axial lines of urban roads.

Illegal occupied land (squatting areas—technocratic approach) (B): These areas grew in several parts of the city, where refugees displaced by the events of the Lebanese civil war (1975–1990) occupied either buildings or entire neighborhoods, that has been abandoned by their owners or tenants (for reasons of security) or occupied large plots of land and transformed them into squatter settlements. The squatter areas usually exist outside the formal legal economic structure of the city of Beirut, such as the Ouzai area, while scattered squatters still exist within the city boundaries in areas such as Ber Hassan and Harsh Tabet.

Foreign refugees' camps (hybrid areas—hybridization) (C): Refugees camps are historically the oldest informal areas of Beirut. Camps were organized for Armenian (1920s), Syriac (1920s), and Palestinian (1948) refugees with the help of international organizations (UNRWA), while Kurds occupied abandoned tenements in the city center. The most famous camps are still the Palestinian camps (e.g., Shatila camp), where the most economic deprivation and misery have existed. Demographic changes are noted in all informal areas, especially since the 1990s, with the arrival of a growing number of non-Lebanese workers (notably Syrian male workers), who live in these areas of the city.

The critical issues are that what are the transactions and struggles behind the creation of urban informality? Does the concept of urban informality differ in Lebanese cities than other cities in the Global South? Does the sociopolitical transformation affect the progress of urban informality? What, if anything, does historiography have to do with the formulation of urban informality? Do politicians have a certain role in supporting the spread of urban informality?

9.4 Political Economic Transitions

The modern state of Lebanon is a unique amalgam of 18 officially recognized religious sects, the product of over a millennium of immigration by Christians and heterodox Muslims from the surrounding Sunni Islamic world and deliberate colonial border demarcation following the collapse of the Ottoman Empire. Political offices in Lebanon have been distributed among its sectarian communities by fixed quotas. Under the terms of the 1943 National Pact, the presidency is reserved for Maronite Christians, the office of prime minister for Sunni Muslims, and the office of parliament speaker for *Shi'a* Muslims (Traboulsi 2007). The following part describes the influence of state involvement in the formulation of urban informality during various periods of Beirut's modern history.

Most informal housing areas in Beirut were established for political, economic, ethnic, and religious reasons. In 1899, after a cholera crisis, a new Quarantine area near *Nahr* (river) in Beirut was constructed. At the beginning of WWI, the site received further refugees and became a main human settlement in Beirut that accommodated 10,500 people of the Armenian community. The great riots of

September 1903 between Greek Orthodox and Sunni groups resulted in splitting the society in Beirut, as an estimated 15,000 Beiruti Christians fled to the mountains in fear of Muslim reprisals (Traboulsi 2007). Since the end of WWII, Beirut has witnessed additional population shifts due to sociopolitical and economic crises[3].

The French molded the ruling class and structured the state by grooming the Maronites for rule and influencing the 1926 Constitution, which, with some modification, remains the basis of the modern state today. Under the French Mandate, serious efforts were initiated by Ecochard to prepare master plans for Beirut. Functionalism and rationalism defined a form of urban zoning in which the segregation of functions became the key concept (Ghorayeb 1998). Perhaps the proposed plan reflected on housing development and social segregation, where the middle-income classes' residential areas were sited in locations far from the elites, so it was the beginning of social exclusion in the city.

After the declaration of Israel as a state in 1948, huge numbers of Palestinians arrived in Beirut. The United Nations Relief and Works Agency (UNRWA) rented some buildings and scattered *Waqf* land plots to shelter these refugees, hence, the beginning of the creation, legally, of foreign camps on the fringe of Beirut City had begun. In 1952, there were six Palestinians camps, one located *at Mar Elias*. In a suburb further south, another two camps were created, Shatellia and Bourg El Bragnahia. Debhia, Geser El Bashia, and Tal El Zahter were three camps developed to the northeast of Beirut. Over time, these camps received more Palestinians and Lebanese who emigrated from surrounding rural areas. Thus, these camps informally developed and expanded.

Arbitrary urbanization soon spread throughout Beirut, and, with the advent of 1952, a sequence plan was set up for Beirut to control the city's administrative boundaries. In the early 1960s, once again, Ecochard set up a plan for Greater Beirut (Nasr and Volait 2003). Despite all these attempts, the plan did not touch the informal areas surrounding the city of Beirut for religious and political reasons, which can be divided into two taxonomies. The first comprises informal housing areas close to industrial zones. These informal areas accommodated people who emigrated from rural areas seeking job opportunities. This type characterized expansion of the existing informal areas located in the eastern sector of the city and were settled close to industrial zones. The second taxonomy involved more Palestinians displaced to Lebanon after defeat in the 1967 war. With the help of UNRWA, the refugees from Palestine settled mainly in scattered camps in northeastern and southern areas of Beirut. Over time, these camps expanded to accommodate people from Lebanese rural areas and from outside Lebanon (*Tall El Zahter* camp became the largest informal area). It was estimated that a total of 600,000 people who were displaced by wars or catastrophes came to Beirut.

[3]These areas formed the main sites for establishing the Dahiya Sharqyyia agglomeration which had received people, mostly Christians, who had escaped from Western Beirut during the civil war and during the two Israeli invasions.

In 1972, Israel marched into South Lebanon, an invasion including 150 southern villages. Nearly 250,000 southern citizens were displaced from homes and lands (Beydoun 1991). A large number of those people moved to the outskirts of Beirut (*Dahiya Janubiyya*). Irregular residential areas swelled up around Greater Beirut, and by 1975 informal communities were growing as a peculiar social case in the cultural texture of Beirut. Such irregular human settlements fell into one of two categories; the first, comprising various social groups of Lebanese and non-Lebanese people, was inside Beirut's administrative borders; the second, reserved for Palestinians some of who were joined by Lebanese families, was in refugees' camps in Beirut's suburbs.

The continuing civil war of 1975 led to a radical change in the demographic fabric of Beirut's overall urban structure. The resulting population shifts and migration due to conflict changed the distribution of the population in and around the city radically. *Dahiya Janubiyya* flourished in southern Beirut as an illegal, anarchical space inhabited by poor *Shiites* led and/or manipulated by *Hezb Allah* [4]. Due to the multiplication of waves of internal displacement and by ethnic conflicts, the demographic structure of the city of Beirut has changed according to the power of who controls what, and who has the power as the ethnic majority or plurality, and their degree of influence on the civil war. The Shiite-Maronite balance that first favored the Maronites shifted gradually towards a Shiite hegemony strengthened by the population displacements from the northeastern suburbs of Beirut, from the *Biqaa,* and from South Lebanon. Also, during the Israeli occupation of the southern part of Lebanon in 1978, a massive wave of people moved towards Beirut, most of them settling in *Dahiya Janubiyya*. The two Shiite parties (*Hezb Allah* and *Amal*) had great influence in shaping *Dahiya Janubiyya*. The Palestinian camps were basically residential tent-clusters set up on land provided by the state or on private property rented by UNRWA.

After the civil war, a dramatic change in the demographic urban fabric of Beirut occurred. Also, the division of the city (eastern and western sectors) created new urban centers to meet the needs and demands of population and market mechanisms. Urban informality formulated an outer informal wall for the city of Beirut consisting of two informal agglomerations; *Dahiya Janubiyya* and *Dahiya Sharqyyia,* separated by an immature pocket which represented the Green Line (demarcation lines) between western and eastern Beirut (see Fig. 9.5). These two agglomerations constituted a defensive human wall for both western and eastern Beirut, and they also reflected the ethnic diversity involving Muslims and Christians.

[4] *Amal* and *Hezb Allah* are two Shiite Muslim political movements. The former emerged in the mid-1970s with the outbreak of the civil war, and the latter emerged in 1982 with the Israeli invasion of Lebanon. In addition to their political and military roles, they provide social, educational, and medical services for Shiites and other Muslim groups inhabiting the urban areas that fall within their spheres of domination. A third party, called the Progressive Socialist Party and headed by Walid Jumblat, as a zaim of Druzes, played a major role in urban informality in Beirut and in the Mountains of Lebanon and was later appointed as the minister of the dislodged.

After 16 years of war, the *Taiif* Accord and the political reforms were agreed upon in 1990 by the various conflicting parties in Lebanon, and this led to the reestablishment of a centralized political authority. The *Taiif* agreement brought many more stakeholders and categories of actors, old and new, into the political game of post-war Lebanese politics. They all had demands and the ability to achieve them, one way or another. In other words, "consensus" had to apply not only in politics, but also on the ground, in the administration, in the institutions, in the diplomatic corps, in privatization, in awarding contracts, and in all other profit-making enterprises (Adwan 2004). These processes resulted in a manifold increase in corruption involving construction/redevelopment, as well as to the spread of urban informality in Lebanon.

The former president Amine Gemayel in his book *Rebuilding Lebanon* (Gemayel 1992) stated the main conditions for recovering from the civil war. His vision was to establish development planning to ensure equal growth in all regions, particularly those with limited natural resources and high birth rates. The main elements of the plan were: first, to bring up to date the master plan for the Beirut area (1983–1986); second, transfer certain economic activities to the regional capitals to relieve congestion in Beirut; third, carry out the integrated development project of pilot villages planned in 1987 but never implemented because of the worsening political situation; fourth, a reconstruction and development plan for Lebanon that defined medium and long-term objectives and the means to reach them (whether by incentive or coercion) so as not to forge a controlled economy. In certain areas, the agricultural land had been abandoned, water resources were poorly used, the electricity supply was faulty, and economic activity was on the decline. After the implementation of the *Solidere* project, the architectural heritage buildings and landmarks of Beirut were overshadowed, even replaced, by high-rise buildings and towers. In the downtown area, the construction of new gated residential communities was taking place, including three comprised of 185 apartments in *Wadi Abu Jmil* alone.

Studies (Lebanese National Report 1996) indicate that there were 45,000 households (of the total number of those dislodged) occupying residences on illegal basis and 120,000 households living under grievous conditions in places unfit for habitation. The government made practical attempts to intervene in the domain of housing through the project of Beirut's southern suburb (*Elissar*), by imposing plans for Beirut's southern inlet and the periphery of the Beirut International Airport. The plan for housing and restoration of dislodged groups, as it appears in Plan 2000 for Reconstruction and Development, met several obstacles. *Dahiya Janubiyya* was considered a main barrier for Beirut's southern part's development. Figure 9.5 illustrates two informal urban agglomerations in Beirut. Postwar reconstruction came without political reconciliation.

The current situation in Lebanon is that the former warlords are today Lebanon's politicians, ministers, and heads of government. Garbage collection has become an urban challenge in Beirut. The Lebanese people participated in the largest protests since the assassination of Hariri in 2005 and the withdrawal of Syrian occupying troops when they demanded not simply a solution to the trash crisis but an end to corruption disguised as sectarianism. Recently, the Beirut explosion of 4 August

2020 exposed a conflict between the society and political class, demanding freedom the country from corruption, ethnicity, sects, and religious autocracy. *All means All,* a declaration by all Lebanese's demonstrations, sought to disarm *Hezb Allah.* However, urban informality is the product of politicians' conflicts, sectarianism structures, civil war, conflict with Israel, international refugees (Palestinian and Syrian), and the international agents' mediation. It is estimated that the ratio of nationals to refugees in Lebanon to about 1: 4, the highest in the world (Fawaz 2017). Thus, this phenomenon correlates with internal and external forces: the former is responsible for sectarianism and economic problems, and the latter is responsible for creating or at least associating, legally, the spread of urban informality. Lebanese territories became battlefields for who can have the upper hand in controlling urban informality and to what extent the local politicians have the power to tackle it.

Relying on the ASUST perspective, as explained in Chap. 1, as multi and continuous processes conceives of transition as an interweaving of six trends. These are policy strategy credibility, consistency of dimensions, policy processes (coherence of processes), systematic change (four perspectives), systematic directions of transitions, and instrument-mix comprehensiveness. The first three trends are currently handled by the international donors headed by the state of France asked for funds for helping Beirut to recover after the Beirut's explosion of August 2020. Donor countries on 9 August asked the Lebanese's state to first help themselves to help them and demanded a clear roadmap for a new political institution to develop the country. On the other hand, the fourth, fifth, and sixth trends are controlled by civil society, NGOs, CBOs, and professionals as subsequently explained. Grassroots participation provides the knowledge and information to effectively use tools to operate on the ground. The other three trends of the ASUST are linked and interrelated with the three pillars and the mechanisms of urban informality.

First, systematic change comprises four perspectives: a complex system of socio-spatial, socioeconomic, socioecological, and sociotechnical perspective. The latter reflects transitions theory in social-science postulates that successful systems are comprised of three main dimensions: socioeconomic, a socioecological perspective, and governance perspectives. These dimensions are in continuous processes and are related and linked to the concept of transitions level, level of intervention, and the role of the three pillars. This inherent complexity requires thinking of urban informality of the Lebanese cities as socioeconomic, the role of three pillars, and political conflicts that must deal with continuous change. Urban informality as complex adaptive systems (Rotmans 2006) is in a continuous process of socioeconomic and socio-spatial transformations and has the potential to be sustainably integrated into the urban context, if a proper policy is applied.

At systematic directions of transitions reflects two dimensions: stakeholders' trends and urban regimes' tasks. It echoes tensions and pressure of urban informality transitions on state and political economy. However, to alleviate these tensions and pressures, innovations in institutional reform and people innovations are essential to achieve transitions. In the case of Beirut, after the explosion of August 2020, citizens, CBOs, NGOs and professionals, come to realize and share the co-created

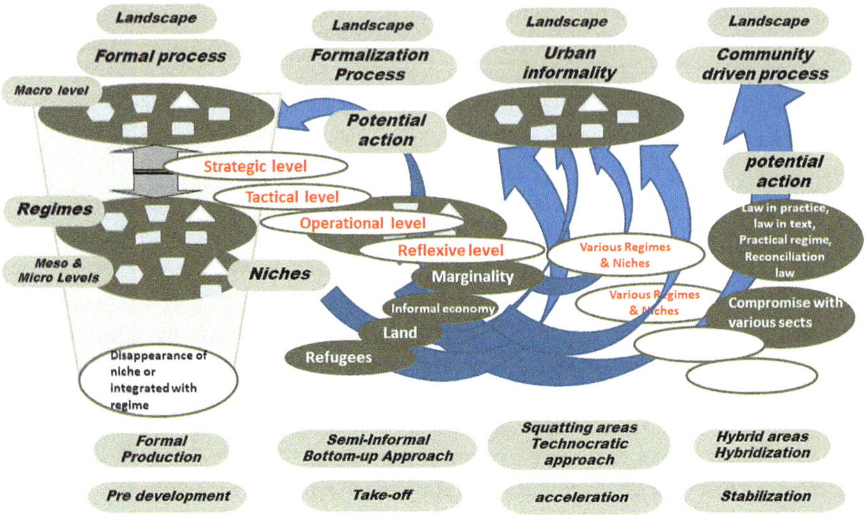

Fig. 9.6 Diagram of the process of management sustainability transitions in urban informality in Lebanon. (Source: Adapted from Koehler et al. 2017; Grin et al. 2010; Geels and Schot 2007; Frantzeskaki et al. 2017a)

socioeconomic transition agenda and are enabling changes in a sustainable way. The processes of transitions are persisting over time require a fundamental change, or spatiotemporal, in handling the development process at three levels. Also, learning by doing is a vital element to avoid misinterpretation of urban transitions.

At comprehensive instrumental mix tendency reflects the transition management processes and housing system economy. It replicates the mechanisms of urban informality through which the housing system links the political economy and the society who are in need of housing. Urban informality transitions, as a sociotechnical landscape, can occur through complex interactions of circular economy of the housing system in which four modes are crucial. These modes are production, coproduction/reproduction, consumption, and distribution of goods and services, interrelated with political economy, state, and society. By enhancing a networked society, as an active urban niche, would stimulate through four issues; increased knowledge, the great acceleration in resource use, rescaling, but also by changing the power of state and grassroots; and by transforming the drivers of change at various socioeconomic-spatial levels.

Figure 9.6 illustrates the process of sustainable transitions in urban informality in Lebanon. The formulation of urban informality in Lebanon is affected formally by politicians and international donors. Indeed, applying the MLP and the MPs to the Lebanese status quo, at the regime level, the main actors of the appearance of urban informality are local and international politicians. However, the phenomenon of urban informality in Lebanon does not follow the mechanisms of the Global South, rather it is established, legally, through top-down processes in which local and international donors play the primary role in the process. It seems that the sociotechnical perspective in relation to urban informality is managed and accelerated on the regime level not from the niche level. In this case, the niche level is integrated and constituted as a part, not isolated, of/from the regime level. Legitimization and the existence of urban informality have political moral support from local and international politicians. Therefore, it appears that the niche level has no impact, or at least does not exist, on the spread of urban informality, rather it is a part of, and interacts, with the regime level. These are brought together in so-called transition arenas or experiments (Frantzeskaki et al. 2018b).

On the landscape level, the development of urban informality transitions has acquired new forces to legitimize their existence in ways different than the procedures or mechanisms applied in the Global South. It could be said that the most influential forces are the role of the international donners who have the upper hand in improving the situation or to eliminating the spread of, or at least to provide the necessary funds to upgrade, urban informality. Therefore, the regime level constitutes the cornerstone to determine, legally, the location, the size, and types of urban informality in Lebanon. However, both parties, national and international, have the main responsibility of the appearance of urban informality in Lebanon. In a later stage, low-income immigrants and displaced Lebanese have played the second role in the urban expansion and transitions of urban informality. On the other hand, transitions management is mainly in the hands of politicians who determine the management process and which areas should be managed. Strategic, tactical, operational, and reflexive levels, as means for transitions management, are handled by the politicians and people who have the power or the willingness to influence urban informality transitions in a sustainable way (see Chaps. 2, 4, 8)), while the urban poor have no say.

However, a lack of genuine opportunities for the urban poor to participate in housing production has led to further feelings of dependence and frustration, negatively affecting people's sense of belonging in the city and the neighborhood. Pressure exerted on people to participate conflicted with their imperative to earn a living, and public servants' expectations of grateful and complicit participation resulted in feelings of frustration. Further, participants' perceptions of nearby neighborhoods with better services, acquired without the necessity to participate, made them feel forgotten. Together, these results illustrate the complexities of citizen participation and social networks (Fawaz 2008) in marginalized neighborhoods. The potential causes for conflict and misunderstanding highlighted in this chapter have relevance for planning strategies and participation policies that seek to include previously marginalized citizens into the social and physical fabric of the city. The demonstrations after the Beirut explosion, in August 2020, against the political bodies showed that there is an urgent need for the involvement of the citizens in development processes to end corruption, as well as, to draw a clear road map for the future. NGOs and social networks proved their willingness to participating in the development process, and they have the potential for transitions, as happened through collective work in cleaning up the areas surrounding the Beirut explosion.

In this sense, urban informality changes can occur through complex interactions, making total control of the Lebanese cities a major challenge, especially through controlled land development as the main component for housing production. Some cities in Lebanon have 50% of their population living in spontaneous areas (Soliman 2012), so the question is how to govern or guide such rapid informal urbanization in a sustainable way. The integration of existing urban informality, including the existing refugees' camps, and the creation of affordable housing within the urban context build on the experience gained in supporting interlinked elements of sustainable transitions that based upon the following cornerstones.

First, the development of city-wide visions that go beyond each project is embedded in the city-regional context. These transitions occurred according to three variables: co-evolution multilevel; multi-phases; and co-design and learning changes in sociotechnical systems. These variables are interacted between complex systems of a sociotechnical niche (e.g., community-driven process, housing production, capital forces, time period that witness transitions, etc.), a sociotechnical regime (e.g., institutions' management, political transitions, planning regulation, land delivery system, etc.), and sociotechnical landscape (e.g., population growth, population mobility, ethnic diversity, social groups, 'radical' change in terms of scope of change traditions, norms etc.).

Second, the transitions process occurs on three levels: macro, meso and micro levels, which constitute a part of, or is interlinked with, the actors involved in transitions process. The micro level echoes the development of locations with basic urban services to provide affordable land plots for housing the new generations. The meso level represents the intervention of government and international donors, market mechanisms, and the role of NGOs in facilitating the integration of urban informality to urban context and enhancing the land-delivery system for low- and middle-income groups. The macro level reflects the output of reaction, action, and interaction between various actors, market forces, citizens, and various variables affecting transitions through which landscapes grow. Third, an integrated and cross-sectoral approach (horizontal and vertical coordination) between the three levels is developed. Various models have come to the fore in this field that aim to explain how transitions unfold and how to govern them (Koehler et al. 2017), and these challenges require system innovation, i.e., deep-structural changes in the sociotechnical configurations underlying the respective sectors (Markard et al. 2012; Van den Bosch 2010). Fourth, new instruments of urban governance, administration, and management, including increased local and international responsibilities and strong local and regional partnerships, are needed. Concentration of resources and funding of selected target areas are encouraged by national and international donors, while bearing in mind that capitalizing on knowledge, exchanging experience and know-how (benchmarking, networking), or learning by doing or doing by learning are vital for the evaluation of the transitions. The accumulation of processes promoting 'lock-in' and path dependency through social networks, actors, and institutions, are the main pathways for such transitions.

9.5 Social Capital and Exclusion

Urban ethnic segregation in Lebanon is a unique model that depends on religion, ethnicity and cultural background. These religious demographics have historically molded the urban fabric and empowered political institutions of the Lebanese territories. Also, these groups maintained extensive social networks, which are reflected in their physical setting and determined the formulation of informal areas according to their ethnic diversity. The urban movements within Beirut have been influenced along with socioeconomic transitions, or may experience political conflict, surrounding the urban fabric of the city. The country has 18 officially recognized religious wings, and most of the population is under an umbrella of two religions: Muslim (Sunni, Shiite, Druze, and Alawites) and Christian (Orthodox, Catholic, and Protestant). It has been calculated that more than 5000 Jews still live in Lebanon (Faghaley 2002). The spatial segregation of the city on the basis of the ethnic, religious, and cultural characteristics of the population is thus not a heritage of a discriminatory past, but rather a characteristic and increasingly common feature of Lebanese society, and to a certain extent, in the regional interest in Lebanese territory.

There were extensive movements of refugees from hunger and war in Lebanon during the civil war and during the two Israeli invasions. The perception of an ethnic diversity that goes beyond the direct impact of immigration involves the spatial concentration of ethnic minorities in Lebanese cities, particularly in Beirut and in specific districts of the city, in which they may even make up the majority of the population (such as *Dahiya Janubiyya*). Unequal development on a regional scale, economic and cultural globalization, ethnic diversity and transport systems favor large shifts of population in Beirut. Also, the exoduses provoked by wars and the pressure from the peoples of nearby countries have led to arbitrary and ad hoc development.

The city of Beirut, therefore, is witnessing an increase in the drifting population, though it is differentiated by ethnic structure and threats of war. The crucial phenomena within Beirut are three different movements of populations. First, the international movements of population (both Muslims and Christians) led to the creation of camps and encouraged international immigrants to settle informally in scattered locations in Beirut (such as *Dahiya Janubiyya* and *Dahiya Sharqyyia*). Next, the mobility of populations of visitors constitutes another trend of the drifting population, mainly comprising students continuing their higher education and using the city and its services without being long-term residents. Finally are the inner population movements, either from rural to urban areas or displaced people from the south of Lebanon who suffered from continuing threats of war where the demographic and physical characteristics of the urban fabric of Beirut city have changed.

Drifting populations give rise to four main problems in the running of Beirut. First, their existence increases the pressure on urban services to an extent which the city cannot handle without at least receiving special aid from higher government levels and international agents. Second, the lack of suitable statistical accounting of those mobile populations and the irregular nature of its movements abjure sound planning of urban services for the city. Third, distortion arises between the people present in the city and the citizens able to deal with the city's problems and administration. Fourth, drifting populations increase the level of consolidation among certain sects within the population that led into the formulation of large informal agglomerations and became difficult to include in plans for future development, unless the approval is obtained from the political leaders and international agents (e.g., Sunni, Shiite, Druze, and Alawites). Finally, the Beirut explosion destroyed a third of its buildings, as some districts were completely annihilated, while others were only partially destroyed, and the city needs between 12–15 billion US$ to recover from this disaster. This explosion has left 300,000 persons homeless, damaging or destroying 50,000–70,000 housing units. About 7000 buildings were completely damaged. Thus, there is an urgent need in a short time to accommodate these people. If 50% of the 70,000 units were rebuilt, the rest could be constructed which would cost around 7.0 billion US$. The state of Lebanon does not have such funds to build this housing. With the weakened and depressed economy and a high rate of inflation, it is expected that many people would create new slum areas on the periphery of the city.

It is argued (Fawaz et al. 2018) that the heterogeneity of the urban area in Beirut puts pressure on the family by making alternative sources of identification available to individuals. Sectarian institutions have exerted pressure on individuals to adopt sectarian identities. Ethnic, national, and political ideological groups have exerted similar pressures. Therefore, since each sect of Beirut's population has its own urban setting, urban informality has followed the same path. The urban setting in many ways reinforced family ties and made them essential. Because of strong family ties, the role of the mediator is a critical one in Middle Eastern culture; therefore, the political leader or Zaim plays a major role in shaping the built environment, and this is his sociopolitical and economic role in the urban setting.

Fawaz argued (Fawaz 2008) that social networks as hierarchical and complex relations are geographically and *historically* situated, and that they result from and alter the strategies of the actors who participate in their production and depend on them to sustain their practices. These social networks function as accumulated capital, enabling developers to strengthen their hold over the production of housing in the neighborhood. For example, *Hayy el-Sellom*, today Lebanon's largest informal settlement accommodates over 100,000 residents, including several generations of rural migrants, refugees displaced by military conflicts in Lebanon, non-Lebanese migrant workers, and others (Fawaz 2008). On other hand, 'the dark side' of social networks, is notably exemplified by their ability to exclude actors considered 'outsiders' from a network or to limit the learning potential and economic exchange with acquaintances. It could be argued that leveraged against informal property rights systems were condemned for confining markets to narrow circles of capital and preventing them from, to use the famous De Soto terminology, 'achieving their capital potentials' (De Soto 2000).

Inequality of income, hunger and war, demonstrations, and discriminatory practices in housing markets lead to a disproportionate concentration of ethnic minorities in certain urban zones within the city of Beirut. For example, a main district for the Jewish community in Beirut is located at *Wadi Abou Gameal*. Each ethnic group has a certain location within the city; subsequently, the urban informality has followed the same pattern during the political and riots crisis, e.g., the riots of 1958 the civil war and the recent riots, after Beirut explosion, against the political class. This form of urban informality has always depended on exceptional circumstances of varying intensity.

Martin (2015) studied the Palestinian camp of *Shatila* in Beirut in which it was found that the transitions “from spaces of exception to ‘campscapes’” are enviable by the time the camp accommodated Palestinians, non-Palestinians, and Lebanese and integrated with surrounding areas. Martin concluded that the refugee camp cannot be uncritically afforded with the space of exception as described by Agamben (1998) for three main reasons. First refugee camps are increasingly becoming permanent solutions to displacement. Second, the exclusion through law reflects a risk by neglecting spaces that deserve equal attention because they are placed outside the ‘normal order’ in other ways. Third, the refugee camp is not the only spatial referent for refugees, but a large proportion of refugees live outside the camp and are integrated within the local society.

On the other hand, access to the local society depends on power and capital in which case there may be integration within a society or there may be formulated a new urban pattern within the urban fabric. In other words, the creation of a new space within the existing spaces or the creation of a state of sect-political within the state (for example *Hezb Allah*). Sect-political is a specific form of political rationality producing new spaces within spaces along economic, political, and sectarian lines as it contradicts what Michel Foucault (2003) ‘biopolitics’ to indicate the inclusion of life in the mechanism of power. It can turn into a space of abjection and, therefore, a technology of power aimed at the elimination of the biological threat. It was Michel Foucault (2003) who first coined the term biopolitics to indicate the inclusion of life in the mechanism of power. Alas, the Lebanese temerity came under the control of sect-political of *Hezb Allah*. In Egypt and in Pakistan, where social relationships with the public sector are deemed essential for access to housing and social services, most informal developers are retired or active military officers or state-accredited topographers working outside the range of their work (Hasan 2001; Soliman 2019)).

It can be noted that the exception does not operate on legal tracks only. As Palestinians are legally excluded from the state's protection, other outcasts are produced along with other kinds of political, social, and economic exclusions. Although this chapter has focused on Beirut city, the Shatila camp that has been officially recognized and, as such, receives relief and assistance, if we are to investigate lives and spaces that are placed at the margins, perhaps researching ‘informal gatherings’ (Palestinian and non-Palestinian) would be equally important. The recognition of a formal status (for both refugees and refugee camps) misses those who are left out and who are not even entitled to humanitarian assistance. Future investigations of the Palestinian lives and camps in Lebanon, e.g., might consider the condition and the politics of informal gatherings (Martin 2015).

9.6 Conclusion

The acceleration of urbanization processes and the emergence of urban informality in Lebanon are largely due to an increase in rural–urban migratory movements which spoils the urban fabric and impacts the natural heritage. Also, the population has shifted during the 17 years of the civil war, and the influx of refugees from neighborhood countries has played a major role in accelerating urban informality. In addition, the recent crisis of the Beirut explosion has created a situation of chaos for sheltering the homeless. Lebanon was suffering a major economic downturn before the explosion. Through these events, the shaping of the urban fabric either formally or informally has spread throughout the countryside and has tended to mushroom linearly along roads, coastlines, and highways and are related to ethnic and state diversity within the country.

The spread of urban informality is the most significant phenomenon shaping the spaces, places, and urban fabric within Beirut. Most of the periphery or former suburbs of Beirut have witnessed arbitrary residential development that has caused (and was caused by) serious socioeconomic and political problems, not only for their residents, but also for the various governments in Lebanon. State power, social segregation, sect-political, and ethnicity have dominated the formulation of urban informality. In many cases, urban movements and urbanization processes, and their discourses, actors, and local and international organizations, have been informally integrated in the structure and practice of local government, either directly or indirectly, through a diversified system of citizen participation, community development, and ethnic membership. As a result, there arises a triple process of urban segregation—on the one hand, that of the various ethnic minorities with respect to the dominant ethnic group, that various refugees with respect to the host community and—on the other hand, that of the different ethnic minorities amongst themselves.

The creation of two urban agglomerations (*Dahiya Janubiyya* and *Dahiya Sharqyyia*) within the city of Beirut has led to a dramatic change in space and place that has shaped the urban environment within the city. This has occurred due to the forces of internal and external political players, population shifts, market mechanisms, and ethnic conflicts. The most significant increases during the 1980s took place in *Dahiya Janubiyya*, basically due to influxes of displaced people from the southern part of Lebanon and the Israeli occupation of the south part.

In the context of economic globalization, the absorption and containment of unplanned zones of urban development have become imperative as a minimal response to the extensive liberalization of the use value of land and to the exchange value of resources and power in the city. The informal areas are finally being considered as integral parts of the urban matrix. Likewise, although economic inequality is influenced by ethnic origin, institutional barriers and social prejudices are much less deeply rooted than in formal areas. Various settlers in Beirut with equally religion-based pasts have thus evolved towards different patterns of spatial segregation and ethnic diversity, and the functions of cultural, institutional and economic factors that favored the mixing of background and social integration in Beirut. What does appear to be established is the tendency towards segregation of ethnic minorities in all parts within the city, and particularly in the peripheral parts of the city.

Based on the modes of confrontation between urban laws, norms, and social exclusion and practices, a critical approach is needed to the integration of the informal housing areas to the urban fabric of a city. Do the (1) day-to-day adjustments, (2) the individual and group arrangements (between occupants, holders of property titles, ethnicity leaders) concerning the right of use of land, and (3) the social integration and constructed property not call into question the often-simplified ideas of the absence of regulation and social homogeneity in the informal and irregular areas?

The early formation and later development of urban informality in Beirut thus depend on how well urban actors are able to mobilize extra-regional and local networks, resources, and knowledge, and involve them in sustained local niche-formation processes. This involves a series of mediation activities between production, reproduction/coproduction, distribution, and consumption of goods and services, between different political priorities and between planning and implementation. Policy-makers, politicians, international donors and practitioners, as a sociotechnical regime, are working to overcome sustainable transitions in urban informality in Lebanon have to acknowledge a potential strategy that accommodates low-income groups, as a sociotechnical niche, of host communities and foreign refugees. Mapping the 'political economy' of (in)formality, in other words explicitly exploring who benefits from ambiguous governance modes in what way, should then become part of the policies, programs, and projects that seek to help refugees and host communities. Ensuring that no one and no place is left behind should be each stakeholder's goal.

This experience showed that even if economists, engineers, and technicians take the real needs of the economic sectors into consideration, the political authority must be closely involved in the definition of development objectives and in the selection of methods of implementation. The Lebanese case has opened a new academic debate on the processes of urban informality transitions in the Middle East because its case is different than the other countries in the Global South.

References

Adwan, C. 2004. *Corruption in Reconstruction: The Cost of National Consensus in Post-War Lebanon*. Washington, DC: CIPE.

Agamben, G. 1998. *Homo Sacer: Sovereign Power and Bare Life*. Stanford: Stanford University.

Beydoun, A. 1991. *Housing Crises and Housing Cooperatives in Lebanon (in Arabic)*. Beirut: Youssef Beydoun Press.

Bourgey, André. 1985. La Guerre et ses Conséquences Géographiques au Liban. *Annales de Géographies, No 521 XCIV e annee*, 1–37.

Castells, M. 2010. *The Rise of the Network Society*. 2nd ed. West Sussex: Wiley.

Cram, F., and D.M. Mertens. 2015. Transformative and Indigenous Frameworks for Multimethod and Mixed Methods Research. In *The Oxford Handbook of Multimethod and Mixed Methods Research Inquiry*, ed. S. Hesse-Biber and R.B. Johnson, 91–109. New York: Oxford University Press.

DAR - IAURIF. 2005. *National Physical Master Plan of the Lebanese Territory*. Beirut: DAR - IAURIF.
De Soto, H. 2000. *The Mestry of Capital: Why Capitalism Triumphs in the Westand Fails Everywhere Else*. London: Blaack Swan.
Dumper, M.R.T., and B.E. Stanley, eds. 2007. *Cities of the Middle East and North Africa: A Historical Encyclopedia*. Santa Barbara: ABC-CLIO, Inc.
Fawaz, M. 2004. *Strategizing for Housing: An Investigation of the Production and Regulation of Low-Income Housing in the Suburbs of Beirut*. Massachusetts Institute of Technology, Department of Urban Studies and Planning, PhD.
———. 2008. An Unusual Clique of City-Makers: Social Networks in the Production of a Neighborhood in Beirut (1950–75). *International Journal of Urban and Regional Research* 32 (3): 565–585.
———. 2017. Planning and the Refugee Crisis: Informality as a Framework of Analysis and Reflection. *Planning Theory* 16 (1): 99–115.
Fawaz, M., M. Krijnen, and D. El Samad. 2018. A Property Framework for Understanding for Understanding Gentrification. *City* 22 (3): 358–374.
Fawyaz, M., and E. Bian. 2003. Informal Housing Areas in Beirut (in Arabic). *Al Mouhandess Journal, Beirut* 17: 36–47.
Foucault, M. 2003. *Society Must Be Defended. Lectures at the College de France, 1975–76*. London: Penguin.
Frantzeskaki, N., V.C. Broto, L. Coenen, and D. Loorbach. 2017a. Urban Sustainability Transitions: The Dynamics and Opportunities of Sustainability Transitions in Cities. In *Urban Sustainability Transitions*, ed. N. Frantzeskaki, V.C. Broto, L. Coenen, and D. Loorbach, 1–19. Oxon: Routledge.
Gaspard, T.K. 2004. *A Political Economy of Lebanon, 1948–2002: The Limits of Laissez-Faire*. Leiden: Koninklijke Brill NV.
Geels, Frank W., and Johan Schot. 2007. Typology of Sociotechnical Transition Pathways. *Research Policy* 36: 399–417.
Gemayel, A. 1992. *Rebuilding Lebanon*. Lanham: University Press of America.
Gendzier, I. 1999. *Notes from the Minifield: United States Intervention in Lebanon and the Middle East 1945–1958*. New York: Westview.
Ghorayeb, M. 1998. The Work and Influence of Michel Ecochard in Lebanon. In *Projecting Beirut: Episodes in the Construction and Reconstruction of a Modern City*, ed. P. Rowe and H. Sarkis, 106–121. London: Prestel-Verlag.
Grin, J., J. Rotmans, and J. Schot. 2010. *Transitions to Sustainable Development: New Directions in the Study of Long Term Transformative Change*. New York: Routledge.
Halabi, Z. 2004. Exclusion and Identity in Lebanon's Palestinian Refugee Camps: A Story of Sustained Conflict. *Environment and Urbanization* 16 (2): 39–48.
Harris, R. 2009. Reflections on Lebanon. In *Lebanon: Liberation, Conflict, and Crisis*, ed. Barry Rubin, 9–24. New York: Palgrave Macmillan.
Hasan, A. 2001. *Working with Communities*. Karachi: City Press.
Janmyr, M. 2018. UNHCR and the Syrian Refugee Response: Negotiating Status and Registration in Lebanon. *The International Journal of Human Rights* 22 (3): 393–419.
Katkhouda, R. 1998. *Affordable Housing in the Suburbs of Beirut: Prospects in Post-War Reconstruction*. Montreal: School of Architecture, McGill University.
Koehler, J., F. Geels, F. Kern, E.. Onsongo, and A. J. Wieczorek. 2017. *A Research Agenda for the Sustainability Transitions Research Network*. Technical. Amsterdam: STRN Working Group, STRN.
Lebanese National Report. 1996. *Lebanese National Report*. Beirut: The Ministry of Environment.
Markard, J., R. Raven, and B. Truffer. 2012. Sustainability Transitions: An Emerging Field of Research and Its Prospects. *Research Policy* 41 (6): 955–967.
Martin, D. 2015. From Spaces of Exception to 'Campscapes': Palestinian Refugee Camps and Informal Settlements in Beirut. *Political Geography* 44: 9–18.
Masri, I. 1999. Development- at What Price? In *Preparing for the Future: Global Pressure and Local Forces*, ed. M. Cohen et al. Washington, DC: the Woodrow Wilson Center Press.

Mertens, D.M. 2009. *Transformative Research and Evaluation*. New York: Guilford.
———. 2012. Transformative Mixed Methods: Addressing Inequities. *American Behavioral Scientist* 20 (10): 1–12.
Nasr, J., and Volait. 2003. *Urbanism Imported or Exported: Native Aspirations and Foreign Plans*. London: Wiley.
Nassara, J., and N. Stel. 2019. Lebanon's Response to the Syrian Refugee Crisis – Institutional Ambiguity as a Governance Strategy. *Political Geography* 70: 44–54.
Rabil, Robert G. 2011. *Religion, National Identity, and Confessional Politics in Lebanon: The Challenge of Islamism*. Hampshire: Palgrave Macmillan.
Rotmans, J. 2006. A Complex Systems Approach for Sustainable Cities. In *Smart Growth and Climate Change*, ed. M. Ruth, 155–180. Cheltenham/Northampton: Edward Elgar.
Rowe, P., and H. Sarkis, eds. 1998. *Projecting Beirut: Episodes in the Construction and Reconstruction of a Modern City*. London: Prestel-Verlag.
Saksouk-Sasso, A., and N. Bekdache. 2015. Private Interest Closing Social Space? A Critical Analysis of Lebanon's Real Estate Sector. In *Lebanon: Towards a Peace Economy*, ed. Jessica Banfield and Victoria Stamadianou, 26–38. London: International Alert.
Salibi, K. 1976. *Crossroads to Civil War*. Caravan: Delmar.
Shwayri, S.T. 2008. From Regional Node to Backwater and Back to Uncertainty: Beirut, 1943–2006. In *The Evolving Arab City: Tradition, Modernity and Urban Development*. ed. Elsheshtawy, Y., 69–98. London and New York: Routledge
Soliman, A. 2004a. *A Possible Way Out: Formalizing Housing Informality in Egyptian Cities*. Lanham: American University Press
———. Tilting at Sphinxes: Locating Urban Informality in Egyptian Cities. In *Urban Informality in an Era of Liberalization: A Transnational Perspective*, ed. Ananya Roy and Nezear Al Sayyad, 171–208. Lanham: Lexington Books.
———. 2004c. Regional Planning Scenarios in South Lebanon: The Challenge of Rural-Urban Interactions in the Era of Liberalization and Globalization. *Habitat International Journal* 27 (1): 385–408.
———. 2008. Diversity of Ethnicity and State Involvement on Urban Informality in Beirut. *Theoretical and Empirical Researches in Urban Management, Year 3, Number 9*: 15–32.
———. 2019. So It's Always a Chance: Community-Led Solutions to New Urban Expansion. In *New Cities and Community Extensions in Egypt and the Middle East*, ed. S. Attia et al., 159–179. Cham: Springer.
Spectrum. 1997. *Housing policy in Lebanon*. Beirut: Spectrum.
Tabet, J. 2001. *Beyrouth, Collection Portrait de Ville*. Paris: Institut Frangais d'Architecture.
The Government of Lebanon and the UN. 2017. *Lebanon Crisis Response Plan 2017–2020*. Beirut: The Government of Lebanon and the United Nations.
The Guardian. 2020. Beirut Families Search for Loved Ones Feared Trapped Under Rubble. *World news* 5: 1–6.
The Ministry of Environment. 2003. *Lebanon State of the Environment Report*. Beirut: The Ministry of Environment.
The Ministry of Social Affairs. 1996. *Population and Housing Statical Facts*. Beirut: The Ministry of Social Affair.
Tierney, J. 2016. *Beirut's Lessons for How Not to Rebuild a Wartorn City*. Washington, DC: Washington Post.
Traboulsi, F. 2007. *A History of Modern Lebanon*. London: Pluto Press.
UNDP. 2009. *Assessment of Capacity Building Needs of NGOs in Lebanon*. Washington, DC: UNDP.
United Nations. 2017. *Habitat III Regional Report: Arab Region*. New York: United Nations.
———. 2018. *Department of Economic and Social Affairs, Population Division*. New York: United Nations.
Van den Bosch S. 2010. *Transition Experiments. Exploring Societal Changes Towards Sustainability*. Ph.D. thesis. Rotterdam: Erasmus University. http://repub.eur.nl/pub/20714.
Wählisch, M., and M. Felsch. 2017. *Lebanon and the Arab Uprisings: In the Eye of the Hurricane*. London: Routledge.

Chapter 10
Hills of Urban Informality in Greater Amman, Jordan

Abstract Jordan is a battlefield country in the Middle East for hosting foreign refugees which has had major influences in setting, or at least transforming, certain socioeconomic and political milieus in the region. Over its recent history, Jordan has passed through several conflicts until it gained stability after WWII, and the conflicts were reflected in urban expansion transitions. As a result of the tremendous socio-economic, socio-spatial, demographic, and political transitions of Jordan, arbitrary urban growth dominates the countryside of the Hashemite Kingdom of Jordan in which urban informality grew.

Jordanian national identity is a multi-scalar Arab supranationalism merging Jordanian and Palestinian identities. Because of the nature of the refugees and the multi-scale of nationalism, any intervention to eliminate or at least to prevent further spread of urban informality in Jordan requires a careful, effective, efficient, and transparent planning strategy. Local inhabitants, refugees, local and national municipalities, and international donors have constituted the stakeholders who can handle this strategy. Urban informality constitutes a political, socio-spatial, and nationalism challenge for the kingdom of Jordan and international agents; therefore, its treatments and transitions differ from those of other countries in the region.

Keywords Urban informality · Nationalism · Urban transitions of Transjordan · Refugees · Twin downtowns · Social segregation and marginality · Management sustainability transitions

10.1 Introduction

The state of Jordan emerged during the WWI era. In 1921, after years of negotiations between Amir Abdullah I on one side and British and French leaders on the other, the newly formed League of Nations accepted the proposal to create the British Mandate of Transjordan from a parcel of the fallen Ottoman Empire. Prior to this period, there was no regional territorial entity referred to as Jordan or Transjordan, but rather it was a part of *Bilad El Sham (Syria, Lebanon, and Jordan)* (Kumaraswamy 2019; Nowar 2006; Alon 2007). With the acceptance of the

A. M. Soliman, *Urban Informality*, https://doi.org/10.1007/978-3-030-68988-9_10

mandate, the British appointed Abdullah I of the Hashemite Dynasty as the nominal ruler of Transjordan. The citizenry of the newly formed capital state comprised little more than 2000–3000 Arab Bedouins and is today a major regional state, with an estimated population of 4.227 million at the start of 2017 (Hashemite Kingdom of Jordan 2017).

In 1946, Transjordan achieved independence, and soon thereafter it changed its name to the Hashemite Kingdom of Jordan. Like many states across the globe, Jordan's national identity was constructed after the state had been created. Jordan's national identity is, of course, multifaceted and fluid. It draws from discourses about the Hashemite Dynasty, Islam, the ancient past, and Bedouin culture, as well as, Palestinian and other Arab nationalities. Domestically, there is a continuing schism and socioeconomic divide between the Transjordanians—the descendants of the inhabitants of the lands which form modern-day Jordan—and the Palestinian-Jordanians, who are the descendants of refugees from Israel and the Occupied Territories who fled after the establishment of Israel in 1948. While Jordan witnessed significant protests during the Arab Uprisings, demonstrations never called for the departure of King Abdallah II, allowing the monarchy room to maneuver its way out of the crisis by implementing a few cosmetic reforms aimed at assuaging dissent.

The ongoing conflicts in Syria and Iraq have had a negative impact on Jordan's socioeconomic and socio-spatial settings and are a potential threat to its stability and security (Ababsa and Abu Hussein 2020). Jordan is one of the largest hosts of refugees in the world, with the influx of 1.4 million Syrian refugees making up 14% of the total population (Hashemite Kingdom of Jordan 2017; Mumtaz et al. 2015). For structural reasons, such as lack of natural resources and cultivable land, the Jordanian economy has been traditionally poor and characterized by a heavy dependence on international aid—5.9% of GDP in 2016—and remittances, which constitute 14.3% of GDP. The economy is heavily dependent on the service sector and especially touristic revenues, which have not been noticeably affected despite post-2011 regional instability (Teti et al. 2018).

Due to these socioeconomic, political, and socio-spatial transformations of the Jordanian territory that are associated with internal migration, an influx of refugees, and regional instability, arbitrary urban growth found fertile ground to grow in the Jordanians' cities, especially the city of Amman. This results in the spreading of urban informality that varies in its typologies between refugees' camps, hybrid areas, and local poor inhabitant areas. The transitions of socioeconomic, political, and socio-spatial transformations that hit the Jordanian territory have a major influence on the formulation of urban informality.

With a focus on Amman city, urban informality within Jordanian territory is examined. Three arguments are discussed. The first is the importance of the multiplicity of state power, nationalism's structure, refugees, and international organizations' influences on urban informality. These divisions encouraged the Jordanians and non-Jordanians to obtain land plots to meet their housing needs at reasonable prices. The second is to understand the extent to which religions and nationalism, along with the political demise of regional instability net mechanisms, affected the urban informality development that is a mapping of the country. To what extent has the variation of nationalities, the pressure of neighborhoods' countries, together with international agents, contributed to shaping the urban fabric within Jordanian cities? Urbanization sprawl, arbitrary urban growth, and the typologies of urban informality in Amman are briefly emphasized to understand the various causes for the appearance and the establishment of urban informality in Jordan.

A transformative mixed methods (Mertens 2012) is used as a philosophical framework that focuses on nationalities in terms of cultural responsiveness. Those dimensions of diversity are associated with power differences, building trusting relationships, and developing mixed methods that are conducive to social change. The transformative paradigm also focuses on the transitions that reside in communities that experience marginalities and oppression on the basis of their nationalities, cultural values, and experiences (Cram and Mertens 2015). This method concentrates on a wide range of appropriate tools, methods, and techniques including literature review, qualitative meta-analysis, qualitative content analysis, on-site empirical investigations, document collection from official administrative units, and multivariate statistical analysis. For data collection, primary literature analysis is accomplished by using secondary data such as Google's maps sources, GIS maps available for Amman city, the national census, and records of the local authorities. The main assumption is that the pressure of the neighboring countries, together with international donors, has contributed to the informality transitions within Jordanian cities in which all the stakeholders have to play a certain role to ensure that emerging urban-informality transitions follow a sustainable path.

The chapter is organized into six sections: the second section observes urban transitions of Transjordan; followed by an overview of urban informality transitions; the fourth part highlights political economy transitions; and the fifth examines social segregation and marginality. The last section concludes that urban informality is the most significant phenomenon shaping the built environment within Jordanian cities, especially Amman city, and that it is caused by the diversity of state power, international donors, and the nationalities structure due to several attributes that could be adjusted in favor of integrating informal areas within the rest of the urban fabric of the cities, melding into a homogeneous built environment.

10.2 Urban Transitions of Transjordan

Transjordan was a part of *Bilad El Sham (Syria, Lebanon, and Jordan)* and had been conquered by several Islamic rulers. Prior to the British Mandate, Transjordan was considered a southern province of Syria (Moore 2004). The history of Jordan's modern merchant class began in 1902 with the development of the *Hijaz* railway linking the area with the Holy Land in Saudi Arabia. Although a mere fragment of the debris of the former Ottoman Empire in 1918, Transjordan became part of the Hashemite King Faisal Ibn al-Hussein's Kingdom of Syria, which was destroyed by the French Empire in 1920. Although Transjordan was artificially created during the implementation of the peace settlement after WWI (as a result of a series of major political accidents), its land was inhabited by some of the most ancient pre-Christian peoples: the Moabites, Edomites, Ammonites, and Nabataeans. On 3 March 1921, in accordance with Arab and Muslim customs and traditions, the Amir Abdullah Ibn al-Hussein was proclaimed the Amir of Transjordan by the people and supported by the British and French rulers. With wisdom, patience, and political acumen, Abdullah was then able to turn the political tide by strict adherence to his political friendship with Britain. For further discussion of the history of Transjordan between 1929 and 1939, see (Kumaraswamy 2019; Nowar 2006; Alon 2007). Transjordan included the current boundary of Jordan and the West Bank, the area located to the west of the River Jordan.

In 1946, Britain ended its mandate over Transjordan, granting full independence to the Kingdom. In the years after the Jordanian state was founded, King Abdullah I took control of Transjordan after liberating it from British rule. The assassination of King Abdullah I had significant implications for the stability of the Hashemite Kingdom and its development as a nation. The accession of Abdullah I's grandson Hussein to the throne in 1953 at the age of 18 did not augur well for the future. Not only did the young king have to establish himself as a credible and capable leader, but he would also have to withstand the dual pressures of British and Arab opinion (Milton-Edwards and Hinchcliffe 2009). However, it was King Hussein who led Jordan's political development, created many of its institutions, and ensured that the Western great powers would view the kingdom as having vital geopolitical and geostrategic importance in both the Cold War and the Middle East peace process. King Hussein played on these concerns and his regime's conservative and anti-communist credentials to solidify ties with the United States, in particular. From its emergence as an independent state, Jordan has held close ties to powerful Western states and has, in fact, depended heavily on foreign aid from these countries to keep the kingdom afloat (Abdul-Hadi 2016). At the beginning of the 1950s, the Kingdom of Jordan shrank to its current boundary and, after the defeat of 1967, the kingdom lost its control of the West Bank of Palestine. Over history, the kingdom hosted refugees from its neighboring countries. After the death of King Hussein in early 1999, King Abdullah II has ruled the country.

Jordan has a prime minister and an elected lower house of parliament (called the Chamber of Deputies); the king appoints the government and the members of the Senate (called the 12th House of Notables). In June 2011, King Abdullah II vowed to relinquish his right to appoint prime ministers. However, in 2011–2012, King Abdullah II fired the Jordanian prime ministers four times in a row—the most recent of these was on 4 October 2012, when he dissolved the parliament for early elections and appointed a new prime minister. Elections on 23 January 2013 were boycotted by opposition groups such as the Muslim Brotherhood and *Hirak*.[1] But an otherwise decent turnout allowed King Abdullah II to declare the elections credible and left the country's largest opposition groups without a strong voice in parliament (Blitz 2014).

However, this delicate political balance is tipping. As prospects for a resolution of the Syrian conflict become increasingly elusive and the number of Syrian refugees in

[1]The Jordanian Youth Movement established after the Arab Uprisings in 2011.

Jordan swells, public disenchantment has turned back towards the Jordanian government. The sharp population increase from the refugees' arrival revealed long-present and deepening fissure in Jordan's political, economic, and social infrastructure. The Syrian and Palestinian influx present Jordan with a political problem. Syrians and Palestinian are highly concentrated in Jordan's most vulnerable communities, and grievances brought to the fore by Syrian refugees have begun to mobilize marginalized Jordanians. As public frustration grows, political conflict is increasingly framed as a struggle against disenfranchisement. This stands in contrast to Jordan's historical political conflicts, which were primarily characterized as struggles between the monarchy and elite interest groups (such as Islamists and Palestinians). The rapid expansion of the Syrian and Iraqi refugee population has accelerated an emerging narrative of the marginalized in the political sphere and has the potential to threaten the stability of the current Jordanian political structure.

The geographical area of the Hashemite Kingdom of Jordan is 89,318 km^2. The total population of the kingdom is 10.05 million (including Syrians and Syrian refugees) (Hashemite Kingdom of Jordan 2017), with an average family size of 5.4 persons, an average density of 73.5 per km^2, and annual population growth of 2.3%. The percentage of the population in urban areas is around 82.6%. The kingdom consists of 12 governorates; distributed as follows: northern governorate: (Irbid, Mafraq, Jarash, and Ajloun), central governorates: (Amman, Zarqa, Balqa', and Madaba), and southern governorate: (Aqaba, Karak, Tafeleh, and Ma'an). The major cities of Jordan are Amman, Zarga, Irbid, Russeifa, Wadi al-Seer, Ajloun, Agaba, Madaba, Salt, and Ar Ramtha (UNDP and GDCD 2008). Currently, "Greater Amman's footprint" has reached 293 square kilometers, twice as large as that of Barcelona, which has a similar population (165 square kilometers for 5.4 million people (Ababsa and Abu Hussein 2020).

As elaborated in Fig. 10.1 and Table 10.1, Amman has witnessed periods of tremendous sociopolitical and socio-spatial transitions since its inception until now (Ababsa and Abu Hussein 2020; Khirfan 2019; Debruyne 2014). The first was the post-independence phase (1918–1948). The city of Amman became the main administration center for Transjordan after WWI. In 1918, Amman's population was less than 3000 inhabitants, with an urban area estimated at 0.321 km^2. Its population was estimated at 10,500 and 45,000 in the 1930s and 1940's, with the city extending over an area of some 2.5 km^2 and 3.5 km^2 respectively (Al Nsour 2016; Potter et al. 2009). Thus, Jordan continued to attract migrants and its political stability attracted successive waves of migrants, internally and externally, to the city from Syria, Palestine, Lebanon, and the newly formed Iraq. Various tribes immigrated from neighboring rural areas into the city. Further religious and ethnic minorities, such as the Kurds and Armenians, also came to live in Amman. In addition, throughout this period, Amman was the focus of domestic immigration. The main factor promoting migration to Amman has been its status as the seat of government. Thus, through the period 1921–1948, Amman experienced gradual and arbitrary socio-spatial expansion. Its population increased around fourfold within nearly 20 years. Subsequently, its urban areas rapidly expanded informally without proper planning. Throughout this period, two principal challenges emerged. The first

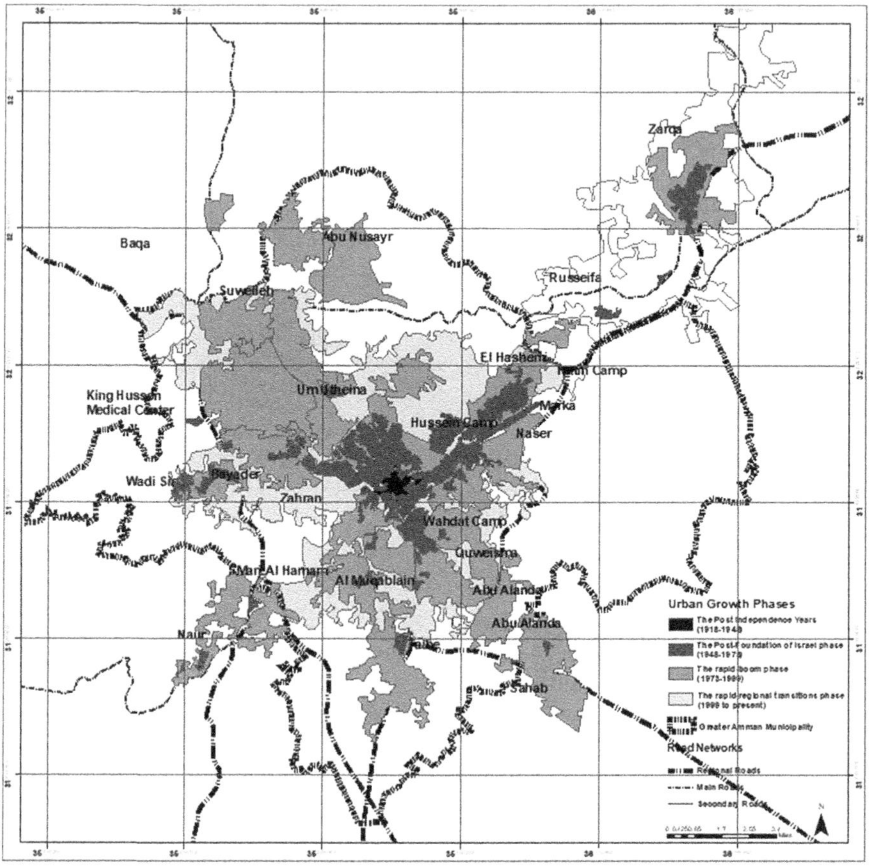

Fig. 10.1 Map showing the phases of urban growth in Amman. (Source: The author, Ababsa and Abu Hussein (2020), GAM, HUDC (2017) and Hashemite Kingdom of Jordan (2014))

concerned urban management agencies' ability to provide social and physical services for a substantially growing population, in particular, the Palestinian refugees. The second concerned the fledgling city's ability to control urban growth whilst ensuring it did not encroach on agricultural land.

Second came the post-foundation-of-Israel phase (1948–1973). But it has been the Palestinians displaced as a result of the foundation of the State of Israel in 1948 who have formed the main wave of migrants to Jordan, in general, and to the city of Amman, in particular. The population of Amman rapidly increased soon after 1948, as a result of the huge influx of Palestinians' refugees. The refugees were mainly accommodated in five hastily prepared camps. Two of these were sited in Amman, namely *Al Hussein* camp in the north of the city and *Al Wehdat* camp in the south (see Fig. 10.1). The camps lacked even the most basic of amenities and services (Kadhim and Rajjal 1988; Municipality of Greater Amman 1993; Khirfan 2017). Amman witnessed arbitrary urban growth around the old core of the city. Due to this

Table 10.1 Data on socioeconomic, political, and reform transitions during the last 70 years in the Kingdom of Jordan

Multi-Level/ Phases of Transitions	Niche		Regime		Landscape
	Economic transitions	Social transitions	Reform transitions	Political transitions	Output & policy
The indepen-dence years (1918–1948)	In 1902 the development of the Hijaz railway to link the area with Holy land in Saudi Arabia. The fledgling Amman's ability to control urban growth whilst ensuring it did not encroach on agri-cultural land	The citizenry of the newly formed capital of Amman was comprised of little more than 2000–3000 Arab Bedouins. The establishment of Palestinian's camps after 1948 war constituted "Misery Pockets" on the periphery of Amman A continuing schism and socio-economic divide between the Transjordanians—the descendants of the inhab-itants of the lands.	Abdullah were then able to turn the political tide by strict adherence to his political friendship with Britain. The period from 1921 to 1948 Amman experi-enced a gradual arbitrary spatial expansion.	The British appointed Abdullah I of the Hash-emite Dynasty, as the nominal ruler of Trans-jordan. Amman became a focus hub for the neighbor-hood's countries.	Amman population was estimated at 10,500 and 45,000 in the 1930s, and 1940's, with the city extending over an area of some 2.5 km^2 and 3.5 km^2 respectively. Religious and ethnic minorities, such as the Kurds and Armenians also came to live in the city of Amman.
Growth phase (1948–1973)	A lack of finance and refugees' migration pre-vent further economic development. The country relied on the international aid to cover the development process.	Another spiral of 'uncontrolled growth', which served to increase its population from 330,000 before the con-flict to around 500,000 shortly after the defeat of 1967 Most of the Jordanians' cities have established based on social diversity according to the place of	Establishment of the Department of Planning in 1952 within the Minis-try of National Economy. In 1955, the Development Plan for Amman was produced.	The foundation of the State of Israel in 1948 have formed the main wave of migrants to Jor-dan The assassination of King Abdullah I had significant implications for the sta-bility of the Hashemite Kingdom and its devel-opment as a nation. The accession of	In 1948, after the estab-lishment of Israel, two camps for Palestinians refugees; Al Hussein and AlWehdat, were constructed in Amman. 1952 plan to guide future urban expansion to accommodate the urgent needs for the internal and the external immigrants to Amman

(continued)

Table 10.1 (continued)

Multi-Level/ Phases of Transitions	Niche		Regime		Landscape
	Economic transitions	Social transitions	Reform transitions	Political transitions	Output & policy
		origin Social tensions occurred between Jordanian and other nationalities.		Abdullah I's grandson Hussein to the throne in 1953 at the age of 18. Second major wave of Palestinians' refugees arrived in Amman after the so-called "Six Day War"	Nine refugee camps were established in the central and northern parts of the country.
Rapid growth phase from 1973 to 1999	Since 1990, Jordan has suffered from the problems of unemployment, poverty and foreign debt. The peace agreement with Israel has reached.	Since 1979, Amman's population, estimated to be 215,000 in 1961, grew to 600,000 in 1974, and 777,800 in 1979, with the city extending over an area of 101 km^2. At the end of the First Gulf War in 1990, about 400,000 people, of both Jordanian or Palestinian origins, returned to Jordan from Kuwait and Iraq and settled in Amman.	Several policies, such as downzoning and building regulations were introduced. The access to the benefits of urbanization has not been equitable.	A regional conflict with Israel through defeating of 1973 war, an oil boom in the region. Lebanese Civil War (1975–1990) displaced people into Jordan.	The establishment of the Municipality of Greater Amman. In 1985 the region had 1 549 500 residents, 59% of the East Bank total The large contingent of migrants from Jordan who worked in the oil-rich states of the region. A dramatic population increase, which was reflected in the urban structure of Amman
Maturity phase from 1999 to the present to present)	Creating job opportunities, reduce poverty, and achieve economic growth, in addition to the establishment of Aqaba	The Jordan Response Plan 2015 (JRP) embeds the refugee response into national development plan	Low-income families were pushed towards living in crowded or sub-standard housings or in informal areas.	The death of King Hussan in 1999, and King Al Abdullah took the role. American invasion of Iraq in 2003, resulted in	The population of Amman amounted to approximately 42.02% of the Kingdom's total of 10,053,000 in 2017

	special economic zone Lack of finance for housing developers and complicated procedures have prevented the lower-income groups from accessing to financial resources. The percentage of poverty in Jordan was 14.4%.	The scarcity of the humanitarian protection space for Syrian refugees and Jordanian host communities has declined by which the development path deteriorated. The current housing supply is not well aligned with the demand, for lower income groups The total number of populations who are living in informal housing areas and camps around 450000 persons. Spatial inequalities are not only a forerunner of social and economic divisions; these, in turn, cause further inequalities and different forms of exclusion and marginalization.	A number of development and duty-free zones were launched in different areas of the Kingdom in 2008. The National Youth Strategy 2005–2010 was developed and has enhanced in 2015. In 2016, the UNDP modifies the JRP into The Regional Refugee & Resilience Plan (3RP) 2019–2020 to offer a strategic, coordination, planning, advocacy, and programming platform for humanitarian and development.	1.4 million Iraqis refugees to Jordan Arab uprising resulted in 2011–2012 wave of protests, Jordan's leadership has promised to accelerate political reforms. The period leading up to the Arab Uprisings a majority of ordinary people were not benefitting from economic growth, inequalities were increasing, and corruption was rife. Approximately 30% of the population are non-Jordanians and more than 1.3 million are Syrians	The Regional Refugee & Resilience Plan (3RP) 2019–2020 to offer a strategic, coordination, planning, advocacy, and programming platform for humanitarian and development The influx of refugees also contributed to the collapse of the housing market in the country and have encouraged or at least associated with the spreading of urban informality A high stock of vacant housing units has left by Iraqis, and these units are unaffordable to lower income groups. The spreading of informal areas in the eastern sector of Amman.

Source: The author

wave of migrants, Jordan established the Department of Planning in 1952 within the Ministry of the National Economy, which was renamed recently as the Ministry of Commerce and Industry. This intervention was seen as the nucleus of socio-economic urban planning. The principal objective of this department was to create plans, frameworks, and strategies to increase the productivity of economic sectors, including agriculture, commerce, and industry based on the available resources and urban growth (Al Nsour 2016). These plans, applied in 1953, were to guide future urban expansion to accommodate the urgent needs of the internal and the external immigrants to the city, as well as to provide various services to the newly established urban areas.

The second major wave of Palestinians' refugees arrived in Amman after the so-called Six Day War', which occurred in 1967 between Israel and the three Arab states of Egypt, Syria, and Jordan. Kadhim and Rajjal (1988). Note that this conflict sent the city into "another spiral of uncontrolled growth", which served to increase its population from 330,000 before the conflict to around 500,000 shortly after it, expanding the city arbitrarily. During this period, nine refugee camps were established in the central and northern parts of the country, and no further refugee camps were developed in Amman. The largest of these was *Baqa,* which is located some 12 km northwest of the city center, and now forms part of the extended metropolitan area (Potter et al. 2009).

Based on the general planning frameworks established in 1953, several guidelines were introduced to control arbitrary urban growth. However, the management of urban growth did not take place because of several barriers, such as a lack of finance and the continuous migration of refugees, but these frameworks formulated the basis for other planning interventions after 1953. In 1955, the Development Plan for Amman was established by foreign consultants in response to the rapid urban growth (Municipality of Greater Amman 1993). It was the first comprehensive physical plan for Amman designed to improve housing, services, and employment. Because of the scarcity of financial resources, lack of planning knowledge, and absence of political patronage, this plan was not implemented on the ground.

Third, the rapid-boom phase (1973–1999). This phase was prejudiced by a regional conflict with Israel through defeat in the 1973 war, an oil boom in the region, and the Lebanese civil war. In 1987, the National Assembly approved the establishment of the Municipality of Greater Amman, combining the previous Municipality of Amman, 13 adjoining municipalities, the new town of *Abu Nuseir*, and 13 villages, councils, and rural areas. Greater Amman covers an area of 528 km^2 of both urban and rural land with a population of around 1.0 million. The Amman-Balqa region is an appropriate regional context for greater Amman. In 1985, the region had 1,549,500 residents, 59% of the East Bank's total. Changes in the population of major settlements between 1979 and 1985 indicate that the population of Greater Amman grew at a slightly faster rate than the population of the region, increasing its share from 57.6% to 58.1%. In 1985, Greater Amman had a population of 900,732 people, with 142,711 households, and an average household size of 6.3 (Municipality of Greater Amman 1993).

The expansion of Amman, from 1973 to 1999, is often referred to as the 'boom years', based on the large contingent of migrants from Jordan who worked in the oil-rich states of the region (Kadhim and Rajjal 1988), many of whom sent back substantial remittances. Throughout this period, the population of Amman continued to be boosted as a result of internal rural–urban migration, the settlement of nomadic tribes, and the wars and crises that affected the region. This 1967 wave of migration combined with additional refugees from the Civil War in Lebanon during 1975, leading to a dramatic population increase that was reflected in the urban structure of Amman. By 1979, the population of Amman had reached 777,855 inhabitants, representing 52.6% of Amman Governorate and about 30% of the nation. By the time of the Population and Housing Census, conducted in 1994, the population reached some 1,307,017, representing an increase of 54.6% since 1979. Amman's population, estimated to be 215,000 in 1961, grew to 600,000 in 1974, and 777,800 in 1979, with the city extending over an area of 101 km^2. This represents an increase of 2277% and was largely due to the second wave of Palestinian migration previously mentioned (Hashemite Kingdom of Jordan 2017).

At the end of the First Gulf War in 1990, about 400,000 people of either Jordanian or Palestinian origin returned to Jordan from Kuwait and Iraq and settled in Amman (Potter et al. 2009). This enormously increased the demand for services, housing, and job opportunities. These pressures and the economic crisis of 1989 led to a decline in the value of the Jordanian dinar. Since 1990, Jordan has suffered from the problems of unemployment, poverty and foreign debt, which can be directly attributed to the third wave of refugees; therefore, an economic reform program was undertaken to address subsequent social-economic problems.

Clearly, the plan of 1955 was a foundation for the urban development of Amman. The city witnessed several seminal transformations in terms of managing urban growth; these transformations have contributed to achieving a relatively sustainable built environment and orderly development. The transformations included instituting Greater Amman Municipality (GAM) as the main agency to manage urban growth, but, once again, laws, acts, and plans were developed by foreign consultants. Many of these policies were applied after the building had already occurred. For example, about 16% of spatial expansion in the year 1979 had already taken place before the Building Regulations Ordinance was adopted as law, so this expansion informally took place on adjacent agricultural land. Several policies, such as downzoning and building regulations, were characterized by uncertainty because they did not meet the needs of low-income groups (Al Nsour 2016).

Finally came the rapid regional-transitions phase (1999 to present). As enumerated in Table 10.2, the last census took place in 2017 in Jordan after the creation of the geographically wider Greater Amman Municipality () area, and by this date, the population of Amman Governorate had increased to some 4,226,700 inhabitants. Thus, at this stage, the population of Amman amounted to approximately 42.02% of the Kingdom's total of 10,053,000 in 2017. This phase has faced the tremendous transitions of receiving two subsequent waves of refugees, from Iraq, Yemen, and Syria.

Table 10.2 Distribution of Jordanians and non-Jordanians in Amman from the censuses of 2004, 2015, and 2017 by the governorate

	2017		2015						2004	
	Population		Jordanians		Non- Jordanians		Total		Population	
Governorates	Number	Percent	Number	Percent	Number	Percent	Number	Percent	Number	Percent
Amman	4,226,700	42.04	2,554,923	38.63	1,452,603	49.78	4,007,526	42.04	1,942,066	38.05
Al-Balqa	518,600	5.16	396,939	6	94,770	3.25	491,709	5.16	346,354	6.79
Al_Zarqa	1,439,500	14.32	923,652	13.97	441,226	15.12	1,364,878	14.32	764,650	14.98
Madaba	199,500	1.98	156,787	2.37	32,405	1.11	189,192	1.98	129,960	2.55
Irbid	1,867,000	18.57	1,316,618	19.91	453,540	15.54	1,770,158	18.57	928,292	18.19
Al_Mafraq	580,000	5.77	314,164	4.75	235,784	8.08	549,948	5.77	244,188	4.78
Jarash	250,000	2.49	167,751	2.54	69,308	2.38	237,059	2.49	153,602	3.01
Ajloun	185,700	1.85	157,162	2.38	18,918	0.65	176,080	1.85	118,725	2.33
Al_Karak	333,900	3.32	272,449	4.12	44,180	1.51	316,629	3.32	204,185	4
Al_ Tafielah	101,600	1.01	90,108	1.36	6183	0.21	96,291	1.01	75,267	1.47
Ma'an	152,000	1.51	127,989	1.94	16,093	0.55	144,082	1.51	94,253	1.85
Al_ Aqaba	198,500	1.97	135,045	2.04	53,115	1.82	188,160	1.97	102,097	2
The Kingdom	10,053,000	100	6,613,587	100	2,918,125	100	9,531,712	100	5,103,639	100

Source: Hashemite Kingdom of Jordan (2014, 2015, 2017)

After the invasion of Iraq in 2003, a wave of refugees arrived in Amman and the population continued to grow at a phenomenal rate. In 2007, the Department of Statistics reported the population of Amman as 2.17 million out of 6.106 million for Jordan as a whole, which equals almost 35.5% of the national population (Hashemite Kingdom of Jordan 2017). In March 2007, the Ministry of the Interior estimated that there were as many as 1.4 million Iraqi nationals currently living in the Greater Amman urban area (Potter et al. 2009). The Syrian refugee influx into Jordan has been massive. As of June 2015, more than 620,000 Syrians were registered with the United Nations Refugee Agency in Jordan. Eighty-four percent of these Syrians live in host communities as opposed to refugee camps (Francis 2015). The most recent wave is composed of 27,000 Yemeni refugees, who arrived since 2014 (Ababsa and Abu Hussein 2020). This huge number of refugees has accelerated the informal urbanization process, as well as spread urban informality, and access to the benefits of urbanization has not been equitable.

The recent and worst wave of refugees occurred after the Syrian crisis in March 2011. According to the government's estimation, Jordan hosts 1.4 million Syrians, of whom 646,700 are refugees and 750,000 lived in the country before the Syrian crisis. Currently, Syrians represent around 14% of the national population. The Jordan Response Plan 2015 (JRP) (UNDP 2016) embeds the refugee response into national development plans, helping to implement sustainable service-delivery systems that meet the needs of both refugees and vulnerable host communities. The most important interventions of the JRP 2015 seeks, to, first, meet the immediate needs of Syrian refugees both in and out of camps, as well as vulnerable Jordanians affected by the crisis. The second aim to restore and reinforce municipal services and infrastructure degraded as a result of the sharp demand increases in critically affected sectors, in particular, solid-waste management, housing, environment, energy, and transport. The third goal is to rapidly expand employment and livelihood opportunities and strengthen the coping capacities of vulnerable Jordanians who have been impacted by the crisis, and finally, to address social imbalances and strengthen social cohesion in Jordanian communities hosting large numbers of refugees (Hashemite Kingdom of Jordan 2014). The main objective of this plan is to ensure urban sustainable transitions for both refugees and of the host communities that satisfy the socioeconomic needs for the refugees without harming the host communities, as well as providing goods and services for the rapid increase of population. On the other hand, this plan does not provide a reasonable housing policy to meet the urgent needs of low-income groups. A part of this plan was a mortgage scheme, but it was beyond the means of the urban poor (Al-Homoud et al. 2009).

In 2016, the UNDP modified the JRP into The Regional Refugee & Resilience Plan (3RP) 2019–2020 to offer a strategic, coordination, planning, advocacy, and programming platforms for humanitarian and development partners to respond to the Syria crisis on the regional level and in host communities. While strategy, planning, and programming are country-led processes, regional coherence is pursued to ensure consistency in responding to planning and implementation, to promote common tools, standards, and innovation, and to enhance advocacy efforts at global and regional levels (UNDP 2016). Two main principles of the 3RP are considered.

The first conducted an exploration of the factors underlying social tensions between Syrian refugees and host communities in Jordan. It found that the main drivers of social tensions include "structural causes that predate the Syrian crisis (such as high levels of poverty, resource scarcity, and lack of municipal capacity to deliver basic services), socioeconomic causes (such as difference in religious, cultural, and social norms between refugee and host communities), and proximate causes (such as decreasing access to affordable quality housing, economic competition over jobs, and the role of international aid in terms of perceptions of fairness, equity, and corruption)." Second, supplying water by truck to the *Za'atari* refugee camp housing more than 80,000 Syrian refugees is an ineffective short-term solution because it is expensive and time-consuming and has harmful environmental impacts.

However, Syrian, Lebanese, Iraqis, Yemeni, and Palestinian refugees have resulted in rapid unorganized expansion in the major cities. This has resulted in crowded sub-standard housing in poorly planned and serviced neighborhoods posing major risks of abuse and conflict, but also high vulnerability to a deteriorating urban environment. The influx of refugees also contributed to the collapse of the housing market in the country and has encouraged or at least been associated with the spread of urban informality on the peripheries of the major Jordanian cities. From a social point of view, the Syrian crisis has increased tensions between Syrian refugees and affected Jordanian host communities in the governorates of *Mafraq, Irbid*, and other cities in the country. The scarcity of the humanitarian protection space for Syrian refugees and Jordanian host communities has declined which further hampers the development path (e.g., increasing the average density of the urban areas, congestion of road capacity, increasing solid waste, scarcity of potable water, etc.). The main question is how to help efforts to improve 3RP and to build resilience within the region. There is an urgent need to find the ways in which resilience in the context of the Syria crisis is understood and measured and to examine how resilience has been addressed in programming and activities under the 3RP.

10.3 Urban Informality Transitions

It is estimated that, in the period 2004–2011, the annual housing need in Jordan was estimated to be 32,000 units on average, whereas supply in the market is about 40,000 units, and the average annual number of units constructed is around 28,589 units (United Nations 2013). Jordan has a relatively high stock of vacant units, (18% before the Syrian crisis), and a low overcrowding ratio that averages 1.3 people/room (Housing Indicators set the standard for sufficient living at <3 persons per room). The ownership-to-rental ratio was 73.1–23.2%. At the lower end of the income ladder (first decile), 44% rent. In the *Mafraq* governorate, rented housing units reached only 12.5% (2010), a fact explained by the strong tribal communities, with more owned and inherited land (United Nations 2013). However, the current housing supply is not well aligned with the demand, in particular, for lower income groups. Various waves of population influx from neighboring countries have affected the Jordanian housing market. However, in 2010 after the stability of Iraq, the demand by the Iraqis decreased, leaving a high stock of vacant housing units, but unaffordable to lower income groups. In other words, Jordan witnessed a housing puzzle: surplus of housing production and high stock of vacant housing units, and at the same time, a lack of affordable housing units for housing the lower-income groups and the urban poor.

Undersupply of low-income housing in Jordan was affected by various factors such as rapid urbanization rates, an influx of Syrian's refugees, inappropriate housing subsidies, and a lack of housing policies and planning. Factors that caused rapid urbanization included political instability in the surrounding countries, major external migrations and flows of refugees in specific years (1948, 1967, 1975, 1982, 1991, 2003, 2011), internal migration from the countryside and the *Badia* (Bedouin land), and economic inflation (Al-Oun 1997; Al-Hasanat 2003). The recent Syrian crisis (2011) has put further pressure on the acceleration of informal urbanization process and increased the size of urban informality in the country, especially, Amman, the capital city. Further, rapid urbanization and housing growth extended planning and zoning ordinances to newly developed areas that were beyond the financial ability of the poor. Thus, low-income families were pushed towards living in crowded or sub-standard housings (Al-Hasanat 2003; Nuseir et al. 2003), or in informal settlements, many of which are sited in the eastern sector of Amman. As illustrated in Fig. 10.2, most of the informal housing areas are in the eastern sector of Amman. Inappropriate housing subsidies, lack of finance for housing developers, and complicated procedures have prevented the lower-income groups from accessing financial resources that can apply to informal areas to meet their housing needs. There existed a lack of mobilization of resources for the purpose of housing at the national level to provide long- term financing to enable all income groups to attain decent housing with an emphasis on low-income segments. This involves the absence of the provision of sustainable incentives systems geared to low-income groups in the domain of housing programs, especially those related to housing production and housing finance. There is a general scarcity of incentives geared to housing developers for boosting the production of housing for low-income people (The Third National Report 2014).

Housing financing in Jordan comes from three sources (Barakat and Tuffaha 1995; Basboos 1995; Al-Omari 2003). First is regular financing through loans from banks and private financial institutions where this source has financed about 18% of the housing units annually. The terms of housing loans provided by banks make it difficult for low-income households to buy a housing unit. Complications include the low amounts of the loans, guarantee requirements, short-payback periods, and the overall loan system. The second source comprises is individual financing through savings, property selling, and money transfers from abroad; this source contributes about 48% of annual housing finance. Third is irregular financing through assistance and loans from family members, relatives, business owners, and real estate and construction material traders, which constitutes about 34% of annual housing financing. In 2002, USAID partnered with the Jordan Mortgage Refinance Company

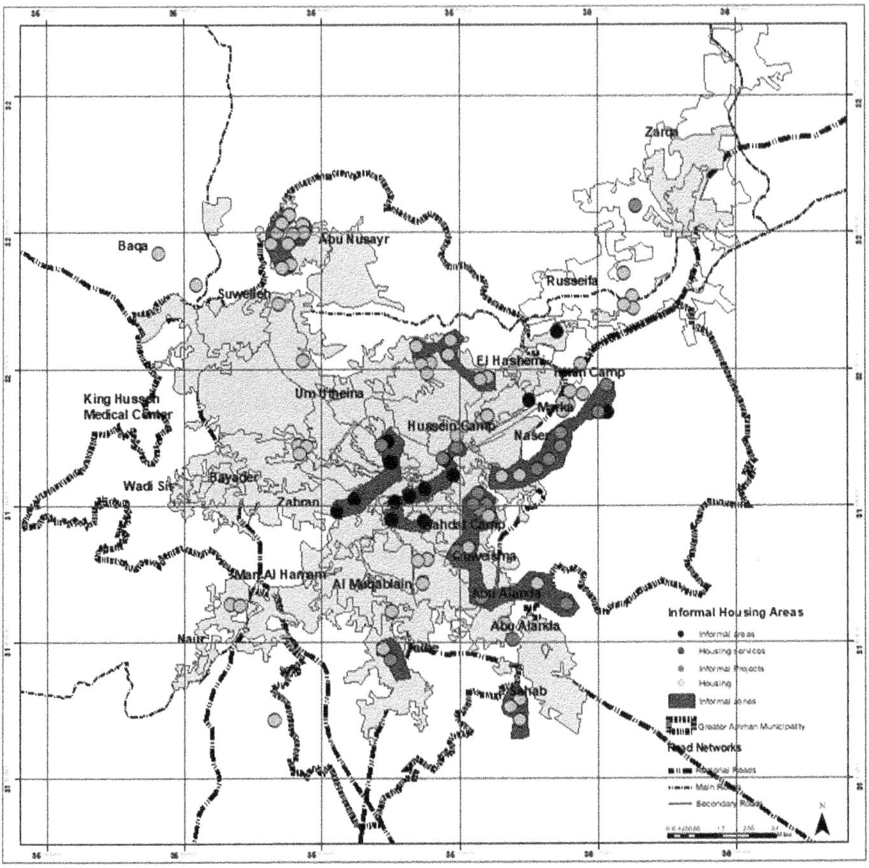

Fig. 10.2 Map showing informal housing areas in Amman. (Source: The author, Ababsa and Abu Hussein (2020) and Hashemite Kingdom of Jordan (2014))

(JMRC) to create the country's first long-term housing bond through the Jordan National Bank (Al-Homoud et al. 2009) (Fig. 10.3).

It is estimated that the total population that lives in informal housing areas and camps is around 450,000 persons (The Third National Report 2014) which constituted around 8.2% of the national population in 2017 (Hashemite Kingdom of Jordan 2017). The definition of 'informal areas' in Jordan remains unclear (Ababsa 2012) and lacks a specific definition. There is also a lack of information that would enable an accurate picture of urban informality to be drawn. In an attempt to define informal settlements, Ababsa (2012) identified two types of informal settlements that differ in terms of the origin of their population: Transjordanian in the first case; and Jordanian of Palestinian origin or only Palestinian (displaced from the Gaza Strip) in the second. Also, this typology relies on a legal criterion related to types of ownership of property. The author also stresses the significant of *sakan 'ashwa'i* which refers almost exclusively to areas inhabited by Palestinian refugees and is not

Fig. 10.3 Photos of informal urban areas in *East Wahdat*, Amman. (Source: The Aga Khan Award for Architecture 1980)

used for informal settlements with rural or Bedouin populations, for which the term 'poor areas' (*ahya' al fuqara*) is preferred. But also ignored are the areas that were established on the periphery of Palestinian camps, which are constructed on treacherous terrain, including floodable wadi basins, steep hillsides, and on agricultural land. These areas accommodated a variety of nationalities from bordering countries (Al Nsour 2016).

Due to the massive wave of Syrian refugees, a new Informal Tended Settlements (ITS) emerged. The UN-Habitat Programme defines ITS (Unicef 2013) in two parts (see Tables 10.3 and 10.4). The first part constitutes residential areas where a group of housing units has been constructed on land to which the occupants have no legal claim, or which they occupy illegally; and second, unplanned settlements having the privilege of some sort of legal land tenure and areas where housing is not in compliance with current planning and housing building regulations (otherwise known as unauthorized housing). Overall, the assessment covered a total of

Table 10.3 Population and household data of ITS on the governorate level

Governorate	Total number of ITS	Total ITS population/ governorate	Average number of people/ITS	Total number of households	Average number of households/ ITS
Balqa	13	1271	97.8	234	18
Irbid	4	341	85.3	64	16
Mafraq	15	1673	111.5	305	20.3

Source: Unicef (2013)

Table 10.4 Governorate-level distributions of land-tenure type and rent cost of ITS

Governorate	Public land tenure (percent of assessed households)	Private land tenure (percent of assessed households)	The average cost of rent per household/month in JOD
Balqa	0	100	43.25
Irbid	26.6	73.4	13.4
Mafraq	17.3	82.7	24.1

Source: Unicef (2013)

32 informal settlements, 603 households, and 3285 individuals across the governorates of Mafraq, Balqa, and Irbid. What has clearly emerged was that demographic indicators vary significantly on both governorate and ITS levels. In terms of absolute population size, informal settlements found in Mafraq were the largest and contained a total of 1673 people across 15 assessed settlements. Irbid, in comparison, displayed by far the lowest concentration of informal settlements (Unicef 2013).

Therefore, as summarized in Table 10.5, housing informality in Amman could be classified into three types, and each type has a sub-typology according to the legality of tenure. The origin of populations is excluded because of a lack of accurate information. These types are: illegal land subdivision (semi-informal areas or *ahya' al fuqara*—housing of the poor), illegal occupied land (squatting areas or Badoun areas), and hybrid areas occupied by Jordanian and non-Jordanian foreign refugees' camps (*sakan 'ashwa'i* or arbitrary housing).

Housing derived from illegal land subdivision (semi-informal areas or *ahya' al fuqara*—housing of the poor) is not developed through established or state-regulated procedures and does not utilize the recognized institutions of housing and housing finance. Nevertheless, such housing is developed on land for which the owner has legal tenure and a formal occupation permit. Semi-informal settlements are generally found in areas of essentially rural character located on the urban fringe that is interspersed with, surrounded by or adjacent to, undeveloped sites or sites that remain in agricultural use. These settlements often develop in advance of the principal lines of urban growth. Their construction is most noticeable during periods of rapid urban expansion, and they generally appear along major roads network (Soliman 2004a, b).

Housing on illegally occupied land (squatting areas or Bedouin areas) were built on steeply sloping ground characterized as the edges of hills *'Jabals'*, and others

Table 10.5 Informal housing typologies in Amman

Main typologies	A- Illegal Land Subdivision (Semi-informal areas) (Transjordan) (ahya' al fuqara) (Bedouin land) Bottom-up Approach		B- Illegal Occupied Land on the peripheries of camps. (Squatting areas and/or unauthorized housing) Mixing between Jordanian and foreigners Technocratic Approach		B- Foreign Refugees Camps (Hybrid areas) (sakan 'ashwa'i) Hybridization	
	Sub typology	Popul.	Sub typology	Popul.	Sub typology	Popul.
Sub typology	**A1- Inside city boundary**	**23,000**	**B1- Inside city boundary**	**16,000**	**C1- Palestinians**	**54,054**
	A1a- Marj Al Hamam	4500	B1a- *Wahdat*	5000	C1a- Al Hussein	23,226
	A1b- Umm Uthayna	6000	B1b- *Jofeh*	2500	C1b- Al Wehdat	27,138
	A1c- Tareq	5000	B1c- *Rimam*	3500	C1c- Talbiya	5000
	A1d- Basman	3000	B1d- *Nuzha*	3000		
	A1e- Wadi Haddada	4500	B1e- *Hia Amir Hassan Hassan*	2000		
	A2-Outside city boundary	**23,100**	**B2-Outside city boundary**	**31,212**	**C2- Outside city boundary**	**66,868**
	A2a- *Taibe*	7500	B2a- Wadi Abdoun	1248	C2a- Marka	30,402
	A2b- *al-Muqablayn*	2500	B2b- Al Hashemi	1692	C2b- Abu Alanda	8000
	A2c- *Marj al-Hamam*	2000	B2c- Lawziya	2040	C2c- Quweisma	10,000
	A2e- *Naur*	3500	B2d- Qaisya	3504	C2d- Naser	6000
	A2f- Hittin	4200	B2e- Nadhif	7146	C2e- Hanikeen	6966
	Hay Tafayla	3400	B2f- Nuzha	2262	C2f- Sahab	5500
			B2g- Musherfa	7836		
			B2j- Hay Tafayla	5484		
	Total Population	**46,100**		**47,212**	**215,544**	**122,232**

Source: Adapted from Soliman (2008) and updated in 2020. The typology is the author's, while the data have been gathered from various sources. It is important to note that the city boundary is defined in this study within a radius of 5 km from the old city center. Details of Palestinian-camp everyday life are illustrated in Ababsa (2012) and Ababsa and Abu Hussein (2020). Informal tented settlements are defined in Unicef (2013). Squatters areas and their characteristics in Amman derived from Al Daly (1999) and Al-Nammari (2013). Areas containing more than 200 households were included. Please note that the total population of informal housing in Amman is 214,234, while the national report (The Third National Report 2014) indicates at the national level a total of 450,000

were located in valley bottoms which are subject to flooding during wet weather, Some of the slopes are so steep that the ground floors of higher-level houses are on the same level and almost touching the roofs of the houses below. Squatter settlements have grown illegally, either on government land or on privately owned land.

Moreover, others have a purchased document known as *'Hejeh'*,[2] not recognized by the Department of Land and Survey (DLS). This category had purchased their plot from the previous 'owner', and it provides only an informal basis for asserting ownership and no legal standing.

Hybrid-settlement areas occupied by Jordanian and non- Jordanian foreign refugees' camps (*sakan 'ashwa'I* or arbitrary housing) were initially legally built on locations rented by UNRWA, but the sizes of individual land plots were below the standard of Jordanian regulation zones. Over time, with increasing pressure for affordable housing and an influx of Syrian refugees, arbitrary residential expansion on the peripheries of the official camps took place. Therefore, this type of informality mixes formal and informal areas that could be considered hybrid areas. This type of settlement includes residential units in formal areas (camps) that have temporarily or permanently acquired degrees of informality. Unlike residential units in informal areas, this type of informality relates to individual dwelling units on a case-by-case basis. In many cases, some units in a formal neighborhood or even an individual building, may be "formal" but others have either been expanded illegally or been modified over time into illegal configurations. Such transformations are most common in the major metropolitan areas in the Middle East.

The great increase in informal housing in the GAM was originally caused by the acceleration of refugees and informal urbanization in Jordan during the 1960s and the massive waves of refugees from Lebanon, Iraq, and Syria. According to Table 10.5, the total population living in informal areas in the GAM is 0.215 million. However, plans for the city never included areas where low-income refugees could be housed. As a result, today more than 10% of the GAM's population (2.17 million) live in areas that were developed informally. Interventions in these areas of the GAM are extremely complex, in part due to their location and the complexity and diversity of housing types described above, and to the diversity of national identities on the other.

Informality is no longer simply the domain of the poor; in fact, it has become a major condition of land ownership and housing for the urban lower-middle and middle classes. This has meant that the urban poor may now have little choice during a housing shortage of obtaining rented units within that sector. The identification of the mechanisms of informal housing development in Jordan indicated that their complexity is a result of a diversity of socioeconomic and socio-spatial characteristics, the role of the international donors, and political situations within the country. Informal areas are now the dominant features of the urban fabric of Jordanian cities. However, as their built environment continues to deteriorate, many questions arise. Should the government allow the continuity of growth of this sector? How should the government control future development of this type of housing? If controls are instituted, what will be the impact of regulation on the urban poor? What consequences would housing controls have for socioeconomic development within the

[2]'Hejeh' or a preliminary contract, is an informal document between the landlord and the buyer regarding how to claim the land ownership; it is not recognized by DLS.

country? How should the government house the urban poor? And, finally as a matter of practical concern, how much can the state do? If this chapter has shown anything, it is that the government alone cannot hope to house the nation's increasing urban population, nor meet the requirements of the refugees.

10.4 Political Economic Transitions

During the last two decades, five major socioeconomic and political transitions have been visited upon the Kingdom of Jordan. These are the financial crisis of 2008, rapid population growth and informal urbanization processes, massive waves of Syrian refugees, transformations of the political arena in the Middle East, and finally, the crises spawned by Covid-19.

First, the financial crisis of 2008 affected economic growth, trade, exports, tourism, and investment, in turn leading to an increase in the budget deficit and public debt. The annual direct cost of hosting Syrian refugees in the kingdom amounts to approximately $US 2 billion, while the annual indirect costs are estimated to be $US 3.1–3.4 billion. This is the case in a situation where Jordan is meant to take action towards the attainment of the Millennium Development Goals (MDGs) and Sustainable Development Goals (SDGs). Overall, considerable achievements were made during the first 10 years, especially in the areas of poverty eradication, maternal and child health, communicable diseases eradication, universal primary education, and environmental sustainability. The percentage of poverty in Jordan was 14.4% in 2010 compared to 13.3% in 2008. This confirms the need to: deal with poverty in its comprehensive manner; use available resources efficiently and effectively in supporting needy families; empower communities to increase their income; and directly support programs for beneficiaries (The Third National Report 2014; Hashemite Kingdom of Jordan 2017). The infant mortality rate was reduced from 34 per 1000 live births in 1990 to 17 per 1000 live births in 2012, while the under-five mortality rate was reduced from 37 per 1000 live births to 21 per 1000 live births over the same period. Maternal mortality rates were significantly lowered from 48 per 100,000 live births in 1992 to 19 per 100,000 live births in 2008. Universal primary education was achieved for boys and girls in 2005 (The Third National Report 2014).

Second, Jordan's population has multiplied sixfold during the period 1952–2017. The size of the population remains on the rise; growing from 5.103 million, 6.6 million, 7.0 million in 2004, 2015, and 2017, respectively (such figures do not include Syrian refugees) (see Table 10.2). At this pace, the population of the Kingdom is expected to reach 8456 million by 2025 (Hashemite Kingdom of Jordan 2017). At present, Jordan is also witnessing a demographic transformation as it heads towards a drop in the population under the age of 15 and an increase in the age group ranging from 15 to 69. Such racing urbanization and development has given rise to many foreseeable adverse outcomes, such as the increasing percentages of poverty and unemployment, the issue of providing decent housing for the lower income category, provision of infrastructure and social services, in addition to the environmental disturbance. Due to the broad geographical area, its topography, and the large and constantly growing population of Amman, the Greater Amman Municipality (GAM) was divided into 27 administrative districts that provide all of the municipal services as stipulated under the municipalities' law and amendments thereof. GAM provides distinctive municipal services with a view to further develop the city of Amman.

Third, over decades, the Kingdom, along with the government, was compelled to scuffle with the many consequences and limitations that clashed with the successive governments' ambitions to implement most of the strategies drafted to develop urban centers. Despite this situation, which still overshadows the region (and has done so for a few decades), and the consecutive compulsory migrations into the Kingdom, especially during the last 3 years (Syrian refugees; where the total population of non-Jordanian residents reached an astounding 2.92 million in 2015—1.4 million of which are Syrians—according to the Higher Population Council) (Hashemite Kingdom of Jordan 2017). This had driven the enormous growth of population and encumbered the Kingdom's infrastructure and limited natural resources, the different governments were able to realize the largest portion of such ambitions on the national level by introducing the Jordan Response Plan 2015 (JRP). However, the consequences and associated limitations have disrupted the implementation of such strategies and has even caused some to be abandoned. The demographic explosion of the refugee population, combined with the existing camp regulations, has had two consequences. First, density and rates of overcrowding have attained extremely high levels, sparking acute social problems and substandard environmental conditions. "UNRWA figures indicate that population density in the camps varied from 70,000 to 103,000 per km^2 in the early 1950s camps; and from 34,000 to 69,000 per km^2 in the post-1967 emergency camps".

Fourth, the Kingdom of Jordan as a part of the Middle East was hit by the Arab Uprising in 2011. As happened in Tunisia, Egypt, Yemen, Syria, and Libya, Jordanians protested asking for the three demands of the Uprising "Bread, Freedom, and Social Justice". These three words mean the protesters wanted to live a decent life with a good standard of living that secured job opportunities and allowed free speech, with which to voice their opinions on various development schemes, to participate in the decisions for their future, and to achieve social justice among the various strata of the society. Jordan witnessed a year of protest, which started on 14 January 2011, organized by the Muslim Brotherhood and the Jordanian Youth Movement "*Hirak*". The demonstrators protested against unemployment and inflation—in particular in food and fuel prices—calling for regime change and accusing King Abdullah II (in power since 1999 and widely regarded as a liberal when he ascended to the throne) of corruption (Sandels 2011). The protests have been routine and persistent ever since, with at least one civilian casualty and an estimated 10,000 people on the streets in November 2012. This crisis entailed a breakdown of the social contract between the state and citizens, a perception of growing inequalities, and a decline in satisfaction with life. Ordinary people became increasingly dissatisfied with their standard of living, with high inflation, with the negative impact of the 2008 global economic crisis, and with a growing food crisis (Hanieh 2013; Teti et al. 2018).

The Arab Spring made the Jordanian public more aware of these issues than ever before. In 2011–2012, protests took place almost every Friday, highlighting popular frustration with unemployment, poverty, and corruption. Importantly, the protests involved a wide spectrum of society (Jordanians, Jordanians of Palestinian origin, and unaffiliated youth movements—*Hirak*). Even members of the Jordanian tribes, traditionally viewed as the monarchy's key support base, were demanding to be treated as citizens rather than subjects. Street protests subsequently ebbed as Jordanians became more cautious following the violent collapse of neighboring countries. Nonetheless, criticism of the government has become increasingly commonplace in private spaces, with many Jordanians frustrated that economic reforms have not yielded long-promised tangible improvements in living standards.

In Jordan, there was both less support for the uprisings and stronger demands for reform rather than regime change. In Jordan, protesters demonstrated for changes in governance but not for the toppling of the king, and there was no regime change. King Abdullah II responded to protests with political and economic concessions, but these left the political system substantively unchanged. Since the 2011–2012 wave of protests, Jordan's leadership has promised to accelerate political reforms. In a series of five discussion papers, the king has asked citizens to participate more effectively in politics by choosing members of parliament according to their policy platforms rather than because of regional or kinship loyalties. He has asserted the need to hold elected officials to account and for the public to remain 'continuously engaged'. However, these discussion papers have remained an academic exercise, lacking corresponding measures to turn theory into policy (AlNasser 2016). As a reaction and to absorb the anger of the protesters, Jordan is committed to the 2030 Agenda and to leaving no one behind. The Government of Jordan opted to prepare its first Voluntary National Review (VNR) and presented it at the High-Level Political Forum during the July 2017 session (The Hashemite Kingdom of Jordan 2017).

Fifth, as Covid-19 swept the world, including Jordan, in 2020, the government has supported the development of sound measures to mitigate the effects of the pandemic on enterprises and workers (Fafo 2020). Among these effects are the decline in cash flow, reduced demand and supply, and the disruption in the value chains because of measures responding to COVID-19. This has affected the progress of the development of urban informality, shrinking the informal economy and eliminating informal construction processes.

However, the Kingdom faces many urban and socioeconomic challenges to tackle the rapid urbanization processes, scarce natural resources and a lack of financial resources, on one side, and the needs of the influx of foreign refugees on the other. First, management of the rapid urban growth is the primary challenge that faces the Kingdom. In the context of his Majesty King Abdullah II's son of Al-Hussein endeavoring to bring about a paradigm shift in the path towards sustainable development in Jordan, a number of development and duty-free zones were launched in various areas of the Kingdom in 2008 and put it into practice in recent years. Such a new approach was aimed at realizing social equality through the distribution of the development outcomes to all governorates and regions, in addition to creating job opportunities, reducing poverty, and achieving economic growth, in addition to the establishment of *Aqaba* Special Economic Zone (ASEZA) in 2011 to manage and regulate the *Aqaba* special economic zone. Furthermore, the Petra development tourism regional authority was established in 2009 and modified in 2015 with the goal of developing the regional economic, social, cultural, and touristic aspects, as well as contributing to developing the local community. The authority is managed and supervised by a council called "the Council of Commissioners".

Second, a relationship between urban and rural development is essential to adapt development as the spinal cord for developing Jordanian society and coming to grips with the reality that no comprehensive development may be achieved without addressing the development of the rural sector and reducing the dichotomies between the countryside and the city, or what is called "integrated development" which begins from the countryside orienting itself towards the city and vice versa. Third, the needs of youth in urban areas caused the insertion of youth issues in the agenda of national priorities, and the National Youth Strategy 2005–2010 was developed and even enhanced in 2015. As prepared by the Higher Youth Council in collaboration with the United Nations Development Program (UNDP) and the United Nations Children's Fund (UNICEF), the strategy comprises a national program to build participatory and an interactive national strategy for youth in Jordan. The key objective of the strategy is to ensure the participation of youth as real actors in various stages of the project, and it considers their capabilities as a key criterion for their participation in the development of the nation.

Fourth, meeting the challenge also encourages integrating gender in urban development plans. Indeed, the gender difference results in a difference in the concerns and priorities of both women and men. This fact requires that the needs of women and men alike should be taken into account in planning, and that the active participation of both in identifying their own needs and decision making by the same proportion as their presence in the community; especially since the percentage of females in Jordan is 47% (in 2017) of the total population (Hashemite Kingdom of Jordan 2017). It is thus imperative to enhance women's participation in urban development plans and to enhance the decision-makers and individuals' awareness on the issue of women participation and their needs in the urban environment because women are better able to identify their needs in urban spaces, and they have a major role in urban informality.

Fifth, in 2008, the National Council for Family Affairs has introduced a participatory methodology to prepare the Jordanian National Strategy for the Elderly. This program was under the heed of His Majesty King Abdullah II and Queen Rania Al Abdullah over the elderly and their Majesties' initiatives in this matter, and to ensure the provision of better health and social care, and improve the services offered to them. This effort was in collaboration with the Ministry of Health, the Ministry of Social Development, the Jordanian Nursing Council, the World Health Organization (WHO), and specialists from the Jordanian universities and NGOs.

Historically, Jordan has been accustomed to sudden influxes of populations fleeing conflict from neighboring countries and seeking safety and security within Jordan's borders. The country has opened its borders to Palestinians, Iraqis, and others in need of shelter many times in the recent past. Since 2011, the kingdom has been called upon again to accept the stream of Syrian refugees fleeing the violence in their homeland. Overall, the number of registered refugees with UNRWA and UNHCR has reached 2.9 million, making Jordan the largest refugee-hosting country worldwide, when compared to the size of its Jordanian population (The Hashemite Kingdom of Jordan 2017). During the past decade, the country has also faced several challenges. Most notably, the tremendous impact of the Syrian crisis and resulting population growth have affected overall development gains and are associated with the acceleration of urban informality in Jordan. According to the 2017 census, Jordan's population stands at 10.05 million (see Table 10.2). Approximately 30% of the population are non-Jordanians, and more than 1.3 million are Syrians, with around 89% of Syrian refugees living in host communities, mostly in the capital Amman, and in the northern governorates. Such circumstances have increased pressure on Jordan's limited resources, especially water resources (Mustafa et al. 2016) and created severe stress on its public services, including health, education, energy, infrastructure, municipal services, and solid-waste management. Water resources constitute a political problem rather than a social problem, and Jordan is currently suffering from water shortages. According to the General Census of Population and Housing for the 2004, the traditional per capita share of water of about ~147 l/day dropped to 89 l/day in 2013.

As the conflict in Syria now enters its ninth year, Jordan hosts 1.4 million Syrians, of whom 646,700 are refugees. Eighty-five percent of refugees live outside camps in some of the poorest areas of the country, and a significant proportion is classified as extremely vulnerable. Approximately 23.5% of all Syrian refugees are women, and almost 53% are children, 18% of whom are under 5 years of age. Providing for their needs has heavily impacted Jordan's public finances, increasing government expenditure on subsidies, public services, housing, and security, and further compounding the negative economic consequences of regional instability. The Jordan Response Plan 2015 (JRP) was introduced to bridge the divide between resilience and humanitarian systems and to reconcile the development programming objectives. It is aimed to meet the immediate needs of host communities and refugees until the year 2030 and is being modified by the Jordanian Policy Paper presented at Habitat III.

A few general insights and urban challenges can be distilled. First, there is an inherent tension between management practices and planning practices in a sustainability transition in Jordan. This is no surprise, given that transition management navigates between urban regimes and institutions, on the one hand, and alternative urbanization visions and practices and international donors, on the other hand. The formulation of the urban informality in Jordan is altered formally by politicians, international donors, and nationalism (Khirfan 2019). Applying the MLP and the MPs to the Jordanian status quo, the niche level is almost integrated with the regime level, and the main actors of the formulation of urban informality are local politicians, various nationalisms, and international donors.

This general planning challenge reenters at the three transition management levels via many pathways. At the strategic level, long-term urban informality objectives might bring opposing visions and interests together; so, it thereby obfuscates underlying conflicts among various actors and stakeholders for urban informality areas and urban life, for both refugees and local citizens. At the tactical level, many international actors and local knowledge schemes seem to dominate the upgrading projects (architects, business interest, policy actors, and international donors) and marginalize residential voices. The specific practices of the operational level has depended on circumventing technical institutional rules (e.g., on housing, safety, financing, water supply) to experiment with urban informality and new spatial plans, mainly for the creative new urban expansion (International Rescue Committee 2018). On the reflexive level, this is about learning from lessons learned and the outcomes of the integration between the regime and niche in urban informality transitions which adjusts various variables.

The application of the ASUST perspective is similar to the Lebanese situation in which the first three trends of policy strategy, namely, credibility, consistency of dimensions, and policy processes (coherence of processes), are affected by the state and the international donors. Donor countries asked for a clear vision to overcome the problem of refugees. On the other hand, the three trends systematic of change (four perspectives), systematic directions of transitions, and instrument mix comprehensiveness are controlled by civil society, NGOs, CBOs, and professionals as subsequently explained. The grassroots participation provides the knowledge and information to develop effective tools to operate on the ground. The three trends of the ASUST are linked and interrelated with the three pillars and the mechanisms of urban informality.

However, the phenomenon of urban informality in Jordan does not follow the mechanisms of the Global South, rather it is established, legally, from top-down processes controlled by the regime in which local and international donors have played the primary role in this phenomenon. As depicted in Fig. 10.4, it seems that the sociotechnical perspective in relation to urban informality is managed and accelerated through integration between the regime level and the niche level. In this case, the niche level is integrated and constitutes a part, not isolated, of/from the regime level. Legitimization of the existence of urban informality has political moral support from local and international politicians. Therefore, it appears that the level of the niche has no impact, or at least does not exist, or constitutes a part of the regime, on the spread or upgrading of urban informality, rather it is a part of, and interacted with, the regime level.

Perhaps most challenging is that the lack of any systemic evaluation at the reflexive level seems to render impossible critical scrutiny and assessment of the proposed objectives and alternative evaluation criteria. Future applications of transition management with a similar historical and socio-spatial context (of urban regeneration) should consider the urban challenge of transition management, especially regarding monitoring and evaluation.

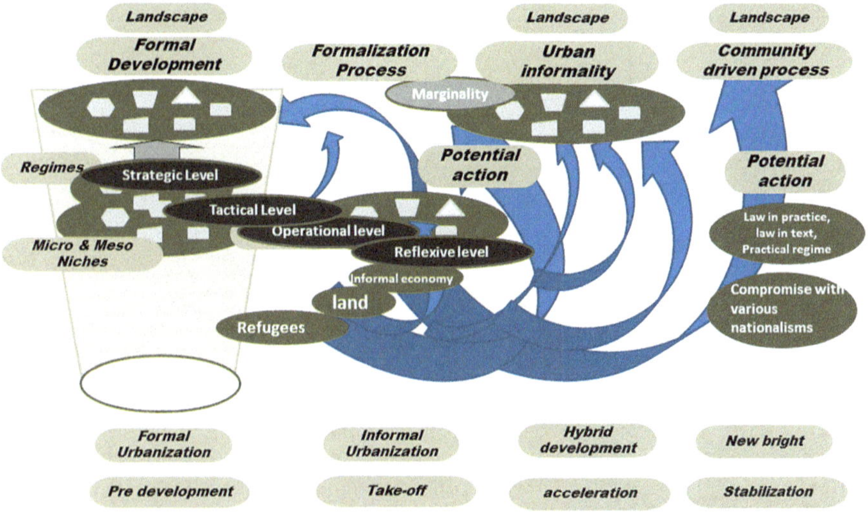

Fig. 10.4 Management sustainability transitions on urban informality in Jordan. (Source: Adapted from Koehler et al. (2017), Grin et al. (2010), Geels and Schot (2007) and Frantzeskaki et al. (2017))

This does not deliver a more 'effective process' or 'better outcomes' per se'. In fact, it might even complicate things and make cooperation and urban governing more complex. What it does do, however, is to question underlying socioeconomic and city socio-environmental assumptions of 'a sustainable' by challenging the taken-for-granted stakeholders (local, regional and international) and frontrunners, abstract long-term sustainability visions, and the gains of local transition projects. Importantly, each and every transition management context is different, which means that such urban issues and concerns differ case by case.

10.5 Social Segregation and Marginality

The Kingdom of Jordan accommodates a variety of nationalities, combining Jordanians, Transjordanians, Palestinians, Iraqis, Syrian, Yemeni, and other nationalities from neighboring countries. These variations of nationalities have reflected on the spatial formulation of the urban fabric of the Jordanian countryside and even more on the decision for managing rapid urban growth. The debate on the Jordan Gates project exemplifies the controversy and diversity of thought among local people and decisions makers (Abdul-Hadi 2016), and even the ways to manage the transitions processes that face the country. The two towers of the Jordan Gates are an urban ruin, a monument to the entrenched system of incentivized development, haphazard administrative decisions, and, above all, to the disregard of public

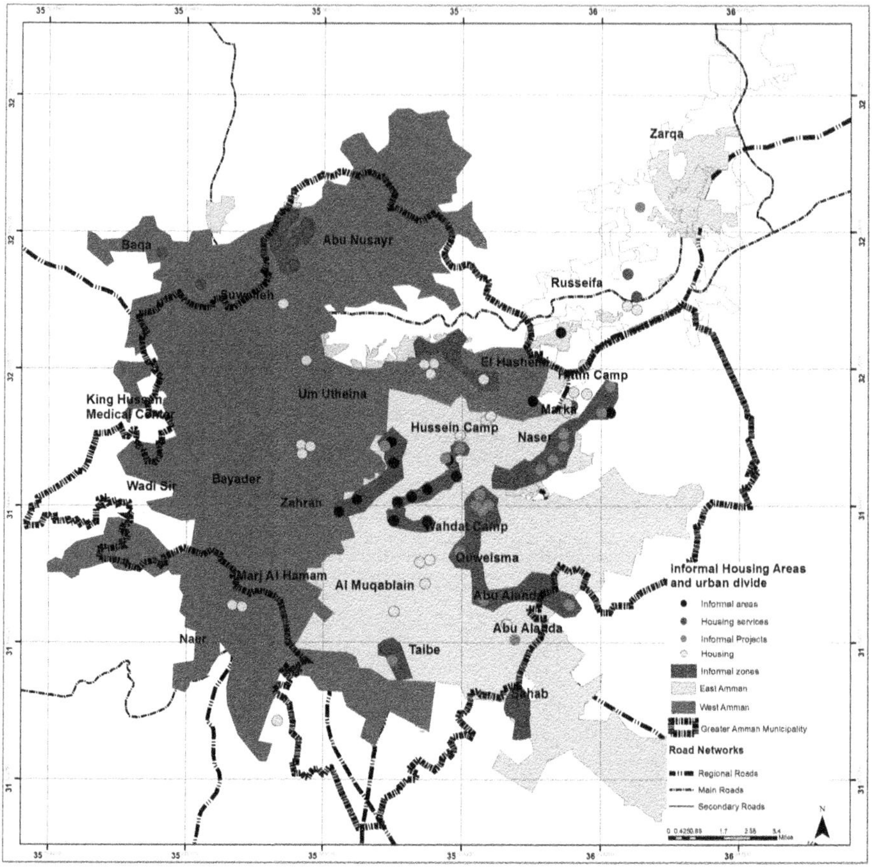

Fig. 10.5 Map of West and East Amman. (Source: The author, adapted from The Hashemite Kingdom of Jordan 2017)

opinion and social welfare. A scholar quoted in the report noted that Jordan Gate "can be seen as one of the worst decisions taken by the Municipality of Amman" (Beauregard and Marpillero-Colomina 2011). This controversy reflects the division of the society into diverse categories in which each stratum of the society has its own role within its own locality.

Amman is a city with twin downtowns (see Fig. 10.5), one of which is the old downtown, called *Wasat Al-Balad*, Arabic for "city center", is the place to go for those on a limited budget. Their shops sell out-of-date designer clothes, handbags, and shoes from brands such as Prada, Gucci, and Louis Vuitton, for as little as one-tenth the price in the country's malls or abroad. The new downtown is west of the old quarters in the central *Al-Abdali* district, where the army and intelligence headquarters were replaced by the high-rise buildings now comprising Amman's central skyline. In Shami's article (Shami 2007), he writes "East Amman is the older part where residential dwellings are on the hillside and commercial areas are distributed in a traditional linear layout along the valleys. West Amman, on the other hand, is less densely populated and more fashionable. Most economic investments are centered in different districts there such as *Abdali, Shmesani, Swedish, and Abdoun*. However, a large physical and social contrast exists between "East and West Amman" (Elsheshtawy 2008; Daher 2008; Abu-Ghazalah 2010; Al-Shawabkeh and Alhaddad 2018).

The result is a growing divergence within Jordanian cities: between the *old* and the *new* part of the urban agglomeration, a "Great Divide" between "establishing", and "emerging" urban spaces. The former space is the new and modern established districts, gated communities of the elite such as Royal Village (Abu-Hamdi 2017); while the latter is the old crowded areas and the *peri-urban informal* settlements of marginal groups. Amman is really made up of the other 'Ammans' within the city, the formal city, and the informal city, found in "the Jordanian countryside", like other cities in the Middle East, Beirut, and Cairo.

Therefore, this spatial segregation divides Amman into two cities in one city, or "the two Ammans" (Potter et al. 2009) (see Fig. 10.5). However, the physical segregation of the city has divided its social structure into two categories: the poor and the rich (Aljafari 2014). Residents talk openly of two Ammans, although in truth there are many: Eastern Amman (which includes the old downtown) is home to the urbanized poor: conservative, more Islamic in its sympathies, and has vast Palestinian refugee camps on its fringe. "Western Amman is a world apart, with leafy residential districts, trendy cafes and bars, impressive art galleries, and young men and women walking openly arm in arm" (Potter et al. 2009). In other words, the city reflects the Western and Eastern cultures, the poor and the rich, the traditional and the modern, legal and illegal, and formal and informal. It is a dual city, a dual culture, and a dual society which results in social segregation among Jordanians.

Spatial inequalities are not only a forerunner of social and economic divisions; they, in turn, cause further inequalities and various forms of exclusion and marginalization. In Jordan's capital, for instance, 97% of households in Western Amman have computers, while in Eastern Amman the proportion is no more than 8%. Not surprisingly, in Western Amman, more than 50% of males and 20% of females earn more than US$ 1400 monthly, compared with only 2% for males and less than 1% for females in Eastern Amman (UN-Habitat 2013). Furthermore, spatial inequalities are reflected by the quality of housing construction, the size of housing units, and the types of building materials used for housing construction. Of course, quality factors contribute in determining the differences in quality as applied by different cultural and social settings (Sweis et al. 2014), but the gap in spatial inequalities should be narrowed to promote social inclusion between various strata of the society. The tensions running through the Palestinian refugee camps in Jordan have resulted in the exercise of three forms of power that are simultaneously at work in the camp, namely, sovereignty, discipline, and government (Oesch 2017). The mutual effects of power are the inclusion of the camp and its dwellers within the territory of the Jordanian state, as well as within the neoliberal city of Amman, although in a particular way, and while maintaining the character of the camp as an excluded humanitarian and temporary space. Thus, the Palestinian refugee camps in Jordan form "a state within the state", a duality of power as in the Lebanese case.

The provision of public services and affordable housing and their perceived quality and equity are the main components affecting how populations judge the responsiveness of their local and national institutions. As indicated, low-income and lower-income groups suffer from access to reasonably-priced housing plots, housing mortgage accessibility, and basic urban services so they are forced to live in informal areas. The large numbers of Syrian refugees, estimated at 613,920 in October 2013, presents an almost 10% increase in the total population of Jordan. This increase translates into a vast immediate demand for housing, roughly estimated at over 86,000 housing units on top of the estimated annual average need of 33,000 units by Jordanians (The Hashemite Kingdom of Jordan 2013).

In the period leading up to the Arab Uprisings, a majority of ordinary people were not benefitting from economic growth, inequalities were increasing, and corruption was rife. During the Uprisings, citizens made clear that they were demonstrating because they wanted a better life, more and better jobs, improved public services, and an end to corruption. It was clear that social justice lay at the heart of citizenship: they were demonstrating against exclusionary practices that denied them the claim to their socioeconomic rights. Certainly, according to Arab Transformations survey data, by 2014 citizens thought that on average their families were worse off than they had been in 2009, with the most noticeable decline in Jordan. On a scale going from 1 to 10, the mean declined in Jordan as just over two-thirds rated their household situation as good as in 2009, only just over a quarter did so in 2014. This brief discussion on households' economic situation illustrates both the challenges and the frustrations that have characterized the post-Uprisings period irrespective of the changes (or lack thereof) that took place on the political and institutional levels (Teti et al. 2018). In Jordan, however, by 2014 citizens were marginally more inclined to think that the government was at least making a reasonable effort to tackle corruption (increasing from 36% to 44%), although it still fell well short of half. This may be because the government had indeed been stepping up the fight against corruption, with amendments to the law in 2012 and 2014, and the Arab Convention against Corruption coming into force in 2013.

Probably even more pervasive than bribes at the level of the ordinary citizen is *Wasta*—the use of connections, influence or clout, which has become deeply embedded in regional societies and is often regarded as a family obligation, as in the most of cities in Middle East. *Wasta* smooths the way to jobs, promotion, bonuses, pay increases, positions of responsibility, university places, and much else in business and government. Those that have *Wasta* can jump the queue and acquire permits, get jobs, obtain favorable rulings from agencies, obtain government contracts, and benefit from government rules that limit competition. Its effect is that who you know is more important and more valuable than what you know or what you can do, and this, as in other forms of corruption, hampers economic development, and impacts business and reduces productivity, both in the public and the private sectors. Contrary to monetary bribes, *Wasta* is based on an economy of favors. It is firmly and widely believed in Egypt, Jordan, and Lebanon that *Wasta* is what gets one employment and, mostly, that no route that does not involve *Wasta* will do so. Between 2011 and 2014, citizens in Egypt and Jordan thought it had become more difficult to get any job at all without *Wasta*, an increase from 18% to 67% in Egypt and 10–74% in Jordan (Teti et al. 2018).

The matter of social cohesion in Jordanian host communities is a complex issue because it is the product of intertwining levels of relationship between community groups/tribes, between Jordanians and their authorities, between Jordanians and Syrian refugees, and also between Syrian refugees and local authorities. Cutting across these relationships are factors of age, gender, income level, and type of living environment (rural/urban). Tension and unrest in Jordanian communities, including in the north, started increasing before the arrival of Syrian refugees. Recent studies have shown that a combined deterioration of purchasing power (inflation, rising unemployment), social equity, and an increasing realization of governance issues (alleged corruption, lack of responsiveness, centralized system) breed frustration among the population and negatively impact the level of confidence that Jordanians have in their institutions (The Hashemite Kingdom of Jordan 2013; see, e.g., Abu-Hamdi 2017). Regional factors greatly shape local perceptions and have direct repercussions in daily life in the Jordanian society, as is the case with the Syria crisis.

Most commonly, tensions that sometimes flare up into open confrontations involve different tribes or disputes between citizens and government, and even more conflicts between citizens and citizens as happened in the *Jabal al-weibdeh* area (Dilworth 2017) and The *Talbiyeh* Camp (Al-Nammari 2013). The increase in violence on university campuses is also a recent phenomenon. However, acknowledging a rise in tensions in Jordanian communities should not overshadow the fact that the great majority of Jordanians continue to feel safe and secure and that the incidence of violent acts remains low compared with other countries in the region. Before the 1967 war, divisions between Jordanian–Jordanians and Palestinian–Jordanians were primarily socioeconomic, but, following the war, Palestinian identity was marginalized in favor of tribalism and narratives of autochthony. This dichotomy resulted in emphasis being placed on tribal autochthony as being truly "Jordanian" in opposition to the Palestinian "other" as a way to mitigate the loss of the ideological strength of pan-Jordanianism following the Six-Day War (Dilworth 2017). In "Amman Is Not a City," Seteney Shami (2007) writes, "the overarching urban narrative in Amman is one that paradoxically emphasizes its fragmentation and deprives the city of claiming a unique urban identity".

On the other hand, humanitarian programs got dealing with Syrian refugees constitutes major national and international efforts to address the Syria crisis within the framework of a coordinated broad-spectrum response. The Jordan Response Plan 2015 (JRP) seeks to bridge the divide between resilience and humanitarian systems and to reconcile the programming objectives, funding mechanisms, and operating systems that often run parallel to each other in addressing short-term people-centered needs, in addition to medium- and longer-term systemic and institutional considerations. The JRP 2015 adopts a resilience-based approach to respond to and mitigate the effects of the Syria crisis on Jordan and Jordanian host communities. The aim of resilience-oriented programming is twofold: first to ensure that shocks and stresses do not lead to long-term deterioration in the well-being of a particular individual, household, system or institution, and second to build capacity development to absorb future shocks and deal appropriately with related stresses (Hashemite Kingdom of Jordan 2014). In addition, housing, a major component of any growth initiative, was not a major substantive focus of the strategic plan of Amman 2025, and the planners could not easily address the refugee situation, a condition over which the municipality has almost no control (Beauregard and Marpillero-Colomina 2011).

10.6 Conclusion

The US-led invasion of Iraq in 2003 drove many thousands of Iraqis into Jordan and created space for the rise of extremist groups. In addition, the decades-old Arab–Israeli conflict to Jordan's west has made the country home to nearly 2.0 million Palestinians. The Hashemite monarchy in Jordan has historically favored East Bank "Transjordanian" tribes and families, rewarding them with positions in the civil service and military that come with job security and benefits, while Jordanians of Palestinian origin tend to dominate the private sector and the informal economy. This is the real situation of the Kingdom of Jordan, as it has many nationalities, each of which has its own traditions and roles with its own locality. Therefore, this variation of nationalities has reflected on the urban fabric of Jordanians' cities and diversification between poor and rich, between Jordanians and non-Jordanians has resulted in a divided urban fabric of the cities into two or three cities: formal, informal, and hybrid cities. This division segregated the society in which many conflicts appeared between citizens and citizens, between citizens and refugees, and between citizens and local municipalities. Therefore, the Kingdom of Jordan has a unique environment in its socioeconomic and spatial structures different than other countries in the Middle East. Pro rata, The Kingdom of Jordan is considered the first country worldwide in hosting the refugees from neighboring countries.

The forced migration caused by the Six Days War in 1967, the civil war in Lebanon (1975–1990), the crisis in Iraq in 2003, and the civil ruin in Syria in 2012 has contributed to social and economic transitions that have affected the housing market and infrastructure services, resulted and caused sudden elevations of prices of houses and land, and undermines the financial capability of the citizens, creating a gap between supply and demand in housing, difficulty in houses its nationals, poor sustainability of housing credit support programs, the lack of small residential land plots, and the spread of urban informality into the hills abutting Jordan's mountains.

Beginning in 1946, the early period of Jordanian independence, most of Jordanians' cities have been established based on social diversity according to the place of origin. The Kingdom of Jordan was fertile land for welcoming refugees from neighboring countries including: the continuing influx of Palestinians refugees (1948 and 1967); the flow of Iraqis refugees in 2003; and the recent Syrian and Yemeni refugees after the Syrian and Yemeni Crises starting in 2011. This huge number of refugees associated with internal rapid urbanization processes and the rapid transitions of the socioeconomics of the oil countries have accelerated arbitrary urban growth in the Jordanian territories as urban informality grew in Jordanians' cities. Thus, urban informality in Jordan has surged out of control, and it can always be linked with the high cost of properties and the strained mortgages for low-income households. This is obvious and reflects the great housing demands associated with migrations due to regional political crises (Palestine, Lebanon, Iraq, Yemen, and Syria) (Al Daly 1999; Dumour and Ababsa 2008).

However, despite the work and effort, urban informality in Jordan is rising at an annual rate of 4.3% (Abed et al. 2015). In addition, it is obvious that most of urban informality's growth is situated in the capital, Amman City. As a result, Amman has developed several crowded informal settlements characterized by high density (based on standards), over 20,000 inhabitants per km^2. Furthermore, locals and residents relocating from countryside and suburbs to the capital Amman for jobs make matters worse. Generally, these settlements were increasingly marginalized because of lack of secured tenure associated with limited access to a decent housing unit, affordable land plots, and poor investments in services at architectural and planning levels (Dumour and Ababsa 2008).

The 'open source' and 'bottom-up' character of some upgrading projects (Sweis et al. 2014) make the 'here and now' indeterminate. The transition dynamics of urban informality and the urban management in Jordan, therefore, relies on the ways in which urban informality is imagined and enacted in the future. This is not only a topical issue for the practice of transition urban management, but also for conceptual work in the field of sustainability transitions in Jordan. The Jordanian case differs in that around one-third of the national population is foreigners, and the rest constitutes various Jordanians' nationalities. The international donors play a great role in the development and transitions processes, and nationalities have a great impact on any decision on development, and finally, Jordan is surrounded by a region of conflicts, mainly Israel, apart from significant water-supply conflicts. Ultimately, the Jordanian case has opened up the analysis by pointing toward the necessity to study the extent to which 'ambiguity' represents a form of governmentality of forced migration (see also Stel 2016), as well as how other types of 'enclosed' spaces generally (e.g., administrative detention centers, migrant worker camps, and so on) produce multiple subjectivities. The Palestinian and Syrian camps represent a spatio-political technology aimed at creating a differentiation in the city, by producing what Oesch (2017) calls a system of variegated citizenship according to which camps dwellers are both included to some extent on the juridical–political and socioeconomic levels, while being excluded administratively and through spatial marginalization.

With the presence of challenges in state-led transitions, it remains to case studies in Amman, the city that practices state-based sustainability principles, to analyze how local dynamics of resources and organizational landscape are integrated into the transition process. These urban-transition management challenges also have broader socio-spatial dimensions that need to be addressed in future transition management practices. Urban informality, as sociotechnical landscapes, are inherently decentered and hierarchical, with centers and peripheries, creating unequal access to green, healthy, and more livable environments. In Amman, there has been a historical division marked by the schism between the wealthier Western parts versus the poorer Eastern parts (often populated by Jordan's poor and recently arrived immigrant families). Arguably, for many residents from the Western parts of Amman, the regeneration of the informal areas (including its refugees' residents) is part of the old city center and its expansion. It is crucial to note that the rise of sustainable development is a work-in-progress. It is a contingent process full of socioeconomic and spatial struggles. The democratic question remains a challenge for transition management and sustainability transitions. Uneven power relationships and socio- spatial developments can simply not be addressed and solved with a guideline or a checklist. They should remind transition researchers and practitioners that urban challenges are here to stay.

References

Ababsa, M. 2012. Public Policies Toward Informal Settlements in Jordan (1965–2010). In *Popular Housing and Urban Land Tenure in the Middle East*, ed. M. Ababsa, B. Dupret, and E. Denis, 259–282. Cairo: The American University in Cairo Press.

Ababsa, M., and A.Z. Abu Hussein. 2020. Metropolitan Amman: Comprehensive Climate Plans. In *Volume II of Greater Than Parts: A Metropolitan Opportunity*, ed. Shagun Mehrotra, Lincoln L. Lewis, Mariana Orloff, and Beth Olberding. Washington, DC: The World Bank.

Abdul-Hadi, A.O.B. 2016. *Nationalism in the Middle East: The Development of Jordanian National Identity Since the Disengagement of 1988*. Durham: Durham University, PhD.

Abed, A., A. Tomah, and D. Dumour. 2015. Assessment of Slums' Upgrading Interventions: Case Study Jabal Al-Natheef, Amman, Jordan. *Innovative Systems Design and Engineering* 6 (6): 1–8.

Abu-Ghazalah, S. 2010. Global Economic Forces' Effect upon Urban Planning: Greater Amman Case Study. *Journal of Urban Planning and Development* 136 (2): 125–34.

Abu-Hamdi, E. 2017. Neoliberalism as a Site-Specific Process: The Aesthetics and Politics of Architecture in Amman, Jordan. *Cities* 60: 102–112.

Alon, Yoav. 2007. *The Making of Jordan: Tribes, Colonialism, and the Modern state*. London: I.B. Tauris & Co Ltd.

Al Daly, J. 1999. *Informal Settlements in Jordan: Upgrading Approaches Adopted and Lessons Learned*. Amman: Housing and Urban Development Corporation.

Al Nsour, J.A. 2016. Managing Urban Growth in the City of Amman, Jordan. *Cities* 50: 93–99.

Al-Hasanat, S. 2003. Housing Information System. *Sixth Annual Seminar on Housing Sector Assessment*. Amman: Housing and Urban Development Corporation, 107–113.

Al-Homoud, M., S. Al-Oun, and A. Al-Hindawi. 2009. The Low-Income Housing Market in Jordan. *International Journal of Housing Markets and Analysis* 2 (3): 233–252.

Aljafari, M. 2014. *Emerging Public Spaces in the City of Amman, Jordan: An Analysis of Everyday Life Practices*. PhD thesis. Dortmund: Dortmund University.

Al-Nammari, F. 2013. Participatory Urban Upgrading and Power: Lessons Learnt from a Pilot Project in Jordan. *Habitat International 39*: 224–231.

AlNasser, H.W. 2016. *New Social Enterprises in Jordan Redefining the Meaning of Civil Society*. London: The Royal Institute of International Affairs.

Al-Omari, Z. 2003. Reality of Housing Sector as the Private Sectors See It. *Sixth Annual Seminar on Housing Sector Assessment*. Amman: Housing and Urban Development Corporation, 131–141.

Al-Oun, S. 1997. *Livestock Marketing in the Jordanian Badia*. PhD dissertation. Durham: University of Durham.

Al-Shawabkeh, R.A. and Alhaddad, I., L. 2018. Project: High Density Mixed Use as an Effective Scheme in Applying Sustainable Urban Design Principles in Amman, Jordan. *International Journal of Humanities and Social Science* 7(10): 57–78.

Barakat, N., and M. Tuffaha. 1995. Private Sector's Role in Housing Limited and Low Income Households. *Housing Sector Assessment Seminar*. Amman: Housing and Urban Development Corporation, 3–12.

Basboos, A. 1995. Housing Financing. *Housing Sector Assessment Seminar*. Amman: Housing and Urban Development Corporation, 2–19.

Beauregard, R.A., and A. Marpillero-Colomina. 2011. More than a Master Plan: Amman 2025. *Cities* 28: 62–69.

Blitz, D. 2014. *The Arab Spring: A Parsimonious Explanation of Recent Contentious Politics*. Nijmegen: Radboud University.

Cram, F., and D.M. Mertens. 2015. Transformative and Indigenous Frameworks for Multimethod and Mixed Methods Research. In *The Oxford Handbook of Multimethod and Mixed Methods Research Inquiry*, ed. S. Hesse-Biber and R.B. Johnson, 91–109. New York: Oxford University Press.

Daher, R. F. 2008. Amman: Disguised Genealogy and Recent Urban Restructuring and Neoliberal Threats. In *The Evolving Arab City: Tradition, Modernity & Urban Development*, ed. Elsheshtawy, Y., 37–68. London and New York: Routledge.

Debruyne, P. 2014. *Spatial rearticulations of statehood: 'Jordan's geographies of power under globalization*. Department of Political and Social Sciences: Ghent University, PhD.

Dilworth, G. 2017. *Jabal Al-Weibdeh: A Counter-Memory of Amman. A Case Study in the Resistance of Memory*. University of Virginia, Independent Study Project (ISP) Collection. 2666.

Dumour, D., and M. Ababsa. 2008. Wadi Abdoun: From Slums to Towers the Upgrading Policy of an Informal Settlements in Amman, Jordan. *Ninth Mediterranean Research Meeting, Workshop no.4 Public Policies and Legal Practice Towards Informal Settlements in the Middle East and Egypt*. Florence: Monticatini Terme, 12–15.

Elsheshtawy, Y. 2008. The Great Divide: Struggling and Emerging Cities in the Arab World. In *The Evolving Arab City: Tradition, Modernity and Urban Development*. ed. Elsheshtawy, Y., 1–26. London and New York: Routledge.

Fafo. 2020. *Impact of the COVID-19 Pandemic on Enterprises in Jordan*. Oslo: Institute for Labour and Social Research.

Francis, Alexandra. 2015. *Jordan's Refugees Crisis*. Washington, DC: Carnegie Endowment for International Peace.

Frantzeskaki, N., V.C. Broto, L. Coenen, and D. Loorbach. 2017. *Urban Sustainability Transitions*. New York: Routledge.

Geels, Frank W., and Johan Schot. 2007. Typology of Sociotechnical Transition Pathways. *Research Policy* 36: 399–417.

Grin, J., J. Rotmans, and J. Schot. 2010. *Transitions to Sustainable Development: New Directions in the Study of Long Term Transformative Change*. New York: Routledge.

Hanieh, A. 2013. *Lineages of Revolt: Issues of Contemporary Capitalism in the Middle East*. Chicago: Haymarket Books.

Hashemite Kingdom of Jordan. 2014. *Jordan Response Plan 2015 for the Syria Crisis*. Amman: Ministry of Planning and International Cooperation.

———. 2015. *General Population and Housing Census 2015*. Amman: The Department of Statistics.

———. 2017. *Jordan in Figures*. Amman: Department of Statistics.

International Rescue Committee. 2018. *From Response to Resilience: Working with Cities and City Plans to Address Urban Displacement, Lessons from Amman and Kampala*. London: International Rescue Committee.

Kadhim, A.M., and Y. Rajjal. 1988. City Profile: Amman. *Cities* 5 (4): 318–325.

Khirfan, L. (ed.). 2017. *Order and Disorder: Urban Governance and the Making of Middle Eastern Cities*. Montreal: McGill-Queens University Press.

———. 2019. Planning and Unplanning Amman: Between Formal Planning and Non-Traditional Agency. In *The Palgrave Handbook of Bottom-up Urbanism*, ed. Konrad Kickert and Mahyar Arefi, 201–218. Cham, Switzerland: Palgrave Macmillan.

Koehler, J., F. Geels, F. Kern, E.. Onsongo, and A. J. Wieczorek. 2017. *A Research Agenda for the Sustainability Transitions Research Network.* Technical. Amsterdam: STRN Working Group, STRN.

Kumaraswamy, P. R. (ed.). 2019. *The Palgrave Handbook of the Hashemite Kingdom of Jordan.* Singapore: Springer Nature Singapore Pte Ltd.

Mertens, D.M. 2012. Transformative Mixed Methods: Addressing Inequities. *American Behavioral Scientist* 20 (10): 1–12.

Milton-Edwards, B., and P. Hinchcliffe. 2009. *Jordan: A Hashemite Legacy.* Oxon: Routledge.

Moore, P.W. 2004. *Doing Business in the Middle East: Politics and Economic Crisis in Jordan and Kuwait.* Cambridge: Cambridge University Press.

Mumtaz, B., I. Heggie, and E. Case. 2015. *Jordan Affordable Housing Programme.* Amman: UN-Habitat.

Municipality of Greater Amman. 1993. Greater Amman: Urban Development. *Cities*, February, 37–49.

Mustafa, D., A. Stamm, and L.M. Scott. 2016. Water User Associations and the Politics of Water in Jordan. *World Development* 79: 164–176.

Nowar, M.A. 2006. *The Development of Trans-Jordan 1929–1939: A History of the Hashemite Kingdom of Jordan.* Reading: Garnet Publishing Limited.

Nuseir, A., M. Tuffaha, and S. Momani. 2003. *Housing Strategies and Its Impact on Housing Sector. Housing and Urban Development Corporation.* Amman: Sixth Annual Seminar on Housing Sector Assessment, 215–222.

Oesch, L. 2017. The Refugee Camp as a Space of Multiple Ambiguities and Subjectivities. *Political Geography* 60: 110–120.

Potter, R.B., D. Darmame, N. Barhamb, and S. Nortcliff. 2009. "Ever-Growing Amman", Jordan: Urban Expansion, Social Polarisation and Contemporary Urban Planning Issues. *Habitat International* 33: 81–92.

Sandels, A. 2011. *JORDAN: Thousands of Demonstrators Protest Food Prices, Denounce Government.* Retrieved from https://latimesblogs.latimes.com/babylonbeyond/2011/01/jordan-protests-food-prices-muslim-brotherhood-tunisia-strike-thousands-government.html. Los Angeles Times. Beirut, January 15.

Shami, S. 2007. Amman Is Not a City. In *Urban Imaginaries. Locating the Modern City*, ed. A. Cinar and T. Bender, 208–235. Minneapolis: University of Minnesota Press.

Soliman, A. 2004a. *A Possible Way Out: Formalizing Housing Informality in Egyptian Cities.* Lanham: American University Press.

———. 2004b. Tilting at Sphinxes: Locating Urban Informality in Egyptian Cities. In *Urban Informality in an Era of Liberalization: A Transnational Perspective*, ed. Ananya Roy and Nezear AlSayyad, 171–208. Lanham: Lexington Books.

———. 2008. Diversity of Ethnicity and State Involvement on Urban Informality in Beirut. *Theoretical and Empirical Researches in Urban Management, Year 3, Number 9*: 15–32.

Stel, Nora. 2016. The Agnotology of Eviction in South Lebanon's Palestinian Gatherings: How Institutional Ambiguity and Deliberate Ignorance Shape Sensitive Spaces. *Antipode* 48 (5): 1400–1419.

Sweis, R.J., R.O. Shanak, A.A. El Samen, and T. Suifan. 2014. Factors Affecting Quality in the Jordanian Housing Sector. *International Journal of Housing Markets and Analysis* 7 (2): 175–188.

Teti, A., P. Abbott, and F. Cavatorta. 2018. *The Arab Uprisings in Egypt, Jordan and Tunisia: Social, Political and Economic Transformations.* Cham: Palgrave Macmillan.

The Aga Khan Award for Architecture. 1980. *East Wahdat Upgrading Programme.* Amman: Urban Development Department.

The Hashemite Kingdom of Jordan. 2013. *Needs Assessment Review of the Impact of the Syrian Crisis on Jordan.* Amman: Ministry of Planning and International Cooperation.

The Third National Report. 2014. *The Third National Report on the Habitat in Cooperation with UN Program for Human Settlement.* New York: United Nations.

UNDP. 2016. *Regional Strategic Overview: Regional Refugee & Resilience Plan (3RP) 2019–2020*. New York: UNDP.
UNDP and GDCD. 2008. *Disaster Risk Management Profile (Amman–Jordan)*. Amman: General Directorate of Civil Defense (GDCD).
UN-Habitat. 2013. *Sate of the World's Cities 2012/2013: Prosperity of Cities*. New York: Routledge.
Unicef. 2013. *Informal Tented Settlements in Jordan: A Multi-Sector Baseline Assessment*. Unicef.
United Nations. 2013. *Needs Assessment Review of the Impact of the Syrian Crisis on Jordan*. Amman: Ministry of Planning and International Cooperation.

Part IV
Epilogue

Chapter 11
A Credible Future

Abstract One of the major objectives of the SDG 11 is rectifying urban informality within the built environment in a sustainable way. Thus, the main objective of this volume has been to highlight the tremendous challenges represented by arbitrary rapid urbanization and the challenges of dwelling in transitions of the Middle East's cities. The region has been affected by crises and rampant inequity demanding that planning professionals throughout the region engage with issues of urban informality. The study is unique as it represents a cross-regional comparative exercise in the Middle East spanning three countries: Egypt, Lebanon, and Jordan and their three capitals, Cairo, Beirut, and Amman.

This chapter summarizes the main findings of the study and draws the main outlines for future intervention to ameliorate or at least to integrate urban informality, in a sustainable way, within the urban context of the Middle Eastern cities. The definitions, taxonomies, and the role of urban informality in different socioeconomic, socio- spatial, and political contexts, as a site of transitions, in relationship to theories and practice strengthen understanding of sustainable transition processes and how they are governed in the context of comparative cross-national perspectives that meet the NUA and the SDGs. It may also enable policymakers, planners, and designers to think about urban informality as a dynamic multi-dimensional phenomenon, or as a new complex pattern of urbanity, or as a new space, or a site of transitions within old spaces embedded in specific contexts.

Keywords A new pattern of urbanity · Governance of sustainability transitions · Complex systems · Sociotechnical perspective · Governance perspective · Arab spring · Capitalist modernity · A site of transitions

11.1 Introduction

Urban informality is a core theme in the study of cities, in particular those that might be termed as in the Global South (Roy and Al-Sayyad 2004; Harris 2018; d'Alençona et al. 2018; Satterthwaite 2014; Banks et al. 2020; Lara-Hernandez et al. 2020). While 'planet of slums' (Davis 2006) pessimism is becoming

A. M. Soliman, *Urban Informality*, https://doi.org/10.1007/978-3-030-68988-9_11

increasingly archaic in academic and policy circles, few would argue that living conditions are satisfactory and change unnecessary (Bunnell and Harris 2012; Gilbert 2009; Sims 2010), and others think that partial intervention leading to sustainability of the homogeneity of the socio-spatial fabrics of urban informality is essential (De Soto 2000; Soliman 2004a, b, 2019, 2020). The variety of those interested in the pursuit of change in this arena is wide. NGOs, CBOs, activists, governments, community groups, academics, artists, international donors, built-environment professionals, and entrepreneurs all find eradication of, or at least upgrading, or integration of, urban informality to be essential for good built environments (Soliman 2020). Most urban authorities in Global South cities are now inundated with products, 'best-practice' guides, plans, and policy statements aiming at supporting self-build housing (Bredenoord et al. 2014; Obremski and Carter 2019; Soliman 2012a, b, c), upgrading informal settlements (Davidson and Payne 1985), Participatory Slum Upgrading Programs (PSUP) (UN-Habitat 2015a, b; Marx and Kelling 2019; Dovey et al. 2020), and reconciliation towards buildings' violators (Soliman 2021). Urban informality becomes "a new complex pattern of urbanity" and/or "a site of transitions" in the Global South by which opens a new academic arena for social science and planning theory in North–South dialogs. It should not treat neither/nor from top-down or bottom-up approaches, rather from a practical and implementable hybrid approach as reflections to the tremendous transitions of socioeconomic, spatiotemporal, political circumstances.

As illustrated in Fig. 1.1, this book focuses on the clarification of three themes: (1) a synopsis of urban informality in the Middle East; (2) ongoing innovative research governance of sustainability transitions; and (3) how to deal with urban informality in the context of sustainability transitions. These themes overlap and crosscut the other two dimensions. The first contains six axes that are: urban sustainability transitions, urbanization process, the paradigm of informality, state power and informality, land rights, and social exclusion/marginality. The second dimension is the informal housing taxonomies and the modes of informal-housing systems of production, coproduction/reproduction, consumption, and distribution of goods and services in the city and its practices. Also, the book tries to introduce what is called the ASUST, shedding light on six perspectives, as explained in the first chapter, as multi and continuous processes conceiving transition as an interference of six trends. These are policy strategy credibility, consistency of dimensions, policy processes (coherence of processes), systematic change (four perspectives), systematic directions of transitions, and instrument mix comprehensiveness. The ASUST echoes the hybrid approach that strive the tremendous transitions in a given environment.

The book has attempted to clarify the importance of urban informality processes on the built environment and how to alter urban informality in the context of ongoing sustainability transitions in the Middle East. Because urban sustainability strategies are playing a key role in the development of cities, this book sheds light on three cities' urban informality within sustainability strategies. The SDG 11 "Tracking Progress towards Inclusive, Safe, Resilient and Sustainable Cities and Human Settlements", Target 1 "By 2030, ensure access for all adequate, safe and affordable housing and basic services and upgrade slums." However, the leaders of the Middle Eastern countries committed to conform with the SDGs until the year 2030 by adopting a target of "*leaving no one behind*."

The book has tried to shed light on the overlap between the three folds, overlapped and crosscutting with the other two dimensions, and intersecting with the six perspectives within the context of a complex system of sustainability transitions of the Middle Eastern cities. It also combines the three themes, two dimensions, the six axes, and the ASUST to illuminate the continuous processes of urban informality transitions in a sustainable way. After the thematic perspective presented in Part 1, Part 2 focuses on an academic basis of the ASUST from various angles of urban sustainability transitions, urbanization in the era of globalization, urban informality transitions, state power and informality, land rights, and marginality and social exclusion. Part 3 targets experimentation and treats case studies of urban informality transitions in Egypt, Lebanon, and Jordan, while the current part focuses on the final findings of the study. The rationale for this structure of the book is that various relevant and distinct characteristics make such organization meaningful. Most importantly, the three case studies spanning Egypt, Lebanon, and Jordan are in relatively similar stages of their urban transitions. Together with remarkable differences in their political, social, socio-spatial, and economic development trajectories, the distinct urbanization processes occurring in the three countries also have a varied imprint on related issues of urban informality, sustainability transitions, and local governance.

This closing chapter is organized into three sections following this introduction. The second section summarizes the relationship between sustainability, the urbanization era, and urban informality transitions. The third section elaborates on the cross-regional scene and urban informality, while the fourth section draws the main outlines for future intervention to move cities toward governance of sustainability transitions on urban informality.

11.2 Sustainability, the Urbanization Era, and Urban Informality Transitions

The basic concepts underlie the sustainability transitions debate that has been going on for a few years now, notably stressing the management of sustainability transitions (Grin et al. 2010; Koehler et al. 2017), and mainly focusing on energy consumption, climate changes, etc., and very rare spatial transitions are examined. There are many different conceptions of sustainability transitions in the literature (Loorbach and Oxenaar 2018), and there are even more understandings of various identifications of governance/management (Frantzeskaki et al. 2017a,b). Our "hub and spikes" provides a structured illustration of the various dimensions attributed to "sustainability transitions" in the literature, especially in urban informality transitions. The linkage between sustainability transitions and urban informality is the cornerstone of this book. Three important points are highlighted in this context. First, the unifying element refers to the sustainability transitions in relation to climate changes, energy consumption, deterioration of natural resources, etc., and summarized the ongoing rapid physical changes in the urban fabric of cities in the Middle East. Second, the various strands in the literature have focused on different facets of the relationship between state intervention, on the one hand, and societal autonomy on the other, ignoring the influence of market forces, either formal or informal or both, social networks, shadow economy, and political transferee on the ongoing transitions of urban informality. Accordingly, the main objective is organized according to a conception of sustainability transitions and how urban informality could become sustainable. Third, it is appropriate to return to the roots of sustainability that show how our approach brings together new insights and/or new wine in old bottles. The intention is to open an academic discussion of the theory and practice of governance of sustainability transitions, the hybrid approach, and how it correlates with urban informality in the Middle East. The intent has been to explore the different dynamics, challenges, and breakthroughs in accelerating sustainability transitions of urban informality across the Middle Eastern cities. Linking the Multilevel Perspectives (MLPs), the Multiphases (MPs), the ASUST, and urban informality, a model of a sociotechnical transition process, as a hybrid approach, has emerged. The present part briefly highlights the six topics that explain the previous discussion on urban informality transitions in the Middle East.

First, the book benchmarks existing debates in the sustainability transitions literature that are inherently inclusive to the Global North (Frantzeskaki et al. 2017a,b, 2018; Bulkeley et al. 2016) as they mix various explicit and implicit dimensions different than the ones in the Global South. Also, it is very rare to question spontaneous urbanization and its negative impact on sustainability transitions. To avoid such analytical problems, the book suggests a three-step analysis to test the configuration of urban informality within sustainability transitions. In the first step, the relationship between urban informality and sustainability transitions is correlated and interacted with prevailing polities.. Only on this basis, a second step could then draw meaningful cross-linkages between governance in the form of institutional structures, stakeholder constellations, sustainability transitions, urban informality and resulting linkage of the final production of the built environment, especially after the introduction of law of reconciliation in Egypt. The third is linking urban informality with the six perspectives of the ASUST which reflects the hybrid approach.

As part of the first step, a sequence of correlations between the transitions level (macro, meso and micro—landscape, regime, and niche), the concept of transitions (socio-spatial, socioeconomic, socioecological, and sociotechnical), devices of transitions (co-evolution, Multilevel Perspective, Multi-phases Perspective and Co-design and learning), levels of intervention (market forces, political economy, circular economy/ shadow economy, government's policies, NGOs, CBOs, and society, values and their expression in life-style), sustainable transitions dimensions (complex systems, sociotechnical perspective, and governance perspective) and the integration of the modes of informal housing systems and its taxonomies with the

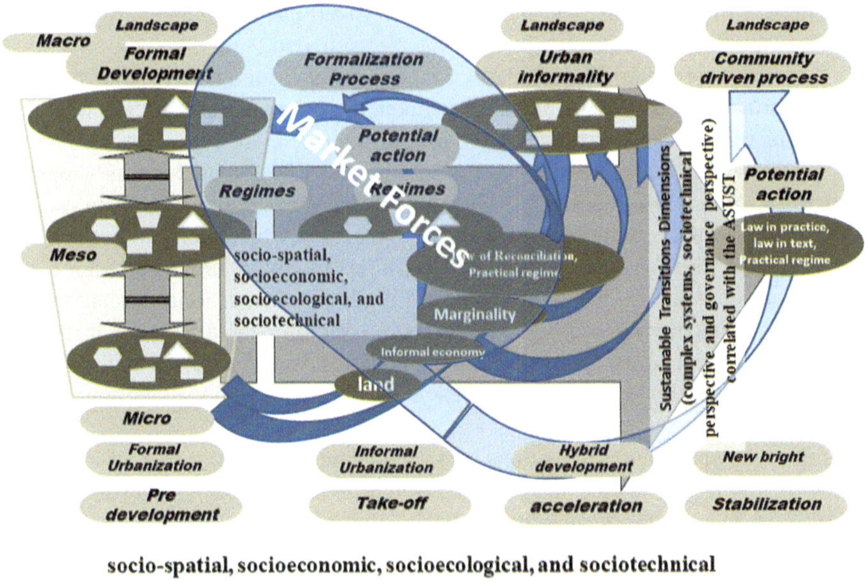

Fig. 11.1 Flowchart depicting four possibilities for urban informality transitions in relation to land issues, the ASUST, and the law of reconciliation. (Source: Soliman 2020)

urban context. This correlation presents a planning process of four modes of sustainability (formal, informal, hybrid, and new lively) and its linkage in the policy dimensions: socio-spatial, spatiotemporal, economic, political, and social networks adaptation. These four dimensions are distinguished along with two variables: the type of instruments applied (legally binding legislation, soft law, or reconciliation law bearing in mind that there are differences between law in text, law in practice, and the reconciliation law) and the approach to implementation (flexible or rigid). Further pathways for the other dimensions could follow suit. The second step focuses on land issues as a main component of residential development and its relationship with the informal economy and marginality. As illustrated in Fig. 11.1, four possibilities have emerged; formal, informal, hybrid, and new lively , and they rely on the MLP and the MP perspectives, each of which has certain treatments and adaptations. These sorts of adaptions have to be within the context of sustainability transitions dynamics and include locked-in/out regimes that are challenged by changing contexts, ecological stress, and societal pressures for change (regimes and niches), as well as experiments and innovations in niches driven by entrepreneurial networks, creative communities, and proactive administrators. But also included are resistance by vested interests, uncertainties about the future among urban populations, political instabilities, circular economy, and the erosion of social services and systems of provision. Treating urban informality as "a site of transitions", various pathways emerge in two directions (vertically and horizontally): first, between niches, regimes, and landscapes; and second, between Multi-phases

Perspective of transitions. The ASUST perspective is integrated and correlated with the four modes of transitions.

Second, most of the Middle East countries suffer from a slow rate of developmental progress and speedy transitions, without the extreme importance of time. The hastening of development progress and its performance in the right place at the right time are the main magnitudes in challenging the rapid growth of urbanization processes to meet the needs of the present and future generations. As long as rapid population growth continues, urbanization processes will remain an unrestrained phenomenon, and urban informality will constitute the future urban pattern for most Middle Eastern cities. It is estimated that Egypt, Lebanon, and Jordan will be welcoming annually around 2.5, 0.03, and 0.26 million people which will result in additional annual housing needs of 0.5, 0.006, and 0.05 million housing units, respectively. Of this, around 50% of the expansion will be under the umbrella of informality.

Informal urbanization transitions, rapid population growth, a lack of an overall urban planning framework, the influx of refugees, and increasing housing demand by the urban poor were among the causes that led to the appearance of urban informality in the Middle East. This phenomenon spreads out onto precipitous hills and agricultural land, renovating or enlarging such sites throughout the cities without either systematic planning or recourse to zoning and regulations. During the last 40 years, it has comprised about 70%, 60%, and 50% of housing production in Egypt, Lebanon, and Jordan, respectively. The result is a growing social and spatial divergence and rapid informal urban transitions in Middle Eastern cities. However, this category of urban informality transitions in the Middle East has not attracted the attention of decision makers nor scholars and will continue over the next few decades unless a proper housing policy is crafted.

Third, urban informality, as "a site of transitions", conceptualizes a range of unregulated/regulated spatial spaces, unsanctioned/sanctioned social networks, legal/illegal economic activities and income-generating strategies, shadow and circular economy, cultural and political mobilization practices, and is most distinct where physically manifested. It is practiced on the ground depending on the amount of capital associated, formally/informally, with an overt, a covert, and hybrid role of the state and various actors, and its speed has fluctuated according to the prevailing socioeconomic, socio-spatial, and political circumstances. Urban informality commonly arises for a combination of many reasons (Harris 2018) and these reasons constitute various pathways in the transitions process.

The persistence of urban informality in the Middle Eastern cities, despite the adaption of various regulatory reform measures and targeted programs to curb them (e.g., formalization of informal settlements by installing basic services and lowering standards, upgrading schemes, assisted self-help housing, PSUP, and reconciliation law etc.), continues to be a puzzle for housing policy in the Middle East. The formulation of urban informality is due to a lack of a national framework for urban planning, an absence of understanding the nature and complexity of urban informality's mechanisms, and an absence of clear land-governance policies. Many countries in the Middle East have a surplus of housing units, vacant housing units (e.g., in Egypt around 12.0 million housing units are empty and closed), and gated housing communities, but most such housing is beyond affordable for the bottom of the social pyramid.

Urban informality as a "site of transitions" and "survival architecture to create permanent facts on the ground" is very dangerous with respect to the destruction not only of the built environment, but also, to the devastation of the social structure of cities in the Middle East and might lead to the destruction of the states in the long term. No doubt, urban informality also leads to a disruption of fertile agricultural land, an urban fabric of cities that cannot exist without a considerable amount of corruption, and favoritism to clients. It has destroyed fertile agricultural land and its spatial and agrarian context (such as the Egyptian case). Urban informality in all its forms, modes, taxonomies, and practices as a habitual national custom has become truly deplorable and utterly condemnable. It has alarmingly become a national phenomenon; this epidemic threatening public order, legal frameworks, and the national pride in its heritage of civil legality and governability, meanings of urbanity, and informality has become not only a way of life in the Middle East, but also a portends future *urbicide* (Shwayri 2012) for the region's survival and a ubiquitous threat to urban existence. Despite that, urban informality is the response to provide housing, as a place for living and work, at affordable cost and to offer a solution to overcome the prevailing inflation, as well as create an opportunity for home ownership and the circular economy. Urban informality undoubtedly creates major servicing problems and additional expenses for the servicing authorities that could have been avoided had a more rational means of organizing land for the poor been available.

Fourth, after the Arab Spring, seven main challenges face the Middle Eastern governments in tackling housing informality in urban areas. The first is that new transitions are taking place within civil society, where NGOs, CBOs and international agencies have come to play a major role in the relationships between the triple trinity of the state, the capital, and the society. The second is the change in the nature of the state, the political economy, and the influence of globalization requiring good governance in which the state becomes the principal enabler of housing production for the bottom strata of society. The third is that public sector management reform and private sector development cannot proceed effectively unless and until they are adapted to the requirements of good urban governance that can be implemented at the municipal level. The fourth is that a new democratic era has emerged in various Middle Eastern countries (the revolt in Egypt and Tunisia) in which civil society, social networks, CBOs, NGOs, and international donors will play a major role in conducting the development process, especially in housing-delivery system. The fifth is that the huge influx of refugees into certain Middle East countries has put further pressure on those countries that makes the situation much worse. The sixth, the pandemic of the Covid-19, has major consequences on the market forces and the housing production, formally or informally. Finally, the recent temporary law of reconciliation in Egypt has created a new taxonomy of informality as "a hybrid formal status" that tolerates the urban fabric of urban informality as it is.

Perhaps the perpetual urban challenge that faces the current transitional circumstances in Middle Eastern countries is the absence of deep understanding of the seriousness of loss of stability, security, retreat of priorities, and the transitions of the economy system. The accomplishment of all targets, enhancing housing delivery systems, and poverty reduction would require a creation of an environment capable of implementing incrementally the priorities of a comprehensive development and to speed the return to a life of natural conditions for the welfare of the societies. A more appropriate balance between rural and urban economic, formal, and informal economy and noneconomic opportunities also appears to be necessary for the amelioration of urban and rural unemployment problems and to slow down the pace of rural–urban migration and regional migration. The challenges facing Middle Eastern governments are the following: to rectify the imbalances between housing demand and supply; to redistribute demographic concentration away from the old urban centers; and to close the gap between what the low-income groups can afford to pay and the minimum cost of a legal plot with basic services. Middle Eastern governments need also to integrate development activities and link housing to larger urban systems of employment and production. They must also decentralize resources to support local enterprise and home building at the local and national levels, providing transparency, accountability, responsibility, and trust in their vision for the future. A closer collaboration or partnership between various stakeholders is required to create the required capital for the development process. Urbanization is not exclusively a challenge for cities alone, rather it is a challenge to face chaos in cities as a result of transitional circumstances in Middle Eastern cities. Sustainable urbanization as an enabler for achieving global prosperity and sustainability should result in added value to a wider development strategy.

Fifth, in contrast, when the state and the market are as weak as in the Middle East, both historically and presently, informal land markets and urban informality flourish. New approaches to land-delivery systems and to urban development in general need to be based on the very social mechanisms and structures whereby the majority of the populations manage their survival. Land-delivery systems need a comprehensive set of policies, or a complete package of policies rather than a piecemeal approach in which one issue is tackled while others are ignored. Housing informality markets and land-conversion procedures need to be formulated in line with socioeconomic reforms and poverty alleviation strategies within the context of servicing the socioeconomic needs of those at the bottom of the pyramid of the society, as well as those at the top. The poor need adjustment mechanisms to participate in the formal market. The top of the pyramid needs more mechanisms to facilitate doing so for the poor. The states of the Middle East have to adapt an approach that supports what people do and to regulate to the benefit of the collective resources that enable affordable housing standards, rather than standards so unfeasible that they leave most of the housing stock unregulated.

The gap between the government policy and the capacity of community members should be narrowed. This is to identify what is of critical importance and what is required for people concerned, as well as what the government can provide and what people can afford. To shorten this gap, proper land-delivery systems associated with the participatory inclusive process should be encouraged. Also, the government should act as an agent/facilitator, not as a contractor, or a developer, for the avail of local citizens and for compatibility and sustainability. Moreover, it should amplify the political goal of rescuing agricultural land and encourage the sustainability of urban development in new virgin locations with various services distant from agricultural areas. Also, simplification of planning and building procedures in new virgin areas must be taken into consideration in law by practice, rather than law by text or applying a spiritual of law to be implemented in practice.

Finally, Target 11.3 of SDGs indicates that, by 2030, enhanced inclusive and sustainable urbanization and capacity for participatory, integrated, and sustainable human settlement planning and management should be in place in all countries. Generally, the expansion of cities is measured by population growth, and, in some cases, where this growth is negative or stagnating or when city boundaries are for some reasons reduced, the ratio would be difficult to interpret in terms of positive or negative change. Most of the Middle Eastern cities are expanding and their boundaries arbitrarily increased two or three-fold of their size six decades ago. Marginality became the main feature of such expansion and accommodated the low-income groups in which social segregation and marginality became noticeable.

A growing divergence exists in Middle Eastern cities, between the new, and the old, modern and traditional, gated and open communities, a "Great Divide" between "establishing", and "emerging" urban spaces, differences between modern and tradition, between white hole and black hole, a great gap between new and old urban fabrics. The former space is the new gated communities of the elites of the society, while the latter is the "Urban Jungle" or urban informality and mirrors the marginal groups. Moreover, the three case studies are characterized by socio-spatial and socioeconomic diversification, and it is manifest in the division of the capital cities into more cities, such as many Cairos within Cairo (Al Sayyad 2006), East Beirut and West Beirut (Fawaz 2017; Shwayri 2008), and East Amman and West Amman (Al Daly 1999; Elsheshtawy 2008; Khirfan 2019; Ababsa 2012).

11.3 Cross-regional Scene and Urban Informality

Looking at the socioeconomic, socio-spatial and political transitions during the last seven decades, the Egyptian state has played various roles advancing urban informality. These roles have pushed low-income and middle-income classes towards the peripheries of the urban centers, as well as excluded them from participating formally in development processes by which informal economies grew and constitute more than 75% of GNP. In these cases, transition management was not prepared to face such transformations and was not able to respond or 'react' even as to an emerging urban problem. Also, the recent law of reconciliation of violators has created a chaotic situation by which the current socio-spatial fabric of urban informality does not improve, so there is a great gap between what are the needs of most of the population, and the official urban strategy by the states in which a situation of *status quo,* laissez-faire, and "a hybrid formal" has dominated the transitions of the urban contexts. This has led to questioning how to develop a situation of transition management for socioeconomically persistent problems such as urban poverty, urban degeneration, urban informality, deprivation, and gentrification.

In Lebanon, the spread of urban informality is the most significant phenomenon shaping the space and urban fabric within Lebanese cities, especially Beirut. Most of the periphery or former suburbs in Beirut have witnessed arbitrary residential development that caused serious socioeconomic and political problems, not only for their residents, but also for the various governments in Lebanon. State power, social segregation, and ethnicity has dominated the formulation of urban informality. In many cases, urban movements and urbanization processes, and their discourses, actors, and organization have been informally integrated within the structure and practices of local government, either directly or indirectly, through a diversified system of citizen participation, community development, politicians' compositions, and ethnicities enrolled. As a result, a double process of urban segregation has been produced—on the one hand, that of the various ethnic minorities with respect to the dominant ethnic group and, on the other hand, that of the different ethnic minorities among themselves. The influx of Syrian refugees and the Beirut explosion constitute a major urban challenge in the Lebanese territory, especially in the Greater Beirut region. The current political system in Lebanon has contributed one way or another in the spread the urban informality, as well as increasing the diversity of the Lebanese urban fabric.

Urban informality in the Kingdom of Jordan differs from other countries in the Global South and in the Middle East. Jordanian national identity is multi-scalar, merging Arab supranationalism with Jordanian and Palestinian identities. It has multiple scales of nationalism (Culcasi 2016) and tribal structures. Urban informality in Jordan is a global issue rather than a local problem. Because of the nature of the international refugees and the multi-scale of nationalism, any intervention to eliminate or at least to prevent further spread of the urban informality or upgrading informal settlements in Jordan requires a concrete and a transparent planning strategy in cooperation with local stakeholders, local and national municipalities, and international donors. This sort of cooperation requires a clear urban-planning strategy based on the ASUST through participatory approaches by involving local and international stakeholders.

In both the Lebanese and Jordanian cases, the refugees' camps need to be reconsidered as a space of multiple ambiguities and subjectivities. Following Oesch's conceptualization (Oesch 2017) of the Palestinian's camp as a 'space of ambiguity', it offers a new analysis of the spatiality of 'the refugee camp and subjectification processes based on their ambiguous character'. And, while recognizing the specificity of camp dwellers in Jordan, e.g., in terms of status, this ambiguity represents a more general principle of governing camps, which certainly unfolds differently in each situation. This ambiguous exercise of power is driven by geopolitical, socio-spatial and socioeconomic considerations. The camps (Palestinian and Syrian) and their dwellers need to be excluded to remain the witnesses of the struggle. But, at the same time, the camps also need to be included and not solely through an exclusion to avoid the complete marginalization of their dwellers that could result in the creation of a subversive population.

Egyptian, Lebanese, and Jordanian experiences have shown that urban informality was resilient and capable of servicing the growing low-income housing needs and other amenities within their localities and was also able to absorb the invasion of refugees. In the Egyptian case, urban informality was accomplished to meet the housing requirements for both the lower strata of Egypt's social pyramid, as well as the influx of refugees (Soliman 2019). In Lebanese and Jordanian cases, urban informality was prepared to absorb very large numbers of refugees and rested on the flexibility and responsiveness of informal housing markets that reacted resiliently and promptly to the spike in demand (Fawaz 2017).

In the three countries, urban informality was resilient and adaptable for socio-economic, socio-spatial and political transitions in various ways, as well as absorbing the mode of the humanitarian crisis. These modes of harnessing and strengthening the flexibilities of urban informality limit their possible negative externalities rather than trying to formalize them. Shifting policies from the emergency-relief mode of immediate service provision that seeks to supplant and replace informal mechanisms to a long-term planning framework would not achieve progress in the long term. Instead, investment in harnessing the resiliency provided by informal channels and mechanisms of housing provision and acquisition will curtail their negative externalities. In other words, to recognize the positive attributes of informality is not to make do with the existence of an active regulatory mechanism, but rather to compose alongside it a more effective planning response to the ongoing demand for housing low-income groups and to meet the requirements of the influx of refugees..

As illustrated in Fig. 11.2, there are differences between the three countries, especially on the niche level. In the Egyptian case, the niche level has major impact on formulating the urban space. The processes of transitions flow by way of the systematic process of the MLP because the development of urban informality started from below on the niche level by converting suitable land for residential development. In the Lebanese case, the niche level is integrated with the regime level and constitutes a part of the structure of the regime, and both are semi-integrated and influence each other. The decision to establish urban informality rests in the hands of the regime influenced by outside forces, e.g., international donors. In Jordan's case, the situation differs from that of the Lebanese since Jordan has a different social structure in which the decision to develop urban informality originates from Jordanian tribes, and the regime supported by international donors.

As illustrated in Fig. 11.3, in all cases land is considered a main component for housing development. Good land governance should cover major areas for policy intervention in the land sector as follows: the legal and institutional framework; land-use planning; and dispute mediation over land ownership. Land governance is a part of governance: it is the spatial component of the act and process within a given environment. The three states should adapt proper land governance as an orderly process by a number of institutions, organized by a set of laws and operated by various actors.

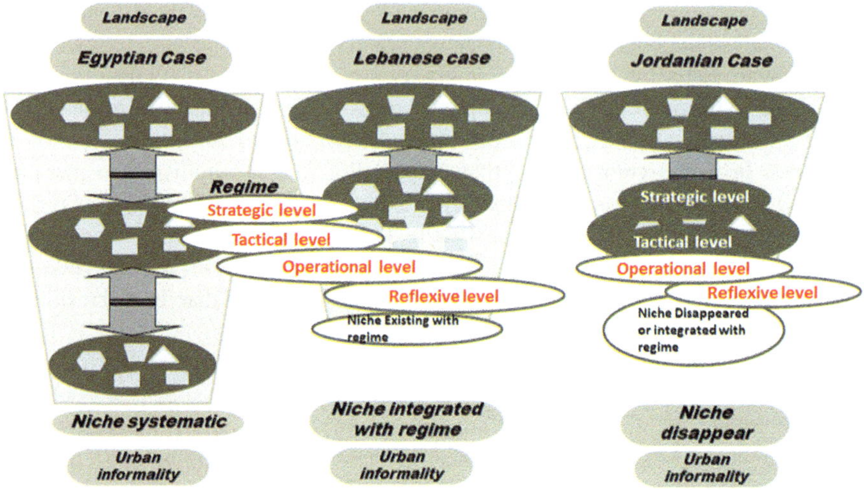

Fig. 11.2 Diagrams explaining the differences between the three case studies. (Source: Adapted from (Grin et al. 2010; Frantzeskaki et al. 2017a,b; Soliman 2020)).

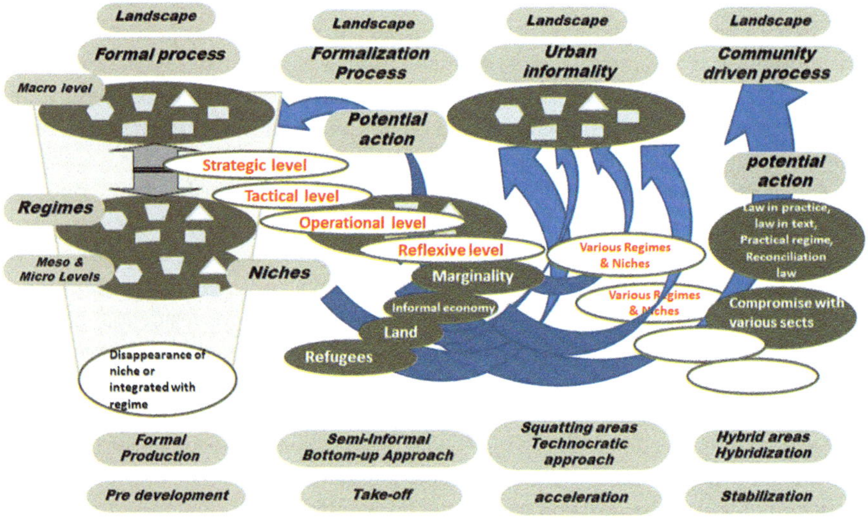

Fig. 11.3 A model that treats urban informality from various perspectives in the three case studies. (Source: Adapted from (Grin et al. 2010; Frantzeskaki et al. 2017a,b; Soliman 2020)).

Good land governance systems should support what people do, what people are willing to do, what people are able to do, how land governance is perceived by people, and to regulate for the benefit of the collective resources which enable a practical and logical land governance system, rather than systems so unfeasible that they leave most urban informality land unregulated. Thus, suggestions that work on the basis of these concepts can permit a more nuanced and appropriate analytical basis for understanding the perception of the three countries' urban dwellers, as well as mechanisms for consolidating urban informality in a sustainable way.

Figure 11.3 illustrates a model that treats urban informality from the ASUST perspective (policy strategy credibility, consistency of dimensions, policy processes, systemic change, systematic directions of transitions, and instrument mix comprehensiveness) in the three case studies. The treatments are varied in relation to the four levels of tactical, strategic, operational, and reflexive, as well as in relation to the MLP and the MPs and potential actions of the community. In the Lebanese and Jordanian cases, compromise with various sects of the society is crucial, while the Egyptian case follows the systematic change of the MLP at different phases of transformation of urban informality. The importance of this model is how to reduce the tension and pressure between the state, market and community in favor of facilitating land-delivery systems for the urban poor in eliminating housing informality or integrating it within urban context. The ASUST is related and integrated with the three informal taxonomies of bottom-up, technocratic, and hybridization approaches.

In general, urban informality can be viewed from three interpretations: institutional, behavioral, and practical. From an institutional point of view, it is essential that housing informality can take various forms; formal and legal shapes, but having in common that they are (to different degrees) hybrid institutions that deal with competing drivers stemming from public, market, and community forces and values. These drivers occur in varying combinations, sometimes reflecting the origins of these organizations and affecting the motivations of the key stakeholders involved. From a behavioral view, urban informality has in common that they (again to different extents) adapt entrepreneurial strategies to fulfill their objectives (versus traditional 'bureaucratic' or 'management-oriented' behavior); conflicts between principles are played out in organizational strategies and day-to-day decisions.

From a practical outlook, the formulation of a given environment in a certain way under certain circumstances is the prime cause for welcoming or not welcoming the formulation of urban informality. This suits any environment of the world if the prevailing socioeconomic, socio-spatial and political circumstances are prepared to encourage people to act informally, thus causing the flourishing of urban informality. These circumstances are varied from one place to another depending on the transitions and challenges that an environment has passed through.

It could be said that any human being on earth is born free and his/her character is shaped due to the surrounding environment accompanying him/her. The changing nature or behavior of human beings is linked with the environment they live in. The main challenge is how to formulate an environment to be capable of providing a better way of life, or at least, how to prepare an environment to encourage people to act formally in a sustainable way. Also, between formal and informal spaces, there is a gray space that might evolve in either informal or formal direction. This gray space reflects the hybrid approach. This explains the prevalence of informality having emphasized its political dynamic, focusing on why residents might flout or why state agencies might tolerate—even welcome—informal arrangements. Therefore, the main task is how to guide this area in a formal sustainable way and avoiding informality. In other words, how to prepare an environment to act formally rather than informally, as well as how to convince the stakeholders to be able and willing to act formally within their environment. On the ground, two or more of these are usually in play, and sometimes all of them, at which point it becomes possible to speak about a planning regime that is defined by informality.

11.4 Discussion

Urban informality, as "a site of transitions", reflects transitional, entrepreneurial, environmental, social, and spatially spaces whose subversive nature is crucial in challenging the 'capitalist modernity' of the Middle East. Sustainability-led thinking has formed part of urban planning dogma for some time, and urban planning is now reciprocally fundamental to the idea of sustainability as urban development provides challenges and opportunities for more sustainable futures (Frantzeskaki et al. 2017a, b). In elaborating on sustainability within planning for human settlements, references are made to stewardship of the natural processes that support human life and equivalently to the social dimensions of urban life, equity, community, and social justice, as well as to the aesthetic and spiritual qualities of the built environment (Friedmann 2005). Planning for more sustainable outcomes is, however, fraught with challenges of future transitions.

Lack of cross-national comparative work that seeks to understand the meaning of urban informality in different socioeconomic, socio-spatial, and political contexts has led to the misunderstanding of the nature of urban informality. Cross-national perspectives, too, seek to understand the governance of sustainability transitions on urban informality in various socioeconomic, socio-spatial, and political contexts. The modes of informal housing systems and its taxonomies, definitions, and the role of urban informality in different socioeconomic and political contexts in relation to theories and practice strengthen the understanding of sustainability transitions processes and how they are governed in the context of comparative cross-national perspectives that meet the NUA. Taxonomies of urban informality may also enable policymakers, planners, and designers to think about urban informality as a multi-dimensional phenomenon embedded in specific contexts. Therefore, taxonomies of urban informality are essential to understand its mechanisms, its transformation as a process and the multi-dimensions of continuous adjustment to changing socioeconomic, socio-spatial, and political circumstances, and to find sustainable ways to integrate these settlements into the urban context. The taxonomy would help to link between theories in relation to rights to the city, inclusion, and justice and practice in relation to processes of implementation of these theories on the ground with different meanings in different socio-political contexts.

Taxonomies of urban informality have been studied in several Global South countries, with emphasis on local circumstances (Soliman 2004a, b; Harris 2018; Devlin 2018; Boanada-Fuchs and Vanessa 2018; Marx and Kelling 2019). In Egypt,

ISDF (ISDF 2016) classified informal areas into two categories, unsafe areas and unplanned areas, the former type having four grades and broken down into twelve sub-taxonomies (ISDF 2016; Khalifa 2011). While the latter constitutes 96% of informal housing in Egypt, there does not appear to be any plan by the responsible authority, except the introduction of the temporary law of reconciliation. However, the current classification are grouped (Soliman 2002, 2004a, b; 2010; Harris 2018; Boanada-Fuchs and Vanessa 2018; Devlin 2018; Antonio et al. 2020; d'Alençona et al. 2018; Marx and Kelling 2019) into three main taxonomies, each of which has six sub-taxonomies.

In the three countries, a taxonomy of urban informality can be formulated into three main categories. This taxonomy divided urban informality into "bottom-up" (semi-informal) on privately owned formerly agricultural land, into "technocratic" (squatting) on formerly state-owned land, and into hybridization (liberal) on land whose ownership is in doubt, or where the construction process has been semi-illegal. This taxonomy relies on six dimensions: economic, legal, technical, organizational, political, and social (Boanada-Fuchs and Vanessa 2018; Soliman 2004a, b; 1996; Marx 2006; Bunnell and Harris 2012; Finn 2014; Iveson 2013; Antonio et al. 2020; Devlin 2018; Marx and Kelling 2019). The study here further identified 18 subtypes within the three main types and divided further within some of the most complex subtypes (see Table 11.1).

The following subsections and accompanying table discuss the three taxonomies and its subtypes and variants. The first taxonomy is bottom-up informality (semi-informal), which is not developed through established or state-regulated procedures and does not utilize the recognized and formal institutions of housing and housing finance. Nevertheless, such housing is developed on land for which the owner has legal tenure and a formal occupation permit. Bottom-up settlements are generally found in areas of essentially rural character located on the urban fringe that are interspersed with, surrounded by, or adjacent to undeveloped sites or land that remains in agricultural use. These settlements often develop in advance of the principal lines of urban growth. Their construction is most noticeable during periods of rapid urban expansion and during socioeconomic and political transitions, and they generally appear along major roads networks. This type is sometimes known in the literature as semi-informal settlement, or a type of illegal agricultural land subdivision (Soliman 2004a, b; Sims 2010). These settlements may be classified into six subtypes; desires, best practice, tactical urbanism, law, overt, and dominant (Marx and Kelling 2019; Harris 2018; Boanada-Fuchs and Vanessa 2018). Each of the three case studies exhibits some degree of these subtypes and is varied in its existence as illustrated in Table 11.1.

The second taxonomy is technocratic informality (squatters) (Antonio et al. 2020; Cirolia 2017). This type is often known as squatter housing on state land. Because this taxonomy overcomes the broader governance frameworks that are complex and reproductive forces in the making of injustices and inadequacy in the city, people squat on public land, and may do so on private land, to house themselves. Squatters are persons who settle on public land without title or who take unauthorized possession of unoccupied premises, such as refugees' camps in Beirut and

Table 11.1 Classification of urban informality in Egypt, Lebanon, and Jordan

Main Taxonomy	Sub taxonomy	City: Cairo-Egypt	City: Beirut-Lebanon	City: Amman-Jordan	Strength	Weakness
A-Bottom-up (Semi informal) (laissez-faire)	A-1 Desires	●	○	●	The nature of property rights, conventual planning, local infrastructure and level of community's services lead other to intervene.	Converts agricultural areas into illegal housing plots and destroys the agricultural production.
	A-2 Best practice	●	⊕	⊕	Settlers are more aware and adapt their setting according to their needs	The state is incapable to cope with community-driven development
	A-3 Tactical Urbanism	●	●	●	Defines informality based on the various mechanisms that adapted to provide land plots at affordable basis	The adaption of development mechanisms depends on place of origin of the settlers, in which encourage rural-urban immigration.
	A4- Law	●	●	●	Settlers have enough knowledge to overcome the law through spiritual law instead of law in text by which the development process at a high rate of speed.	The residents transfer the law in text into law in practice or soft law to overcome their legal conflicts, and this make a huge dispute over legality of tenure
	A5- Overt	●	●	●	It spreads urban informality through cooperation between official institutions and community	The state under certain circumstances supports informality and in another prevent this phenomenon.
	A6- Dominant	●	○	○	Social consolidation and social networks basis are playing important role in formulating the informality	The official institution cannot cope with the speed of development process of urban informality
B-Technocratic (squatting) (Status quo)	B1-Needs	●	●	●	It promotes new and creative ways in which urban users through informal tactics can spur and foster urban innovation by increasing the competitiveness of cities that contributes to economic development.	Such informal uses vary from fulfilling basic needs like housing, ass use value, to a more leisurely activities like artistic activities.
	B2-Cry & demand	●	●	●	Encouraging citizens to appropriate the urban realm through creative and informal practices in providing housing at the lost cost.	governments have cleared some informal areas, rehousing a few residents, while trying to tame and regulate others, but to limited effect
	B3-DIY Urbanism	⊕	⊕	○	A conceptualization of politics to encourage self-help production as the cheapest and speediest way to provide housing at the lowest cost.	Land as a main component of housing considers the main challenge for such process.
	B4- Condition	●	⊕	●	Allows settlers to act free to provide housing and implies for those living 'in it' and, in the best cases, to empower them in their settings.	Arbitrary conditions imposed upon differential spaces in the urban context might destroy the built environment.
	B5- Latent	●	⊕	⊕	Latency exists everywhere and on an almost limitless range of issues; after all, there are no restraints on the character of urban development at the poles.	The shadow of policy either with international donors or local institutions played an important role in the formulation of urban informality.
	B6- Diffuse	●	●		The status of a Status quo policy of regulation and policy mattered, and in a variety of ways, including a range of land use, building, and occupancy regulations.	Planners and other municipal officials often turn a blind eye to sustain status quo of the situation. Such a strategy can itself become almost formal.
C-Hybridisation Liberal (A blind Eye)	C1- Necessities	●	●	●	Emerges from the practices of categorization by the state and/or planning derived by the neoliberalism	Status quo might encourage further informal development on the periphery of cities.
	C2- Everyday Urbanism	●	○	⊕	Endeavors to produce, only that it should be constructed in a bottom-up, citizen-driven manner.	Unplanned urbanism causes further deterioration of the built environment.
	C3- Currency (urban commons)	●	●	●	Takes a relational perspective, emphasising flux, changes and dynamic conditions, while still recognising regularity as a driving force for development.	The local dynamics and relations of the context and specially place towards informality, rather those ones coming from neoliberal impulses.
	C4- Embedded	●	⊕	⊕	Collective activity often produces a cooperation among settlers in an organization way, commonly within a particular social group.	In areas with various sects or with various nationalisms prevent collective driven process.
	C5- Authenticity	●	●	●	The existing areas reflect the real social struggle for shelters.	Some politicians ignore the reality of the existing of the informality and might lead into further informality.
	C6- Covert (a hybrid formal)	●	●	●	Some practices may remain invisible to the state for years or even decades. Issuing law of reconciliation for buildings' violators.	The invisibility of informality might encourage newcomers to follow suite. Creating a new taxonomy of informality.

○: Low, ⊕: Moderate, ●: High

Source: The taxonomy of urban informality produced by the author and relies on various sources; (Soliman 2004a, b; Boanada-Fuchs and Vanessa 2018; Bunnell and Harris 2012; Harris 2018; Devlin 2018; Devlin Forthcoming)

Amman. Squatting, known as *wada'yad* (hand claim), has meant and continues to mean different things in different cultures and at different times in each culture's existence. It is a concept that is neither absolute nor static, but which evolves over time. Squatting is thus a *"Cultural Construct"*; more specifically, it is a political construct. This urban informality may be classified into six subtypes; needs, cry and demand (Lefebvre 1968), DIY (Do It Yourself) urbanism, condition, latent, and diffuse (Marx and Kelling 2019; Harris 2018; Boanada-Fuchs and Vanessa 2018). Each of the three case studied exhibits some degree of these subtypes and is varied in its occurrence as illustrated in Table 11.1.

The third type is hybridization informality, which includes residential units in formal areas that have temporarily or permanently acquired degrees of informality. This type involves housing built on public or private land that may have originated as formal housing, but over time it has been transformed one way or another and is now informal. It includes the recent adjustment for violators, as a law of reconciliation, for people who constructed illegally residential units or buildings. This type is known as ex-formal housing. Indeed, some properties currently in this category may have passed back and forth between formal and informal status several times (Soliman 2004a, b). Unlike residential units in informal areas, this type of informality relates to individual dwelling units in scattered areas on a case-by-case basis. In many cases, some units or areas in a formal neighborhood, or even an individual building, may be "formal", but others have either been added illegally or been modified over time into illegal configurations. Such transformations reflect "a hybrid formal" and are most common in the major metropolitan areas; they comprise six subtypes: status quo, everyday urbanism, currency, embedded, authenticity, and covert (Marx and Kelling 2019; Harris 2018; Boanada-Fuchs and Vanessa 2018; Antonio et al. 2020). Each of the three case studies exhibits some degree of those sub-taxonomies according to their local circumstances. The degree of occurrence in each country is classified into three levels; high, modest, and weak, as summarized in Table 11.1.

Implementing principles of governance of sustainability transitions (Frantzeskaki et al. 2017a,b) on urban informality at the local and national levels are essential. A model of integration and interlinked between the ASUST perspectives, various taxonomy of urban informality, the three pillars, and the three thresholds is reached. However, the challenge for very healthy cities is how to transform existing urban informality towards environmental sustainability without compromising the socio-economic benefits that they provide.

As illustrated in Fig. 11.4, the three taxonomies of urban informality—bottom-up, technocratic, and hybridization—are in a dynamic process, and may change their categories over time from one taxonomy to another, also depending on people's behavior and action in each alteration. These alterations represent various pathways for sustainability transitions process and are related to the level of the involvement of the three pillars, the ASUST perspective, and the degree of the three thresholds of people's transitions. All of which are explained in the following.

The first taxonomy of housing as a bottom-up approach reflects semi-informal housing production built on privately owned agricultural land and has some sort of legality of tenure. It does not follow state-regulated procedures and does not utilize the recognized institutions of housing and housing finance but mostly relies on the shadow economy. Most of it builds illegally without obtaining building license nor does it comply with building and planning regulations. By contrast to the bottom-up type, the second type of housing taxonomy is technocratic taxonomy. It is usually established on desert or hills sites relatively far from the city center. The major disadvantage here is that the cost of access to job opportunities is greater. The major advantage, however, is that the land may be obtained cheaply, or at no cost at all. Such settlements are usually located adjacent to a main road which acts as a strong edge limiting future expansion, and their streets are narrow and mainly used by pedestrians. Such settlements generally offer only limited basic services, and they may lack all social and public services. They have generally been established by collective invasions by local citizens, from Bedouin regions, or refugees. Hence, social networks (Fawaz 2017) are very important within them and, indeed, often have played a major role in the original land-invasion process.

The development of hybridization has largely resulted from the absence of planning control and the weakness of the building and planning regulations to deal with the housing needs of low-income groups. However, understanding the built environment of hybridization development may assist both government officials and private professionals today to understand the transformation process by which citizens have adapted to change their status from formal into informal and vice versa, and the adaption of law to meet their transformation. This taxonomy might open a debate to help understand the mechanisms, tools, and methods to properly value this sector so that these assets may be harnessed to promote development and alleviate poverty.

To adjust or modify the existing condition, especially after the introduction of the temporary law of reconciliation in Egypt, for each taxonomy and pathways of urban informality, one has to understand exactly three thresholds of people's transitions; *what people are willing to do, what people are able to do, and what people actually do and what people are able to do* in their own settings. The first threshold concerns ***the willingness of people's reaction*** towards their settings that affect how planning happens. This leads to the bottom-up approach which reflects the relationship between society and capital that has resulted in the level of distribution and consumption of goods and services. These deserve more attention than they have received. Once willingness has been introduced, the crossing of thresholds depends on the numbers of people involved in informal practices, their physical location, their amount of capital consumed on goods and services, and their manner of social organization. A second threshold concerns the emergence of embedded social legitimacy and ***what people are prepared to do*** towards developments with practical significance for those involved in the process of development and also, the degree of people's potentialities for planning practices and wider politics. This echoes the technocratic approach and the interaction between the capital and society which determines the level of production and reproduction of goods and services for satisfying the level of demand of housing and other social amenities. A third threshold concerns ***what people do and what people are able to do*** and represents

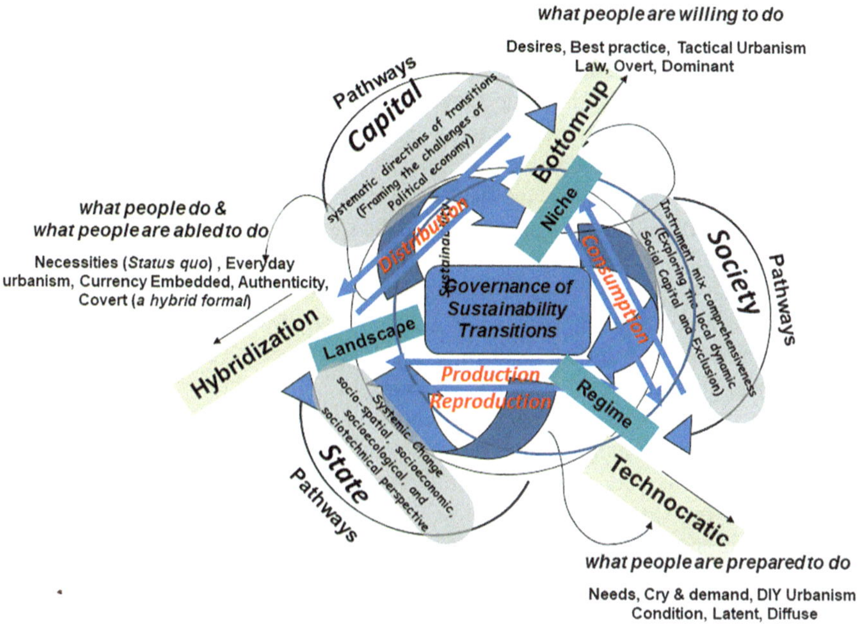

Fig. 11.4 The ASUST, the three pillars, the three thresholds and the three housing taxonomies in relation to governance of sustainability transitions on urban informality within the three cities. (Source: The author)

the hybridization approach in which regular/irregular practices become covert, an unavoidable matter for debate and policy. It also represents the interaction between the state and capital and determines the level of distribution of goods and services. This is the least predictable step: It depends on the prevalence and physical visibility of changing legal development into illegal development and vice-versa, but on much else besides.

Generally, because of the differences in socioeconomic, socio-spatial and political characteristics of the three cities, the variation of the three pillars, the three thresholds, the four modes of informal housing systems, and the three taxonomies of urban informality is due to the correlation of the ASUST perspective, the local circumstances, and the type of the built environment. On the other hand, as explained in Chap. 1, the three perspectives of the ASUST are integrated through systemic change (four dimensions), systematic directions of transitions, and instrument mix comprehensiveness, while the other three perspectives are attributed to the state policies. Otherwise, the socio-spatial characteristics of urban informality in the three cities differ, except for the geographical features and the layout of housing plots. In Cairo, housing blocks are generally two–three floors taller than in Beirut and Amman. The size of housing plots differs in the three cities, while the housing mechanisms are nearly the same. The everyday ways of life vary across the three cities, and the amount of money spent on housing is nearly the same when compared to each household's income level.

Linking the three thresholds, the three pillars, the six perspectives of the ASUST, the modes of housing, and the three taxonomies of urban informality with three levels of: the Multi-Level perspectives (MLPs), Multi phases (MPs) and Multi Socio-spatial perspectives (MSSs), the process of transition would change its levels due to the local circumstances and the variations of pathways. As illustrated in Fig. 11.5, urban informality represents a model in the form of a 3D outlook to link MLPs, MPs, and MSSs. It is a dynamic and flexible process that reflects "a site of transitions", to cope with the continuous transitions of a given environment.

The horizontal level represents the MPs consists of three phases: emergence, diffusion, and reconfiguration, reflecting the three taxonomies of bottom-up, technocratic, and hybridization. The vertical level is the MLPs which echo the three level of a sociotechnical perspective of niche, regime, and landscape. The perpendicular on the horizontal and vertical levels represents the MSSs level and comprises the three stages of scattered, collective, and consolidation, which mirrors the modes of informal housing systems of production, coproduction/reproduction, consumption, and distribution. The vertical and horizontal levels reflect the integration and interlinkage between the process of the MLPs and the MPs in which various pathways are merged while the perpendicular level of the MSSs is integrated with both previous levels at the vertical and horizontal levels. The intersection between the horizontal level of the MLPs and the perpendicular level of the MSSs crosses at three points of emergence-scattered, diffusion-collective, and reconfiguration-consolidation, and these spread vertically to integrate with the levels of the MLPs of niche, regime and landscape. In between, there are various pathways crossing horizontal, and vertical, and diagonal levels at the three points already mentioned.

The reproductive stability of these configurations is dependent, on the one hand, on their internal networked relations based on a multiplicity of supporting and aligned actors, and on the other hand, on the potential overlap of boundaries of arenas, which open for conflicts and possible reconfigurations. Actors and grassroots are said to have not confined themselves to working on the niche level: they also sought visionary changes of politics, value systems, and societal organization, and thus acted to recruit among the public and politicians to achieve changes to the landscape framing the urban informality. The challenge of transition is to combine an analytical understanding of the stability and path-dependent dynamics of dominant societal configurations of production, coproduction/reproduction, distribution, and consumption of goods and services in which the three taxonomies generate a process-oriented understanding of the situated actors' possibilities of engaging in transitional processes. While the regime concept and the MLPs seem well suited to cater to the configuration of traditional sociotechnical production or supply systems in which core institutions are easily identified, it is not very helpful in describing spheres of consumption and normalized everyday practices.

Niches might be integrated within a regime (as the case of Lebanon) or niche might despair (as the case of Jordan) and be replaced by the regime. The MLPs helps to specify and probe such underlying assumptions, and to analyze the complex

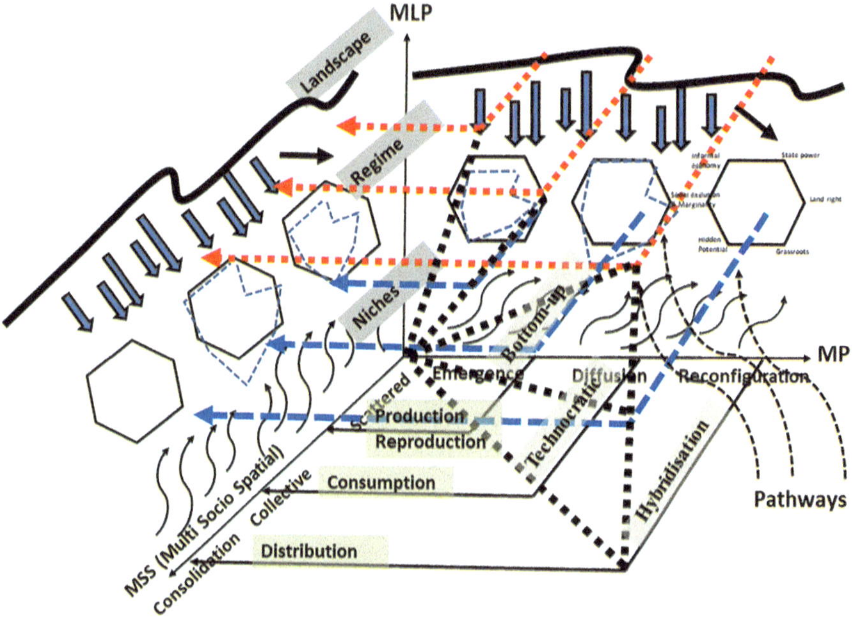

Fig. 11.5 3D perspective of the MLPs, MPs, and MSSs in relation to urban informality, as "a site of transitions" reflects the three thresholds, three modes of informal housing systems, and its three taxonomies in relation to governance of sustainability transitions on urban informality within the three cities. (Source: The author)

diversity of the (ascribed/constructed) roles of different sectors and actors in multi-level transition dynamics. An important element for doing so is the lock-out of 'civil society'. In much of the transition's literature so far, 'civil society' is mostly characterized in terms of 'grassroots innovation' and/or 'niche-activities'. The MLPs helps us to see civil society beyond such a niche-based focus because it also acknowledges the third sector as "a different articulation of the social into economic life by often powerful socioeconomic organizations and not only the petty under-scaled or undercapitalized local initiatives of moaning communities". While explicitly acknowledging the significant contribution of grassroots community activism, it is also important to recognize that the third sector encompasses more than that, including—in fact—rather powerful regime structures such as labor unions, large cooperatives, and religious networks (Avelino and Wittmayer 2017). As urban informality constitutes one of the main challenges of Middle Eastern cities to be tackled within the next few years, the urban development fields associated with the quality of life in residential areas should be formulated within the context of urban sustainable transitions that are resilient, competitive, and capable of providing better life standards.

It is hoped that this book will contribute to an exchange of knowledge between the Global North and South on urban-planning innovation and how it is possible to use approaches from the Global North that apply in the Global South/Middle East.

Three concerns are critical for further research. First is the need for recognizing conflicts and pressure as an important condition for change to help actors and grassroots to navigate. The second is the observation that actors and grassroots engage at all levels in society including visions, institutions, and innovations. The final one is the need to address the role of academic theories and to advise regarding governance of transition processes in which they function as entrenched actors. The fourth observation is to make sure to link between theories and practices to achieve proper sustainable transitions environment. Last and not least, is the hope that this book opens a new arena of scientific discussion between urban informality and sustainability transition perspectives, between conceptual and empirical, and between structural and practical. As this volume has pointed out in numerous places, there are still many important research and policy questions to be answered in moving cities toward governance of sustainability transitions in urban informality. The pathways identified throughout are largely tentative and require much additional research and practical demonstration. We have little understanding, for instance, of what the minimum nature of urban informality, or the nature of its mechanisms, is that is required by urbanites to grasp the lessons of "the new way of urbanity", or "a site of transitions" of urban informality. Much additional research is needed to better understand, moreover, the full package of resilient benefits that is provided by urban informality and even the different basic ways in which a better way of life can become manifest.

References

Ababsa, M. 2012. Public Policies Toward Informal Settlements in Jordan (1965–2010). In *Popular Housing and Urban Land Tenure in the Middle East*, ed. M. Ababsa, B. Dupret, and E. Denis, 259–282. Cairo: The American University in Cairo Press.

Al Daly, J. 1999. *Informal Settlements in Jordan: Upgrading Approaches Adopted and Lessons Learned*. Amman: Housing and Urban Development Corporation.

Al Sayyad, N. 2006. Whose Cairo? In *Cosmopolitan: Politics, Culture, and Urban Space in the New Globalized Middle East*, ed. D. Singerman and P. Amar, 539–542. Cairo: American Unversity Press in Cairo.

Antonio, J., C.M. Lara-Hernandez, C.M. Coulter, and A. Melis. 2020. Temporary Appropriation and Urban Informality: Exploring the Subtle Distinction. *Cities* 99. https://doi.org/10.1016/j.cities.2020.102626.

Avelino, F., and J. Wittmayer. 2017. A Multi-Actor Perspective on Urban Sustainability Transitions. In *Urban Sustainability Transitions*, ed. N. Frantzeskaki, V.C. Broto, L. Coenen, and D. Loorbach, 272–284. Oxon: Routledge.

Banks, N., M. Lombard, and D. Mitlin. 2020. Urban Informality as a Site of Critical Analysis. *The Journal of Development Studies* 56 (2): 223–238.

Boanada-Fuchs, A., and F.B. Vanessa. 2018. Towards a Taxonomic Understanding of Informality. *IDPR* 40 (4): 397–420.

Bredenoord, J., P.V. Lindert, and P. Smets, eds. 2014. *Affordble Housing in the Urban Global South: Seeking Sustainable Solutions*. Oxon: Routledge.

Bulkeley, H., L. Coenen, N. Frantzeskaki, C. Hartmann, A. Kronsell, L. Mai, and Y.V. Palgan. 2016. Urban Living Labs: Governing urban Sustainability Transitions. *Current Opinion in Environmental Sustainability* 22: 13–17.

Bunnell, T., and A. Harris. 2012. Reviewing Informality: Perspectives from Urban Asia. *International Development Planning Review* 34 (4): 339–348.
Cirolia, L.R. 2017. Overcoming the Disjunctures: Competing Discourses on Informal Settlements in South Africa. *IDPR* 39 (4): 443–459.
Culcasi, K. 2016. Warm Nationalism: Mapping and Imagining the Jordanian Nation. *Political Geography* 54: 7–20.
d'Alençona, P.A., H. Smith, and E.A. de Andres. 2018. Interrogating Informality: Conceptualisations, Practices and Policies in the Light of the New Urban Agenda. *Habitat International* 75: 59–66.
Davidson, F., and G. Payne. 1985. *Urban Project Manual: A Guide to Preparing Upgrading and New Development Projects Accessible to Low Income Groups*. Liverpool: Liverpool University Press.
Davis, M. 2006. *Planet of Slums*. London: Verso.
De Soto, H. 2000. *The Mestry of Capital: Why Capitalism Triumphs in the Westand Fails Everywhere Else*. London: Blaack Swan.
Devlin, R.T. 2018. Asking 'Third World questions' of First World informality: Using Southern Theory to Parse Needs from Desires in an Analysis of Informal Urbanism of the Global North. *Planning Theory* 17 (4): 568–587.
———. Forthcoming. Temporary Appropriation and Informality: A Nuanced Relationship. *Cities*.
Dovey, K., et al. 2020. Viewpoint Informal Settlement Is Not a Euphemism for 'Slum': What's at Stake Beyond the Language? *IDPR*. https://doi.org/10.3828/idpr.2020.14.
Elsheshtawy, Y. 2008. The Great Divide: Struggling and Emerging Cities in the Arab World. In *The Evolving Arab City: Tradition, Modernity and Urban Development*. ed. Elsheshtawy, Y., 1–26. London and New York: Routledge.
Fawaz, M. 2017. Planning and the Refugee Crisis: Informality as a Framework of Analysis and Reflection. *Planning Theory* 16 (1): 99–115.
Finn, D. 2014. DIY Urbanism: Implications for Cities. *Journal of Urbanism* 7 (4): 381–398.
Frantzeskaki, N., V.C. Broto, L. Coenen, and D. Loorbach. 2017a. Urban Sustainability Transitions: The Dynamics and Opportunities of Sustainablity Transitions in Cities. In *Urban Sustainability Transitions*, ed. N. Frantzeskaki, V.C. Broto, L. Coenen, and D. Loorbach, 1–19. Oxon: Routledge.
———. 2017b. *Urban Sustainability Transitions*. New York: Routledge.
Frantzeskaki, N., M. Bach, K. Holscher, and F. Avelino. 2018. *Co-Creating Sustainable Urban Futures: A Primer on Applying Transition Management in Cities*. Cham: Springer.
Friedmann, J. 2005. Globalization and the Emerging Culture of Planning. *Progress in Planning* 64: 183–234.
Gilbert, A. 2009. Extreme Thinking About Slums and Slum Dwellers: A Critique. *SAIS Review of International Affairs* 29 (1): 35–48.
Grin, J., J. Rotmans, and J. Schot. 2010. *Transitions to Sustainable Development: New Directions in the Study of Long Term Transformative Change*. New York: Routledge.
Harris, R. 2018. Modes of Informal Urban Development: A Global Phenomenon. *Journal of Planning Literature* 33 (3): 267–286.
ISDF. 2016. *Slums Areas in Egypt*. Cairo: ISDF.
Iveson, K. 2013. Cities Within the City: Do-It-Yourself Urbanism and the Right to the City. *International Journal of Urban and Regional Research* 37 (3): 941–956.
Khalifa, M.A. 2011. Redefining Slums in Egypt: Unplanned Versus Unsafe Areas. *Habitat International* 35: 40–49.
Khirfan, L. 2019. Planning and Unplanning Amman: Between Formal Planning and Non-Traditional Agency. In *The Palgrave Handbook of Bottom-up Urbanism*, ed. Konrad Kickert and Mahyar Arefi, 201–218. Cham, Switzerland: Palgrave Macmillan.
Koehler, J., F. Geels, F. Kern, E.. Onsongo, and A. J. Wieczorek. 2017. *A Research Agenda for the Sustainability Transitions Research Network*. Technical. Amsterdam: STRN Working Group, STRN.

Lara-Hernandez, et al. 2020. Temporary Appropriation and Urban Informality: Exploring the Subtle Distinction. *Cities* 99. https://doi.org/10.1016/j.cities.2020.102626.
Lefebvre, H.. 1968. *The Right to the City*. The Anarchist Library.
Loorbach, D., and S. Oxenaar. 2018. *Counting on Nature: Transitions to a Natural Capital Positive Economy by Creating an Enabling Environment for Natural Capital Approaches*. Rotterdam: Dutch Research Institute for Transitions (DRIFT), Erasmus University.
Marx, C. 2006. Conceptualising the Economy' to Make Urban Land Markets Work for the Poor. *The Urban Land Seminar, November 2006*. Muldersdrift: NA. NA.
Marx, Colin, and Emily Kelling. 2019. Knowing Urban Informalities. *Urban Studies* 56 (3): 494–509.
Obremski, Helena, and Claudia Carter. 2019. Can Self-Build Housing Improve Social Sustainability Within Low-Income Groups? *TPR* 90 (2): 167–193.
Oesch, L. 2017. The Refugee Camp as a Space of Multiple Ambiguities and Subjectivities. *Political Geography* 60: 110–120.
Roy, A., and N. Al-Sayyad. 2004. *Urban informality: Transnational perspectives from the Middle East, Latin America, and South Asia*. Lanham: Lexington Books.
Satterthwaite, D. 2014. Guiding the Goals: Empowering Local Actors. *SAIS Review of International Affairs* 34(2, Summer-Fall): 51–61.
Shwayri, S.T. 2008. From Regional Node to Backwater and Back to Uncertainty: Beirut, 1943–2006. In *The Evolving Arab City: Tradition, Modernity and Urban Development*, Ed. Elsheshtawy, Y, 69–98. London and New York: Routledge.
———. 2012. Modern Warfare and the Theorization of the Middle Eastern City. In *Urban Theory Beyond the West*, ed. T. Ednesor and M. Jayne, 259–270. London: Routledge.
Sims, D. 2010. *Understanding Cairo: The Logic of a City Out of Control*. Cairo: America University Press in Cairo.
Soliman, A. 1996. Legitimizing Informal Housing: Accommodating Low-Income Groups in Alexandria, Egypt. *Environment and Urbanization Journal* 8 (1): 183–193.
———. 2002. Typology of Informal Housing in Egyptian Cities: Taking Account of Diversity. *IDPR* 24 (2): 177–201.
———. 2004a. *A Possible Way Out: Formalizing Housing Informality in Egyptian Cities*. Lanham: American University Press.
———. 2004b. Tilting at Sphinxes: Locating Urban Informality in Egyptian Cities. In *Urban Informality in an Era of Liberalization: A Transnational Perspective*, ed. Ananya Roy and Nezear AlSayyad, 171–208. Lanham: Lexington Books.
———. 2010. Rethinking Urban Informality and the Planning Process in Egypt. *IDPR Journal* 32 (2): 119–143.
———. 2012a. Housing and the State in the Middle East. In *International Encyclopedia of Housing and Home*, ed. Susan J. Smith, 346–354. Amsterdam: Elsevier.
———. 2012b. The Egyptian Episode of Self-Build Housing. *Habitat International Journal* 36 (2): 226–236.
———. 2012c. Building Bridges with the Grassroots: Housing Formalization Process in Egyptian Cities. *Journal of Housing and the Built Environment* 27 (2): 241–260.
———. 2019. So It's Always a Chance: Community-Led Solutions to New Urban Expansion. In *New Cities and Community Extensions in Egypt and the Middle East*, ed. S. Attia et al., 159–179. Cham: Springer.
———. 2020. Understanding Sustainability Transitions for Urban Informality in the Middle East. *International Journal of Urban Sustainable Development*. https://doi.org/10.1080/19463138.2020.1728692.
———. 2021. Informal Cairo: The Making of Urban Fabric. In *The Routledge Handbook on Cairo*, ed. N.A. Al Sayyad. New York: Routledge.
UN-Habitat. 2015a. *Slum Almanac 2015/2016: Tracking Improvement in the Lives of Slum Dwellers*. Nairobi: UN-Habitat.
———. 2015b. *UN-Habitat Global Activities Report 2015: Increasing Synergy for Greater National Ownership*. Nairobi: United Nations Human Settlements Programme (UN-Habitat), 31.

Bibliography

Ababsa, M. 2012. Public Policies Toward Informal Settlements in Jordan (1965–2010). In *Popular Housing and Urban Land Tenure in the Middle East*, ed. M. Ababsa, B. Dupret, and E. Denis, 259–282. Cairo: The American University in Cairo Press.

Ababsa, M., and A.Z. Abu Hussein. 2020. Metropolitan Amman: Comprehensive Climate Plans. In *Volume II of Greater Than Parts: A Metropolitan Opportunity*, ed. Shagun Mehrotra, Lincoln L. Lewis, Mariana Orloff, and Beth Olberding. Washington, DC: The World Bank.

Abdul-Hadi, A.O.B. 2016. *Nationalism in the Middle East: The Development of Jordanian National Identity Since the Disengagement of 1988*. Durham: Durham University, PhD.

Abed, A., A. Tomah, and D. Dumour. 2015. Assessment of Slums' Upgrading Interventions: Case Study Jabal Al-Natheef, Amman, Jordan. *Innovative Systems Design and Engineering* 6 (6): 1–8.

Abrams, C. 1964. *Man's Struggle for Shelter in an Urbanizing World*. Cambridge, MA: MIT Press.

Abu-Ghazalah, S. 2010. Global Economic Forces' Effect upon Urban Planning: Greater Amman Case Study. Journal of Urban Planning and Development 136 (2): 125–34.

Abu-Hamdi, E. 2017. Neoliberalism as a Site-Specific Process: The Aesthetics and Politics of Architecture in Amman, Jordan. *Cities* 60: 102–112.

Abu-Lughod, J. 1965. A Tale of Two Cities: The Origins of Modern Cairo. *Comparative Studies in Society and History* 12 (1): 429–457.

———. 1969. Varieties of Urban Experience: Contrast, Coexistence and Coalescence in Cairo. In *Middle East Cities*, ed. I.M. Lapidus, 159–187. Stanford: California University Press.

———. 1971. *Cairo: 1001 Years of the City Victorious*. Princeton: The Princeton University Press.

Abu-Lughod, J., and Richard Hay Jr., eds. 1977. *Third World Urbanization*. New York: Methuen.

Abusaada, H., and A. Elshater. 2020. COVID-19's Challenges to Urbanism: Social Distancing and the Phenomenon of Boredom in Urban Spaces. *Journal of Urbanism: International Research on Placemaking and Urban Sustainability*. https://doi.org/10.1080/17549175.2, https://doi.org/10.1007/s11524-020-00438-6.

Adler, P., and S. Kwon. 2002. Social Capital Respective or a New Concept. *Academy of Management Review* 27 (1): 17–40.

Adwan, C. 2004. *Corruption in Reconstruction: The Cost of National Consensus in Post-War Lebanon*. Washington, DC: CIPE.

Agamben, G. 1998. *Homo Sacer: Sovereign Power and Bare Life*. Stanford: Stanford University.

Al Daly, J. 1999. *Informal Settlements in Jordan: Upgrading Approaches Adopted and Lessons Learned*. Amman: Housing and Urban Development Corporation.

Al Naber, M., and F. Nolle. 2016. The Politics of Accessing Desert Land in Jordan. *Land Use policy*: 492–503.

A. M. Soliman, *Urban Informality*, https://doi.org/10.1007/978-3-030-68988-9

Al Nsour, J.A. 2016. Managing Urban Growth in the City of Amman, Jordan. *Cities* 50: 93–99.

Al Sayyad, N. 2004. Urban Informality as a "New" Way of Life. In *Urban Informality: Transnational Perspectives from the Middle East, Latin America, and South Asia*, ed. A. Roy and N. Al Sayyad, 7–30. Lanham: Lexington Books.

———. 2006. Whose Cairo? In *Cosmopolitan: Politics, Culture, and Urban Space in the New Globalized Middle East*, ed. D. Singerman and P. Amar, 539–542. Cairo: American University Press in Cairo.

Al-Ahram. 2010a. *Egyptian Newspaper*. April 7th.

———. 2010b. *Egyptian Newspaper*. May 8th.

Alexander, C., et al. 1975. *The Oregon Experiment*. Mountain View: The Center for Environmental Structure.

Al-Hasanat, S. 2003. Housing Information System. *Sixth Annual Seminar on Housing Sector Assessment*. Amman: Housing and Urban Development Corporation, 107–113.

Al-Homoud, M., S. Al-Oun, and A. Al-Hindawi. 2009. The Low-Income Housing Market in Jordan. *International Journal of Housing Markets and Analysis* 2 (3): 233–252.

Aljafari, M. 2014. *Emerging Public Spaces in the City of Amman, Jordan: An Analysis of Everyday Life Practices*. PhD thesis. Dortmund: Dortmund University.

Allen, A. 2014. Peri-Urbanisation and the Political Ecology of Differential Sustainability. In *A Routledge Handbook on Cities of the Global South*, ed. S. Oldfield and S. Parnell, 21–36. London/New York: Routledge.

Al-Nammari, F. 2013. Participatory Urban Upgrading and Power: Lessons Learnt from a Pilot Project in Jordan. *Habitat International 39*: 224–231.

AlNasser, H.W. 2016. *New Social Enterprises in Jordan Redefining the Meaning of Civil Society*. London: The Royal Institute of International Affairs.

Al-Omari, Z. 2003. Reality of Housing Sector as the Private Sectors See It. *Sixth Annual Seminar on Housing Sector Assessment*. Amman: Housing and Urban Development Corporation, 131–141.

Al-Oun, S. 1997. *Livestock Marketing in the Jordanian Badia*. PhD dissertation. Durham: University of Durham.

Al-Oun, S., M. Al-Homoud, and J. Abdel Jawad. 2010. Fitness Between House Choices versus Market Production in Jordan. *22nd International Housing Research Conference, Istanbul, 4–7 July*.

Al-Shawabkeh, R.A. and Alhaddad, I., L. 2018. Project: High Density Mixed Use as an Effective Scheme in Applying Sustainable Urban Design Principles in Amman, Jordan. *International Journal of Humanities and Social Science* 7 (10): 57–78.

Alon, Yoav. 2007. *The Making of Jordan: Tribes, Colonialism, and the Modern state*. London: I.B. Tauris & Co Ltd.

Amin, A. 1976. *Unequal Development: An Essay on the Social Formations of Peripheral Capitalism*. Sussex: The Harvester Press Limited.

Amin, S. 1980. *Class and Nation: Historically and in the Current Crisis*. London: Heinemann.

———. 2006. *A Life Looking Forward: Memoirs of an Independent Marxist*. London: Zed Books Ltd.

———. 2011. *Ending the Crisis of Capitalism or Ending Capitalism?* Oxford: Pambazuka Press.

Amineh, M. Parvizi. 2007. *The Greater Middle East in Global Politics : Social Science Perspectives on the Changing Geography of the World Politics*. Leiden, The Netherlands: Brill.

Andonova, L.B., et al. 2009. Transnational Climate Governance. *Global Environmental Politics* 9 (2): 52–73.

Angel, S., R. Archer, S. Tanphiphat, and E. Weglin. 1985. *Land for Housing the Poor*. Singapore: Selected Books.

Angel, S., J. Parent, D. Clivco, and A. Blei. 2011. *Making Room for a Planet of Cities*. Cambridge: Lincoln Institute of Land Policy.

Angotti, T. 2013. Urban Latin America: Violence, Enclaves and Struggles for Land. *Latin American Perspectives* 40: 5–20.

Anholt, R., and G. Sinatti. 2020. Under the Guise of Resilience: The EU Approach to Migration and Forced Displacement in Jordan and Lebanon. *Contemporary Security Policy* 41 (2): 311–335.

Antonio, J., C.M. Lara-Hernandez, C.M. Coulter, and A. Melis. 2020. Temporary Appropriation and Urban Informality: Exploring the Subtle Distinction. *Cities* 99. https://doi.org/10.1016/j.cities.2020.102626.

Archer, R. 1999. The Potential of Land Pooling/Readjustment to Provide Land for Low-Cost Housing in Developing Countries. In *Making Common Ground: Public-Private Partnerships in Land for Housing*, ed. G. Payne, 113–133. London: Intermediate Tech.

Arrighi, G., and J.S. Saul. 1973. *Essays on the Political Economy in Africa*. London: Monthly Review Press.

Attia, Sahar. 2019. Al Alamein New City, a Sustainability Battle to Win. In *New Cities and Community Extensions in Egypt and the Middle East: Visions and Challenges*, ed. Attia Sahar et al., 1–18. Cham: Springer.

Avelino, F., and J. Wittmayer. 2017. A Multi-Actor Perspective on Urban Sustainability Transitions. In *Urban Sustainability Transitions*, ed. N. Frantzeskaki, V.C. Broto, L. Coenen, and D. Loorbach, 272–284. Oxon: Routledge.

Avelino, F., J. Grin, B. Pel, and S. Jhagroe. 2016. The Politics of Sustainability Transitions. *Journal of Environmental Policy & Planning* 18 (5): 557–567.

Badawy, M.A.S. 1999. *Islamic Associations in Cairo*. Unpublished MA thesis. Cairo: American University in Cairo.

Bai, X., R. McAllister, R. Beaty, and B. Taylor. 2010. Urban Policy and Governance in a Global Environment: Complex Systems, Scale Mismatches and Public Participation. *Current Opinion in Environmental Sustainability* 2: 1–7.

Banks, N., M. Lombard, and D. Mitlin. 2020. Urban Informality as a Site of Critical Analysis. *The Journal of Development Studies* 56 (2): 223–238.

Barakat, N., and M. Tuffaha. 1995. Private Sector's Role in Housing Limited and Low Income Households. *Housing Sector Assessment Seminar*. Amman: Housing and Urban Development Corporation, 3–12.

Barker, I., and R. Linden. 1985. *Community Crime Prevention*. Ottawa: Ministry of Solicitor General Canada.

Basboos, A. 1995. Housing Financing. *Housing Sector Assessment Seminar*. Amman: Housing and Urban Development Corporation, 2–19.

Baser, H., and P. Morgan. 2008. *Capacity, Change and Performance. European Centre for Development Policy Management*. Dordrecht: HE Maastricht.

Bayat, A. 2000. From "Dangerous Classes" to "Quiet Rebels" Politics of Urban Subaltern in the Global South. *International Sociology* 15 (3): 533–557.

———. 2004. Globalization and the Politics of the Informals in the Global South. In *Urban Informality: Transnational Perspectives from the Middle East, Latin America, and South Asia*, ed. Ananya Roy and Nezar AlSayyad, 79–102. Lanham: Lexington Books.

———. 2009. Middle Eastern Megacities: Social Exclusion, Popular Movements and the Quiet Encroachment of the Urban Poor. In *Megacities: The Politics of Urban Exclusion and Violence in the Global South*, ed. Kees Koonings and Dirk Kruijt, 82–106. London: Zed Books Ltd.

———. 2010. *Life as Politics: How Ordinary People Change the Middle East*. Amsterdam: Amsterdam University Press.

———. 2012. Marginality: Curse or Cure. In *Marginality and Exclusion in Egypt*, ed. Ray Bush and Habib Ayeb, 14–27. Cairo: American University Press in Cairo.

Beall, J., and S. Fox. 2009. *Cities and Development*. London: Routledge.

Beauregard, R.A., and A. Marpillero-Colomina. 2011. More than a Master Plan: Amman 2025. *Cities* 28: 62–69.

Bell, W. 1982. *Tilting at Windmills: Considerations on the Nature of the State*. London: University College London, Department of Geography, occasional Paper No. 40.

Bell, S.T. 2007. Deep-Level Composition Variables as Predictors of Team Performance: A Meta-Analysis. *Journal of Applied Psychology* 92 (3): 595–615.

Beydoun, A. 1991. *Housing Crises and Housing Cooperatives in Lebanon (in Arabic)*. Beirut: Youssef Beydoun Press.
Bicknell, J., D. Dodman, and D. Satterthwaite. 2009. *Adapting Cities to Climate Change: Understanding and Addressing the Development Challenges*. London: Earthscan.
Binz, C., and B. Truffer. 2008. Leapfrogging in Infrastructure-Identifying Transition Trajectories Towards Decentralized Urban Water Management Systems in China. *DRUID Conference Paper*. Copenhagen: DRUID, 1–26.
Blitz, D. 2014. *The Arab Spring: A Parsimonious Explanation of Recent Contentious Politics*. Nijmegen: Radboud University.
Blokland, H. 2016. *Pluralism, Democracy and Political Knowledge: Robert A. Dahl and His Critics on Modern Politics*. New York: Routledge.
Blokland, T., and M. Savage. 2008. Social Capital and Networked Urbanism. In *Networked Urbanism: Social Capital in the City*, ed. T. Blokland and M. Savage, 1–20. Hampshire: Ashgate Publishing Limited.
Boanada-Fuchs, A., and F.B. Vanessa. 2018. Towards a Taxonomic Understanding of Informality. *IDPR* 40 (4): 397–420.
Bolay, Jean-Claude, et al. 2016. *Learning from the Slums for the Development of Emerging Cities*. Cham: Springer.
Bourdieu, P. 1977. *Outline of Theory of Practice*. Cambridge, UK: The University of Cambridge.
Bourgey, André. 1985. La Guerre et ses Conséquences Géographiques au Liban. *Annales de Géographies, No 521 XCIV e annee*, 1–37.
Brand, U. 2005. Order and Regulation: Global Governance as a Hegemonic Discourse of International Politics? *Review of International Political Economy* 12 (1): 155–176.
Bredenoord, J., P.V. Lindert, and P. Smets, eds. 2014. *Affordble Housing in the Urban Global South: Seeking Sustainable Solutions*. Oxon: Routledge.
Breman, J. 2019. *Capitalism, Inequality and Labour in India*. Cambridge, UK: Cambridge University Press.
———. 2020. Informality: The Bane of the Laboring Poor Under Globalized Capitalism. In *The Informal Economy Revisited: Examining the Past, Envisioning the Future*, ed. M. Chen and F. Carré, 31–37. Oxon: Routledge.
Bridge, G., S. Bouzarovski, M. Bradshaw, and N. Eyre. 2013. Geographies of Energy Transition: Space, Place and the Low-Carbon Economy. *Energy Policy* 53: 331–340.
Bromely, R. 1978. The Urban Informal Sector: Why Is It Worth Discussing? *World Development* 6: 1033–1039.
———. 1979. *The Urban Informal Sector: Critical Perspectives on Employment and Housing Policies*. Headington Hill Hall: Pergamon Press Ltd.
———. 2004. Power, Property and Poverty: Why De Soto's Mystery of Capital Cannot Be Solved. In *Urban Informality: Transnational Perspectives from the Middle East, Latin America, and South Asia*, ed. A. Roy and N. Al Sayyad, 271–288. Lanham: Lexington Books.
Brundtland, G.H. 1987. *The World commission on Environment and Development: Our Common Future*. Technical Report. New York: United Nations.
Buckley, R., and J. Kalarickal, eds. 2006. *Thirty Years of World Bank Shelter Lending What Have We Learned?* Washington, DC: World Bank, The International Bank for Reconstruction and Development.
Bulkeley, H., L. Coenen, N. Frantzeskaki, C. Hartmann, A. Kronsell, L. Mai, and Y.V. Palgan. 2016. Urban Living Labs: Governing urban Sustainability Transitions. *Current Opinion in Environmental Sustainability* 22: 13–17.
Bums, L.S., and J. Friedmann. 1985. *The Art of Planning: Selected Essays of Harvey S. Pedoff*. New York: Plenum Press.
Bunnell, T., and A. Harris. 2012. Reviewing Informality: Perspectives from Urban Asia. *International Development Planning Review* 34 (4): 339–348.
Burgess, R. 1982. Petty Critique of John Turner's. In *Self-Help Housing: A Critique*, ed. P. Ward, 55–97. London: Mansell Publishing Co.

Burgess, R., M. Carmona, and T. Kolstee. 1997. *The Challenge of Sustainable Cities: Neoliberalism and Urban Strategies in Developing Countries*. London: Zed Books Ltd.

Bush, R., and H. Ayeb. 2012. Introduction: Marginality and Exclusion in Egypt and the Middle East. In *Marginality and Exclusion in Egypt*, ed. R. Bush and H. Ayeb, 3–13. Cairo: American University in Cairo Press.

Butterwick, S., and R.L. Lawrence. 2009. Creating Alternative Realities: Arts-Based Approaches to Transformative Learning. In *Transformative Learning in Practice: Insights from Community, Workplace, and Higher Education*, ed. J. Mezirow, E.W. Taylor, et al., 35–45. San Francisco: Jossey-Bass.

Cacioly, Claudio, Jr. 2000. Can Urban Management Deliver the Sustainable City? Guided Densification in Brazil Versus Informal Compactness in Egypt. In *Compact Cities: Sustainable Urban Forms for Developing Countries*, ed. Mike Jenks and Rod Burgess, 127–140. London: Spon Press.

Cammett, M., I. Diwan, A. Richards, and J. Waterbury. 2015. *A Political Economy of the Middle East*. Boulder: Westview Press.

Canclini, N.G. 1995. *Hybrid Cultures Strategies for Entering and Leaving Modernity*. Minneapolis: University of Minnesota Press.

CAPMAS. 2016a. *General Statistics for Population and Housing: Population Census*. Cairo: Central Agency for Public Mobilization and Statistics.

———. 2016b. *A Study for Development and Upgrade Informal Areas in Egypt (in Arabic)*. Cairo: CAPMAS.

Carnoy, M., M. Castells, S.S. Cohen, and F.H. Cardoso. 1993. *The New Global Economy in the Information Age: Reflections on Our Changing World*. Pennsylvania: The Pennsylvania State University.

Caroline, M., and S. Yatzimirsky. 2014. *Megacity Slums: Social Exclusion, Space and Urban Policies in Brazil and India*. London: Imperial College Press.

Castells, M. 1975. Advanced Capitalism, Collective Consumption, and Urban Contradictions: New Sources of Inequality and New Models for Change. In *Stress and Contradiction in Modern Capitalism*, ed. L. Lindberg, R. Alford, C. Crouch, and C. Offe. Cambridge: Lexington Books.

———. 1977. *The Urban Question*. London: Edward Arnold (Publishers) Ltd.

———. 1983. *The City and Grassroots: A Cross-Cultural Theory of Urban Social Movements*. London: Edward Arnold (Publishers) Ltd.

———. 1997. *The Power of Identity*. Malden: Blackwell Publishers Inc.

———. 2000. Materials for an Exploratory Theory of the Network Society. *British Journal of Sociology* 51 (1): 5–24.

———. 2010. *The Rise of the Network Society*. 2nd ed. West Sussex: Wiley.

———. 2015. *Networks of Outrage and Hope: Social Movements in the Internet Age*. Cambridge: Polity Press.

Castells, M., and P. Hall. 2004. *Technopoles of the World: The Making of Twenty First Century Industrial Complexes* (electronic edition). Oxon: Routledge.

CDR. 2016. *Habitat III Lebanese National Report*. Final Report. Grand Serail/Beirut/Lebanon: Council for Development and Reconstruction.

Chadwick, G. 1971. *A System View of Planning: Towards a Theory of Urban and Regional Process*. Oxford/Headington Hill Hall: Pergamon Press Ltd.

Chang, Rui-dong, J. Zuo, Z. Zhao, V. Soebarto, G. Zillante, and X. Gan. 2017. Approaches for Transitions Towards Sustainable Development: Status Quo and Challenges. *Sustainable Development*. https://doi.org/10.1002/sd.1661.

Chase, J.L., M. Crawford, and J. Kaliski, eds. 2008. *Everyday Urbanism*. New York: Monacelli Press.

Chen, M. 2006. Rethinking the Informal Economy: Linkages with the Formal Economy and the Formal Regulatory Environment. In *Linking the Formal and Informal Economy: Concepts and Policies*, ed. R. Kanbur, E. Ostrom, and B. Guba-Khasnobis, 75–92. Oxford: Oxford University Press.

Chen, Marty, and Françoise Carré. 2020. *The Informal Economy revisited: Examining the Past, Envisioning the Future*. Oxon: Routledge.
Cirolia, L.R. 2017. Overcoming the Disjunctures: Competing Discourses on Informal Settlements in South Africa. *IDPR* 39 (4): 443–459.
Clark, D. 2003. *Urban World: Global City*. 2nd ed. London: Routledge.
Coase, R.H. 1960. The Problem of Social Cost. *Journal of Law and Economics*: 1–44.
Coenen, L., P. Benneworth, and B. Truffer. 2012. Toward a Spatial Perspective on Sustainability Transitions. *Research Policy* 41: 968–979.
Cohen, M. 2009. Aid, Density, and Urban Form: Anticipating Dakar. In *Urban Land Markets: Improving Land Management for Successful Urbanization*, ed. S.V. Lall, M. Freire, B. Yuen, R. Rajack, and J.-J. Helluin, 385–398. Dordrecht/Heidelberg: Springer.
Collier, D. 1976. *Squatters and Oligarchs*. Baltimore: Johns Hopkins University Press.
Collier, P. 2002. Social Capital and Poverty: A Microeconomic Perspective. In *The Role of Social Capital in Development: An Empirical Assessment*, ed. T. Van Bastelaer and C. Grootaert. Cambridge: Cambridge University Press.
Colsaet, A., Y. Laurans, and H. Levrel. 2018. What Drives Land Take and Urban Land Expansion? A Systematic Review. *Land Use Policy* 79: 339–349.
Cooke, P. 2008. Clean Tech and an Analysis of the Platform Nature of Life Sciences: Further Reflections Upon Platform Policies. *European Planning Studies* 16: 375–393.
Cornelius, W. 1975. *Politics and Migrant Poor in Mexico City*. Stanford: Stanford University Press.
Corsín Jiménez, A. 2014. The Right to Infrastructure: A Prototype for Open Source Urbanism. *Environment and Planning D: Society & Space* 32: 342–362.
Cram, F., and D.M. Mertens. 2015. Transformative and Indigenous Frameworks for Multimethod and Mixed Methods Research. In *The Oxford Handbook of Multimethod and Mixed Methods Research Inquiry*, ed. S. Hesse-Biber and R.B. Johnson, 91–109. New York: Oxford University Press.
Crowther, D., S. Seifi, and A. Moyeen. 2018. Responsibility and Governance in Achieving Sustainability. In *Approaches to Global Sustainability, Markets, and Governance*, ed. D. Crowther, S. Seifi, and A. Moyeen, 1–18. Singapore: Springer.
Culcasi, K. 2016. Warm Nationalism: Mapping and Imagining the Jordanian Nation. *Political Geography* 54: 7–20.
d'Alençon, P.A., L. Moretto, and H. Smith. 2016. *N-AERUS – Recommendations for the New Urban Agenda*. Working Paper Group, Cities Alliance.
d'Alençona, P.A., H. Smith, and E.A. de Andres. 2018. Interrogating Informality: Conceptualisations, Practices and Policies in the Light of the New Urban Agenda. *Habitat International* 75: 59–66.
Daher, R. F. 2008. Amman: Disguised Genealogy and Recent Urban Restructuring and Neoliberal Threats. In *The Evolving Arab City: Tradition, Modernity & Urban Development*, ed. Elsheshtawy, Y., 37–68. London and New York: Routledge.
Dahl, R.A. 1961. *Who Governs? Democracy and Power in an American City*. Clinton: Yale University.
Daily Star. 2004. *Lebanese Newspaper*. May 28.
DAR - IAURIF. 2005. *National Physical Master Plan of the Lebanese Territory*. Beirut: DAR - IAURIF.
Davidson, F., and G. Payne. 1985. *Urban Project Manual: A Guide to Preparing Upgrading and New Development Projects Accessible to Low Income Groups*. Liverpool: Liverpool University Press.
Davis, M. 2006. *Planet of Slums*. London: Verso.
de Haan, Arjan. 2007. *Reclaiming Social Policy: Globalization, Social Exclusion, and New Poverty Reduction Strategies*. New York: Palgrave Macmillan.
de Satgé, R., and W. Vanessa. 2018. *Urban Planning in the Global South*. Cham: Palgrave Macmillan.

De Soto, H. 1989. *The Other Path: The Invisible Revolution in the Third World*. London: I.B. Taurus.

———. 2000. *The Mestry of Capital: Why Capitalism Triumphs in the Westand Fails Everywhere Else*. London: Black Swan.

Debruyne, P. 2014. *Spatial Rearticulations of Statehood: 'Jordan's Geographies of Power Under Globalization*. Department of Political and Social Sciences: Ghent University, PhD.

DEFID. 1999. *Literature Review on Peri-Urban Natural Resource Conceptualization and Management Approaches*. London: DEFID, project no. R6949.

Deininger, K., H. Selod, and A. Burns. 2012. *Land Governance Assessment Framework, Identifying and Monitoring Good Practice in the Land Sector*. Washington, DC: The world Bank.

Delgado, M.C. 2015. Contentious Copts: The Emergence, Success, and Decline of the Maspero Youth Movement in Egypt. In *Contentious Politics in the Middle East: Popular Resistance and Marginalized Activism: beyond the Arab Uprisings*, ed. Fawaz A. Gerges, 259–280. New York: Palgrave Macmillan.

Dempsey, N., and M. Jenks. 2010. The Future of the Compact City. *Built Environment* 36 (1): 5–8.

Denis, E. 2006. Cairo as Neo-Liberal Capital? From Walled City to Gated Communities. In *Cairo Cosmopolitan: Politics, Culture, and Urban Space in the New Globalized Middle East*, ed. D. Singerman and P. Amar, 47–71. Cairo: The American University in Cairo Press.

———. 2012. The Commodification of the Ashwa'ryyat: Urban Land, Housing Market Unification, and de Soto's Interventions in Egypt. In *Popular Housing and Urban Land Tenure in the Middle East*, ed. M. Ababsa, M. Dupret, and E. Denis, 227–258. Cairo: American University.

Dennis, R. 2005. *Marginality, Power, and Social Structure: Issues in Race, Class, and Gender*. London: Elsevier Ltd.

Devlin, R.T. 2011. 'An Area That Governs Itself': Informality, Uncertainty and the Management of Street Vending in New York City. *Planning Theory* 10 (1): 53–65.

———. 2018. Asking 'Third World Questions' of First World informality: Using Southern Theory to Parse Needs from Desires in an Analysis of Informal Urbanism of the Global North. *Planning Theory* 17 (4): 568–587.

———. Forthcoming. Temporary Appropriation and Informality: A Nuanced Relationship. *Cities*.

di Milano, Politecnico. 2013. Cities to be Tamed? Standards and Alternatives in the transformation of the urban South. *The Journal of Urbanism* 1(26). *Cities to be Tamed?* Milan.

Diana Reckien, D., and J. Karecha. 2008. Sprawl in European Cities: The Comparative Background. In *Urban Sprawl in Europe: Landscapes, Land-Use Change and Policy*, ed. Lila Leontidou, Chris Couch, and Gerhard Petschel-Held, 39–67. Oxford: Blackwell Publishing Ltd.

Dilworth, G. 2017. *Jabal Al-Weibdeh: A Counter-Memory of Amman. A Case Study in the Resistance of Memory*. University of Virginia, Independent Study Project (ISP) Collection. 2666.

Dixon, Tim, et al. 2014. Introduction. In *Urban Retrofitting for Sustainability Mapping the transition to 2050*, ed. Malcolm Eames, Miriam Hunt, Simon Lannon, and Tim Dixon, 1–16. Abingdon/Oxon: Routledge.

Doebele, W. 1987. The Evolution of Concepts of Urban Land Tenure in Developing Countries. *Habitat International* 11 (1): 7–22.

———. 1994. Urban Land and Macroeconomic Development: Moving from 'Access for Poor' to Urban Productivity. In *Methodology for Land and Housing Market Analysis*, ed. G. Jones and P.M. Ward, 44–54. Cambridge, MA: Lincoln Land Institute of Land Policy.

Dominelli, L., and M. Moosa-Mitha. 2014. *Reconfiguring Citizenship: Social Exclusion and Diversity Within Inclusive Citizenship Practice*. Surrey: Ashgate Publishing Limited.

Dorman, W. J. 2007. *The Politics of Neglect: The Egyptian State in Cairo, 1974–198*. PhD. thesis, School of Oriental and African Studies, University of London.

Dorman, W.J. 2012. Exclusion and Informality: The Praetorian Politics of Land Management in Cairo, Egypt. *International Journal of Urban and Regional Research*. https://doi.org/10.1111/j.1468-2427.2012.01202.x1-27.

———. 2013. Exclusion and Informality: The Praetorian Politics of Land Management in Cairo (Egypt). *International Journal of Urban and Regional Research.*

Dovey, K., et al. 2020. Viewpoint Informal Settlement Is Not a Euphemism for 'Slum': What's at Stake Beyond the Language? *IDPR.* https://doi.org/10.3828/idpr.2020.14.

Drakakis-Smith, D. 1979. Low-Cost Housing Provision in the Third World: Some Theoretical and Practical Alternatives. In *Housing in Third World Countries*, ed. H.S. Murison and J.P. Lea, 14–22. London: The Macmillan Press Ltd.

———. 2000. *Third World Cities.* 2nd ed. London: Routledge.

Drakakis-Smith, D., and W.B. Fisher. 1975. *Housing in Ankara, Occasional Paper Number 7 (New Series).* Durham: Department of Geography, University of Durham.

Duffy, H. 1995. *Competitive Cities: Succeeding in the Global Economy.* London: E & FN Spon.

Dumour, D., and M. Ababsa. 2008. Wadi Abdoun: From Slums to Towers the Upgrading Policy of an Informal Settlements in Amman, Jordan. *Ninth Mediterranean Research Meeting, Workshop no.4 Public Policies and Legal Practice Towards Informal Settlements in the Middle East and Egypt.* Florence: Monticatini Terme, 12–15.

Dumper, M.R.T., and B.E. Stanley, eds. 2007. *Cities of the Middle East and North Africa: A Historical Encyclopedia.* Santa Barbara: ABC-CLIO, Inc.

Duquennois, A., and P. Newman. 2009. *Linking the Green and Brown Agendas: A Case Study on Cairo, Egypt.* Technical. Nairobi: UN-Habitat, Unpublished case study prepared for the Global Report on Human Settlements.

Durand-lasserve, A. 2005. Land for Housing the Poor in African Cities: Are Neo-Customary Processes an Effective Alternative to Formal Systems? In *Urban Futures, Economic Growth and Poverty Reduction*, ed. N. Hamdi. London: ITDG Publishing.

Dwyer, D.J. 1975. *People and Housing in Third World Cities.* London: Longman.

Eckstein, Susan. 1988. *The Poverty of Revolution: The State and the Urban Poor in Mexico. (Revised edition 1988.).* Princeton: Princeton University Press.

Edelenbos, J. 2005. Institutional Implications of Interactive Governance: Insights from Dutch Practice. *Governance* 18 (1): 111–134.

Edensor, E., and M. Jayne, eds. 2011. *Urban Theory Beyond the West: A World of Cities.* London: Taylor & Francis.

EEA. 2019. *Sustainability Transitions: Policy and Practice.* Technical. Luxembourg: European Environment Agency.

Egypt, Arab Republic of, and UNDP. 2007. *Project 00056471.* New York: Untied Nation Development Program.

Egypt Human Development Report. 1996. *Egypt Human Development Report.* Cairo: UNDP.

Ehteshami, A. 2007. *Globalization and Geopolitics in the Middle East: Old Games New Rules.* Oxon: Routledge.

Elkington, J. 1997. *Cannibals with Forks: The Triple Bottom Line of 21st Century Business.* Oxford: Capstone.

Elmouelhi, H. 2019. New administrative Capital - Cairo: Power, Urban Development and Social Injustice – the Official Egyptian Model of Neoliberalism. In *Neoliberale Urbanisierung Stadtentwicklungsprozesse in derarabischen Welt*, ed. Al-Hamarneh, A., Margraff, J., Scharfenort, N., Bielefeld, Germany: Transcript Verlag.

Elsheshtawy, Y. (ed.). 2004. *Planning Middle Eastern Cities : An Urban Kaleidoscope in a Globalizing World.* London: Routledge.

Elsheshtawy, Y. 2008. The Great Divide: Struggling and Emerging Cities in the Arab World. In *The Evolving Arab City: Tradition, Modernity and Urban Development*, ed. Elsheshtawy, Y., 1–26. London and New York: Routledge.

Elzen, B., F. Geels, and K. Green. 2004. *System Innovation and the Transition to Sustainability: Theory, Evidence and Policy.* Gloucestershire: Edward Elgar Publishing Limited.

Elzen, B., M. Barbier, M. Cerf, and J. Grin. 2012. Stimulating Transitions Towards Sustainable Farming Systems. In *Farming Systems Research into the 21st Century: The New Dynamic*, ed. I. Darnhofer, D. Gibbon, and B. Dedieu, 431–455. New York: Springer.

Engles, F. 1970. *The Housing Question*. Moscow: Progress Publishers.
ESCWA. 2003. *Survey of Economic and Social Developments in the ESCWA Region*. New York: United Naations.
Etingoff, K. 2017. *Sustainable Cities: Urban Planning Challenges and Policy*. Waretown: Apple Academic Press, Inc.
Evans, P.B., and M. Timberlake. 1980. Dependence, Inequality, and the Growth of the Tertiary: A Comparative Analysis of Less Developed Countries. *American Sociological Review* 45: 531–552.
Fafo. 2020. *Impact of the COVID-19 Pandemic on Enterprises in Jordan*. Oslo: Institute for Labour and Social Research.
Faghaley, K. 2002. *Encyclopedia of Lebanon (in Arabic)*. Vol. 1. Beirut: International Publisher.
Faludi, A. 1970. The Planmng Environment and the Meaning of Planning. *Regional Studies* 4: 1–9.
———. 1973. *A Reader in Planning Theory*. Oxford/Headington Hill Hall: Pergamon Press Ltd.
FAO. 2001. *Urban and Peri-Urban Agriculture*. Rome: Food and Agriculture Organization of the United Nations.
Farha, L. 2017. *Global-Commodity-Housing-Has-Lost-Its-Currency Universal*. http://citiscope.org/story/2017/-.
Fathy, H. 1973. *Architecture for the Poor: An Experiment in Rural Egypt*. Chicago: The University of Chicago Press.
Fawaz, M. 2004. *Strategizing for Housing: An Investigation of the Production and Regulation of Low-Income Housing in the Suburbs of Beirut*. Massachusetts Institute of Technology, Department of Urban Studies and Planning, PhD.
———. 2008. An Unusual Clique of City-Makers: Social Networks in the Production of a Neighborhood in Beirut (1950–75). *International Journal of Urban and Regional Research* 32 (3): 565–585.
———. 2009. Neoliberal Urbanity and the Right to the City: A View from Beirut's Periphery. *Development and Change* 40 (5): 827–852.
———. 2017. Planning and the Refugee Crisis: Informality as a Framework of Analysis and Reflection. *Planning Theory* 16 (1): 99–115.
Fawaz, M., and I. Peillen. 2003. *Understanding Slums: Case Studies for the Global Report on Human Settlements 2003: Urban Slums Reports: The Case of Beirut, Lebanon*. Nairobi: Un-Habitat.
Fawaz, M., M. Krijnen, and D. El Samad. 2018. A Property Framework for Understanding for Understanding Gentrification. *City* 22 (3): 358–374.
Fawyaz, M., and E. Bian. 2003. Informal Housing Areas in Beirut (in Arabic). *Al Mouhandess Journal, Beirut* 17: 36–47.
Fergany, N. 2016. *Arab Revolution in the 21st Century? Lessons from Egypt and Tunisia*. Hampshire: Palgrave Macmillan.
Ferguson, B., and P. Smets. 2010. Finance for Incremental Housing; Current Status and Prospects for Expansion. *Habitat International* 34: 288–298.
Fernandes, E., and A. Varley, eds. 1998. *Illegal Cities: Law and Urban Change in Developing Countries*. London: Zed Books Ltd.
Fiedler, J. 2014. *Urbanisation, Unlimited: A Thematic Journey*. Cham: Springer.
Finn, D. 2014. DIY Urbanism: Implications for Cities. *Journal of Urbanism* 7 (4): 381–398.
Fischhbach, M. 2000. *State, Society, and Land in Jordan*. Boston: Brill.
Forrest, R., S. Yee Koh, and B. Wissink. 2017. *Cities and the Super-Rich: Real Estate, Elite Practices, and Urban Political Economies*. New York: Palgrave Macmillan.
Forrester, K. 2019. *In the Shadow of Justice: Postwar Liberalism and the Remaking of Political Philosophy*. Oxfordshire: Princeton University Press.
Foucault, M. 2003. *Society Must Be Defended. Lectures at the College de France, 1975–76*. London: Penguin.
Francis, Alexandra. 2015. *Jordan's Refugees Crisis*. Washington, DC: Carnegie Endowment for International Peace.

Frantzeskaki, N., and H. de Haan. 2009. Transitions: Two Steps from Theory to Policy. *Futures* 41 (9): 593–606.

Frantzeskaki, N., and D. Loorbach. 2012. Governing Societal Transitions to Sustainability. *International Journal of Sustainable Development* 15 (1/2).

Frantzeskaki, N., V.C. Broto, L. Coenen, and D. Loorbach. 2017a. Urban Sustainability Transitions: The Dynamics and Opportunities of Sustainability Transitions in Cities. In *Urban Sustainability Transitions*, ed. N. Frantzeskaki, V.C. Broto, L. Coenen, and D. Loorbach, 1–19. Oxon: Routledge.

———. 2017b. *Urban Sustainability Transitions*. New York: Routledge.

Frantzeskaki, N., M. Bach, K. Holscher, and F. Avelino. 2018a. *Co-Creating Sustainable Urban Futures: A Primer on Applying Transition Management in Cities*. Cham: Springer.

———. 2018b. Introducing Sustainability Transitions Thinking in Urban Contexts. In *Co-creating Sustainable Urban Future: A Primer on Applying Transition Management in Cities*, ed. Matthew Bach, Katharina Holscher, and Flor Avelino Niki Frantzeskaki, 63–79. Cham: Springer.

Frantzeskaki, N., M. Bach, and P. Mguni. 2018c. Understanding the Urban Context and Its Challenges. In *Co-Creating Sustainable Urban Futures: A Primer on Applying Transition Management in Cities*, ed. Niki Frantzeskaki et al. Tokyo/Heidelberg/New York/ Dordrecht/London: Springer.

Frantzeskaki, N., M. Bach, K. H€olscher, and F. Avelino. 2018d. Transition Management in and for Cities: Introducing a New Governance Approach to Address Urban Challenges. In *Co-creating Sustainable Urban Futures: A Primer on Applying Transition Management in Cities*, ed. N. Frantzeskaki, K. H€olscher, M. Bach, and F. Avelino, 1–40. Cham: Springer.

Friedmann, J. 2005. Globalization and the Emerging Culture of Planning. *Progress in Planning* 64: 183–234.

———. 2011. *Insurgencies: Essays in Planning Theory*. Oxon: Routledge.

———. 2014. Austrians-in-the-World. Conversations and Debates About Planning and Development. *European Spatial Research and Policy* 21 (1): 11–22.

Friedrich, E., B. Hillier, and A. Chiaradia. 2009. Anti-Social Behaviour and Urban Configuration Using Space Syntax to Understand Spatial Patterns of Socio-Environmental Disorder. In *Proceedings of 7th International Space Syntax Symposium*, ed. D. Koch, L. Marcus, and J. Steen, vol. 34, 1–16. Stockholm: Royal Institute of Technology (KTH).

Fuenfschilling, L., and B. Truffer. 2014. The Structuration of Socio-Technical Regimes – Conceptual Foundations from Institutional Theory. *Research Policy* 43 (4): 772–791.

Galal, A. 2004. *Potential Winners and Losers from Business Formalization*. Working Paper Series 95. Cairo: The Egyptian Center for Economic Studies.

Gans, H. 1962. *The Urban Villagers; Group and Class in the Life of Italian-Americans*. New York: Free Press of Glencoe.

———. 1971. The West End: An Urban Village. In *Internal Structure of the City*, ed. L.S. Bourne, 300–308. Oxford: Oxford University Press.

Gaonkar, D.P. 2001. *Alternative Modernities*. Durham: Duke University Press.

Gaspard, T.K. 2004. *A Political Economy of Lebanon, 1948–2002: The Limits of Laissez-Faire*. Leiden: Koninklijke Brill NV.

Gause, G. 2011. Why Middle East Studies Missed the Arab Spring? *Foreign Affairs* 90 (4): 81–90.

Geddes, P. 1915. *An Introduction of the Town Planning Movement and to the Study for Civics*. London: Williams & Norgate.

Geels, F. 2002. Technological Transitions as Evolutionary Reconfiguration Processes: A Multi-Level Perspective Case-Study. *Research Policy* 31 (8/9): 1257–1274.

———. 2004. From Sectoral Systems of Innovation to Socio-Technical Systems: Insights About Dynamics and Change from Sociology and Institutional Theory. *Research Policy* 33: 897–920.

———. 2005. The Dynamics of Transitions in Socio-Technical Systems: A Mulit-Level Analysis of the Transition Pathway from Horse-Drawn Carriages To Automobiles. *Technology Analysis and Strategic Management* 17 (4): 445–476.

———. 2010. Ontologies, Socio-Technical Transitions (to Sustainability), and the Multilevel Perspective. *Research Policy* 39: 495–510.

———. 2014. Regime Resistance Against Low-Carbon Transitions: Introducing Politics and Power into the Multi-Level Perspective Theory. *Culture Society* 31 (5): 21–40.

———. 2020. Micro-Foundations of the Multi-Level Perspective on Socio-Technical Transitions: Developing a Multi-Dimensional Model of Agency Through Crossovers Between Social Constructivism, Evolutionary Economics and Neo Institutional Theory. *Technological Forecasting and Social Change* 152. https://doi.org/10.1016/j.techfore.2019.119894.

Geels, F., and J.J. Deuten. 2006. Local and Global Dynamics in Technological Development: A Socio-Cognitive Perspective on Knowledge Flows and Lessons from Reinforced Concrete. *Science and Public Policy* 33: 265–275.

Geels, Frank W., and Johan Schot. 2007. Typology of Sociotechnical Transition Pathways. *Research Policy* 36: 399–417.

———. 2010. The Dynamics of socio-Technical Transitions: A Sociotechnical Perspective. In *Transitions to Sustainable Development; New Directions in the Study of Long Term Transformative Change*, ed. J. Grin, J. Rotmans, and J. Schot, 11–93. New York: Routledge.

Geels, Frank W., M.P. Jacobsson, and S. Hekkert. 2008. The Dynamics of Sustainable Innovation Journeys. *Technology Analysis & Strategic Management* 20 (5): 521–536.

Geels, F.W., F. Kern, G. Fuchs, N. Hinderer, K.G. Nele, J. Mylan, M. Neukirch, and S. Wassermann. 2016a. The Enactment of Socio-Technical Transition Pathways: A Reformulated Typology and a Comparative Multi-Level Analysis of the German and UK Low-Carbon Electricity Transitions (1990–2014). *Research Policy* 45: 896–913.

Geels, F.W., F. Berkhout, and D.P. van Vuuren. 2016b. Bridging Analytical Approaches for low-Carbon Transitions. *Nature Climate Change* 6: 576–583.

Geels, et al. 2017. *A Research Agenda for the Sustainability Transitions Research Network*. Amsterdam: STRN.

Gelder, Jean-Louis Van. 2009. Legal Tenure Security, Perceived Tenure Security and Housing Improvement in Buenos Aires: An Attempt Towards Integration. *IJURR* 33: 126–146.

Gemayel, A. 1992. *Rebuilding Lebanon*. Lanham: University Press of America.

Gendzier, I. 1999. *Notes from the Minifield: United States Intervention in Lebanon and the Middle East 1945–1958*. New York: Westview.

Genus, A., and A.-M. Coles. 2008. Rethinking the Multi-Level Perspective of Technological Transitions. *Research Policy* 37 (9): 1436–1445.

Gerges, F.A. 2015. *Contentious Politics in the Middle East: Popular Resistance and Marginalized Activism Beyond the Arab Uprisings*. New York: Palgrave Macmillan.

Gharipour, M. 2016. Urban Landscape: Public Space and Environment in Cities of the Contemporary Middle East. In *Contemporary Urban Landscapes of the Middle East*, ed. Mohammad Gharipour, 1–21. Oxon: Routledge.

Ghorayeb, M. 1998. The Work and Influence of Michel Ecochard in Lebanon. In *Projecting Beirut: Episodes in the Construction and Reconstruction of a Modern City*, ed. P. Rowe and H. Sarkis, 106–121. London: Prestel-Verlag.

Gilbert, A. 2009. Extreme Thinking About Slums and Slum Dwellers: A Critique. *SAIS Review of International Affairs* 29 (1): 35–48.

———. 2012. De Soto's the Mystery of Capital: Reflection on the Book's Public Impact. *IDPR* 34: 5–16.

Gilbert, A., and J. Gugler. 1982. *Cities, Poverty and Development: Urbanisation in the Third World*. Oxford: Oxford University Press.

Gilbert, A., and P. Ward. 1985. *Housing, the State and the Poor: Policy and Practice in Three Latin American Cities*. Cambridge, MA: Cambridge University Press.

GLTN's Arab Land Initiative. 2021. *Cairo: Second Arab Land Initiative Conference*.

Goodfellow, T. 2020. Political Informality: Deals, Trust Networks, and the Negotiation of Value in the Urban Realm. *The Journal of Development Studies* 56 (2): 278–294.

GOPP. 2012. *The Detailed Plan of El Matariya Area in Cairo City*. Cairo: The General Organization for Physical Planning (GOPP).

Gough, I., G. Wood, A. Barrientos, and P. Bevan. 2008. *Insecurity and Welfare Regimes in Asia, Africa and Latin America: Social Policy in Development Contexts*. New York: Cambridge University Press.

Goulet, D. 2006. *Development Ethics at Work*. Oxon: Routledge.

Grashoff, and Yang. 2020. Towards Critique and Differentiation: Comparative Research on Informal Housing. In *Comparative Approaches to Informal Housing Around the Globe*, ed. U. Grashoff, 1–21. London: UCL Press, University College London.

Grin, J., and A. Loeber. 2007. Theories of Learning. Agency, Structure and Change. In *Handbook of Public Policy Analysis. Theory, Politics, and Methods*, ed. Gerald J. Miller, Mara S. Sidney, and Frank Fischer, 201–222. New York: CRC Press.

Grin, J., J. Rotmans, and J. Schot. 2010. *Transitions to Sustainable Development: New Directions in the Study of Long Term Transformative Change*. New York: Routledge.

GTZ. 2004. *New Practice of Participatory Local Development in Egypt's Urban Areas. Policy Paper, Ministry of Planning, Egypt*. Cairo: German Technical and Financial Cooperation.

Gugler, J., and W. Flanagan. 1978. *Urbanization and Social Change in West Africa*. Cambridge: Cambridge University Press.

Guha-Khasnobis, B., R. Kanbur, and E. Ostrom. 2006. Beyond Formality and Informality. In *Linking the Formal and Informal Economy: Concepts and Policies*, ed. Ravi Kanbur, Elinor Ostrom, and Basudeb Guha-Khasnobis, 1–18. Oxford: Oxford University Press.

Guibrunet, L. 2017. *The Contribution of the Informal Economy to Urban Sustainability: Case Study of Waste Management in Tepito, Mexico City*. PhD thesis. London: University College London.

Gunderson, L.H., and C.S. Holling. 2002. *Understanding Transformations in Human and Natural Systems*. Washington, DC: Island Press.

Gupta, J., K. Pfeffer, H. Verrest, and M. Ros-Tonen, eds. 2015. *Geographies of Urban Governance: Advanced Theories, Methods and Practices*. Cham: Springer.

Guyer, J.I. 2011. Describing Urban 'No Man's Land' in Africa. *Africa* 81 (3): 474–492.

H€olscher, K. 2018. So What? Transition Management as a Transformative Approach to Support Governance Capacities in Cities. In *Co-creating Sustainable Urban Futures: A Primer on Applying Transition Management in Cities*, ed. N. Frantzeskaki, K. H€olscher, M. Bach, and F. Avelino, 375–396. Cham: Springer.

Hadjitheodorou, N.H. 1981. An Institutional Framework for the Upgrading Process of Manshiet Nasser in Cairo, Egypt. *Ekistics* 48 (286): 71–78.

Halabi, Z. 2004. Exclusion and Identity in Lebanon's Palestinian Refugee Camps: A Story of Sustained Conflict. *Environment and Urbanization* 16 (2): 39–48.

Hall, P. 1998. *Cities in Civilization: Culture, Technology and Urban Order*. London: Weidenfeld and Nicolson.

———. 2014. *Cities of Tomorrow: An Intellectual History of Urban Planning and Design Since 1880*. 4th ed. London: Blackwell.

Hall, P., and U. Pfeiffer. 2000. *Urban Future 21: A Global Agenda for Twenty First Century Cities*. Lonodon: Routledge.

Hamdi, N. 2004. *Small Change: About the Art of Practice and Limits of Planning in Cities*. London: Earthscan.

Hanieh, A. 2013. *Lineages of Revolt: Issues of Contemporary Capitalism in the Middle East*. Chicago: Haymarket Books.

Harms, H. 1982. Historical Perspectives on the Practice and Politics of Self-Help Housing. In *Self-Help Housing: A Critique*, ed. P. Ward, 17–53. London: Mansell.

Harris, R. 2002. A Double Irony: The Originality and Influence of John F.C. Turner. *Habitat International*: 48–56.

———. 2009. Reflections on Lebanon. In *Lebanon: Liberation, Conflict, and Crisis*, ed. Barry Rubin, 9–24. New York: Palgrave Macmillan.

———. 2014. Urban Land Market: A Southern Exposoure. In *The Routledge Handbook on Cities of the Global South*, ed. S. Parnell and S. Oldfield, 9–20. Oxon: Routledge.
———. 2018. Modes of Informal Urban Development: A Global Phenomenon. *Journal of Planning Literature* 33 (3): 267–286.
Harrison, P. 1979. *Inside the Third World*. Harmondsworth: Penguin Books.
Harriss-White, B. 2019. Waste, Social Order, and Physical Disorder in Small-Town India. *The Journal of Development Studies*. https://doi.org/10.1080/00220388.2019.1577386.
Hart, K. 1973. Informal Income Opportunities and Urban Employment in Ghana. *Journal of Modern African Studies* 11 (1): 61–89.
Harvey, D. 1973. *Social Justice and the City*. Oxford: Blackwell Publishers.
———. 1985. *The Urbanization of Capital: Studies in the History and Theory of Capitalist Urbanization 2*. Oxford: Basil Blackwell Ltd.
———. 2009. *Cosmopolitanism and the Geographies of Freedom*. New York: Columbia University Press.
———. 2012. *Rebel Cities: From the Right to the City to the Urban Revolution*. London: Verso.
———. 2016. *The Ways of the World*. London: Profile Books Ltd.
Hasan, A. 2001. *Working with Communities*. Karachi: City Press.
Hashemite Kingdom of Jordan. 2014. *Jordan Response Plan 2015 for the Syria Crisis*. Amman: Ministry of Planning and International Cooperation.
———. 2015. *General Population and Housing Census 2015*. Amman: The Department of Statistics.
———. 2017. *Jordan in Figures*. Amman: Department of Statistics.
Hatina, Meir. 2007. *Identity Politics in the Middle East Liberal Thought and Islamic Challenge in Egypt*. London: Tauris Academic Studies.
Healey, P. 2003. Collaborative Planning in Perspective. *Planning Theory* 2 (2): 101–123.
Henderson, J.V. 2009. The Effect of Residential Land Market Regulations on Urban Welfare. In *Urban Land Markets: Improving Land Management for Successful Urbanization*, ed. S.V. Lall et al., 25–50. Dordrecht/Heidelberg: Springer.
Hendriks, B. 2008. The Social and Economic Impacts of Peri-Urban Access to Land and Secure Tenure for the Poor: The Case of Nairobi, Kenya. *IDPR* 30 (1): 27–66.
Hendriks, C.M. 2009. Policy Design Without Democracy? Making Democratic Sense of Transition Management. *Policy Sciences* 42 (4): 341–368.
Hoogma, R., R. Kemp, J. Schot, and B. Truffer. 2002. *Experimenting for Sustainable Transport: The Approach of Strategic Niche Management*. London: Spon Press.
Hopkins, H. 1969. *Egypt, the Crucible-the Unfinished Revolution of the Arab World*. London: Martin Secker & Warburg Limited.
Hosseinioon. 2018. Measuring Urban Resilience to Natural Disasters for Iranian Cities: Challenges and Key Concepts. In *Urban Disaster Resilience and Security: Addressing Risks in Societies*, ed. F. Fiedrich and A. Fekete, 71–90. Cham: Springer.
———. 2019. Urban Resilience and Informality: Effects of Formalisation in Golestan, Iran. In *Urban Resilience for Risk and Adaptation Governance: Theory and Practice*, ed. G.B. Morató, O.C.N. Tollin, and M.R.C. Jordi, 11–128. Cham: Springer.
Hou, J. 2010. (Not) Your Everyday Public Space. In *Insurgent Public Space: Guerilla Urbanism and the Remaking of Contemporary Cities*, ed. J. Hou, 1–44. New York: Routledge.
Iaquinta, D., and A. Drescher. 2001. More than the Spatial Fringe: An Application of the Periurban Typology to Planning and Management of Natural Resources. In *Rural–Urban Encounters: Managing the Environment of the Peri-Urban Interface*. London: Development Planning Unit; University College.
IIHS. 2016. *Slum Almanac 2015/2016: Tracking Improvement in the Lives of Slum Dwellers*. New Delhi: Indian Institute for Human Settlements.
Ilbert, R. 1984. Heliopolis: Colonial Enterprise and Town Planning Success? *The Expanding Metropolis Coping with the Urban Growth of Cairo*. Cairo: Proceeding of seminar nine in the series architecture transformations in the Islamic World, November 11–15, 36–42.

ILO. 1972. *Employment, Income and Inequality: A Strategy for Increasing Productivity*. Geneva: International Labor Office.

Ingram, J. 2015. Framing Niche-Regime Linkage as Adaptation: An Analysis of Learning and Innovation Networks for Sustainable Agriculture Across Europe. *Journal of Rural Studies* 40: 59–75.

International Rescue Committee. 2018. *From Response to Resilience: Working with Cities and City Plans to Address Urban Displacement, Lessons from Amman and Kampala*. London: International Rescue Committee.

IRFED, Mission. 1963. *Besjimr et PossibiiA de DiAAloppent dm Liban*. Beirut: Ministdre du Plan.

ISDF. 2016. *Slums Areas in Egypt*. Cairo: ISDF.

Ismail, S. 2014. The Politics of the Urban Everyday in Cairo: Infrastructures of Oppositional Action. In *The Routledge Handbook on Cities of the Global South*, ed. S. Oldfield and S. Parnell, 269–280. London: Routledge.

Iveson, K. 2013. Cities Within the City: Do-It-Yourself Urbanism and the Right to the City. *International Journal of Urban and Regional Research* 37 (3): 941–956.

Jackson, L. 2010. *Influences on Financial Decision Making Among the Rural Poor in Bangladesh*. Unpublished Ph.D. thesis. Sydney: School of Marketing, The University of Western Sydney.

Jacobs, J. 1961. *The Death and Life of Great American Cities*. New York: Vintage Books, a division of Random House Inc.

Jamshidi, M. 2014. *The Future of the Arab Spring Civic Entrepreneurship in Politics, Art, and Technology Startups*. Oxford: Elsevier Inc.

Janmyr, M. 2018. UNHCR and the Syrian Refugee Response: Negotiating Status and Registration in Lebanon. *The International Journal of Human Rights* 22 (3): 393–419.

Jenkins, P. 2004. Beyond the Formal/Informal Dichotomy: Access to Land in Maputo, Mozambique. In *Reconsidering Informality Perspectives from Urban Africa*, ed. Karen Tranberg Hansen and Mariken Vaa, 210–226. Madrid: Grafilur Artes Gráficas.

———. 2013. *Urbanization, Urbanism and Urbanity in an African City: Home Spaces and House Cultures*. New York: Palgrave Macmillan.

Jenks, M., and R. Burgess. 2000. *Compact Cities: Sustainable Urban Forms for Developing Countries*. London: Spon Press.

Jessop, Bob. 2002. Liberalism, Neoliberalism and Urban Governance: A State Theoretical Perspective. *Antipode* 34 (3): 452–472.

———. 2010. What Follows Neo-liberalism? The Deepening Contradictions of US Domination and the Struggle for a New Global Order. In *Political Economy and Global Capitalism: The 21st Century, Present and Future*, ed. Bob Jessop, Richard Westra, and Robert Albritton, 67–88. London: Anthem Press.

———. 2013. *The Capitalist State: Marxist Theories and Methods*. Oxford: Martin Robertson & Company Ltd.

Johnstone, M. 1983. Housing Policy and the Urban Poor in Peninsular Malaysia. *Third World Planning Review* 5: 249–272.

Joint AESOP/ACSP Congress. 2013. Planning for Resilient Cities and Regions. In *AESOP/ACSP 5th Joint Congress 2013*. Dublin: University College Dublin.

Jørgensen, U. 2012. Mapping and Navigating Transitions – The Multi-Level Perspective Compared with Arenas of Development. *Research Policy* 41: 996–1010.

Juppenlatz, M. 1970. *Cities in Transformation: The Urban Squatter Problem of the Developing World*. Queensland: University of Queensland Press.

Kabisch, N., H. Korn, J. Stadler, and A. Bonn, eds. 2017. *Nature-based Solutions to Climate Change Adaptation in Urban Areas: Linkages Between Science, Policy and Practice*. Cham: Springer.

Kadhim, A.M., and Y. Rajjal. 1988. City Profile: Amman. *Cities* 5 (4): 318–325.

Kanger, L., and J. Schot. 2018. Deep Transitions: Theorizing the Long-Term Patterns of Socio Technical Change. *Environmental Innovation and Societal Transitions*. https://doi.org/10.1016/j.eist.2018.07.006.

———. 2019. Deep Transitions: Theorizing the Long-Term Patterns of Sociotechnical Change. *Environmental Innovation and Societal Transitions* 32: 7–21.
Kanie, N., and F. Biermann, eds. 2017. *Governing Through Goals: Sustainable Development Goals as Governance Innovation*. London: The MIT Press.
Karaman, O., et al. 2020. Plot by Plot: Plotting Urbanism as an Ordinary Process of Urbanisation. *Antipode*: 1–30.
Katkhouda, R. 1998. *Affordable Housing in the Suburbs of Beirut: Prospects in Post-War Reconstruction*. Montreal: School of Architecture, McGill University.
Kattaa, M., and S. Al Cheikh Hussein. 2009. Women Entrepreneurs Facing the Informality in Rural and Remote Areas in Syria. *European Journal of Social Sciences* 11 (4): 624–642.
Kauffman, Craig M. 2017. *Grassroots Global Governance: Local Watershed Management Experiments and the Evolution of Sustainable Development*. Oxford: Oxford University Press.
Keivani, R. 2018. *The Concept of Densification in the Middle Eastern Cities*. Proposed research. Oxford: Oxford Brookes University.
Kemp, R., J. Schot, and R. Hoogma. 1998. Regime Shifts to Sustainability Through Processes of Niche Formation: The Approach of Strategic Niche Management. *Technology Analysis & Strategic Management* 10: 175–196.
Kemp, R., D. Loorbach, and J. Rotmans. 2005. Transition Management as a Model for Managing Processes of Co-Evolution Towards Sustainable Development. *The International Journal of Sustainable Development and World Ecology*: 1–15.
Kemp, R., J. Rotmans, and D. Loorbach. 2007a. Assessing the Dutch Energy Transition Policy: How Does It Deal with Dilemmas of Managing Transitions? *Journal of Environment Policy Planning* 9 (3–4): 315–331.
Kemp, René, Derk Loorbach, and Jan Rotmans. 2007b. Transition Management as a Model for Managing Processes of Co-evolution Towards Sustainable Development. *International Journal of Sustainable Development & World Ecology* 14: 1–15.
Kendall, S.H. 1976. The Barangay as a Community in the Philippines. *Ekistics* 42 (242): 16–20.
Khalidi, R. 1998. The "Middle East" as a Framework of Analysis: Re-mapping a Region in the Era of Globalization. *Comparative Studies of South Asia, Africa and the Middle East* 18 (1): 74–81
Khalifa, M.A. 2011. Redefining Slums in Egypt: Unplanned Versus Unsafe Areas. *Habitat International* 35: 40–49.
Kim, S.J., and S.B. Merriam. 2011. Methodological Challenges in Studying Transformative Learning. In *Adult Education Research Conference*, 364–370. Toronto: New Prairie Press.
King, K. 2009. Workplace Performance–PLUS: Empowerment and Voice Through Professional Development and Democratic Processes in Health Care Training. *Performance Improvement Quarterly* 21 (4): 55–74.
Kipper, R. 2009. Cairo: A Broader View. In *Cairo's Informal Areas: Between Urban Challenges and Hidden Potentials*, ed. Regina Kipper and Marton Fischer, 13–15. Lisbon: Norprint SA.
Kipper, R., and M. Fischer. 2009. *Cairo's Informal Areas: Between Urban Challenges and Hidden Potentials*. Cairo: GTZ.
Kirchherr, J., D. Reike, and M. Hekkert. 2017. Conceptualizing the Circular Economy: An Analysis of 114 Definition. *Resources, Conservation & Recycling* 127: 221–232.
Khirfan, L. (ed.). 2017. *Order and Disorder: Urban Governance and the Making of Middle Eastern Cities*. Montreal: McGill-Queens University Press.
———. 2019. Planning and Unplanning Amman: Between Formal Planning and Non-traditional Agency. In *The Palgrave Handbook of Bottom-Up Urbanism*, ed. M. Arefi and C. Kickert, 201–218. Cham, Switzerland: Springer International Publishing.
Kjaer, A.M. 2004. *Governance*. Cambridge: Polity Press.
Knight, F.H. 1921. *Risk, Uncertainty, and Profit*. Boston: Houghton.
Koehler, J., F. Geels, F. Kern, E. Onsongo, and A. J. Wieczorek. 2017. *A Research Agenda for the Sustainability Transitions Research Network*. Technical. Amsterdam: STRN Working Group, STRN.

Köhler, J., F.W. Geels, and F. Kern. 2019. An Agenda for Sustainability Transitions Research: State of the Art and Future Directions. *Environmental Innovation and Societal Transitions*. https://doi.org/10.1016/j.eist.2019.01.004.

Kooiman, J. 2003. *Governing as Governance*. London: SAGE.

Koonings, K., and D. Kruijt. 2009. *Megacities: The Politics of Urban Exclusion and Violence in the Global South*. London: Zed Books.

Kreibich, V. 2012. The Mode of Informal Urbanization: Reconciling Social and statutory Regulation in Urban Land Management. In *Urban Informalities: Reflections on the Formal and Informal*, ed. M. Waibel and C. McFarlane. Surrey: Ashgate.

Kuhl, S. 2009. Capacity Development as the Model for Development Aid Organizations. *Development and Change* 40: 551–577.

Kumaraswamy, P.R. (ed.). 2019. *The Palgrave Handbook of the Hashemite Kingdom of Jordan*. Singapore: Springer Nature Singapore Pte Ltd.

LaFreniere, G.F. 2007. *The Decline of Nature: Environmental History and the Western Worldview*. Bethesda/Dublin/London: Academic.

Laguerre, M.S. 1994. *The Informal City*. New York: St. Martin's Press, Inc.

Lall, S.V., M. Freire, B. Yuen, R. Rajack, and J.J. Helluin, eds. 2009. *Urban Land Markets Improving Land Management for Successful Urbanization*. London: Springer.

Lamba, H.S. 2010. Understanding the Ideological Roots of Our Global Crises: A Pre-Requisite for Radical Change. *Futures* 42 (10): 1079–1087.

Landman, K., and M. Napier. 2010. Waiting for a House or Building Your Own? Reconsidering State Provision, Aided and Unaided Self-Help in South Africa. *Habitat International* 34: 299–305.

Landy, F., and M.C. Saglio-Yatzimirsky. 2014. Conclusion: Spatial Justice, Exclusion and Urban Policies in Brazil and India. In *Megacity Slums: Social Exclusion, Space and Urban Policies in Brazil and India*, ed. Marie-Caroline Saglio-Yatzimirsky and Frédéric Landy, 409–426. Covent Garden/London: Imperial College Press.

Laquian, A.A. 1968. *Slums Are for People: The Barrio Magsaysay Pilot Project in the Philippine Urban Community Development*. Manila: Bustamente Press.

Lara-Hernandez, J.A., and A. Melis. 2018. Understanding the Temporary Appropriation in Relationship to Social Sustainability. *Sustainable Cities and Societies*: 1–14.

Lara-Hernandez, et al. 2020. Temporary Appropriation and Urban Informality: Exploring the Subtle Distinction. *Cities* 99. https://doi.org/10.1016/j.cities.2020.102626.

Layton, R.A. 2009. On Economic Growth, Marketing Systems, and the Quality of Life. *Journal of Macro Mark* 29 (4): 349–362.

Lebanese National Report. 1996. *Lebanese National Report*. Beirut: The Ministry of Environment.

Leduka, R. 2004. The Law and Access to Land for Housing in Maseru, Lesotho. In *Reconsidering Informality Perspectives from Urban Africa*, ed. K.T. Hansen and M. Vaa, 176–192. Spain: Grafilur Artes Gráficas.

Leeds, A. 1969. The Significant Variables Determining the Character of squatter Settlements. *America Latina* 12: 44–86.

———. 1972. *Forms of Squatment Political Organisation: The Politics of Control in Brazil*. M. A. thesis. Austin: University of Texas.

Lefebvre, H. 1968. *The Right to the City*. The Anarchist Library.

———. 1971. *Everyday Life in the Modern World*. New York: Harper & Row Publishers.

———. 1976. *The Survival of Capitalism: Reproduction of Relation of Production*. New York: Allison and Busby.

———. 1991. *The Production of Space*. Maiden: Blackwell Publishing.

———. 1996. *Writings on Cities*, trans. and ed. E. Kofman and E. Lebas. Oxford: Basil Blackwell.

Leftwich, A. 2010. *Redefining Politics. People, Resources and Power*. New York: Routledge.

Lewis, W. Arthur. 1954. Economic Development with Unlimited Supplies of Labor. *Manchester School of Economics and Social Studies* 22 (2): 139–191.

———. 1966. Culture of Poverty. *Scientific American* 215 (4): 19–25.

Lipton, M. 1984. Family, Fungibility and Formality: Rural Advantages of Informal Non-Farm Enterprise versus the Urban-Formal State. In *Human Resources, Employment and Development, vol. 5: Developing Countries*, ed. S. Amin, 189–242. London: MacMillan, for International Economic Association.

Lloyd, P. 1979. *Slums of Hope? Shanty Towns of the Third World.* Manchester: Manchester University Press.

Lomnitz, Larissa. 1977. *Como sobreviven los marginados. Mexico DF: Siglo XXI. English Translation Networks and Marginality: Life in a Mexican Shantytown.* New York: Academic.

Loorbach, D. 2007. *Transition Management New Mode of Governance for Sustainable Development.* Ph.D. thesis. Grifthoek/Utrecht: International Books.

———. 2010. Transition Management for Sustainable Development: A Prescriptive Complexity-Based Governance Framework. *Governance: An International Journal of Policy Administration and Institutions* 23 (1): 161–183.

Loorbach, interview by A. Soliman. 2020. *Series Editor* (March 15).

Loorbach, D., and S. Oxenaar. 2018. *Counting on Nature: Transitions to a Natural Capital Positive Economy by Creating an Enabling Environment for Natural Capital Approaches.* Rotterdam: Dutch Research Institute for Transitions (DRIFT), Erasmus University.

Loorbach, D., and J. Rotmans. 2010. The Practice of Transition Management: Examples and Lessons from Four Distinct Cases. *Futures* 42 (3): 237–246.

Loorbach, D., J. Wittmayer, H. Shiroyama, J. Fujino, and S. Mizuguchi. 2016. *Governance of Urban Sustainability Transitions: European and Asian Experiences.* Tokyo: Springer.

Loorbach, D., N. Frantzeskaki, and F. Avelino. 2017. Sustainability Transitions Research: Transforming Science and Practise for Societal Change. *Annual Review of Environment and Resources* 42: 599–626.

Lydon, M., and A. Garcia. 2015. *Tactical Urbanism: Short-Term Action for Long-Term Change.* Washington, DC: Island Press.

Lynch, Kenneth. 1960. *The Image of the City.* Cambridge, MA: MIT Press.

———. 2005. *Rural–Urban Interaction in Developing Countries.* London: Routledge.

Maconachie, R. 2007. *Urban Growth and Land Degradation in Developing Cities: Change and Challenges in Kano, Nigeria.* Hampshire: Ashgate Publishing Limited.

Madsen, S.R., and B.J. Cook. 2010. Transformative Learning: UAE, Women, and Higher Education. *Journal of Global Responsibility* 1 (1): 127–148.

Maheshwari, B., P. Vijay Singh, and B. Thoradeniya. 2016. Balanced Urban Development: Is It a Myth or Reality? In *Balanced Urban Development: Options and Strategies for Liveable Cities*, ed. P. Basant Maheshwari, Singh Thoradeniya, and Bhadranie Vijay, 3–16. Cham: Springer.

Malmvig, H. 2014. Free Us from Power: Governmentality, Counter-Conduct, and Simulation in European Democracy and Reform Promotion in the Arab World. *International Political Sociology* 8: 293–310.

Mangin. 1967. Latin American Squatter Settlements. A Problem and a Solution. *Latin American Research Review* 2: 65–98.

Mangin, W., and J.F.C. Turner. 1963. Urbanisation Case History in Peru. *Architectural Design* 339: 366–374.

Mansoob, M.S. 2002. *Globalization, Marginalization and Development.* London: Routledge.

March, J.G., and J.P. Olson. 1995. *Democratic Governance.* New York: Free Press.

Markard, J., R. Raven, and B. Truffer. 2012. Sustainability Transitions: An Emerging Field of Research and Its Prospects. *Research Policy* 41 (6): 955–967.

Marrengane, N., and S. Croese, eds. 2021. *Reframing the Urban Challenge in Africa: Knowledge Co-production from the South.* Oxon: Routledge.

Marshall, F., and J. Dolley. 2018. Transformative Innovation in Peri-Urban Asia. *Research Policy.* https://doi.org/10.1016/j.respol.2018.10.007.

Martin, R.J. 1969. The Ecology of Squatter Settlement. *Architecture Review* 145 (865): 213–214.

———. 1974. The Architecture of Under-Development, or the Route to Self-Determination in Design. *Architectural Design* 44: 626–643.

Martin, J. 2008. *The Poulantzas Reader: Marxism, Law and the State*. London: VERSO.

Martin, D. 2015. From Spaces of Exception to 'Campscapes': Palestinian Refugee Camps and Informal Settlements in Beirut. *Political Geography* 44: 9–18.

Marx, C. 2006. Conceptualising the Economy' to Make Urban Land Markets Work for the Poor. *The Urban Land Seminar, November 2006*. Muldersdrift: NA. NA.

Marx, Colin, and Emily Kelling. 2019. Knowing Urban Informalities. *Urban Studies* 56 (3): 494–509.

Masri, I. 1999. Development- at What Price? In *Preparing for the Future: Global Pressure and Local Forces*, ed. M. Cohen et al. Washington, DC: the Woodrow Wilson Center Press.

McAuslan, P. 1985. *Urban Land and Shelter for the Poor*. London: Earthscan.

———. 2003. *Bringing the Law Back In: Essays in Land, Law and Development*. Aldershot: Ashgate.

———. 2013. *Land Law Reform in Eastern Africa, Traditional or Transformative? A Critical Review of 50 Years of Land Law Reform in Eastern Africa,1961–2011*. Oxon: Routledge.

McFarlane, C., and M. Waibel, eds. 2012. *Urban Informalities: Reflections on the Formal and Informal*. London: Ashgate.

McGee, T. 1982. Labour Mobility in Fragmented Labour Market, the Role of Circulatory Migration in Rural-Urban Relations in Asia. In *Towards a Political Economy of Urbanization in Third World Countries*, ed. H. Safa, 47–66. Delhi: Oxford University Press.

Meadowcroft, J. 2000. Sustainable Development: A New(ish) Idea for a New Century? *Political Studies* 48: 270–387.

———. 2007. Who is in Charge here? Governance for Sustainable Development in a Complex World. *Journal of Environmental Policy and Planning* 9 (3): 299–314.

Meagher, K. 2016. Taxing Times: Taxation, Divided Societies and the Informal Economy in Northern Nigeria. *The Journal of Development Studies*. https://doi.org/10.1080/00220388.2016.1262026.

Mehta, B., and A. Dastur. 2008. *Approaches to Urban Slums: A Multimedia Sourcebook on Adaptive and Proactive Strategies*. Washington, DC: World Bank.

MENA. 2005. *The Macroeconomic and Sectoral Performance of Housing Supply Policies in Selected MENA Countries: A Comparative Analysis*. Dutch Trust Fund for Low Income Housing. Beirout: MENA Region.

Mertens, D.M. 2009. *Transformative Research and Evaluation*. New York: Guilford.

———. 2012. Transformative Mixed Methods: Addressing Inequities. *American Behavioral Scientist* 20 (10): 1–12.

Mguni, P. 2015. *Sustainability Transitions in the Developing World: Exploring the Potential for Integrating Sustainable Urban Drainage Systems in Sub-Saharan Cities*. Copenhagen: PhD submitted to University of Copenhagen.

Miliband, R. 1969. *The State in the Capitalist Society*. New York: Basic Books, Inc., Publishers.

Miller, C. 2006. Upper Egyptian Regionally Based Communities in Cairo Traditional or Modern Forms of Urbanization? In *Cosmopolitan Cairo: Politics, Culture, and urban Space in the New Global Middle East*, ed. D. Singerman and P. Amar, 375–397. Cairo: The American University Press in Cairo.

Miltin, D., and J. Thompson. 1995. Participatory Approaches in Urban Areas: Strengthening Civil Society or Reinforcing the Status Quo? *Environment and Urbanization* 7: 231–250.

Milton-Edwards, B., and P. Hinchcliffe. 2009. *Jordan: A Hashemite Legacy*. Oxon: Routledge.

Mitlin, D., and D. Satterthwaite. 2004. *Empowering Squatter Citizen: Local Government, Civil Society, and Urban Poverty Reduction*. London: Earthscan.

Mitlin, D., and A. Walnycki. 2019. Informality as Experimentation: Water Utilities' Strategies for Cost Recovery and Their Consequences for Universal Access. *The Journal of Development Studies*.https://doi.org/10.1080/00220388.2019.157738.

Mofeed, R., and N. Elgendy. 2016. Re-Presenting the Egyptian Informal Urbanism in Films Between 1994–2014. *International Journal of Humanities and Social Sciences* 10 (6): 2044–2050.
Mohan, B. 2011. *Development, Poverty of Culture, and Social Policy*. London: Palgrave Macmillan.
Monkelbaan, J. 2019. *Governance for the Sustainable Development Goals Exploring an Integrative Framework of Theories, Tools, and Competencies*. Singapore: Singapore Pte Ltd./Springer.
Moore, P.W. 2004. *Doing Business in the Middle East: Politics and Economic Crisis in Jordan and Kuwait*. Cambridge: Cambridge University Press.
Moore, J.W. 2015. *Capitalism in the Web of Life: Ecology and the Accumulation of Capital*. London: Verso.
Moreno, Jose Antonio. 1970. *Barrios in Arms; Revolution in Santo Domingo*. Pittsburg: University of Pittsburg Press.
Moser, Carline. 1994. The Informal Sector Debate, Part 1: 1970–1983. In *Contrapunto: The Informal Sector Debate in Latin America*, ed. C. Rakowski, 11–29. Albany: State University of New York Press.
———. 1998. The Asset Vulnerability Framework: Reassessing Urban Poverty Reduction Strategies. *World Development* 26 (1): 1–19.
Mountjoy, A. 1978. *The Third World: Problems and Perspectives*. London: Macmillan.
Mukhija, V., and A. Loukaitou-Sideris. 2014. Introduction. In *The Informal American City: Beyond Taco Trucks and Day Labor*, ed. V. Mukhija and A. Loukaitou-Sideris, 1–17. Cambridge: MIT Press.
Mumford, Lewis. 1962. *The City in History*. London: Faber.
Mumtaz, B., I. Heggie, and E. Case. 2015. *Jordan Affordable Housing Programme*. Amman: UN-Habitat.
Munck, R. 2005. *Globalization and Social Exclusion. A Transformationalist Perspective*. Bloomfield: Kumarian Press.
Municipality of Greater Amman. 1993. Greater Amman: Urban Development. *Cities*, February, 37–49.
Murshed, S.M. 2002. *Globalization, Marginalization and Development*. London: Routledge.
Mustafa, D., A. Stamm, and L.M. Scott. 2016. Water User Associations and the Politics of Water in Jordan. *World Development* 79: 164–176.
Nada, M. and Sims, D. 2020. *Assessment of Land Governance in Egypt*. Cairo: The World Bank
Napier, M. 2009. Making Urban Land Markets Work Better in South African Cities and Towns: Arguing the Basis for Access by the Poor. In *Urban Land Markets Improving Land Management for Successful Urbanization*, ed. S.V. Lall, 71–100. Dordrecht/Heidelberg: Springer.
Nasr, J., and Volait. 2003. *Urbanism Imported or Exported: Native Aspirations and Foreign Plans*. London: Wiley.
Nassara, J., and N. Stel. 2019. Lebanon's Response to the Syrian Refugee Crisis – Institutional Ambiguity as a Governance Strategy. *Political Geography* 70: 44–54.
Nelson, J. 1969. *Migrants, Urban Poverty and Instability in Developing Nations*. Technical. Cambridge, MA: Harvard University, Occasional Papers in International Affairs, No 22.
———. 1970. The Urban Poorp Disruption or Political Integration in Third World Cities? *World Politics* 22 (3): 393–414.
———. 1979. *Access to Power: Politics and the Urban Poor in Developing Nations*. Princeton: Princeton University Press.
Nicklin, S, and B. Cornwell. 2019. *A Better World Volume 5: Actions and Commitments to the Sustainable Development Goals*. Towards the Achievement of SDG 11. London: Tudor Rose.
Nordstrom, O.A., and L. Steier. 2015. Social Capital: A Review of its Dimensions and Promise for Future Family Enterprise Research. *International Journal of Entrepreneurial Behavior and Research* 21 (6): 801–813.
Nowar, M.A. 2006. *The Development of Trans-Jordan 1929–1939: A History of the Hashemite Kingdom of Jordan*. Reading: Garnet Publishing Limited.

Nunan, F., and N. Devas. 2004. Accessing Land and Services: Exclusion or Entitlement? In *Urban Governance, Voice and Poverty in the Developing World*, ed. Nick Devas, 164–185. London: Earthscan Publications Ltd.

Nuseir, A., M. Tuffaha, and S. Momani. 2003. *Housing Strategies and Its Impact on Housing Sector. Housing and Urban Development Corporation*. Amman: Sixth Annual Seminar on Housing Sector Assessment, 215–222.

O'Ballance, E. 1998. *Civil War in Lebanon, 1975–92*. Hampshire: Palgrave Macmillan Ltd.

O'Brien, K. 2015. Political Agency: The Key to Tackling Climate Change. *Science* 350: 1170–1171.

O'Connor, J. 1973. *The Fiscal Crisis of the State*. New York: St James Press.

Obeng-Odoom, F. 2012. Land Reform in Africa: Theory, Practice, and Outcome. *Habitat International* 36 (1): 161–170.

———. 2016. Property in the Commons: Origins and Paradigms. *Review of Radical Political Economics* 48 (1): 9–19.

Obremski, Helena, and Claudia Carter. 2019. Can Self-Build Housing Improve Social Sustainability Within Low-Income Groups? *TPR* 90 (2): 167–193.

OECD. 2016. *The Governance of Inclusive Growth*. Paris: OECD Publishing.

Oesch, L. 2017. The Refugee Camp as a Space of Multiple Ambiguities and Subjectivities. *Political Geography* 60: 110–120.

Osman, T. 2010. *Egypt on the Brink: From Nasser to Mubarak*. New Haven/London: Yale University Press.

Owen, R. 2004. *State, Power and Politics in the Making of the Modern Middle East*. London: Routledge.

Palmer, D., S. Fricska, and B. Wehrmann. 2009. *Towards Improved Land Governance: Land Tenure Working Paper 11*. Nairobi: (FAO of UN for human settlement programme).

Pamuk, A. 2000. Informal Institutional Arrangements in Credit, Land Markets and Infrastructure Delivery in Trinidad. *IJURR* 24: 379–396.

Paris, C. 1982. *Critical Reading in Planning Theory*. Oxford/Headington Hill Hall: Pergamon Press Ltd.

Park, R. 1967. The City: Suggestions for Investigation of Human Behavior in the Urban Environment. In *The City*, ed. R.E. Park, E.W. Burgess, and R.D. McKenzie, 1–46. Chicago: The University of Chicago Press.

Parnell, S., and S. Oldfield. 2014. *The Routledge Handbook on Cities of the Global South*. Abingdon/Oxon: Routledge.

Parvin, A. 2013. Architecture (and the Other 99%): Open-Source Architecture and Design Commons. *Architectural Design* 83 (6): 90–95.

Payne, G. 1974. Functions of Informality: Squatter Settlements in Delhi. *Ekistics* 34 (224): 63–66.

———. 1977. *Urban Housing in the Third World*. London: Leonard Hill.

———. 1999. Introduction. In *Making Common Ground: Public/Private Partnerships in Land for Housing*, ed. G. Payne, 1–16. London: Intermediate Technology Press Ltd.

———., ed. 2002. *Land, Rights and Innovation: Improving Security for the Urban Poor*. London: ITDG.

Payne, G., A. Durand-Lasserve, and C. Rakodi. 2009. Social and Economic Impacts of Land Titling Programs in Urban and Periurban Areas: A Short Review of the Literature. In *Urban Land Markets Improving Land Management for Successful Urbanization*, ed. S.V. Lall, M. Freire, B. Yuen, R. Rajack, and J.J. Helluin, 133–162. London: Springer.

Peattie, L. 1974. The Concept of Marginality as Applied to Squatter Settlements. In *Latin American Urban Research* 4: 101–109, ed. W. Cornelius and F. Trueblood.

———. 1979. Housing Policy in Developing Countries: Two Puzzles. *World Development* 79: 1017–1022.

———. 1987. Shelter Development and the Poor. In *Shelter, Settlement and Development*, ed. L. Rodwin, 263–280. Boston: Allen & Unwin.

———. 1996. Urban Research in the 1990s. In *Preparing for the Urban Future: Global Pressures and Local Forces*, ed. M. Cohen, B. Ruble, J. Tulchin, and A. Garland, 371–391. Washington, DC: The Woodrow Wilson Center Press.
Peattie, L., and J. Aldrete-Haas. 1981. Marginal Settlements in Developing Countries: Research Advocacy of Policy and Evolution of Programs. *Annual Review of Sociology* 7: 157–175.
Pel, B., and R. Kemp. 2020. Between Innovation and Restoration; Towards a Critical-Historicizing Understanding of Social Innovation Niches. *Technology Analysis & Strategic Management*. https://doi.org/10.1080/09537325.2020.1750588.
Perlman. 1976. *The Myth of Marginality: Urban Poverty and Politics in Rio de Janeiro*. Stanford: California University Press.
Perlman, Janice. 2004. Marginality: From Myth to Reality in the Favelas of Rio de Janeiro, 1969–2002. In *Urban Informality: Transnational Perspectives from the Middle East, Latin America, and South Asia*, ed. Ananya Roy and Al Sayyad, 105–146. Lanham: Lexington Books.
Piketty, T. 2014. *Capital in the Twenty-First Century*. London: Th e Belknap Press of Harvard University Press.
Pittfor, A. 2019. System Change, Not Climate Change. *Internationa Politices and Society*.https://www.ips-journal.eu/regions/global/article/show/system-change-not-climate-change-3712/.
Pomponi, F., and A. Moncaster. 2017. Circular Economy for the Built Environment: A Research Framework. *Journal of Cleaner Production* 143: 710–718.
Portes, A. 1972. Rationality in the Slums: An Essay in Interpretive Sociology. *Comparative Studies in Society and History* 14 (3).
Portes, A., and J. Walton. 1976. *Urban Latin America: The Political Condition from Above and Below*. Austin: University of Texas Press.
Potter, R.B., D. Darmame, N. Barhamb, and S. Nortcliff. 2009. "Ever-Growing Amman", Jordan: Urban Expansion, Social Polarisation and Contemporary Urban Planning Issues. *Habitat International* 33: 81–92.
Poulantzas, N. 1979. *State, Power Socialism*. London: Verso.
Poustie, M., N. Frantzeskaki, and R. Brown. 2016. A Transition Scenario for Leapfrogging to a Sustainable Urban Water Future in Port Vila, Vanuatu. *Technological Forecasting and Social Change* 105: 129–139.
Pritchard, M. 2013. Land, Power and Peace: Tenure Formalization, Agricultural Reform, and Livelihood Insecurity in Rural Rwanda. *Land Use Policy* 30: 186–196.
Rabil, Robert G. 2011. *Religion, National Identity, and Confessional Politics in Lebanon: The Challenge of Islamism*. Hampshire: Palgrave Macmillan.
Rahman, A., and J. Westley. 2001. The Challenge of Ending Rural Poverty. *Development Policy Review, Volume* 19 (4): 553–562.
Rakodi, C. 1997. *The Urban Challenge in Africa: Growth and Management of Its Large Cities*. Tokyo: United Nations University Press.
Rakowski, C.A. 1994a. Introduction: Land and Housing Development in Middle Eastern and North Africa Cities. *International Journal of Urban and Regional Research* 12 (1): 1–7.
———., ed. 1994b. *Contrapunto: The Informal Sector Debate in Latin America*. Albany: Sunny Press.
Rapoport, A. 1977. *Human Aspects of Urban Form: Towards a Man – Environment Approach to Urban Form and Design*. Oxford/Headington Hill Hall: Pergamon Press Ltd.
Rashed, T., D. Stow, and D. Fugate. 2005. Measuring Temporal Compositions of Urban Morphology Through Spectral Mixture Analysis: Toward a Soft Approach to Change Analysis in Crowded Cities. *International Journal of Remote Sensing* 26: 699–707.
Raven, R.P.J.M., E. Heiskanen, R. Lovio, M. Hodson, and B. Brohmann. 2008. The Contribution of Local Experiments and Negotiation Processes to Field-Level Learning in Emerging (Niche) Technologies: Meta-analysis of 27 New Energy Projects in Europe. *Bulletin of Science, Technology and Society* 28: 464–477.
Rawls, J. 1971. *A Theory of Justice*. Cambridge, MA: Belknap Press.
Ray, T. 1969. *The Politics of the Barrios of Venezuala*. Berkeley: University of California Press.

Richardson, L., C. Durose, and B. Perry. 2018. Coproducing Urban Governance. *Politics and Governance* 6 (1): 145–149.
Rip, A., and R. Kemp. 1998. Technological Change. In *Human Choice and Climate Change, Volume 2*, ed. S. Rayner and E.L. Malone, 327–399. Columbus: Battelle Press.
Roberts, B. 1973. *Organizing Strangers: Poor Families in Guatemala City*. Austin: University of Texas Press.
———. 1978. *Cities of Peasants: The Political Economy of Urbanization in the Third World*. Beverly Hills: Sage Publication.
Robinson, J. 2006. *Ordinary Cities: Between Modernity and Development*. London: Routledge.
Robinson, Daron Acemoglu, and A. Jams. 2012. *Why Nations Fail: The Origins of Power, Prosperity, and Poverty*. London: Profile Books Ltd.
Rocco, R., and J.V. Ballegooijen, eds. 2018. *The Routledge Handbook on Informal Urbanization*. New York: Routledge.
Rodgers, D., J. Beall, and R. Kanbur. 2012. Re-thinking the Latin American City. In *Latin American Urban Development into the 21st Century: Towards a Renewed Perspective on the City*, ed. J. Beall, R. Kanbur, and D. Rodgers, 3–36. Basingstoke: Palgrave Macmillan.
Rotmans, J. 2006. A Complex Systems Approach for Sustainable Cities. In *Smart Growth and Climate Change*, ed. M. Ruth, 155–180. Cheltenham/Northampton: Edward Elgar.
Rotmans, J., R. Kemp, and M. van Asselt. 2001. More Evolution Than Revolution: Transition Management in Public Policy. *Foresight* 3 (1): 15–31.
Rowe, P., and H. Sarkis, eds. 1998. *Projecting Beirut: Episodes in the Construction and Reconstruction of a Modern City*. London: Prestel-Verlag.
Roy, A. 2004. The Gentleman's City: Urban Informality in the Calcutta of New Communism. In *Urban informality: Transnational Perspectives from the Middle East, Latin America, and South Asia*, ed. A. Roy and N. Al Sayyad, 147–170. Lanham: Lexington Books.
———. 2005. Urban Informality: Toward an Epistemology of Planning. *Journal of the American Planning Association* 71 (2): 147–158.
———. 2009. The 21st-Century Metropolis: New Geographies of Theory. *Regional Studies* 43 (6): 819–830.
———. 2011a. Postcolonial Urbanism: Speed, Hysteria, Mass Dreams. In *Worlding Cities : Asian Experiments and the Art of Being Global*, ed. A. Roy and A. Ong, 307–335. Chichester: Blackwell Publishing Limited.
———. 2011b. Slumdog Cities: Rethinking Subaltern Urbanism. *International Journal of Urban and Regional Research* 35 (2): 223–238.
———. 2014. Worlding the South: Toward a Post-colonial Urban Theory. In *The Routledge Handbook on Cities of the Global South*, ed. S. Parnell and S. Oldfield, 9–20. London: Routledge.
Roy, A., and N. Al-Sayyad. 2004. *Urban informality: Transnational perspectives from the Middle East, Latin America, and South Asia*. Lanham: Lexington Books.
Runkel, C. 2009. *The Role of Urban Land Titling in Slum Improvement the Case of Cairo*. Berlin: Institut für Stadtund Regionalplanung, Technische Universität Berlin, Diploma Thesis.
Sachs, J. 2005. *The End of Poverty. How We Can Make It Happen in Our Lifetime*. London: Penguin Books.
Safa, Helen. 1970. The Poor Are Like Everyone Else, Oscar. *Psychology Today* 4 (4): 26–32.
Sait, S., and H. Lim. 2006. *Land, Law and Islam: Property and Human Rights in the Muslim World*. London: Zed Books Ltd.
Saksouk-Sasso, A., and N. Bekdache. 2015. Private Interest Closing Social Space? A Critical Analysis of Lebanon's Real Estate Sector. In *Lebanon: Towards a Peace Economy*, ed. Jessica Banfield and Victoria Stamadianou, 26–38. London: International Alert.
Salibi, K. 1976. *Crossroads to Civil War*. Caravan: Delmar.
Salmon, T.C., and M.F. Imber. 2008. *Issues in International Relations*. Oxford: Routledge.
Samman, K. 2006. *Cities of God and Nationalism: Mecca, Jerusalem, and Rome as Contested World*. Paradigm: Boulder.

Sampson, R.J., S.W. Raudenbush, and F. Earls. 1997. Neighbourhoods and Violent Crime: A Multilevel Model of collective Efficacy. *Science* 277: 918–924.

Sandels, A. 2011. *JORDAN: Thousands of Demonstrators Protest Food Prices, Denounce Government.* Retrieved from https://latimesblogs.latimes.com/babylonbeyond/2011/01/jordan-protests-food-prices-muslim-brotherhood-tunisia-strike-thousands-government.html. Los Angeles Times. Beirut, January 15.

Sandercock, L. 1998. *Toward Cosmopolis*. New York: Wiley.

Satterthwaite, D. 2014. Guiding the Goals: Empowering Local Actors. *SAIS Review of International Affairs* 34(2, Summer-Fall): 51–61.

———. 2020. Editorial: Getting Housing Back onto the Development Agenda: The Many Roles of Housing and the Many Services It Should Provide Its Inhabitants. *Environment and Urbanization* 32 (1): 3–18. https://doi.org/10.1177/0956247820905212.

Satterthwaite, D., D. Archer, S. Colenbrader, and D. Dodman. 2020. Building Resilience to Climate Change in Informal Settlements. *One Earth* 21: 143–156.

Schot, J. 1998. The Usefulness of Evolutionary Models for Explaining Innovation: The Case of the Netherlands in the Nineteenth Century. *History of Technology* 14: 173–200.

———. 2016. Confronting the Second Deep Transition Through the Historical Imagination. *Technological Culture* 57 (2): 445–456.

Schot, J.W., and F. Geels. 2008. Strategic Niche Management and Sustainable Innovation Journeys: Theory, Findings, Research Agenda, and Policy. *Technology Analysis and Strategic Management* 20 (5): 537–554.

Schot, J., and L. Kanger. 2016. *Deep Transitions: Emergence, Acceleration, Stabilization and Directionality Science Policy Research Unit (SPRU)*. Sussex: University of Sussex.

Schot, J., and W.E. Steinmueller. 2016. *Framing Innovation Policy for Transformative Change: Innovation Policy 3.0 - Draft Version 2 (Draft Working Paper).* Brighton: University of Sussex, Science Policy Research Unit (SPRU).

Scott, J. 1986. Everyday Form of Peasant Resistance. *Journal of Peasant Studies* 13 (2): 5–35.

Seely, J. 1959. The Slum: Its Nature, Use and Users. *Journal of the American Institute of Planners* 25 (19): 7–11.

———. 1971. The Slum: Its Nature, Use and Users. In *Internal Structure of the City*, ed. L.S. Bourne, 464–474. Oxford: Oxford University Pres.

Sela, A. 2014. The PLO at Fifty: A Historical Perspective. *Contemporary Review of the Middle East* 1 (3): 269–333.

Shami, S. 2007. Amman Is Not a City. In *Urban Imaginaries. Locating the Modern City*, ed. A. Cinar and T. Bender, 208–235. Minneapolis: University of Minnesota Press.

Shirazi, M., and R. Keivani, eds. 2019. *Urban Social Sustainability: Theory, Policy and Practice*. New York/Abingdon: Routledge.

Shwayri, S.T. 2012. Modern Warfare and the Theorization of the Middle Eastern City. In *Urban Theory Beyond the West*, ed. T. Ednesor and M. Jayne, 259–270. London: Routledge.

Sibley, D. 1995. *Geographies of Exclusion: Society and Difference in the West*. London: Routledge.

Silberstein, P. 1969. Favela Living: A Personal Solution to Larger Problems. *America Latina* 12 (3): 183–200.

Silva, P. 2016. Tactical Urbanism: Towards an Evolutionary Cities' Approach? *Environment and Planning B: Planning & Design* 43 (6): 1040–1051.

Silver, H. 2010. Divided Cities in the Middle East. *Community and Development* 9: 357–345.

Simon, D. 2008. Urban Environments: Issues on the Peri-Urban Fringe. *Annual Review of Environment and Resources* 33: 167–185.

Sims, D. 2010. *Understanding Cairo: The Logic of a City Out of Control*. Cairo: America University Press in Cairo.

———. 2014. *Egypt's Desert Dreams: Development or Disaster*. Cairo: The American University in Cairo Press.

———. 2019. *The Distorting Luxury of Infinite Space: Egypt's New Capital, New Cities, and New Industrial Areas*. Cairo: American university Press in Cairo.

Singerman, D. 1995. *Avenues of Participation: Family, Politics, and Networks in Urban Quarters of Cairo*. Princeton: Princeton University Press.

———. 2011. The Siege of Imbaba, Egypt's Internal "Other", and the Criminalization of Politics. In *Cairo Contested: Governance, Urban Space, and Global Modernity*, ed. Diane Singerman, 111–144. Cairo: The American University in Cairo Press.

Singerman, D., and P. Amar. 2006. *Cairo Cosmopolitan: Politics, Culture, and Urban Space in the New Globalized Middle East*. Cairo: The American University in Cairo Press.

Smith, A., and R. Raven. 2012. What Is Protective Space? Reconsidering Niches in Transitions to Sustainability. *Research Policy* 41: 1025–1036.

Smith, A., J. Vob, and J. Grin. 2010. Innovation Studies and Sustainability Transitions: The Allure of the Multi-level Perspective and Its Challenges. *Research Policy* 39 (4): 435–448.

Smith, A., M. Fressoli, D. Abrol, E. Arond, and A. Ely. 2017. *Grassroots Innovation Movements*. London: Routledge.

SoJo, C. 2001. *Social Development in Latin America: Issues for Public Policy*. Technical report. Washington, DC: World Bank.

Soliman, A. 1988. Housing Mechanisms in Egypt: A Critique. *Journal of Housing and Environmental* 4 (1): 31–50.

———. 1991. *Approaches to Urban Low-Income Housing: Between Theoriticality and Practicality*. The Third International Congress of the Major Metropolises, October 14–19, 1990. Melbourne: The Proceedings of Metropolis'90, 123–131.

———. 1996. Legitimizing Informal Housing: Accommodating Low-Income Groups in Alexandria, Egypt. *Environment and Urbanization Journal* 8 (1): 183–193.

———. 1999. Public/Private Partnerships in Three Egyptian Cities. In *Making Common Ground: Public/Private Partnerships in Land Provision for Housing the Urban Poor*, ed. G. Payne. London: Intermediate Technology Publications (for DFID).

———. 2002. Typology of Informal Housing in Egyptian Cities: Taking Account of Diversity. *IDPR* 24 (2): 177–201.

———. 2004a. *A Possible Way Out: Formalizing Housing Informality in Egyptian Cities*. Lanham: American University Press.

———. 2004b. Tilting at Sphinxes: Locating Urban Informality in Egyptian Cities. In *Urban Informality in an Era of Liberalization: A Transnational Perspective*, ed. Ananya Roy and Nezear AlSayyad, 171–208. Lanham: Lexington Books.

———. 2004c. Regional Planning Scenarios in South Lebanon: The Challenge of Rural-Urban Interactions in the Era of Liberalization and Globalization. *Habitat International Journal* 27 (1): 385–408.

———. 2007. *Urban Informality in Egyptian Cities: Coping with Diversity*. Fourth Urban Research symposium 2007, Urban Land Use and Land Markets, May. Washington, DC: World Bank.

———. 2008. Diversity of Ethnicity and State Involvement on Urban Informality in Beirut. *Theoretical and Empirical Researches in Urban Management, Year 3, Number 9*: 15–32.

———. 2010. Rethinking Urban Informality and the Planning Process in Egypt. *IDPR Journal* 32 (2): 119–143.

———. 2012a. Housing and the State in the Middle East. In *International Encyclopedia of Housing and Home*, ed. Susan J. Smith, 346–354. Amsterdam: Elsevier.

———. 2012b. The Egyptian Episode of Self-Build Housing. *Habitat International Journal* 36 (2): 226–236.

———. 2012c. Building Bridges with the Grassroots: Housing Formalization Process in Egyptian Cities. *Journal of Housing and the Built Environment* 27 (2): 241–260.

———. 2013a. *Living With/in Ashwaiyyat in Egypt: Is It a New Innovation of Urbanity?* The Flexible City, International Symposium. Oxford: Oxford University, 24–25 October.

———. 2013b. *The Poor in Search of Shelter: An Examination of Squatter Settlements in Alexandria, Egypt*. Deutschland: Lap Lambert Academic Publishing.

———. 2015a. Collective Planning Process: A Driving Seat for Formalising Urban Informality in Egypt. *International Journal of Urban Sustainable Development*, June 1–24.

———. 2015b. The Crucible of Housing the Bottom of the Pyramid in Egypt. In *Housing the Urban Poor in Developing Countries*, ed. B. Aldrich and R. Sandhu, 165–189. Jaipur: Rawat Publications.

———. 2017. Land Readjustment as a Mechanism for New Urban Land Expansion in Egypt: Experimenting Participatory Inclusive Processes. *International Journal of Urban Sustainable Development* 9 (3): 313–332.

———. 2019. So It's Always a Chance: Community-Led Solutions to New Urban Expansion. In *New Cities and Community Extensions in Egypt and the Middle East*, ed. S. Attia et al., 159–179. Cham: Springer.

———. 2020. Understanding Sustainability Transitions for Urban Informality in the Middle East. *International Journal of Urban Sustainable Development*. https://doi.org/10.1080/19463138.2020.1728692.

———. 2021a. Informal Cairo: The Making of Urban Fabric. In *The Routledge Handbook on Cairo*, ed. N.A. Al Sayyad. New York: Routledge.

———. 2021b. *Rethinking Urban Sustainability Transitions on land readjustment in Egypt, paper presented on 24th February in Second Arab Land Conference, 22–24 February*. Cairo: Egypt.

Sovacool, B. 2016. How Long Will It Take? Conceptualizing the Temporal Dynamics of Energy Transitions. *Energy Research & Social Science* 13: 202–221.

Spaans, M., and B. Waterhout. 2017. Building Up Resilience in Cities Worldwide: Rotterdam as Participant in the 100 Resilient Cities Programme. *Cities* 61: 109–116.

Spectrum. 1997. *Housing policy in Lebanon*. Beirut: Spectrum.

Stel, Nora. 2016. The Agnotology of Eviction in South Lebanon's Palestinian Gatherings: How Institutional Ambiguity and Deliberate Ignorance Shape Sensitive Spaces. *Antipode* 48 (5): 1400–1419.

Stilwell, F., and K. Jordan. 2004. The Political Economy of Land: Putting Henry George in His Place. *The Journal of Australian Political Economy* 54 (December): 119–134.

Stirling, A. 2011. Pluralising Progress: From Integrative Transitions to Transformative Diversity. *Environmental Innovation and Societal Transitions* 1 (1): 82–88.

Stokes, C. 1962. A Theory of Slums. *Land Economic* 38 (3): 187–197.

Stokes, S.C., T. Dunning, M. Nazareno, and V. Brusco. 2013. *Brokers, Voters, and Clientelism: The Puzzle of Distributive Politics*. Cambridge, MA: Cambridge University Press.

Stren, R. 2002. Local Governance and the Development of Associational Life: An Exploration. In *Democratic Governance and Urban Sustainability*, ed. J.S. Tulchin, D.H. Varat, and B.A. Ruble, 9–28. Washington, DC: Woodrow Wilson International Center for Scholars.

Subrahmanyam, G. 2016. *Addressing Informality in Egypt*. Tunis: African Development Bank, AFDB, Working Paper.

Sudra, T. 1976. *Low Income Housing System in Mexico City*. Un-published Ph.D. Dissertation, Cambridge, MA: Massachusetts Institute of Technology.

Sugranyes, A., and C. Mathivet. 2010. *Cities for All: Proposals and Experiences Towards the Right to the City*. Santiago: Habitat International Coalition (HIC).

Sunam, R. 2015. The Significance of Foreign Labour Migration and Land for Poverty Reduction in Nepal. In *Poverty Reduction Policies and Practices in Developing Asia*, ed. Almas Heshmati, Esfandiar Maasoumi, and Guanghua Wan, 218–234. New Delhi: Springer.

Sweis, R.J., R.O. Shanak, A.A. El Samen, and T. Suifan. 2014. Factors Affecting Quality in the Jordanian Housing Sector. *International Journal of Housing Markets and Analysis* 7 (2): 175–188.

Swilling, Mark, and Maarten Hajer. 2017. *Governance of Urban Transitions: Towards Sustainable Resource Efficient Urban Infrastructures. Environmental Research Letters* 12 (2017): 125007.

Swilling, M.J., K. Musephine, and J. Wakeford. 2015. Developmental States and Sustainability Transitions: Prospects of a Just Transition in South Africa. *Journal of Environmental Policy and Planning* 5 (5): 1–23.

Tabet, J. 2001. *Beyrouth, Collection Portrait de Ville*. Paris: Institut Frangais d'Architecture.

Talani, L. S. 2014. *The Arab Spring in the Global Political Economy*. Hampshire, UK: Palgrave Macmillan.

Taylor, P.J., R.J. Johnnson, and M.J. Watts. 1995. *Geographies of Global Change: Remapping the World in the Late Twentieth Century*. Oxford: Blackwell.

Teti, A., P. Abbott, and F. Cavatorta. 2018. *The Arab Uprisings in Egypt, Jordan and Tunisia: Social, Political and Economic Transformations*. Cham: Palgrave Macmillan.

The Aga Khan Award for Architecture. 1980. *East Wahdat Upgrading Programme*. Amman: Urban Development Department.

The Government of Lebanon and the UN. 2017. *Lebanon Crisis Response Plan 2017–2020*. Beirut: The Government of Lebanon and the United Nations.

The Guardian. 2015. *A New Cairo: Egypt Plans £30bn Purpose-Built Capital in Desert*. London: The Guardian.

———. 2020. Beirut Families Search for Loved Ones Feared Trapped Under Rubble. *World news* 5: 1–6.

The Hashemite Kingdom of Jordan. 2013. *Needs Assessment Review of the Impact of the Syrian Crisis on Jordan*. Amman: Ministry of Planning and International Cooperation.

———. 2017. *Jordan's Way to Sustainable Development: First National Voluntary Review on the Implementation of the 2030 Agenda*. New York: High-Level Political Forum on Sustainable Development.

The Ministry of Environment. 2003. *Lebanon State of the Environment Report*. Beirut: The Ministry of Environment.

The Ministry of Social Affairs. 1996. *Population and Housing Statical Facts*. Beirut: The Ministry of Social Affair.

The Third National Report. 2014. *The Third National Report on the Habitat in Cooperation with UN Program for Human Settlement*. New York: United Nations.

The World Bank. 2008. *Approaches to Urban Slums*. Washington, DC: The International Bank for Reconstruction and Development/The World Bank.

Tidball, K.G., N. Frantzeskaki, and T. Elmqvist. 2016. Traps! An Introduction to Expanding Thinking on Persistent Maladaptive States in Pursuit of Resilience. *Sustainability Science* 11: 861.

Tierney, J. 2016. *Beirut's Lessons for How Not to Rebuild a Wartorn City*. Washington, DC: Washington Post.

Tipple, G. 1999. User-Initiated Transformations of Government-Built Housing Stocks: Lessons from Developing Countries. *Journal of Urban Technology* 6 (3): 17–35.

———. 2000. *Extending Themselves: User-Initiated Transformations of Government-Built Housing in Developing Countries*. Liverpool: Liverpool University Press.

Todaro, M.P., and S.C. Smith. 2012. *Economic Development*. 11th ed. New York: Addison-Wesley.

Torres, H., H. Alves, and M. De Oleiera. 2007. São Paulo Peri-Urban Dynamics: Some Social Causes and Environmental Consequences. *Environment & Urbanization* 19 (1) (World Bank): 207–223.

Toth, O. 2009. *Strategic Plan for an Egyptian Village: A Policy Analysis of the Loss of Agricultural Land in Egypt*. MSc thesis. Wagen: The School of Planning of the College of Design, Architecture, Art and Planning.

Traboulsi, F. 2007. *A History of Modern Lebanon*. London: Pluto Press.

Truffer, B., and L. Coenen. 2012. Environmental Innovation and Sustainability Transitions in Regional Studies. *Regional Studies* 46 (1): 1–21.

Truffer, B., J.T. Murphy, and R. Raven. 2015. The Geography of Sustainability Transitions: Contours of an Emerging Theme. *Environmental Innovation and Societal Transitions* 17: 63–72.

Tulchin, J.S., D.H. Varat, and B.A. Ruble. 2002. *Democratic Governance and Urban Sustainability*. Washington, DC: Woodrow Wilson International Center for Scholars.

Tunas, D., and A. Peresthu. 2010. The Self-Help Housing in Indonesia: The Only Option for the Poor? *Habitat International* 34: 315–322.

Turner, J. 1967. Barriers and Channels for Housing Development in Modernising Countries. *Journal of the American Institute of Planners* 33: 167–181.

———. 1968. The Squatter Settlement: Architecture That Works. *Architectural Design* 38: 355–360.

———. 1976. *Housing by People*. London: Marion Boyars Publishers Ltd.

———. 1991. Releasing the Potentials of Self-Help Housing. In *Metropolis in Ascendancy: Housing and Population*, ed. D. Ingemann, 31–37. Melbourne: Housing and Construction Victoria.

———. 2015. Explaining the (Non-) Encampment of Syrian Refugees: Security, Class and the Labour Market in Lebanon and Jordan. *Mediterranean Politics* 20 (3): 386–404.

Turner, J., and R. Fichter. 1972. *Freedom to Build: Dwellers Control of the Housing process*. New York: Macmillan.

Turnheim, B., F. Berkhout, F. Geels, A. Hof, A. Mcmeekin, and B. Nykvist. 2015. Evaluating sustainability Transitions Pathways: Bridging Analytical. *Global Environmental Change* 35: 239–253.

UN. 2016a. *Habitat III Issue Papers: 20 – Housing*. New York: UN.

———. 2016b. *The New Urban Agenda*. New York: UN.

———. 2018. *Department of Economic and Social Affairs, Population Division, Online Edition*. Population. New York: United Nations.

———. 2019a. *United Nations, Department of Economic and Social Affairs, Population Division, World Population Prospects 2019, Volume II: Demographic Profiles*. New York: United Nations.

———. 2019b. *Department of Economic and Social Affairs, Population Division*. World Urbanization Prospects: The 2018 Revision (ST/ESA/SER.A/420). New York: United Nations.

———. 2020. *Department of Economic and Social Affairs, Population Division, Online Edition*. New York: United Nations Publications.

UNDP. 2003. *World Economic and Social Survey, Various Issues*. Technical Report. Nairobi: United Nations Publication.

———. 2009. *Assessment of Capacity Building Needs of NGOs in Lebanon*. Washington, DC: UNDP.

———. 2016. *Regional Strategic Overview: Regional Refugee & Resilience Plan (3RP) 2019–2020*. New York: UNDP.

UNDP, and CLEF. 2008. *Making the Law Work for Everyone: Report of the Legal Empowerment of Poor*. Washington, DC: United Nations development program.

UNDP and GDCD. 2008. *Disaster Risk Management Profile (Amman–Jordan)*. Amman: General Directorate of Civil Defense (GDCD).

UNESCO. 2019. *The United Nations World Water Development Report 2019: Leaving No One Behind*. World Water Assessment Programme. Paris: The United Nations Educational, Scientific and Cultural Organization.

UN-Habitat. 2003. *The Challenge of Slums: Global Report on Human Settlements*. London: Earthscan Publications Ltd.

———. 2006. *Slums Upgrading Processes and Challenges in Asia-Pacific*. Paper Presented in Asia-Pacific Ministerial Conference, 13–16 December. New Delhi: Asia-Pacific Ministerial Conference.

———. 2008. *The State of African Cities 2008: A framework for Addressing Urban Challenges in Africa*. Nairobi: UN-HABITAT.

———. 2009. *Urban World Bridging the Urban Divide, Why Cities Must Built Equally*. Technical. Nairobi: Un-Habitat.

———. 2010. *Cities and Citizens' Series Bridging the Urban Divide: Sao Paulo a Tale of Two Cities*. Urban Planning. Nairobi: Un-Habitat.

———. 2011a. *Islamic Principles and Land: Opportunities for Engagement*. Nairobi: UN-HABITAT.
———. 2011b. *Labanon Urban Profile: A Task Review Report October 2011*. Beirut: UN-HABITAT.
———. 2011c. *Lebanon Urban profile: Adisk Review Report*. Nairobi: UN-HABITAT.
———. 2012. *Tapping the Potential: The Role of Grassroots in Land Policy Implementation*. Technical report. Nairobi: UN-Habitat.
———. 2013. *Sate of the World's Cities 2012/2013: Prosperity of Cities*. New York: Routledge.
———. 2015a. *Slum Almanac 2015/2016: Tracking Improvement in the Lives of Slum Dwellers*. Nairobi: UN-Habitat.
———. 2015b. *UN-Habitat Global Activities Report 2015: Increasing Synergy for Greater National Ownership*. Nairobi: United Nations Human Settlements Programme (UN-Habitat), 31.
———. 2016a. *Urbanization and Development: Emerging Futures*. World Cities Report 2016. Nairobi: United Nations Human Settlements Programme (UN-Habitat).
———. 2016b. *The New Urban Agenda*. Nairobi: UN-Habitat.
———. 2016c. *Evaluation of Participatory Review of Egyptian Planning and Related Urban Development Legislation to Support Sustainable Urban Development, April 2016*. Nairobi: UN-Habitat.
———. 2018a. *DG 11 Synthesis Report 2018: Tracking Progress Towards Inclusive, Safe, Resilient and Sustainable Cities and Human Settlements*. SDG 11 Synthesis Report High Level Political Forum 2018. Nairobi: Un-Habitat.
———. 2018b. *Prosperity for All: Enhancing the Informal Economy Through Participatory Slum Upgrading*. Nairobi: UN-Habitat.
———. 2018c. *Addressing the most Vulnerable First -Pro-Poor Climate Action in Informal Settlements*. Thematic Guide. Nairobi: UN-Habitat.
———. 2018d. *National Urban Policies Program in Lebanon: Diagnosis Report 2018*. Nairobi: Un-Habitat.
———. 2020. *UN-Habitat COVID-19: Response Plan*. Technical. Nairobi: UN-Habitat.
UN-Habitat and Arab Republic of Egypt. 2015. *Egypt Housing Profile*. Cairo: Arab Republic of Egypt.
Unicef. 2013. *Informal Tented Settlements in Jordan: A Multi-Sector Baseline Assessment*. Unicef.
United Nations. 2005. *Second National Report on the MDGs in the Syrian Arab Republic*. Damascus: United Nations.
———. 2013. *Needs Assessment Review of the Impact of the Syrian Crisis on Jordan*. Amman: Ministry of Planning and International Cooperation.
———. 2017. *Habitat III Regional Report: Arab Region*. New York: United Nations.
———. 2018. *Department of Economic and Social Affairs, Population Division*. New York: United Nations.
Van den Bergh, J.C.J.M. 2013. Economic-Financial Crisis and Sustainability Transition: Introduction to the Special Issue. *Environmental Innovation for Societal Transitions* 6: 1–8.
Van den Bosch S. 2010. *Transition Experiments. Exploring Societal Changes Towards Sustainability*. Ph.D. thesis. Rotterdam: Erasmus University. http://repub.eur.nl/pub/20714.
Vasenev, V.I., E. Dovletyarova, Z. Cheng, and R. Valentini, eds. 2018. *Megacities 2050: Environmental Consequences of Urbanization*. Cham: Springer.
Vellema, S. 2011. *Transformation and Sustainability in Agriculture: Connecting Practice with Social Theory*. Wageningen: Wageningen Academic Publishers.
Viotti, E.B. 2001. *National Learning Systems: A New Approach on Technical Change in Late Industrializing Economies and Evidences from the Cases of Brazil and South Korea. Science*. Cambridge, MA: Center for International Development, Harvard University, Technology and Innovation Discussion Paper No. 12.
Voss, J.-P., and B. Bornemann. 2011. The Politics of Reflexive Governance: Challenges for Designing Adaptive Management and Transition Management. *Ecology & Society* 16 (2): 9.

Voss, J.P., D. Bauknecht, and R. Kemp, eds. 2006. *Reflexive Governance for Sustainable*. London: Edward Elgar Publishing.
Voß, Jan-Peter, A. Smith, and J. Grin. 2009. Designing Long-Term Policy: Rethinking Transition. *Policy Science* 42: 275–302.
VROM. 2001. *Where There's a Will There's a World*. The Hague: Ministry of Housing: Spatial Planning Environment (VROM).
Wählisch, M., and M. Felsch. 2017. *Lebanon and the Arab Uprisings: In the Eye of the Hurricane*. London: Routledge.
Ward, B. 1976. *The Home of Man*. Harmondsworth: Penguin Books Ltd.
———. 1982a. Informal Housing: Conventional Wisdoms Reap-Praised. *Built Environment* 8 (2): 84–94.
———. 1982b. *Self-Help Housing: A Critique*. London: Alexandrine Press, Mansell Publishing Co.
———. 1989. *Progress for a Small Planet*. London: Penguin.
Ward, P. 2003. *From the Marginality of the 1960s, to the "New Poverty" of Today: A LARR Research Forum*. Workshop. Austin: University of Texas at Austin.
———. 2004. Informality of Housing Production at the Urban-Rural Interface: The "Not-so Strange Case" of the Texas Colonias. In *Urban Informality: Transnational Perspectives from the Middle East, Latin America, and South Asia*, ed. A. Roy and N. Al Sayyad, 243–270. Lanham: Lexington Books.
Ward, B., and R. Dubos. 1972. *Only One Earth: The Care and Nauntance of a Small Planet*. Stockholm: United Nations conference on the human environment.
Waterbury, J. 1973. *Cairo: Third World Metropolis, Part three, Housing and Shelter*. Cairo: American Universities Field Staff Papers, Northeast Africa Series, 189, No. 8 (September).
Watson, V. 2009. Seeing from the South: Refocusing Urban Planning on the Globe's Central Urban Issues. *Urban Studies* 46 (11): 2259–2275.
Wheaton, W.C. 1998. Land Use and Density in Cities with Congestion. *Journal of Urban Economics* 2 (43): 558–572.
Wheeler, S. 1998. Planning Sustainable and Livable Cities. In *The City Reader*, ed. R.T. LeGates and F. Stout. London/New York: Routledge.
Wieczorek, Anna J. 2018. Sustainability Transitions in Developing Countries: Major Insights and Their Implications for Research and Policy. *Environmental Science and Policy* 84: 204–216.
Wignaraja, K., ed. 2009. *Capacity Development*. Nairobi: A UNDP Primer. Available at: www.undp.org/capacity
Wikan, U. 1980. *Life Among the Poor in Cairo*. London: Tavistock Publications Ltd.
Willems, W. 2019. 'The Politics of Things': Digital Media, Urban Space, and the Materiality of Publics. *Media, Culture & Society* 41 (8): 1192–1209.
Wittmayer, J., F. Steenbergen, and N. Frantzeskaki. 2018. Transition Management: Guiding Principles and applications. In *Co-creating Sustainable Urban Futures: A Primer on Applying Transition Management in Cities*, ed. N. Frantzeskaki, M. Bach, and P. Mguni, 81–102. Tokyo/Heidelberg/New York/Dordrecht/London: Springer.
Wittmayer, J., J. Backhaus, F. Avelino, B. Pel, T. Strasser, I. Kunze, and L. Zuijderwik. 2019. Narratives of Change: How Social Innovation Initiatives Construct Societal Transformation. *Futures* 112. https://doi.org/10.1016/j.futures.2019.06.005.
Wolfram, M. 2018. Urban Planning and Transition Management: Rationalities, Instruments and Dialectics. In *Co-creating Sustainable Urban Futures: A Primer on Applying Transition Management in Cities*, ed. N. Frantzeskaki et al., 103–125. Berlin: Springer.
Woolcock, M., and D. Narayan. 2000. Social Capital: Implications for Development Theory, Research, and Policy. *World Bank Research Observer* 15(2): (World Bank Res) 225–249.
World Bank. 2000. *Cities in Transition: World Bank Urban and Local Government Strategy. Urban Planning*. Washington, DC: The International Bank for Reconstruction.
Yacobi, H., and R. Shechter. 2005. Rethinking Cities in the Middle East: Political Economy, Planning, and the Lived Space. *The Journal of Architecture*: 499–515.

Yap, K., and K. De Wandeler. 2010. Self-Help Housing in Bangkok. *Habitat International* 34: 332–341.

Yeh, S., and L.L. Laquian. 1979. *Housing Asia's Millions: Problems, Policies and Prospects for Low-Cost Housing in Southeast Asia*. Ottawa: International Development and Research Centre.

Yifiachel, O., and H. Yakobi. 2004. Control Resistance, And Informality: Urban Ethnocracy in Beer-Sheva, Israel. In *Urban Informality: Transnational Perspectives from the Middle East, Latin America, and South Asia*, ed. A. Roy and N. Al Sayyad, 209–239. Lanham: Lexington books.

Yiftachel, O. 2009. Theoretical Notes on 'Gray Cities': The Coming of Urban Apartheid? *Planning Theory* 8 (1): 88–100.

Author Index

A. M. Soliman, *Urban Informality*, https://doi.org/10.1007/978-3-030-68988-9

Subject Index

A. M. Soliman, *Urban Informality*, https://doi.org/10.1007/978-3-030-68988-9

D

R

S

T

City Index

A. M. Soliman, *Urban Informality*, https://doi.org/10.1007/978-3-030-68988-9

The manufacturer's authorised representative in the EU is Springer Nature Customer Service Centre GmbH, Europaplatz 3, 69115 Heidelberg, Germany. If you have any concerns regarding our products, please contact ProductSafety@springernature.com

Printed and bound by CPI Group (UK) Ltd, Croydon, CR0 4YY
30/09/2026
02225491-0017